ROUTLEDGE HANDBOOK OF WAR, LAW AND TECHNOLOGY

This volume provides an authoritative, cutting-edge resource on the characteristics of both technological and social change in warfare in the twenty-first century, and the challenges such change presents to international law.

The character of contemporary warfare has recently undergone significant transformation in several important respects: the nature of the actors, the changing technological capabilities available to them, and the sites and spaces in which war is fought. These changes have augmented the phenomenon of non-obvious warfare, making understanding warfare one of the key challenges. Such developments have been accompanied by significant flux and uncertainty in the international legal sphere. This handbook brings together a unique blend of expertise, combining scholars and practitioners in science and technology, international law, strategy and policy, in order properly to understand and identify the chief characteristics and features of a range of innovative developments, means and processes in the context of obvious and non-obvious warfare. The handbook has six thematic sections:

- Law, war and technology
- Cyber warfare
- Autonomy, robotics and drones
- Synthetic biology
- New frontiers
- International perspectives.

This interdisciplinary blend and the novel, rich and insightful contribution that it makes across various fields will make this volume a crucial research tool and guide for practitioners, scholars and students of war studies, security studies, technology and design, ethics, international relations and international law.

James Gow is Professor of International Peace and Security and Co-Director of the War Crimes Research Group at King's College London, UK.

Ernst Dijxhoorn is Assistant Professor in the Institute of Security and Global Affairs (ISGA) at Leiden University, the Netherlands.

Rachel Kerr is Reader in International Relations and Contemporary War and Co-Director of the War Crimes Research Group at King's College London, UK.

Guglielmo Verdirame is Professor of International Law at the Department of War Studies and the Dickson Poon School of Law, King's College London, UK.

ROUTLEDGE HANDBOOK OF WAR, LAW AND TECHNOLOGY

Edited by
James Gow, Ernst Dijxhoorn, Rachel Kerr and Guglielmo Verdirame

Routledge
Taylor & Francis Group
LONDON AND NEW YORK

First published 2019 by Routledge
2 Park Square, Milton Park, Abingdon, Oxon OX14 4RN

and by Routledge
605 Third Avenue, New York, NY 10017

First issued in paperback 2021

Routledge is an imprint of the Taylor & Francis Group, an informa business

Publisher's Note
The publisher has gone to great lengths to ensure the quality of this reprint but points out that some imperfections in the original copies may be apparent.

British Library Cataloguing-in-Publication Data
A catalogue record for this book is available from the British Library

Library of Congress Cataloging-in-Publication Data
Names: Gow, James, editor. | Dijxhoorn, Ernst. editor. | Kerr, Rachel, editor. | Verdirame, Guglielmo, editor.
Title: Routledge handbook of war, law and technology / edited by James Gow, Ernst Dijxhoorn, Rachel Kerr and Guglielmo Verdirame.
Description: Abingdon, Oxon ; New York, NY : Routledge, 2019. | Includes bibliographical references and index.
Identifiers: LCCN 2018038044 (print) | LCCN 2018039685 (ebook) | ISBN 9781351619981 (Web PDF) | ISBN 9781351619974 (ePub) | ISBN 9781351619967 (Mobi) | ISBN 9781138084551 (hardback) | ISBN 9781315111759 (e-book)
Subjects: LCSH: War (International law) | War–Technological innovations.
Classification: LCC KZ6385 (ebook) | LCC KZ6385 .R686 2019 (print) | DDC 341.6–dc23
LC record available at https://lccn.loc.gov/2018038044

ISBN 13: 978-1-03-209247-8 (pbk)
ISBN 13: 978-1-138-08455-1 (hbk)

Typeset in Bembo
by Wearset Ltd, Boldon, Tyne and Wear

CONTENTS

FIGURES

TABLES

CONTRIBUTORS

Kenneth Anderson is a professor at Washington College of Law, American University; a visiting fellow of the Hoover Institution; and a non-resident senior fellow of the Brookings Institution. He writes on international law, the laws of war, weapons and technology, and national security; he is the author, with Benjamin Wittes, of *Speaking the Law: The Obama Administration's Addresses on National Security Law* (Hoover Institute Press, 2015).

Air Commodore **Bill Boothby** (Retd) served for 30 years in the Royal Air Force Legal Branch, retiring as Deputy Director of Legal Services in July 2011. In 2009 he took a Doctorate at the Europa Universität Viadrina, Frankfurt (Oder) in Germany and published *Weapons and the Law of Armed Conflict* through OUP (now in its 2nd Edition) in the same year. His second book, *The Law of Targeting*, appeared with the same publisher in 2012. He has been a member of Groups of Experts that addressed Direct Participation in Hostilities and that produced the *HPCR Manual of the Law of Air and Missile Warfare*, the 2013 *Tallinn Manual on the Law of Cyber Warfare* and the *Leuven Manual on Peace Operations Law*. His third book, addressing *Conflict Law*, was published in 2014. In March 2018, with Professor Wolff Heintschel von Heinegg, he published with CUP a Detailed Commentary on the *US Department of Defense Law of War Manual* and his edited volume on *New Technologies and the Law in War and Peace* was published by CUP in December 2018. In October 2018 he was appointed Adjunct Professor at La Trobe University. He teaches, among other places, at the Australian National University, at the University of Southern Denmark and at the Geneva Centre for Security Policy.

Bleddyn Bowen is a Lecturer in International Relations, University of Leicester. He was previously a Lecturer in Defence Studies at the Defence Studies Department, King's College London, and a Teaching Fellow at the Department of International Politics, Aberystwyth University. He holds a PhD in International Politics from Aberystwyth University. In Spring 2014, he was a Visiting Scholar at the Space Policy Institute in Washington, DC.

Ernst Dijxhoorn is Assistant Professor in the Institute of Security and Global Affairs (ISGA) at Leiden University. He was previously research associate and lecturer at King's College London, where he worked on Economic and Social Research Council-funded projects on Science and Technology and Militancy and Violence in West Africa. He is author of *Quasi-states, Critical Legitimacy and International Criminal Courts* (2017).

Matteo Bencic Habian is a trainee lawyer at Bonelli Erede Pappalardo LLP and holds an MA in International Peace and Security from King's College London.

Paweł Frankowski is Assistant Professor in the Institute of Political Science and International Relations at the Jagiellonian University, Poland.

Georg Gassauer was chief operating officer at Train of Hope, Vienna, and oversaw the transit of over 150,000 refugees through Vienna between August 2015 and January 2016. Currently, he is an independent researcher associated with Princeton University's Liechtenstein Institute on Self Determination.

Tony Gillespie is an engineer with research and management experience in government, industry, and academia. He retired from Dstl (Defence Science and Technology Laboratory) in 2014 as Avionics and Mission Systems Fellow, having worked on many technical and legal aspects of radar and autonomous systems. Prior to this, he established and ran a microwave research department for BAESYSTEMS before becoming a Chief Project Engineer. His career started as a student apprentice, followed by a PhD and ten years as a radio astronomer, building advanced instrumentation and observing with it. Tony is currently a Visiting Professor at University College London and was elected a Fellow of the Royal Academy of Engineering in 2014.

James Gow is Professor of International Peace and Security and Co-Director of the War Crimes Research Group at King's College London. He is a non-resident scholar with the Liechtenstein Institute, Princeton University. From 2013–16, Gow held a Leverhulme Trust Major Research Fellowship. He has served as an expert adviser and an expert witness for the Office of the Prosecutor at the UN International Criminal Tribunal for the former Yugoslavia (1994–98), and as an Expert Adviser to UK Secretaries of State for Defence. Gow has held visiting positions at the University of Sheffield, the Woodrow Wilson International Center for Scholars in Washington, Columbia University, and Princeton University. His numerous publications include *The Art of Power: Freedman on Conflict, War and War Crimes*, *Prosecuting War Crimes: Lessons and Legacies of the International Criminal Tribunal for the former Yugoslavia* and *Security, Democracy and War Crimes* (as co-author), all in 2013, and *War, Image and Legitimacy* (2007), *The Serbian Project and Its Adversaries: a Strategy of War Crimes* (2003) and *Triumph of the Lack of Will: International Diplomacy and the Yugoslav War* (1997).

Cecilie Hellestveit is a Fellow at the Norwegian Academy of International Law (NAIL). She holds a PhD in international laws of armed conflict. Hellestveit has been a Fellow at the Peace Research Institute, Oslo (PRIO), the Norwegian Centre for Human Rights (University of Oslo), and International Law and Policy Institute (ILPI). She has also has been affiliated with the Atlantic Council in Washington DC and Max Planck Institute in Heidelberg. In 2008–10 she served as special rapporteur on the conduct of hostilities in military operations to the International Society for Military Law and the Laws of War. Hellestveit is currently a member of the Council on Ethics for the Norwegian Petroleum Fund.

Maziar Homayounnejad researches targeting law and weapons law aspects of new weapon systems. He recently completed a PhD at the Dickson Poon School of Law, King's College London, entitled 'Lethal Autonomous Weapon Systems Under the Law of Armed Conflict'.

Oscar Jonsson is director of the Stockholm Free World Forum (Frivärld) during 2019, a Swedish foreign and security policy think tank, and associated researcher at the Swedish Defence

University. Oscar has worked as subject-matter expert in the Policy and Plans department at the Swedish Armed Forces Headquarters. Oscar is the author of *The Russian Understanding of War: Blurring the Boundaries of War and Peace* (GUP 2019) and holds a PhD from the Department of War Studies, King's College London.

Rachel Kerr is Reader in International Relations and Contemporary War in the Department of War Studies at King's College London and co-Director of the War Crimes Research Group at King's with James Gow. She is the author of *The International Criminal Tribunal for the Former Yugoslavia: Law, Diplomacy and Politics* (OUP, 2004); *Peace and Justice: Seeking Accountability after War* (Polity, 2007), with Eirin Mobekk, *The Military on Trial: The British Army in Iraq* (Wolf Legal Publishers, 2008), and co-edited *Prosecuting War Crimes: Lessons and Legacies of 20 Years of the International Criminal Tribunal for the Former Yugoslavia* (Routledge, 2013), with James Gow and Zoran Pajic. In 2009–10, Dr Kerr was a Fellow at the Woodrow Wilson International Center for Scholars in Washington, DC, and in 2011–13, she was a Visiting Research Associate at the Centre for International Policy Studies, University of Ottawa, Canada.

Elaine Korzak leads the Cyber Initiative and is Assistant Professor of Cyber Security at the Middlebury Institute of International Studies, Monterey, after being a W. Glenn Campbell and Rita Ricardo-Campbell National Fellow at the Hoover Institution. She was previously a pre-doctoral and postdoctoral cybersecurity fellow at CISAC. She has a PhD in War Studies and an MA in International Peace and Security from King's College London and an LL.M in Public International Law from the London School of Economics and Political Science (LSE). She has held posts in various governmental and non-governmental institutions (both national and international) where she has worked on disarmament and international security issues.

Anne-Marie le Gloannec (deceased) was Director of Research at the International Research Centre of Sciences Po in Paris, and an Associate Researcher at the European Policy Center in Brussels. She previously served as Deputy Director of the Berlin-based Centre Marc Block from 1997 to 2002 and taught at the Johns Hopkins University in Bologna, the Université de Paris 1, the Freie Universität Berlin, the Luiss Guido-Carli University in Rome as well as the Universities of Viadrina, Stuttgart, and Cologne. From 1984 to 2005 she held a fellowship at the Woodrow Wilson International Center for Scholars in Washington, DC and from May to June 2015 she was a fellow at the Nobel Institute in Oslo. She authored several books and was a regular contributor to the French newspapers *L'Express* and *Le Figaro*. Her last book *Continent by Default. The European Union and the Demise of Regional Order* was posthumously released by Sciences Po and Cornell University Press in September 2017.

Filippa Lentzos is a Senior Research Fellow at King's College London, with a joint appointment in the Department of Global Health & Social Medicine and the Department of War Studies. She is also an Associate Senior Researcher within Armament and Disarmament at the Stockholm International Peace Research Institute (SIPRI), a biosecurity columnist at the Bulletin of the Atomic Scientists, and the NGO Coordinator for the Biological Weapons Convention. Lentzos was the social science lead on the first synthetic biology centre established in the UK.

Christopher Lowe OBE, FREng, FInstP, FRSC, is Emeritus Professor of Biotechnology at the University of Cambridge. The principal focus of his research programme over 40 years was the high-value/low-volume sectors of pharmaceuticals, fine chemicals, and diagnostics. The work is characterised by not only being highly inter- and multi-disciplinary, but also covering the entire range from pure science to strategic applied science, some of which has significant commercial applications.

Jack McDonald is a Lecturer in War Studies at the Department of War Studies, King's College London. He holds a PhD in War Studies from King's College London, as well as an MA in International Peace & Security. Prior to taking up a lectureship at King's, he worked as a policy researcher, and prior to that as a Teaching & Research Fellow in the Department. He was a Research Associate on the ESRC-funded project, *SNT Really Makes Reality* from 2013–14.

Thrishantha Nanayakkara is Director of Morphological Computation and Learning Lab, Imperial College, London. His research in controllable stiffness robots tries to understand how physical circuits in the body and environment contribute to solve computational problems to do with efficient survival in unstructured environments.

Sir David Omand was the first UK Security and Intelligence Coordinator, responsible to the Prime Minister for the professional health of the intelligence community, national counter-terrorism strategy, and 'homeland security'. He served for seven years on the Joint Intelligence Committee. He was Permanent Secretary of the Home Office from 1997 to 2000, and before that Director of GCHQ (the UK Sigint Agency). Previously, in the Ministry of Defence as Deputy Under Secretary of State for Policy, he was particularly concerned with long-term strategy, with the British military contribution in restoring peace in the former Yugoslavia and the recasting of British nuclear deterrence policy at the end of the Cold War. He was Principal Private Secretary to the Defence Secretary during the Falklands conflict, and served for three years in NATO Brussels as the UK Defence Counsellor. He has been a visiting Professor in the Department of War Studies since 2005.

Richard E. Overill is a Senior Lecturer in Computer Science in the Department of informatics, King's College London. Since 1996 his principal research interests have centred on cyber security and digital forensics resulting in 58 interdisciplinary publications. He is a Chartered Mathematician, a Chartered Scientist and a Chartered Engineer.

Brian Rappert is Professor of Science, Technology, and Public Affairs at the University of Exeter. His long-term concern lies with the social and ethical dilemmas associated with scientific and technical expertise and attempts to enhance the humanitarian restrictions governing the conduct of war. Recent publications include *The Dis-eases of Secrecy: Tracing History, Memory, and Justice* (Jacinda, 2018), *Absence in Science, Security and Policy from Research Agendas to Global Strategy* (Palgrave, 2016), *Sensing absence: How to see what isn't there in the study of science and security* (Palgrave, 2015) and *How to Look Good in War* (Pluto, 2012).

Marco Roscini is Professor of International Law at the Westminster Law School. Prof. Roscini has a PhD from the University of Rome 'La Sapienza' and was previously a Research Fellow in International Law at the University of Verona School of Law. He lectured in international security law (*jus ad bellum*, law of armed conflict, and disarmament law) at University College London (UCL), King's College London, Queen Mary University of London, and the Ecole des Relations Internationales in Paris. Prof. Roscini has published widely in the field of international security law. He is the author of *Le zone denuclearizzate* (Nuclear weapon-free zones,

Giappichelli, 2003) and of *Cyber Operations and the Use of Force in International Law* (OUP, 2014). He is also the co-editor of *Non-proliferation Law as a Special Regime* (CUP, 2012).

Ariane Tabatabai is a senior associate with the Proliferation Prevention Program at the Center for Strategic and International Studies (CSIS) and the director of curriculum and a visiting assistant professor at the Georgetown University Edmund A. Walsh School of Foreign Service. Dr Tabatabai is also an international civilian consultant for NATO; a columnist for the *Bulletin of the Atomic Scientists*; a Truman National Security Fellow; and a 2017–18 postdoctoral fellow at the Harvard Kennedy School's Belfer Center for Science and International Affairs, where she was previously an associate in the International Security Program and the Project on Managing the Atom in 2014–15 and a Stanton Nuclear Security Fellow in 2013–14. Previously, Dr Tabatabai was a non-resident scholar with the James Martin Center for Nonproliferation Studies at the Monterey Institute.

Fleur Richard-Tixier is a Junior Consultant at CEIS, European Office. She holds an MA in Geopolitics and International Relations, Catholic Institute, Paris.

Guglielmo Verdirame is Professor of International Law at King's College London. He was previously a Lecturer in the Faculty of Law at the University of Cambridge and a Fellow of the Lauterpacht Centre for International Law (2003–11); a Junior Research Fellow at Merton College, Oxford (2000–03); and a Research Officer at the Refugee Studies Centre at the University of Oxford (1997–98). He has also held a visiting appointment at Harvard Law School (2007) and was Director of Studies for Public International Law at the Hague Academy of International Law (2006). He is the author of *The UN and Human Rights: Who Guards the Guardians?* (2011, CUP), Winner of the Biennial ACUNS Book Award, and of *Rights in Exile* (Berghahn Books, 2005).

Matthew C. Waxman is the Liviu Librescu Professor of Law and the faculty chair of the National Security Law Program at Columbia Law School. Before joining the Law School faculty, he served in senior positions at the State Department, the Department of Defense, and the National Security Council. Waxman was a Fulbright Scholar to the United Kingdom, where he studied international relations and military history. He is a member of the Council on Foreign Relations, where he also serves as Adjunct Senior Fellow for Law and Foreign Policy, and he is the co-chair of the Cybersecurity Center at the Columbia Data Science Institute. He holds a J.D. from Yale Law School.

PREFACE AND ACKNOWLEDGEMENTS

This book has its origins in a major ESRC-funded research project, led by Guglielmo Verdirame, James Gow, and Rachel Kerr: *SNT Really Makes Reality: Technological Innovation, Non-Obvious Warfare and the Challenges to International Law* (ESRC Ref: ES/K011413/1) which was conducted at King's College London and ran from 2013 to 2015. The project was funded as part of the Research Councils UK (RCUK)'s £2.1 million *Science and Security* Programme, run jointly by the Economic and Social Research Council (ESRC), the Defence Science and Technology Laboratory (DSTL) and the Arts and Humanities Research Council (AHRC). The Science and Security Programme was part of the wider RCUK Global Uncertainties Programme, which ran from 2008 to 2018. Our thanks must first therefore go to RCUK and the DSTL, and specifically to the DSTL lead and scientific experts assigned to the project, Mark Ashforth and Tony Gillespie, as well as Lou Martingale, who had DSTL responsibility for the programme. We would also like to acknowledge Jack McDonald who was the initial Research Associate on the project, before Ernst Dijxhoorn took over and saw the project through to the end. It is also important to acknowledge various individuals who made contributions with research papers and in other ways, who are not, for various reasons, authors in the final volume, but who all made vital contributions to our research and understanding along the way: Sir Daniel Bethlehem, former principal Legal Advisor to the UK Foreign and Commonwealth Office, Jason Reifler, Christopher Coker at the LSE, Sarah Soliman at RAND, Clément Guitton, Lola Frost, Leverhulme Artist-in-Residence in the Department of War Studies, and our former colleague at King's, Professor Thomas Rid, SAIS Johns Hopkins University. Finally, we need to thank the members of our Advisory Group, who included Tony Gillespie and Mark Ashforth: Sam Selvadurai (Policy Unit, FCO), Robert Fox (Defence Editor, *Evening Standard*), Rodney Dixon (Barrister, Temple Garden Chambers), Tony Coll (Tony Coll Associates) and Ben Wilkinson (Policy Institute, King's College London). It is also necessary to thank Steve Hill, former Deputy Director (Cyber and Government Security), National Security Secretariat, Cabinet Office), who featured relatively late in the project and process, but whose input was invaluable on some questions. Our thanks too to Gordon Burck for editorial assistance in preparing the bulk of the chapters, and Mia El-Husseini for helping us get to the finish line.

Our aim in the research was to investigate the characteristics of technological and social change in the context of obvious and non-obvious warfare in the twenty-first century, and the challenge such change presents ethically and to international law. There was, and remains, a clear need for this as a result of significant change in the character of contemporary warfare, relating specifically to the nature of the actors and the changing technological capabilities available to them, coupled with significant flux and uncertainty in the international legal sphere. We sought in the project to identify the range of scientific and technical innovations that presented the most acute challenges, and investigate the legal and ethical dimensions of those challenges in terms of international humanitarian law and international human rights law. It was imperative, we felt, to bring together technology experts, with practical and operational understanding, international legal experts, and experts in strategy, in order properly to understand and identify the chief technical characteristics and features of a range of innovative developments, and to reflect upon the ethical and legal dimensions of their exploitation in the context of obvious and non-obvious warfare. This was a major achievement and we did so in a series of workshops held at King's College London, starting in 2013. Many of the contributions to this handbook were commissioned for those workshops and formed the basis of our discussions, while others were commissioned but were not part of the workshops, with their work being reviewed at distance.

We also identified gaps in our coverage, especially in relation to emerging issues, less well covered in the literature, such as those relating to potential military applications of space technology, and potential challenges and opportunities of ensuring accountability where the law was breached. We also noted gaps identified in peer review of the book in progress, guided by the ever patient and supportive Andrew Humphrys, Senior Editor at Routledge – to whom our thanks must also go. We therefore commissioned a series of additional contributions. This part of the project benefited from the support of the British International Studies Association International Law Working Group, which sponsored a further workshop and also panels at its annual convention. All of this, of course, prolonged the process of getting the handbook ready. But it was clearly the right course of action and worth the delay and additional effort, in terms of offering a fuller picture and marking out new territories and avenues for future research. Our thanks therefore to all of our contributors, and especially to the original participants for bearing with us patiently. These thanks must go posthumously with prayers to the wonderful Anne-Marie le Gloannec, who died after a battle with cancer on 26 April 2017.

And finally, we – James, especially – must thank Gabriel for instigating the original project, and also all our near, dear and loved ones for putting up with all that it takes to bring a project such as this to fruition.

RK, AJWG, EEAD and GV
London, July 2018

1

INTRODUCTION

Technological innovation, non-obvious warfare and challenges to international law

Rachel Kerr

We are experiencing a period of immense upheaval in all spheres of human existence. In this context, it is not surprising that war too is undergoing rapid and dramatic change. Over the last few decades, received images of conventional war based on highly organised and trained forces engaged in a 'duel' have become almost entirely outdated. Whether characterised in terms of so-called 'new wars',[1] Revolutions in Military Affairs,[2] hybrid wars,[3] virtual wars,[4] human wars,[5] spectator-sport wars[6] and wars among the people,[7] the image of war has shifted radically from the twentieth century experience of major inter-state war to a twenty-first century dominated by wars within as well as among states and involving a complex mix of state and non-state actors. These wars are often categorised in terms of what they are not rather than what they are, including 'non-state',[8] 'non-obvious' and 'non-linear' wars, such as those in Crimea and eastern Ukraine.[9]

At the same time, the context in which technological and scientific innovation is occurring is itself rife with new ambiguities. Media of all kinds, both traditional and the new social media, play a dominant role in the ways in which military operations are perceived and supported (or not) by publics who consume a constant stream of information and commentary – some of it in the form of powerful and shocking visual imagery streamed around the world even as events are unfolding. Meanwhile, violations of international law give rise not only to state responsibility, but also to individual criminal responsibility and there has been a proliferation of mechanisms for accountability, including the establishment of a permanent International Criminal Court, operational from July 2002. Legal scrutiny and argumentation has given rise to what some have called 'Lawfare'[10] and others the 'judicialisation of war',[11] demonstrated most pertinently in the still ongoing 'fury of litigation' that followed the ill-fated 2003 Iraq War.

In many respects, international law and politics have been found wanting. The war in Syria demonstrated the limits of international action when the UN Security Council is divided. In the context of the so-called 'war on terror', it was argued that international law was at once too weak – not up to the challenge of preventing atrocities – and outdated, or 'quaint' in the face of the new challenges wrought by a globalised terrorist threat. A major area of contention concerns the range of new technologies that have entered the battlefield in the last decades, which challenge both our understanding of the use of kinetic force and the state-centric paradigm central to the laws of war. We are at the beginning of a curve of intense technological change; unsurprisingly, predicting where these processes will take us is largely a speculative exercise.

Some technological developments, 'cyber war' for example, challenge our existing categories so profoundly that one wonders whether they still come under the broad umbrella of 'war'.[12] Other technologies, such as synthetic biology, have not yet entered the battlefield but their destructive potential is such that prevention seems at first glance to be the only sensible strategy.

These technologies are transforming the modern battlefield. A central concern for international lawyers and military practitioners alike is whether and how existing laws governing armed conflict – the Law of Armed Conflict (LOAC) or International Humanitarian Law (IHL) – are applicable to new and emerging technologies in the context of contemporary war. What is the effect of these developments on the international law of war – and on the ethical ideas that animate it? Do these changes call for more than evolving interpretations of existing principles and rules? Is there anything inherently new or different about the range of new technologies in a military context and the dilemmas they raise, or are they merely an extension of the challenges we have already identified. In other words, do they cause new problems or simply exacerbate old ones?

The book

This handbook investigates the characteristics of technological and scientific innovations in the context of obvious and non-obvious warfare, exploring their legal, ethical, and strategic dimensions. The questions it examines are important and complex. They have often been considered within the confines of specialist disciplines with little genuine cross-departmental interaction. By contrast, this handbook attempts to create a new and common ground for scholarly debate among lawyers, policy specialists, and the science and technology community, with experts from different fields invited to develop reflections that transgress traditional academic boundaries.

The interdisciplinary nature of the book marks it out from the various books on the law of armed conflict and new military technologies published over the last few years. It is not the aim of this book to map out the legal issues around new technologies in a systematic way. The objective is, rather, to stimulate new thinking about these developments, and address the big strategic, technological, legal and policy questions behind them. As with every genuinely interdisciplinary project, the editors faced a challenge – both substantive and presentational. The project was greatly assisted by the fact that editors and contributors – together with other participants – presented drafts of their chapters in two workshops and by the skill and flexibility of the individual authors – all experts in their respective fields. The workshops, and other aspects of the research that led to this book, were funded through a grant from the UK Research Councils and the Defence, Science and Technology Laboratory (DSTL), part of the UK Ministry of Defence's Science and Security Programme. The result (we hope readers will agree) is a book that provides insight into the key challenges posed by the advent of new technologies in the shifting contexts of obvious and non-obvious wars.

The handbook is organised in six parts:

- I Law, War and Technology
- II Cyber Warfare
- III Autonomy, Robotics and Drones
- IV Synthetic Biology
- V International Perspectives
- VI New Frontiers

Part I examines the context in which discussions about technological change are taking place, focusing on the existing international legal framework, on the process of political contestation in the creation of new law, and on the legal and technical framework in which new means of warfare are developed. In Chapter 2, James Gow and Ernst Dijxhoorn review changes in the strategic and legal landscape in which technological change must be accommodated and understood – referred to elsewhere in terms of a *Transformation of Strategic Affairs*.[13] Contemporary war is different in that wars are no longer fought for decisive victory by two, or more, state-armies meeting on the battlefield; almost all wars now involve non-state actors alongside or against the state and are fought 'among the people'.[14] In this context, the key to success is the ability to create and maintain legitimacy and every decision has the potential to have strategic impact. Complicating the picture further is the power of strategic communication (including misinformation) via all types of media, deployed by states and non-state actors, and amplifying the reach of the latter, and an increasingly blurred picture of who is waging war, how and when. Modalities of technological change, discussed in elsewhere in the Handbook, must therefore be understood against this background.

In Chapter 3, Bill Boothby presents an historical overview of the legal and institutional framework governing the introduction of new weapons technology. As Boothby makes clear, these are not developed, procured and fielded in a legal vacuum. International law includes rules that prohibit certain weapons entirely in armed conflict, while others are the subject of restrictions as to the circumstances in which they can lawfully be employed. Critical to this body of weapons law are the two established principles of *humanity* and *military necessity*. The rules of the law regulating weapons represent the balance that states have struck between these conflicting interests, a balance that will vary from weapon to weapon depending on how states perceive the military need associated with the weapon and how they interpret the humanitarian concerns that have motivated the legal provision in question. He concludes by considering the particular problems posed by the development of automated weapons, where at the present time, a human remains in the loop but any developments lead to fully autonomous systems, it will need to be determined that the machine can successfully negotiate the complex decision-making process involved. In Chapter 4, Tony Gillespie, formerly of DSTL, turns the question around and considers what the review process looks like from the point of view of a technologist and how, in practice, states comply with Article 36 of Additional Protocol I to the Geneva Conventions, which obliges them to review new means and methods of war during the procurement process. Gillespie continues the focus on autonomy of the previous chapter, and examines the ever-more extensive use of automated decision-support tools in the military command and control (C2) chain. Gillespie argues that the use of a weapon cannot be separated from the surrounding system, but at the same time, there must be clear boundaries to the weapon under review otherwise review risks becoming an open-ended process. Finally, in this section, Brian Rappert steps back and considers the underlying principles and motivations behind the regulation of new weapons technologies. Scientific and technological developments are often accompanied by legal, political, and social concerns. In the context of armed conflict, this is manifested in moral and humanitarian apprehension about the 'harm' caused by new weapons technologies. Autonomous weapons systems, direct energy weapons, and cyber warfare are but a few of the areas that have generated such concerns in recent years. In the past, these concerns have been accommodated within the framework of international humanitarian law, and the debate has been predicated on the balance of military necessity and humanity that sits at the core of that body of law. As Boothby demonstrates in his chapter, the balancing of these principles has underpinned regulation of armed conflict, evidenced in a number of specific legal rules. Rappert proposes a radical shift, however, to an alternative normative framework that takes as its starting point not whether new technologies should be argued 'out' but rather that the case needs to be made for allowing them 'in'.

In Parts II, III and IV, the contributors examine a range of new technologies and the challenges they pose to international law. In Chapters 6 and 7, Elaine Korzak and James Gow address the problem of cyber warfare, interrogating first the concepts of armed conflict and armed attack to determine on what basis and under which circumstances the law of armed conflict applies to computer network attacks (CNAs) and then, where it does apply, how are principles of distinction and proportionality applied in situations characterised by ambiguity and attrition. Overall, they find that, while significant progress has been made on finding ways to apply the law to cyber warfare, the unique features of these new types of attacks, particularly their non-kinetic mode of operation, their range of possible effects, and their perceived anonymity, create significant difficulties for the application of international law. Even if issues of distinction and proportionality can be resolved, and there are difficulties there, as Korzak and Gow outline in Chapter 7, the problem of attribution for computer network attacks may yet make such determinations moot since there is no possibility of pursing legal recourse without an identified perpetrator. In Chapter 8, Marco Roscini delves deeper into issues around proportionality in relation to targeting in cyber operations, arguing that such operations offer opportunities as well as challenges. Roscini compares two parameters (incidental damage on civilians and civilian property on the one hand, and the attacker's concrete and direct military advantage on the other) of different nature but of equal standing in specific attacks, asking what does 'damage' and 'military advantage' mean in the cyber context?

Chapter 9 shifts the focus to 'big' data and digital intelligence. As Sir David Omand argues, the actions of Edward Snowden pushed the legal and moral issues associated with intelligence collection in the digital age to the forefront of public debate. Given the origin of the material, much of the debate centred on the United States, but the methods of 'bulk access' to internet data and analysis of metadata have raised privacy operations elsewhere as well. Omand's chapter focuses on the UK's regulatory systems, depicted as 'broken' by critics, and demonstrates that in fact, the British political and legal frameworks that regulate surveillance and intelligence activities place important limitations on the collection of, and access to, data by UK civil servants both in the UK and in the wider world. Given that intelligence support to military and counter-terrorism operations will require digital means in the information age, the regulatory systems that constrain intelligence gathering activity are necessary for ongoing legitimacy. The final chapter in this section, by James Gow (Chapter 10) returns to the domain of cyber warfare. Amid the discussions about how existing law might be applied, or where new law might be needed, two highly important factors have largely been ignored. The first of these is the relationship between offence and defence in cyber warfare, where governments are in the business of building robust defensive infrastructure but the advantage lies squarely with the individual able to launch an offensive attack. The second concerns the human factor; that is, for all the legal and technical discussion that there might be, the key to cyber success or failure largely rests at the level of the individual. Minor human errors or unilateral acts of destruction can have major consequences.

Part III returns to the highly contentious issues around the use of automated, autonomous and semi-autonomous weapons systems, briefly discussed in Chapter 3. Thrish Nanayakkara opens the section with an explanation of the technology of automation and how autonomy and rationality are conceptualised in the world of robotics. Contrary to traditional belief that humans are autonomous beings who can take rational decisions, it is argued that rationality itself consists of conditioned processes that take uncertain trajectories. Therefore, conditioning and empowerment through training and technological augmentation of information fed to the process of decision-making is very important for those who operate semi-autonomous robots that provide sensory feedback to humans to take decisions. This imposes challenges both in the domain of incorporating ethical guidelines in the training process, and in terms of providing flawless technological support. Moreover, autonomy can emerge in any system regardless of whether there is a software code to control actions or not. Therefore, when imposing bounds on the autonomy of a machine to map a situation to a potentially harmful action, mechanisms should be set in place to limit the entire behaviour as an embodied entity rather than a software code alone.

Even in the case of counter action for hacking, focusing on software alone can lead to potentially dangerous outcomes. In Chapter 12, Jack McDonald critiques the contemporary debate on autonomous weapon systems. Analysing the different positions about humans 'on the loop' or 'in the loop', McDonald argues that the debate ignores the role of rules of engagement and military structures in targeting decisions. The real challenge of autonomous systems is their existing and potential role in augmenting human decisions that result in uses of lethal force. The integration of autonomous systems capable of producing or enhancing 'intelligence' therefore poses a huge challenge to the concept of command responsibility. In Chapter 13, Matthew Waxman and Kenneth Anderson interrogate the legal and policy challenges of armed 'drones' and autonomous weapon systems. Waxman and Anderson mount a strong case that armed UAVs and autonomous weapon systems *can* be effectively regulated under the well-established law of armed conflict framework applicable to all weapon systems. A combination of factors, however, including rapid technological development and low levels of transparency in how these emergent weapon systems are used or will be used, contributes to scepticism about whether existing international law is sufficient and makes it difficult for the international community to reach consensus understandings about *how* to interpret and apply existing law to these systems. For strategic reasons therefore, they counsel in favour of greater policy and operational transparency by the United States and some of its close allies. In Chapter 14, Maziar Homayounnejad and Richard Overill push further on the ethical framework underpinning objections to 'killer robots'. He interrogates the arguments for the increased use of non-human (robotic) warriors, premised on the assumption that robots lack common human failings and weaknesses – both deliberate, born of hatred, anger, and frustration, or as a result of physical tiredness, hunger and duress. Robots, in this view, 'do not rape' (or commit other war crimes). However, as the authors argue, robots *can* rape, in that they could be programmed to do so, or to inflict other equivalent harm. Given that rape, sexual assault, and other forms of inhuman treatment, including torture, have frequently been deployed as strategic as well as tactical weapons in contemporary wars, there is no reason to believe that deploying robots, as such, would enhance prevention of such crimes. Indeed, it may even be more likely given that robots, as well as lacking human failings, also lack human virtue. The final chapter in this section returns to a more narrowly technical perspective. Tony Gillespie responds to critiques of autonomous weapon systems in reports of non-governmental organisations, such as Human Rights Watch's *Losing Humanity, The Case Against Killer Robots*. Full autonomy, he argues, has no clear definition in the context of weapon systems but is part of a continuum with high and low levels of automation. *All* weapon systems currently have a human in both their command and control chains. What drives the level of automation is the time constant for human intervention. If we move on the spectrum toward greater autonomy, we can learn from developments in the civilian sphere, which are likely these days to be ahead of the military curve in any case.

The third major area of technological innovation considered in this handbook is the emergent field of synthetic biology, the subject of Part IV. In Chapter 16, Christopher Lowe considers the ethical, political and legal concerns about rapid developments in biotechnology and in particular its potential military application. This field of biology mirrors modern engineering by designing novel life functions and forms with a predictable box of materials and parts. At present, whole genome synthesis requires multiple capabilities in software, hardware and wetware to be brought together and integrated in a well-funded laboratory environment, but that situation could change as synthetic biology becomes 'lego-ised' and de-skilled, with much of the genomic data available on the internet and many of the biobricks and process kits becoming commercially available. Lowe warns us that with reductions in the costs of DNA sequencing, synthesis and chemical genomics, coupled with the universality of the internet and 'kit-ification' of biological recipes, this type of research is no longer the preserve of government-supported academic institutions or large corporations, but is now in the domain of biohackers who are able to conduct research in their kitchens or garages. The following chapters consider the legal implications of this new technology. In Chapter 17, Filippa Lentzos and Cecilie Hellestveit analyse the security threat posed by efforts to engineer biology and the question of whether it is containable under the existing law prohibiting the development and use of biological weapons. In the remaining two chapters, Guglielmo Verdirame and Matteo Bencic Habian consider the implications of 'dual-use' scientific exploration. In Chapter 18, Matteo Bencic Habian sets the advent of the new biotechnologies with military application in historical context. In a sense, the law, as ever, sought to respond to new threats, with attempts to exploit new scientific discoveries with the advent of the study of virology in the nineteenth century for military use followed some years later by international law seeking to outlaw its development and use in war. Matteo Bencic Habian uses this framework to consider responses to current and potential future threats emerging from the developments in biotechnologies, and, crucially, the increased ease with which non-state actors and terrorist groups might be able to obtain and develop expertise to use them to devastating effect. Next, in Chapter 19, Verdirame and Matteo Bencic Habian extend the discussion to ask 'who is in charge' of regulating synthetic biology? On the one hand, there are those, mainly scientists, who believe that the government lacks the expertise necessary not only to regulate this field but especially to implement any such regulation. The scientific community should instead 'self-regulate'. On the other hand, there are those, to be found mainly among national security experts, who point out that scientists do not have a proper appreciation of the wider strategic context, including the capabilities of states and non-state actors; the self-regulation by the scientific community would almost certainly create a security gap. Verdirame and Matteo Bencic Habian propose instead the establishment of a more effective, flexible, collaborative, approach to regulation based on both bottom-up self-regulation and top-down regulation.

In Part V, we turn our attention to some of the 'New Frontiers' of technological and scientific innovation and their operational implications. In Chapter 20, Bleddyn Bowen considers the prospect of 'Space Wars', and investigates the prospects for a Code of Conduct and the Treaty of the Prevention of Placements of Weapons in Outer Space (PPWT), and their inherent problems in the context of space weapons proliferation. Bowen argues that the flaw in the current framework is that it seeks to regulate space security as a separate domain of activity, whereas on the one hand, it cannot be considered distinct from global and security considerations on the ground, nor can there be a neat distinction between military and civilian considerations – rather space exploration and exploitation is truly 'dual-use'. In both these senses, space warfare is not a qualitatively different domain, but simply the continuation of war by other means. Paweł Frankowski, in Chapter 21, continues the dual-use theme, and considers the consequences for global security of private companies operating in space. As Frankowski explains, the rising importance of geo-intelligence, space surveillance and telecommunication for global security together with new kinds of security challenges and vulnerabilities, such as environmental problems in outer space, pose new challenges to security and to the legal framework. This is compounded by the significant role that private profit-oriented companies play in the new security environment in the US in particular, changing the landscape for both law and practice. In this context, the new market for subcontractors in space applications raises important questions about the growing dependence on private resources in a traditional sphere of state activity – security – provided from and through outer space. In Chapter 22, James Gow and Georg Gassauer discuss the novel use and application of biometric technologies, such as iris recognition, to register and track refugee movements in the wake of the Syria crisis. Whilst such technologies offer significant advantages, they are not, as Gow and Gassauer demonstrate, without their practical, ethical, and legal challenges.

In Chapters 23 and 24, James Gow and Ernst Dijxhoorn explore 'what soldiers think' of these developments. These chapters present the findings of focus group research with groups of military personnel. For all its novelty and the challenging qualities it brings, research subjects embraced cyber warfare as simply another part of warfare. While soldiers underlined repeatability as a quality distinguishing conventional weapons from their cyber counterparts – confirming that there were crucial differences – the cyber capabilities were simply an addition to the realm of weapons used in warfare, and issues were generally interpreted through the lens of conventional armed conflict and the contribution that the cyber arms made. There was strong agreement that old concepts of what constitutes *use of force* might be relevant and it was reasonable to maintain existing frameworks, but also that there was scope for considering that destructive power of cyberspace operations in other than military realms and against civilian objects might also constitute 'armed attack'. The discussions were most difficult in relation to the potential for weaponised synthetic biology. Weapons that, idealistically, could remove incidental harm and offer great force protection raised major ethical concerns among respondents, concerned that, if using such a weapon, the potential harm to civilians and others might outweigh the protection offered. This, in itself, reflected and concerned the profound psychological impact that the prospect of using genetic weapons introduced. It was judged that this effect would not only affect the military in the battlespace – wherever the theatre was to be defined – but also society, as a whole. In contrasting views, some respondents strongly favoured adoption of genetic weapons, in view of the potential discriminatory benefit they might bring, while others called for a complete ban, even before any deployable weapon had emerged. This call for an outright ban ran counter to the sanguine sense that existing normative frameworks could well embrace both cyber warfare and autonomy. It also confirmed that the strength of division regarding prospective synthetic biological weapons, including the sense that new rules might be required, sets this area of innovation in relation to warfare and the very obviously 'invisible' non-obvious warfare it represents apart from other domains of novel weaponisation, such as autonomy and cyber warfare. In the final Chapter (25), Maziar Homayounnejad reflects on the potential possibilities and challenges of future war crimes prosecution for transgressions in the areas of technological innovation outlined in the book. Aside from the legal challenges, Homaynoujenad considers the challenges and opportunities posed by new forms of digital evidence, focusing on four areas in which forensic investigation of war crimes, or international crimes, involving cyber technologies could be possible: the use of black box recorders; access control data; code stylometry; and keystroke analysis.

The last Part, VI, considers international perspectives. In Chapter 26, Oscar Jonsson interrogates how the Russian view of information warfare is changing in light of technology. Russia's notably broader understanding of information warfare, dating from the 1990s, comprises both an information-technical and an information-psychological aspect. Conceptually, this broader approach enabled Russia to deal with the changing information landscape following the rapid increase of social media. Western concerns have grown ever since the Russian cyber attacks in Estonia and Georgia. In the Ukrainian conflict, meanwhile, we have seen a low level of cyber attacks but a full-intensity information-psychological conflict. In support of this effort, Russia has successfully updated its approach to the current information environment with paid bloggers and commentators on internet forums. Internally, Vladimir Putin has steadily consolidated control of domestic media, both traditional and social, to a critical degree. Externally, *Russia Today* has established itself as the main network in the US and UK broadcasting the Kremlin message. Jonsson argues therefore that the Russian strategy on cyber warfare can be best understood in connection with the wider Russian approach to the use of information in the conflict and security arena. Next, Ariane Tabatabai considers questions of ethics and legality in Islamic thought as applied to unconventional warfare and technological innovation. She examines both the Sunni and Shi'a approaches to the laws regulating warfare and discusses their implementation by various Muslim actors, including state and non-state players. Tabatabai argues that while the nuances that exist in Western debates surrounding the legality of means and methods of warfare is absent in Islamic jurisprudence, the general prescriptions of the faith are very similar to the rules and regulations of international humanitarian law.

The following chapters focus on the perspectives of the other members of the P5. In Chapter 28, Anne-Marie le Gloannec and Fleur Richard-Tixier examine debate on cyber war in the French political and military establishment, focusing on two sets of questions: 1. Do politicians and policy-makers deem it possible to regulate cyber warfare through international law? Has France made any contribution to this – within the P5 or in other *fora* (NATO, ESDP, others)?; and 2. Are strategic and legal doctrines about cyber war being developed – something comparable, for example, to the doctrines on nuclear restraint (MAD), which emerged in the early 1960s? The question of capabilities is connected to this debate. What do we know about French cyber war capabilities? And where does France fit in in its relations to states such as the US and the UK? And what is the attitude of the establishment to these developments? In Chapters 29 and 30, Elaine Korzak examines the position of the US, UK, Russia and China through the lens of discussions at the UN, which have seen the US and the UK line up on one side, and Russia and China on the other. Russian efforts to gain a new international cyber treaty have not materialised. Instead, a broader debate on norms of responsible state behaviour, favoured by the US and the UK, has emerged. In the course of discussions, all four states have acknowledged the applicability of international law to state conduct in cyberspace and attention has shifted towards questions of implementation. Korzak argues that the emerging interpretative approaches of the US–UK and Russia–China axes with regard to the implementation of international humanitarian law leave open the possibility that additional norms may be created, but, for the most part, interpretative approaches in the context of international law on the use of force reveal divergence – limiting prospects for the development of dedicated norms. The international debate and development of states' interpretative approaches illustrate the complexity of challenges prompted by the emergence of cyber attacks. An assessment of the adequacy of international law in regulating this new type of warfare will necessarily need to go beyond a debate over the need for a new international legal treaty along the lines of known weapons conventions. However, the nascent stages of states' views with regard to the legal challenges created by cyber attacks indicate that any development of dedicated norms in this area will be subject to numerous factors and their complex interplay. Positions with regard to proportionality have not yet crystallised, and there is consequently greater room for the adoption of new rules by consent.

This sense of being in an era of considerable flux and change resounds through all of the contributions to this handbook. As stated at the outset, we are on the curve of massive change taking place in the legal sphere, in response both to technological and scientific innovation and potential military application of these new developments, and in the context of the changing character of contemporary war. It is our hope and intention that this handbook provides a solid way into thinking about these changes underway and their implications as they unfold.

Notes

1 Mary Kaldor, *New and Old Wars: Organised Violence in a Global Era*, 3rd edn, Cambridge: Polity, 2012; Herfried Munkler, *The New Wars*, Cambridge: Polity, 2005.
2 Eliot Cohen, 'A Revolution in Warfare', *Foreign Affairs* , Vol. 75, 1996, pp. 37–54.
3 Frank Hoffman, *Conflict in the 21st Century: The rise of Hybrid Wars*, Arlington, VA: Potomac Institute for Policy Studies, 2007.
4 Michael Ignatieff, *Virtual War: Kosovo and Beyond*, London: Chatto & Windus, 2000.
5 Christopher Coker, *Humane Warfare: the new ethics of postmodern war*, London: Routledge, 2002.
6 Colin McInnes, 'Spectator Sport Warfare', *Contemporary Security Policy*, Vol. 20, 1999, pp. 142–65.
7 Lieutenant General Sir Rupert Smith, *The Utility of Force: The Art of War in the Modern World*, London: Allen Lane, 2005.
8 Therése Pettersson and Peter Wallensteen, 'Armed conflicts, 1946–2014', *Journal of Peace Research*, Vol. 52, No. 4 (2015), pp. 536–50.
9 Martin Libicki, 'The Specter of Non-Obvious Warfare', *Strategic Studies Quarterly*, 2012, pp. 88–101.
10 David Kennedy coined the term 'Lawfare' to refer to the way in which law has become a vernacular for legitimation of war. David Kennedy, *Of War and Law*, Princeton: Princeton University Press, 2006.
11 Gerry Simpson, *Law, War and Crime: War Crimes, Trials and the Reinvention of International Law*, Cambridge: Polity, 2007.
12 Thomas Rid, *Cyber War Will Not Take Place,* London: Hurst, 2013.
13 Lawrence Freedman, *The Transformation of Strategic Affairs*, London: IISS, 2006.
14 Smith, *The Utility of Force* (see note 7 above).

PART I

Law, war and technology

2
OBVIOUS AND NON-OBVIOUS
The changing character of warfare

Ernst Dijxhoorn and James Gow

The character of warfare has been in a period of change for over two decades. In many instances, the received images of conventional warfare, based on the highly organised and trained forces engaged in the Second World War, have become outmoded and inappropriate. The nature of both the actors in contemporary warfare and of the changing technological capabilities available to those waging war fundamentally challenge existing international law and the state-centric paradigm of the use of armed force involving some degree of kinetic force – that is, energy transfer through blast and fragmentation.[1] While states remain a focal point for conflict in the world – all warfare is, on one level or another, conducted in relation to states – there is no war, whether obvious or non-obvious, in the contemporary era, which does not involve non-state actors – whether these are proxies for states (including private military security companies), territorial or non-territorial insurgent movements, terrorist, national or transnational movements, or coalitions of states (or states and other types of non-state actors).[2] Similarly, a range of new technical means, most often and obviously epitomised by cyber technology and notions of cyber attacks, and their potential application and exploitation, cannot easily be accommodated within the existing legal framework, if at all.[3] Cyber warfare dominated thinking in the context of non-obvious warfare – that is, modes of warfare in which the identity of actors, the character of particular actions, or the very fact of warfare are either unknown or ambiguous.[4] Yet, other notable instances include, but not exclusively, space warfare, drone warfare, autonomous weapons, and the chilling prospects of synthetic biological weapons. What has been described as a revolution in military affairs has been ushered in by technological advancements that are out of line with established international law. Characteristics such as swiftness, non-kinetic nature, anonymity (or plausible deniability), and distance, have proven difficult to accommodate with a legal paradigm based on a state-centred concept of armed force involving some degree of kinetic force, in the context of conventional 'Geneva Convention' warfare. This chapter reviews the changes in the strategic and legal landscape in which technological change must be accommodated and understood – referred to elsewhere in terms of a *Transformation of Strategic Affairs*.[5]

The changing character of warfare

The shift in the main features and dominant forms of warfare over the last decades of the twentieth century and first decades of the twenty-first century occurred under the influence of various factors. The changing international order,[6] itself both an effect and a cause of changes in statehood,[7] changed the character of warfare. Conventional warfare as seen in the early twentieth century was characterised by regular armed forces of states engaging with each other; by the end of that same century this way of waging war – by applying as much armed force as possible to another state's centre of gravity – made place for a way of waging war in which other factors dominate. Between the Napoleonic era and the Cold War the means and methods for armed forces to wage 'total war' developed in which the key objective was to apply mass force to destroy the power centre of the enemy in order to overcome them. Yet, in the twenty-first century, instead of seeking to destroy the enemy by applying greater amounts of destructive force, the key purpose of armed force became the achievement of a quality or a condition, rather than an objective, physical demand.[8]

William S. Lind, and then Thomas X. Hammes, called this new paradigm 'Fourth Generation Warfare'.[9] Others have characterised it as '3-block warfare',[10] or labelled it as a shift to 'hybrid war'.[11] What 'Fourth Generation Warfare', '3-block warfare' and 'hybrid war' have in common is that all notions recognise the relative complexity of war, and the salience of politics, and the will of population groups as being central to war. As a result, in contemporary conflict, brute force alone is no longer enough to win a war. In fact, the application of brute force might make it harder to win, so that if armed force is used it should be calibrated judiciously to fit with the political and social contexts of the operation. Lawrence Freedman described this phenomenon, where narrative replaced sheer brute force as the decisive element in warfare, as the *Transformation of Strategic Affairs*.[12] Following Freedman's logic, Joseph Nye summarised this as: 'It is not whose army wins, but whose story wins'.[13]

The focus of many theories of contemporary warfare is on how major combat with decisive victory as objective has been replaced by warfare in which the use of technological developments, linked to cultural aspects of warfare, and applied in a network oriented, diverse environment can lead to the accumulation of effects that cause the enemy to collapse. This led to effects-based warfare, in which the message became as important as the missile; and in which the missile only served its purpose if it sent the right signal. Rupert Smith, in his book *The Utility of Force*, distinguishes between six key characteristics of contemporary warfare that differ from the previous paradigm, which he calls 'industrial' warfare; it is fought not for victory but to create political or strategic conditions, in contemporary war non-state actors are central, and the key to war is the struggle for the will of 'the people', because war is fought amongst the people.[14] All of these characteristics, and especially all these characteristics combined, mean that the ends, demands, means and character of warfare have changed.[15]

At the same time that the particular manifestations of war changed due to changes of actors, objectives, and the technological means available to attain those objectives, the essence of war remained the same. The character of warfare might differ in different contexts or periods, yet the fundamental nature of war is necessary and eternal. Regardless of how and by whom the art of war is practised, it is always practised for a political purpose.[16] Only because of its political purpose can war be considered a legitimate means to settle political disputes when other means to do so fail. War always involves a dedicated social organisation for the management of restrained coercive violence; armed forces that are trained and disciplined in the management and application of violence. That the application of violence is restrained is central in this because it demonstrates that even the special condition of war is subject to conventions and norms. While the content of these rules and conventions might change, there are always laws of war that make a distinction possible between deeds that are not permissible under any circumstance, not even war, and acts that, under some circumstances, are permitted only in the extraordinary condition of war.

This shift of the character of warfare can be explained by reference to the primary and secondary 'trinities' identified by Carl von Clausewitz.[17] These constitute the eternal essence of warfare. Clausewitz's primary trinity, building on the idea that war was a political phenomenon where armed force was applied to reach decisions, was governed by the interplay of reason, chance (or probabilities) and passion. This captures the very nature of war. Along with its political purpose is the involvement of armed forces (no matter how force itself is put to work) and conventions. Clausewitz's secondary trinity comprised the government, the armed forces and the people. This links directly to contemporary notions of war.[18] Indeed, it provides a lucid perspective on the changed character of warfare while retaining a hold on the essence of war. Thus, those who argue that Clausewitz is no longer pertinent, or that Clausewitzian-type warfare had become obsolete as the trinity no longer applied are mistaken.[19] Although Martin van Creveld was right in that the conflicts typical for the nineteenth and twentieth century in which one state applies mass force against another state became less likely to occur, he was wrong in claiming that the Clausewitzian trinity had become irrelevant.[20] Clausewitz maintained that warfare is the outcome of a trial of physical strength and a tussle of will: and both were always, and remain, present in warfare, even if that tussle is fought with different means.

In the era of 'industrial warfare' military success or failure was judged by who attained victory in a trial of strength, and the power this gained the victor. The parallel conflicts fought during the Cold War had already revealed a crucial transition in the character of, and belligerents in, warfare.[21] In the 1990s it had become apparent that the balance had shifted to the other end of the scale and that a struggle for power and influence – for the will of the people – took precedence over winning battles to take, hold or destroy something with military means. While the balance between these elements had reversed, that balance still applied to both the primary and secondary Clausewitzian Trinities; to reason, chance and passion and to political leaders, armed forces and peoples. The secondary Clausewitzian trinity was still relevant to every single war fought: even as the character of war changed, political elites, armed forces, and communities remained the relevant variables in any armed conflict. In this context of a shift towards winning the struggle of wills in order to win the trial of strength, the outcome of any war, in large part, was defined by the need to 'win' through success at the level of 'hearts and minds'. As Smith correctly asserts, the battle for 'hearts and minds' has gone from being a support activity in military operations to being their central purpose.[22]

In a battle for hearts and minds, the set piece battles involving large groups of armed forces lost most of their relevance but regular and professional armed forces did not. In fact, there might even be a greater demand for the skills of well-trained professional soldiers as those who apply coercive violence – not least because they have to deal with new subtleties and challenges in the context of contemporary conflict. In the socio-political conditions of the current world, the type of armed force applied in, for instance, the two World Wars will not work. Yet, even with the advanced technical capabilities available in the twenty-first century, armed force, by its very nature, is a blunt instrument and therefore the skill and precision of those applying it where necessary is at a premium. Especially because of the potentially grave consequences if something goes wrong in the application of armed force, those chances have to be minimised.[23] In contemporary armed conflict the application of kinetic force – physical force that generates blast and fragmentation – to destroy life or property, despite technological advancement, requires unprecedented guile to be effective in its use.

The need to use force effectively in a manner that is not about physical control or destruction but about creating a condition or set of conditions was demonstrated by the 2003 invasion of Iraq. In that instance the phase of traditional combat typical for 'industrial' warfare fought with the aim of conquering and holding territory was short and relatively successful. This led to the famous photo opportunity in May 2003 of US President George W. Bush on the USS *Abraham*

Lincoln, standing beneath a 'mission accomplished' banner.[24] The years that followed made it painfully clear that this phase was a prelude to a long and intensive armed conflict – fought in a manner that reflects the dominant mode of warfare of the twenty-first century. The problems of legitimacy surrounding the Global War on Terror and especially the Iraq War were further manifestations of this changing character of war.

Legitimacy and success in contemporary warfare

As Gow pointed out in *War and War Crimes*, the Iraq War also demonstrated the changing relevance of war crimes in how war is waged and emerging legal issues at the heart of any conceptualisation of war. This could first be seen in the wars of the early 1990s, when satellite television networks made it possible to broadcast mass murder almost real-time into the living rooms of Western audiences, the most salient example being the war in the former Yugoslavia. This led to what Frits Kalshoven and Liesbeth Zegveld called 'a shift from concern to condemnation' in the thinking on human rights violations,[25] and it gave rise to the establishment of the first modern war crimes tribunal. Or, as Smith describes it, it meant that in contemporary armed conflict force is judged by its morality and legality.[26]

The means for achieving full military victory became limited because the morality of force is defined by the legality of force. War crimes accusations – and, indeed, prosecutions – have come ever more to address the normal conduct of military affairs, rather than the exceptional and clearly illegal. This means that the calculus of armed force, the kinds of judgements made by those ordering and applying armed force changed. Especially, but not exclusively, this applied to the members of armed forces of States Members to the Rome Statute for the International Criminal Court international law considerations are vital. The establishment of a permanent International Criminal Court means that violations of international law give rise not only to state responsibility, but also to individual criminal responsibility. Changing legal and judicial environments that contributed to the salience of war crimes accusations and issues of wrong and right in the conduct of contemporary military operations had an impact on strategy and other kinds of considerations that have to be made in the conduct of warfare. Some have called this phenomenon, demonstrated most pertinently in the still ongoing 'fury of litigation' that followed the ill-fated 2003 Iraq War, 'Lawfare', and others the 'judicialisation of war'.[27]

In modern warfare, the aim is, as noted above, what Smith calls 'a set of conditions'. Rather than attaining a concrete physical objective contemporary strategy has as its central aim to achieve a quality. Attaining a physical objective might sometimes be a requirement to be successful in creating that quality, but it is that quality or condition that is the aim of contemporary warfare. In order to be successful in creating this condition, an appreciation of different levels of warfare is needed. While the traditional hierarchy of levels of warfare – political, strategic, operational and tactical – is still present in modern war these levels do not operate as hierarchically as expected, and while the levels of warfare are defined in relation to each other, different levels do not influence decision-making at other levels in a straightforward manner.

Smith stresses the importance for those ordering the use of armed force to consider the overall aim of a mission when deciding on the level and degree with which force is applied – especially as force needs to serve the purpose of creating a political or strategic condition and the wrong level of force or degree of armed force could undermine the overall objective to create that condition. This is further complicated by the fact that contemporary war is fought in a context where non-state actors are prevalent. This is most notable at the sub-, quasi- or non-state level but the concerns of multi-state international organisations, coalitions and partnerships also have to be taken into account even if they do not impact directly on the ability to use force.

Moreover, the key is the struggle for the will of 'the people' because war is fought amongst the people. However, as Smith notes, the 'people' include also the 'global theatre' of third parties who watch wars, express views on them, and even influence their aims.[28] It is this multi-dimensional arena that distinguishes contemporary war from previous paradigms of war.

In contemporary war legitimacy is the key to success, yet it is also harder than ever to create and maintain legitimacy,[29] not least because legitimacy has to be sought in various constituencies and the ability to create and maintain legitimacy is affected by multiple constituencies simultaneously. This problem of multiple constituencies is a reflection of the globalised, internationalised and interconnected world in which fighting armed conflicts is no longer the sole domain of state entities. Although the state remains central in many ways, the realities of armed conflict mean that the involvement of non-state actors, whether of sub-state groups within a country, such as Hezbollah, or a transnational network, such as al Qa'ida, or of multinational coalitions, alliances or partnerships, such as NATO, prevent the individual state and its armed forces from acting freely, if at all, or unilaterally. As a result, rather than in the case of two state armies straightforwardly engaging in battle with each other, those engaged in contemporary warfare have to take multiple constituencies into account.

While often imagined as Clausewitz's duel between two parties, warfare always involves alliances and coalitions and is rarely binary in practice, and rarely has been. In this binary form, the notions of legitimacy and success as victory can be related to the famous Clausewitzian secondary 'trinity' of government, armed forces and the people at home.[30] The key aspects of nineteenth-century statehood are reflected in this trinity, but the importance of legitimacy and of messages, narratives or 'hearts and minds' that are central in contemporary warfare can also be linked to the 'trinity'. Understanding of the secondary Clausewitzian trinity in the contemporary context is vital to legitimacy as it still operates in every single conflict situation; the relationships between political leaders, their armed instruments and the communities on whom they depend, or whose support they seek, remain relevant in every armed conflict.[31]

The trinity became even more significant because of its factorial expansion into a more complex, multidimensional trinity. Decisions have to be gauged against the perspectives of multiple trinities, not only within the entities' own society, but also in that of allies and opponents, and the global audience. The expansion of the trinity was characterised by Gow as the Multidimensional Trinity Cubed (Plus) – (which can be expressed by the emblem Trinity3(+), or T3(+)).[32] Firstly, a battle of wills has to be fought on the home front, in each case, comprising political leaders, armed forces and people. Secondly, each aspect of the opponent's triangle of political leaders, armed forces and people needs to be influenced, as well as all of them at the same time. Thirdly, there are multiple global audiences, all subject to the same information and the same images, and all affecting the environment in which any strategic campaign is conducted.[33]

As an extra dimension to the complex conditions of alliances and support in contemporary warfare, there are multiple audiences for, or potential reservoirs of, support (or challenges to legitimacy), and there is a need to be aware of transnational communities cutting across the boundaries of states. With this comes the need to recognise how these transnational communities might be affected, and how their ideas about what is legitimate and what not are influenced by both the messages sent by missiles and bombs and those sent via Facebook and on Twitter. In the contemporary world, individuals have access to people and ideas around the world and communities are no longer based on a geographical proximity, but on shared interests.

Legitimacy holds the key to success in contemporary warfare. Various constituencies are important in a context of strategic-tactical compression. Tactical-level action needs to be coordinated, not only within its own country's context, but also with those of its allies because it potentially affects the legitimacy of other contingents, and of the operation as a whole. In situations of complexity and ambiguity, the same set of relationships applies on all sides. The third category, comprising multiple trinities that are not directly engaged in a conflict, but which form an international public, what Smith calls audiences in the 'global theatre', also needs to be influenced. This means that tactical and operational actions have the potential to be significant at the strategic level.[34] To be successful it is essential to take appropriate action at any particular level of the conflict to achieve the desired effect at the overall strategic level.

Scientific innovations and the revolution in information and communication technologies gave rise to the globalisation of economies and financial institutions. These innovations also enabled the development of highly advanced kinetic weapon systems for state armies (often designed with the previous paradigm of industrial warfare in mind), some of which are discussed in this volume. Finally, science and technology originally not designed for use in warfare, or even to have a kinetic effect, is used in the struggle for wills that modern war has become. Both non- and quasi-state entities that are engaged in armed conflicts, and various state actors as well, learned how to make effective use of modern technology to contribute to establishing the condition or conditions that are the objectives in contemporary conflict. In an interconnected environment, victory can only be achieved by superior use of all available networks and effectively deployed force to send the desired messages to the multiple constituencies relevant in the conflict.[35] It is important how that force is best utilised and deployed to be successful in ensuring legitimacy for the use of armed force and for the organisation in the relevant constituencies. Success, even in conflict, is not a zero-sum game. Legitimacy is not a zero-sum game in the sense that when one entity loses it its opponent automatically gains legitimacy and thereby success, but if the narrative of one belligerent is more successful in a constituency, its opponent will likely lose legitimacy In a struggle of wills it is therefore often opportune to deploy a strategy that hampers the ability of the opponent in a conflict to create and maintain legitimacy. In contemporary armed conflict this is not just done by missiles, but by deploying all available means and networks and a combination of the use or threat of classic armed force with other means, whether they are sabotage, economic means, or other obvious or non-obvious means.

In conclusion, the context in which technological and scientific innovation is occurring – in which the character of warfare has changed, even if the essence of warfare remained unchanged – is itself rife with ambiguities. Yet, despite the changed manifestations of war, the scientific innovations and technological capabilities discussed in this book, when used in warfare, are deployed for political purpose. This deployment ideally involves designated social organisations trained at deploying that force with some restraint, whether they can be called 'armed forces' in a classic sense or not. And finally, this force is subject to conventions and norms, even if it is not always immediately apparent exactly what these norms and conventions are.

Conclusion

Contemporary war is different in that wars are no longer fought for decisive victory by two, or more, state armies meeting on the battlefield, but to create political or strategic conditions. Although states still wage war, modern armed conflicts typically also involve at least one belligerent that is not a state; these are called terrorists, insurgents or rebels, Moreover, modern conflicts often involve coalitions such as NATO or ad hoc coalitions 'of the willing'; regional or supranational organisations. The key to war is the struggle for the will of 'the people', and it is fought amongst the people. This means that the ability to create and maintain legitimacy is the key to success in contemporary armed conflict. But, as explained above, those who order armed force have to take into account how that influences various constituencies. As a result every decision has the potential to have an impact at the strategic level; there is a strategic-tactical compression.

Being successful in creating and maintaining legitimacy is dependent on the clever use of armed force. But the means to achieve all-out victory are limited because the morality of force is judged by the legality of it. Legal issues are at the heart of any decision at every level of warfare, not in the least because international criminal prosecutions gave rise to individual criminal responsibility. Moreover, success in creating a condition or set of conditions that are the aim of modern war is not just dependent on the clever use of armed force but of all means at the disposal of a party: an unlimited access to almost real-time (mis)information through both traditional and the new (social) media has a huge impact on the means to create and maintain legitimacy for military operations and the use of armed force. This is particularly so as this legitimacy has to be sought in various publics who consume, and base their judgement on, a constant stream of information and commentary – some of it in the form of powerful and shocking visual imagery streamed around the world even as events are unfolding. Finally, as discussed in this book, in contemporary war it might be more or less obscured *who* is waging war in the first place, as with the (Russian) 'Little Green Men' taking over the Crimean peninsula, or in the case of using cyber capabilities or drone strikes.[36] In an era in which information is easily shared and available to many, disinformation can be deployed as a weapon, maybe more than ever before. For all these actors: states, supranational, multinational, and quasi-state entities, success in contemporary armed conflict depends largely on legitimacy: the ability to create and maintain it, in various constituencies simultaneously. 'Hybrid' 'non-obvious', or Fourth generation warfare means that an entity has to use every means at that entity's disposal; the use or threat of 'classic' armed force combined with other means, including cyber capabilities and economic measures as means of applying pressure, and providing competing or confusing narratives to prevail in a struggle for wills.

Notes

1 Martin van Creveld, *On Future War*, London: Brassey's, 1991; Chris Hables Gray, *Postmodern War: The New Politics of Conflict*, London Routledge, 1997; Lieutenant General Sir Rupert Smith, *The Utility of Force: The Art of War in the Modern World*, London: Allen Lane, 2005; James Gow, *Defending the West*, Cambridge: Polity, 2005.
2 Smith, *The Utility of Force*, Gow, *Defending the West.*
3 James Gow & Rachel Kerr, 'Law and War in the Global War on Terror', in A. Hehir, N. Kuhrt and A. Mumford (eds.) *International law, security and ethics: policy challenges in the post-9/11 world*, London: Routledge, 2011 pp. 61–78; Philip Bobbitt, *The Shield of Achilles: War, Peace and the Course of History*, Penguin: London, 2003; Philip Bobbitt, *Terror and Consent: The Wars for the Twenty-first Century*, London: Allen Lane, 2008.
4 Martin C. Libicki, 'The Specter of Non-Obvious Warfare', *Strategic Studies Quarterly*, Fall 2012, pp. 88–101.
5 Lawrence Freedman, *The Transformation of Strategic Affairs*, Adelphi Paper 379, London: Routledge for the International Institute of Strategic Studies (IISS), 2006.
6 Gow, *Defending the West* (see note 1 above).
7 Bobbitt, *The Shield of Achilles* (see note 3 above).
8 Smith, *The Utility of Force*, Ch. 7 (see note 1 above).
9 William S. Lind, 'Understanding Fourth Generation War', *Military Review*, September-October 2004, pp. 12–16; Thomas X Hammes, *The Sling and the Stone: On War in the 21st Century*, St Paul: Zenith Press, 2004, p. 208; Thomas X Hammes, 'War Evolves into the Fourth Generation', *Contemporary Security Policy*, Vol. 26, No. 2, 2005.
10 Gen. Charles Krulak, 'The Strategic Corporal: Leadership in the Three Block War', *Marines Magazine*, January 1999.
11 Margaret S. Bond, *Hybrid War: A New Paradigm for Stability Operations in Failing States*, Carlisle, PA: US Army War College, 30 March, 2007; Frank G. Hoffman, *Conflict in the 21st. Century: The Rise of Hybrid Wars*, Arlington, VA: The Potomac Institute, December 2007.
12 Lawrence Freedman, *The Transformation of Strategic Affairs.*

13 Joseph Nye, 'Soft Power and the Struggle Against Terrorism', Lecture, The Royal Institute of International Affairs, Chatham House, London, 5 May 2005., See also: Joseph Nye, *The Paradox of American Power: Why the World's Only Superpower Can't Go It Alone*, Oxford: Oxford University Press, 2002; Joseph Nye, 'US Power and Strategy After Iraq', *Foreign Affairs*, Vol. 82, No. 4, July-August 2003.
14 Smith, *The Utility of Force*, Ch. 7 (see note 1 above).
15 Milena Michalski & James Gow, *War, Image and Legitimacy: Viewing Contemporary Conflict*, New York: Routledge, 2007 p. 200.
16 James Gow, *War and War Crimes*, London: Hurst & Co, 2013.
17 Carl von Clausewitz, *On War*, trans. J. J. Graham, 'Introduction' and 'Notes' by Colonel F. N. Maude, C. B (Late R. E.) and. 'Introduction to the New Edition', by Jan Willem Honig, New York: Barnes and Noble, 2004.
18 This is Clausewitz's secondary trinity, which is a reflective derivative of the primary trinity of reason (linked mainly to government), chance (linked mainly to the military) and passion (linked mainly to the people). See Beatrice Heuser, *Reading Clausewitz*, London: Pimlico, 2002, pp. 53–54.
19 Such as Mary Kaldor and Martin van Creveld in their notion of 'post-Clausewitzian' warfare. Mary Kaldor, *New and Old Wars: Organized Violence in a Global Era*, Cambridge: Polity, 2001; Martin van Creveld, *On Future War*.
20 Van Creveld, *On Future War* (see note 1 above).
21 Ernst Dijxhoorn, *Quasi-State Entities and International Criminal Justice*, Abingdon: Routledge, 2017, p. 6.
22 Smith, *Utility of Force*, p. 278 (see note 1 above).
23 Even when applied by professionals with advanced weapons, the application of armed force can go horribly wrong; the bombing of the Chinese Embassy in Belgrade in 1999, or the Médecins Sans Frontières hospital in Kunduz in 2005 both had severe political and diplomatic consequences But this may be demonstrated in the case of the soldiers manning the BUK installation that shot down Malaysian Airlines Flight 17 on 17 July 2014. This – most likely an unintentional consequence of the misapplication of force – not only led to the tragic loss of 298 civilians but also complicated Russian foreign relations significantly; and the ensuing investigation hampers the Russian ability plausibly to deny military involvement in Ukraine and potentially its ability to wage non-obvious war. Joint Investigation Team MH17, Update in Criminal Investigation MH17 Disaster, Press Statement 24 May 2018. www.om.nl/onderwerpen/mh17-crash/.
24 While President Bush stood on that aircraft carrier off the coast of California he claimed that: 'Major combat operations in Iraq have ended. In the battle of Iraq, the United States and our allies have prevailed'. CNN, *Bush makes historic speech aboard warship*, 2 May 2003. http://edition.cnn.com/2003/US/05/01/bush.transcript/ (accessed 8 June 2018).
25 Michalski and Gow, *War, Image and Legitimacy* (see note 15 above), p. 118; Frits Kalshoven and Liesbeth Zegveld, *Constraints on the Waging of War*, Geneva: ICRC, 2001, p. 185.
26 Smith, *The Utility of Force*, p. 268 (see note 1 above).
27 Gerry Simpson, *Law, War and Crime: War Crimes, Trials and the Reinvention of International Law*, Cambridge: Polity, 2007.
28 Smith, *The Utility of Force*, p. 289 (see note 1 above).
29 It could be argued that in some ways legitimacy has always been key to military success. Philip Bobbitt has argued that law, or legitimacy, and strategy have always been two sides of the victory coin – law has historically been invoked to confirm military success, while armed force is needed to enforce legal claims. Bobbitt, *Shield of Achilles*, p. 226 (see note 3 above). Notions of 'will', 'hearts and minds' and, ultimately, in effect, legitimacy, were recognised as relevant to counter-insurgency operations in the twentieth century; the same types of issues are relevant to the dominant forms of warfare, reflecting the nature of all types of conflict, in the contemporary era. On counter-insurgency and its evolution see, for some of the better examples, John MacKinlay, *The Insurgent Archipelago*, London and New York: Hurst/Columbia University Press, 2010; Thomas Rid and Thomas Keaney, eds., *Understanding Counterinsurgency: Doctrine, Operations and Challenges*, Abingdon: Routledge, 2010; and David Kilcullen, *Counterinsurgency*, London: Hurst and Co., 2010. For an unusual and particularly reflective account of the issues see M. L. R. Smith, 'Whose Hearts and Whose Minds: The Curious Case of Global Counterinsurgency', *Journal of Strategic Studies*, February 2010.
30 See the discussion in note 18 above; This is Clausewitz's secondary trinity, which is a reflective derivative of the primary trinity of reason (linked mainly to government), chance (linked mainly to the military) and passion (linked mainly to the people). See Beatrice Heuser, *Reading Clausewitz*, London: Pimlico, 2002: pp. 53–54; John Stone's subtle analysis of technology and the trinities casts useful light

on the latter and their relevance to contemporary armed conflict; see also Stone, 'Clausewitz's Trinity' and Contemporary Conflict', *Civil Wars*, Vol. 9, No. 3.

31 Gow, *War and War Crimes* (see note 16 above).

32 Michalski & Gow, *War, Image and Legitimacy* (see note 15 above), p. 201.

33 Ibid.

34 While this idea was already labelled 'the strategic corporal' by US Marine General Charles Krulak in the 1990s, it did not penetrate other parts of the US military, and painful lessons had to be learned during the Iraq and Afghanistan Wars 'The Strategic Corporal: Leadership in the Three Block War', *Marines Magazine*, Vol. 83, Issue 1, January 1999; Terry Terriff, Frans Osinga and Theo Farrell, eds., *A Transformation Gap?: American Innovations and European Military Change*, Stanford: Stanford University Press, 2010.

35 Hammes, *The Sling and the Stone*, p. 208 (see note 9 above).

36 Libicki, 'Non-Obvious Warfare' (see note 4 above).

3

WEAPONS LAW, WEAPON REVIEWS AND NEW TECHNOLOGIES

Bill Boothby

New weapons technologies are not developed, procured and fielded in a legal vacuum. International law includes rules that prohibit certain weapons entirely in armed conflict, while others are the subject of restrictions as to the circumstances in which they can lawfully be employed. Critical to this body of weapons law are the two established, customary legal notions, namely humanity and military necessity. These principles do not constitute rules in their own right but are best seen as the mutually counterbalancing principles that underpin the whole of the law of armed conflict, and thus the law relating to weapons. The rules of the law regulating weapons represent the balance that states have struck between conflicting interests; a balance that will vary from weapon to weapon depending on how states perceive the military need associated with the weapon and how they interpret the humanitarian concerns that have motivated the legal provision in question.

Several of the legal rules we are about to discuss refer to 'weapons', 'methods', and 'means' of warfare and it is important to understand the meaning of these terms, none of which is defined in any treaty. The ordinary meaning of a 'weapon' is 'a thing designed or used for inflicting bodily harm or physical damage; a means of gaining an advantage or defending oneself'.[1] One commentator referred to 'an offensive capability that can be applied to a military object or enemy combatant';[2] so it seems sensible to conclude that a weapon is a device, munition, implement, substance, object, or piece of equipment that is used, designed, or intended to be used for these purposes.

That notion will extend to all arms, munitions, materiel, instruments, mechanisms, or devices that have an intended effect of injuring, damaging, destroying, or disabling personnel or property. This implies that a weapon system comprises the weapon or munition itself and associated elements required for its operation having a directly injuring or damaging effect on people or property. Examples include all munitions and technologies such as projectiles, small arms, mines, explosives, and all other devices and technologies that are physically destructive or injury-producing.[3]

'Methods' and 'means' include 'weapons in the widest sense, as well as the way in which they are used'.[4] So, methods and means of warfare mean, respectively, tactics, whether lawful or unlawful, and weapons and weapon systems.

This means that the legal rules that will shortly be considered prohibit or limit the use of things that are designed, intended, or used to harm persons or to damage property. If the purpose in acquiring weapons is to gain a military advantage or to enable certain kinds of attack to be undertaken, such as killing or injuring enemy personnel or damaging or destroying his property, weapon systems will also include platforms and equipment which will not as such cause damage or injury, but which form an essential part of the system that does have that effect.

How weapons law emerged

Dr Francis Lieber[5] prepared a Code issued to the Union side in the American Civil War in which he explained military necessity as the necessity of those measures which are indispensable for securing the ends of the war. He considered that military necessity should not permit cruelty, such as the infliction of suffering for its own sake, or the use of poison.[6] Although the Lieber Code was never adopted by states and is therefore not regarded as a source of international law as such, it does represent an informed assessment of law as it then existed. The prohibition of the use of poison is also reflected in Article 23(a), Hague Regulations 1907,[7] and is certainly a rule that applies to all States and in all armed conflicts, international and non-international.

The idea that the infliction of suffering for its own sake is prohibited later developed into one of the two core principles of the law of weaponry which prohibits weapons that by nature cause superfluous injury or unnecessary suffering. We will discuss this principle in a later section of this chapter.

A Declaration adopted in St Petersburg in 1868 prohibited 'the employment … of any projectile of a weight below 400 grammes, which is either explosive or charged with fulminating or inflammable substances'.[8] It is fair to say that the weight limit prescribed by the treaty no longer reflects the law.

Customary law does, however, continue to prohibit the use of explosive or incendiary bullets designed exclusively for use against personnel. This is because a solid round would achieve the relevant military purpose, so using an explosive round against such targets will cause injury for which there is no military necessity.[9]

The Peace Conferences conducted in The Hague in 1899 and 1907 took the development of weapons law further. Declaration II prohibited the use of 'projectiles the sole object of which is the diffusion of asphyxiating or deleterious gases',[10] but only applied to the states party in the case of a war between two or more of them and would not apply if a state that was not a party took part in the conflict, and the Declaration did not prohibit weapons which combined gas diffusion with some other object, such as blast or fragmentation. The Declaration did not therefore prevent the gas attacks on the trenches during the First World War that resulted in so many casualties and so much suffering. It was in the aftermath of this experience that the Geneva Protocol of 1925 was adopted. Many decades later, the Chemical Weapons Convention of 1993 and the Biological Weapons Convention of 1972 addressed these mass casualty weapons in an altogether more comprehensive way.

The third Declaration of 1899 prohibited 'the use of bullets which expand or flatten easily in the human body, such as bullets with a hard envelope which does not entirely cover the core or is pierced with incisions'.[11] It also only applied in the case of a war between two or more states that are party to the treaty and would not apply if a non-contracting state entered the conflict. The prohibition now has customary law status, meaning that it binds all states whether or not they are party to the treaty. The customary rule prohibits the use in armed conflicts between states of bullets that are designed, in the intended circumstances of use, to expand or flatten easily in the human body.

The position in relation to armed conflicts within a state is more complex. Consider, for example, a hostage situation in which a hostage-taker must be instantly disabled for the protection of civilians, or consider a situation in which the ricochet to be expected from a high velocity round will imperil civilians in the vicinity of the target. The use of expanding ammunition in such relatively limited non-international armed conflict situations would cause the target additional suffering for which there is a military purpose, namely the protection of the relevant civilians, so the superfluous injury/unnecessary suffering rule would not, arguably, be breached in those limited kinds of circumstance and, therefore, neither would international law.[12]

Superfluous injury and unnecessary suffering

At this point we should conclude our brief account of the early emergence of the law of weaponry and start to get to grips with its two core principles. The first of these prohibits injury that lacks military utility, an idea first expressed in the modern era in the Preamble to the St Petersburg Declaration 1868:

> Considering that the progress of civilisation should have the effect of alleviating as much as possible the calamities of war;
>
> That the only legitimate object which states should endeavour to accomplish during war is to weaken the military forces of the enemy;
>
> That for this purpose it is sufficient to disable the greatest possible number of men;
>
> That this object would be exceeded by the employment of arms which uselessly aggravate the sufferings of disabled men, or render their death inevitable;
>
> That the employment of such arms would therefore be contrary to the laws of humanity.[13]

The Brussels Declaration and the Oxford Manual contain similar statements,[14] but it was in Article 23(e) of the Hague Regulations of 1907 that the notion was first expressed as a substantive rule of law, as follows: 'It is especially forbidden to employ arms, projectiles or material calculated to cause unnecessary suffering.'

Seventy years later, in API, the modern formulation of the rule provides:

> It is prohibited to employ weapons, projectiles and material and methods of warfare of a nature to cause superfluous injury or unnecessary suffering.[15]

This customary rule, which binds all states in all types of armed conflict, is central in importance to the law of weaponry. The practical application of the rule, however, involves a comparison of inherently dissimilar phenomena. The elements to be compared are, it is suggested, the degree and relative extent of the suffering or injury that the use of the weapon will inevitably occasion and the generic military advantage or utility to be anticipated from the employment of the given weapon in its intended circumstances of use. The terms 'superfluous' and 'unnecessary' confirm the comparative nature of the test that is to be applied, leading to the conclusion that a weapon is likely to breach the rule if it may be expected to cause injury on a scale significantly greater than that to be expected of alternative weapons that yield the same generic military advantage or utility.

Such a comparison process only makes sense if it is the employment of the weapon in its normal, designed circumstances of use that is evaluated.[16] Perfectly lawful weapons are capable of being misused and of having unacceptable or even unlawful effects as a result of such misuse. The proper basis for judgement is therefore how the weapon behaves when used in its normal circumstances, within its designed range, employing its intended power setting or velocity, and when being directed at its intended category of target.

So the principle can be expressed as follows:

> the legitimacy of a weapon, by reference to the superfluous injury and unnecessary suffering principle, must be determined by comparing the nature and scale of the generic military advantage to be anticipated from the weapon in the application for which it is designed to be used, with the pattern of injury and suffering associated with the normal intended use of the weapon.[17]

It is for states to evaluate which weapons breach the rule and to assess new weapons and weapon technologies accordingly.

Indiscriminate weapons rule

The second fundamental customary principle of the law of weaponry prohibits weapons whose nature it is to be indiscriminate. During the period before 1974 a rule that indiscriminate attacks are prohibited certainly existed, but there was no treaty text specifically addressing weapons that by nature are indiscriminate. An authoritative commentator noted in 1975 that when the negotiations of what was to become API commenced, not all experts were prepared to acknowledge that a rule prohibiting identifiable 'indiscriminate weapons' had 'acquired the status of a rule of positive international law'.[18] API prohibited indiscriminate attacks[19] which it defined as including attacks 'which employ a method or means of combat which cannot be directed at a specific military objective; or ... which employ a method or means of combat the effects of which cannot be limited' as required by the Protocol and which consequently are of a nature to strike military objectives and civilians or civilian objects without distinction. The clear references in the rule to means of combat render the pre-existing indiscriminate attacks rule into a rule that applies specifically and explicitly to weapons.

The V2 rocket used to attack Southern England from September 1944,[20] its predecessor the V1 rocket, and some Scud rockets have been cited as examples of weapons that would have breached the rule.

Geneva Gas Protocol 1925

The perceived inadequacies of the Hague Declaration IV of 1899 were addressed in 1925 with the adoption of the Geneva Gas Protocol.[21] This treaty prohibited the use in war of asphyxiating, poisonous, or other gases and of all analogous liquids, materials, or devices, and it extended the prohibition to the use of bacteriological methods of warfare. The 'sole object' language, which had provided the loophole in the 1899 text, was not repeated in the 1925 Protocol.

France, the UK and US became party to the Protocol on the stated understanding that they would only remain bound by the prohibitions so long as the adverse party in an armed conflict did not use the prohibited weapons. These 'no first use' arrangements remained relevant until the adoption of the Chemical and Biological Weapons Conventions.[22]

Weapons and the environment

The next relevant piece of conventional law came some 50 years later with the adoption of the Environmental Modification Convention of 1976 (ENMOD).[23] This treaty prohibited military or other hostile use of environmental modification techniques if these would have widespread, long-lasting or severe effects as the primary means of destruction, damage or injury to another state party.[24] This provision will therefore need to be considered if a weapon uses the environment as an instrument for causing the stated degree of damage to another state that is party to the treaty.

The other environmental protection rule within the law of armed conflict is to be found in Articles 35(3) and 55 of API. Article 35(3) prohibits the employment of methods or means of warfare which are intended, or may be expected, to cause widespread, long-term, and severe damage to the natural environment. Article 55 builds on this rule by imposing a requirement to take environmental care and by expressly prohibiting methods or means that are intended or may be expected to cause the prohibited damage and thereby to prejudice the health or survival of the population. Such a requirement to take care will clearly apply to the selection and design of weapons. The API rules are, however, concerned with collateral damage to the environment, as opposed to its use as a weapon.

Conventional Weapons Convention of 1980

Further progress in developing the law of weaponry was achieved with the adoption of the Convention on Prohibitions or Restrictions on the Use of Certain Conventional Weapons which May be Deemed to be Excessively Injurious or to have Indiscriminate Effects, which we shall refer to as the CCW.[25] This Convention facilitates the negotiation and adoption of Protocols that address particular types of weapon. Some of the provisions in these Protocols merely restate the customary principles that we have already discussed or apply to the particular kind of weapon the established targeting rules that apply as a matter of law to all attacks. The following summary will focus on the provisions that seem to develop the law of weaponry.

CCW Protocols prohibit the following specific weapons:

Protocol 1 prohibits the use of 'any weapon the primary effect of which is to injure by fragments which in the human body escape detection by x-rays'.[26]

Protocol II[27] and Amended Protocol II[28] address the use in armed conflict of mines,[29] booby-traps, and other devices.[30] Protocol II defines booby-traps as 'any device or material which is designed, constructed or adapted to kill or injure and which functions unexpectedly when a person disturbs or approaches an apparently harmless object or performs an apparently safe act'.[31] Under Article 6 of Protocol II it is prohibited to use booby-traps in the form of an apparently harmless portable object if they are specifically designed and constructed to contain explosive material and to detonate when they are disturbed or approached.[32] The use of booby-traps is also prohibited if they are in any way attached to or associated with:

- internationally recognised protective emblems, signs or signals;
- sick, wounded or dead persons;
- burial or cremation sites or graves;
- medical facilities, medical equipment, medical supplies or medical transportation;
- children's toys or other portable objects or products specially designed for the feeding, health, hygiene, clothing or education of children;
- food or drink;
- kitchen utensils or appliances except in military establishments, military locations or military supply depots;
- objects clearly of a religious nature;

- historic monuments, works of art or places of worship which constitute the cultural or spiritual heritage of peoples;
- animals or their carcasses.[33]

Amended Protocol II prohibits:

- mines, booby-traps or other devices which employ a mechanism or device specifically designed to detonate the munition by the presence of commonly available mine detectors as a result of their magnetic or other non-contact influence during normal use in detection operations;[34]
- a self-deactivating mine equipped with an anti-handling device that is designed in such a manner that the anti-handling device is capable of functioning after the mine has ceased to be capable of functioning;[35]
- anti-personnel mines that do not incorporate in their construction a material or device that enables the mine to be detected by commonly available technical mine detection equipment and provides a response signal equivalent to a signal from 8 grammes or more of iron in a single coherent mass;[36]
- remotely delivered anti-personnel mines which do not comply with the following requirements:[37] their design and construction must be such that no more than ten percent of activated mines will fail to self-destruct within 30 days after emplacement, and each mine must have a back-up self-deactivation feature designed and constructed so that, in combination with the self-destruction mechanism, no more than one in 1000 activated mines will function as a mine 120 days after emplacement;[38]
- remotely delivered mines other than anti-personnel mines, unless, to the extent feasible, they are equipped with an effective self-destruction or self-neutralisation mechanism and have a back-up self-deactivation feature, which is designed so that the mine will no longer function as a mine when the mine no longer serves the military purpose for which it was placed in position;[39]
- and booby-traps or other devices in the form of apparently harmless portable objects which are specifically designed and constructed to contain explosive material.[40]

Protocol IV to CCW, which owes much to the general disapproval of blinding as a method of warfare, provides:

> It is prohibited to employ laser-weapons specifically designed, as their sole combat function or as one of their combat functions, to cause permanent blindness to unenhanced vision, that is, to the naked eye or to the eye with corrective eyesight devices.[41]

At the time of writing, it can be concluded that a customary rule in the terms of Protocol I to CCW is in the process of emerging and that a customary rule in the precise terms of article 1 of Protocol IV has emerged.

Bacteriological or biological weapons

The 1925 Geneva Gas Protocol only prohibited the use of bacteriological or biological weapons. Article I of the Biological Weapons Convention[42] takes matters considerably further, in that the states party undertakes never in any circumstances to develop, produce, stockpile, or otherwise acquire or retain:

1 Microbial or other biological agents or toxins whatever their origin or method of production, of types and in quantities that have no justification for prophylactic, protective or other peaceful purposes;
2 Weapons, equipment or means of delivery designed to use such agents or toxins for hostile purposes or in armed conflict.

The Convention does not explicitly prohibit 'use', but at the Fourth Review Conference in 1996 it was agreed among the States party that Article 1 has the effect of prohibiting the use of such weapons.[43]

The comprehensive terms of the prohibitions mean that they apply to all classes of conflict. Very many states are party to the treaty, including almost all militarily significant states, and the consistent practice of states supports the conclusion that the prohibition on use of such weapons is now a customary rule of international law which therefore binds all states, whether or not they are party to this treaty.[44] It is also probably the case that the treaty's prohibition on the possession, stockpiling, transfer and development of such weapons is also customary in nature. So it is unlawful for any state, whether it is party to the Biological Weapons Convention or not, to plan, prepare for, equip itself for, or undertake a biological attack.

While attempts have been made at two Review Conferences to agree upon a verification mechanism for the Convention, it has not, to date, proved possible to adopt such a provision.

Chemical weapons

The possession of chemical weapons was also not prohibited by the 1925 Gas Protocol. The adoption in 1993 of the Chemical Weapons Convention,[45] which applies to all classes of conflict, therefore takes the law forward by providing that the participating states agree never under any circumstances:

- To develop, produce, otherwise acquire, stockpile or retain chemical weapons or transfer, directly or indirectly, chemical weapons to anyone;
- To use chemical weapons;
- To engage in any military preparations to use chemical weapons;
- To engage in any military operations to use chemical weapons;
- To assist, encourage or induce, in any way, anyone to engage in any activity prohibited to a State party under the Convention.[46]

'Chemical weapons' means, together or separately:

a Toxic chemicals and their precursors, except where intended for purposes not prohibited under this Convention, as long as the types and quantities are consistent with such purposes;
b Munitions and devices, specifically designed to cause death or other harm through the toxic properties of those toxic chemicals ... which would be released as a result of the employment of such munitions and devices;
c Any equipment specifically designed for use directly in connection with the employment of munitions and devices specified in sub-paragraph (b).[47]

A toxic chemical is any chemical which, through its chemical action on life processes, can cause death, temporary incapacitation or permanent harm to humans or animals. This includes all such chemicals, regardless of their origin or of their method of production, and regardless of whether they are produced in facilities, in munitions, or elsewhere.[48]

A 'precursor' is any chemical reactant which takes part, at any stage, in the production, by whatever method, of a toxic chemical, including any key component of a binary or multi-component chemical system.[49]

If the chemical is intended for purposes which are not prohibited under the Convention and if the amount held is consistent with such innocent purposes, its possession is lawful. Purposes not prohibited under the Convention are:

a Industrial, agricultural, research, medical, pharmaceutical or other peaceful purposes;
b Protective purposes, namely those purposes directly related to protection against toxic chemicals and to protection against chemical weapons;
c Military purposes not connected with the use of chemical weapons and not dependent on the use of the toxic properties of chemicals as a method of warfare;
d Law enforcement, including domestic riot control purposes.[50]

The net effect of these provisions is, quite simply, to prohibit chemical warfare as that term is colloquially understood. The treaty applies in all classes of conflict and the prohibition on use of chemical weapons is clearly now a rule of customary law,[51] with the result that all states, irrespective of their participation in the Convention, are prohibited from using such weapons.

Riot control agents may not be used as a method of warfare, but they may be used for law enforcement, including domestic riot-control purposes. Riot control agents are 'chemicals not listed in a Schedule to the Treaty which can produce rapidly in humans sensory irritation or disabling physical effects which disappear within a short time following termination of exposure'.[52] In the context of modern armed conflict, it is not of course always clear where law enforcement ends and armed conflict begins.

Anti-personnel landmines

Consensus is required before a CCW Protocol can be adopted,[53] and no consensus could be achieved in favour of an Anti-Personnel Landmine (APL) ban. It was therefore an ad hoc process that led to the adoption of the Ottawa Convention,[54] which provides:

Each state party undertakes never under any circumstances:

a To use anti-personnel mines;
b To develop, produce, otherwise acquire, stockpile, retain or transfer to anyone, directly or indirectly, anti-personnel mines;
c To assist, encourage or induce, in any way, anyone to engage in any activity prohibited to a State party under this Convention.

An anti-personnel mine is a mine designed to be exploded by the presence, proximity, or contact of a person and that will incapacitate, injure or kill one or more persons. Mines designed to be detonated by the presence, proximity, or contact of a vehicle as opposed to a person and that are equipped with anti-handling devices are not considered anti-personnel mines as a result of being so equipped.[55]

The treaty prohibitions apply in all classes of conflict, but have not yet achieved customary law status; they are therefore binding only on states party to the treaty.

Cluster munitions

The Convention on Cluster Munitions, adopted at a meeting in Dublin in 2008, obliges states party never under any circumstances:

- to use cluster munitions,
- to develop, produce, otherwise acquire, stockpile, retain or transfer to anyone, directly or indirectly, cluster munitions,
- to assist, encourage or induce anyone to engage in any activity prohibited to a State party under the Convention.[56]

A cluster munition means a conventional munition that is designed to disperse or release explosive sub-munitions each weighing less than 20 kilograms and includes those explosive sub-munitions. It does not mean the following:

a A munition or sub-munition designed to dispense flares, smoke, pyrotechnics or chaff, or a munition designed exclusively for an air defence role;
b A munition or sub-munition designed to produce electrical or electronic effects;
c A munition that, in order to avoid indiscriminate area effects and the risks posed by unexploded sub-munitions, has all of the following characteristics:
   i Each munition contains fewer than 10 explosive sub-munitions;
   ii Each explosive sub-munition weighs more than 4 kilograms;
   iii Each explosive sub-munition is designed to detect and engage a single target object;
   iv Each explosive sub-munition is equipped with an electronic self-destruction mechanism;
   v Each explosive sub-munition is equipped with an electronic self-deactivating feature.[57]

'Explosive sub-munitions',[58] 'self-destruction mechanism', and 'self-deactivating mechanism' are defined terms. Article 21 addresses the interoperability issues that may arise for states party engaging in military cooperation and operations with states not party to the Convention.

The rules in the Convention apply to states party only, but in all classes of conflict.

Non-international armed conflict

The scope of application of the Conventional Weapons Convention and its annexed Protocols was extended in 2001[59] to 'situations referred to in Article 3 common to the Geneva Conventions of 12 August 1949'.[60] The Protocols have that extended scope in relation to states that ratify that extension of 2001. APII always did apply to such conflicts.[61]

The Biological Weapons, Chemical Weapons, Ottawa, and Cluster Munitions Conventions apply to all classes of conflict, as do the customary law on superfluous injury/unnecessary suffering and indiscriminate weapons principles and the customary rules of weapons law, such as the prohibition of poisons or poisoned weapons. The position in relation to expanding bullets was discussed earlier.

Treaty law restrictions on the use of certain weapons

Protocol II and Amended Protocol II to the CCW include restrictions on the use of mines, booby-traps and other devices the details of which lie outside the scope of this chapter. Protocol IV[62] also placed restrictions on the use of certain weapons that are not specifically prohibited.

Protocol III[63] to the CCW defines an incendiary weapon as

> any weapon or munition which is primarily designed to set fire to objects or to cause burn injury to persons through the action of flame, heat, or a combination thereof, produced by a chemical reaction of a substance delivered on the target.[64] [It is prohibited] in all circumstances to make any military objective located within a concentration of civilians[65] the object of attack by air-delivered incendiary weapons.[66]

It is also prohibited to make military objectives located within a concentration of civilians the object of attack using incendiary weapons other than air-delivered incendiary weapons

> except when such military objective is clearly separated from the concentration of civilians and all feasible precautions are taken with a view to limiting the incendiary effects to the military objective and to avoiding, and in any event to minimizing, incidental loss of civilian life, injury to civilians and damage to civilian objects.[67]

The Protocol also prohibits making forests or other kinds of plant cover the object of attack using incendiaries except when such natural elements are used to cover, conceal, or camouflage combatants or other military objectives or have themselves become military objectives.[68]

Non-lethal weapons

NATO policy refers to non-lethal weapons as 'weapons which are explicitly designed and developed to incapacitate or repel personnel, with a low probability of fatality or permanent injury, or to disable equipment with minimal undesired damage or impact on the environment'.[69]

The important point to note is that the 'non-lethal' character of a weapon does not affect the applicability to it of the principles and rules discussed so far in this chapter.

Weapon reviews

States that are party to API[70] are required

> [i]n the study, development, acquisition or adoption of a new weapon, means or method of warfare [...] to determine whether its employment would, in some or all circumstances, be prohibited by th[e] Protocol or by any other rule of international law applicable to the High Contracting Party.[71]

If, therefore, a weapon, means or method is being studied, then the weapons review duty applies. It will be a matter for national judgement when general technology research becomes the study of a weapon.[72]

'Development' involves the application of materials, equipment, and other elements to form a weapon and includes the improvement, refinement, and probably the testing of the prototype weapons with a view to achieving optimal performance.[73] 'Acquisition' involves obtaining weapons from commercial undertakings and/or from other states and 'adoption' involves a state or its armed forces deciding to use a weapon or method of warfare in military operations.

Customary law requires all states to review new weapons to determine whether they comply with the law that applies to the relevant state. The final paragraph of the 1868 St Petersburg

Declaration, Article 1 of Hague Convention IV of 1907, the AMW Manual,[74] and the *Tallinn Manual* all lead to this conclusion.[75] The ICRC concludes that '[t]he requirement that the legality of all new weapons, means and methods of warfare be systematically assessed is arguably one that applies to *all* States, regardless of whether or not they are party to Additional Protocol 1.[76] However, relatively few states are known systematically to ensure the review of all new weapons.[77]

The law is not prescriptive as to the form that a weapon review must take, does not lay down any required procedure, and does not oblige states to disclose the contents of their reviews. Advice to an appropriate commander may be sufficient.[78]

The legal review must apply the rules of weapons law that apply to the state in question. These will be the two customary principles discussed in sections 3 and 4, the customary rules that have been identified, and the treaty prohibitions and restrictions to which the relevant state is party. The important point to note is that it is the principles and rules of current law that must be applied to new weapons and weapon technologies. The peculiarities of a new weapon technology are, therefore, no basis for arguing that an established weapons law rule does not have to be complied with.

Autonomous weapons

We have discussed the principles and rules of the law of weaponry and the obligation legally placed on all States to review new weapons. This section considers how these principles and rules apply to an emerging novel technology, namely autonomous weapons.

A weapon systems is 'man on the loop' if it is capable of automated or autonomous operation but is supervised by an operator who has the capability to intervene and override an attack decision that the automated or autonomous decision-making process makes. Contrast that to 'man in the loop' weapon systems in which the human operator decides which target is to be engaged and undertakes the attack by initiating the firing mechanism using the remote-control facility built into the weapon system.[79]

The weapon reviewer is concerned with the law applicable to the weapon system as such and not with the legality of a particular attack. He will therefore want to know whether a 'man on the loop' weapon system is capable of use in compliance with targeting law, including the distinction, discrimination, and proportionality principles and the precautions rules. If an individual empowered to countermand unsatisfactory machine-made attack decisions is enabled properly to supervise the autonomous or automated attack decision-making and to intervene when it is appropriate to do so, the autonomous or automated nature of the initial decision-making facility is unlikely to raise international weapons law concerns;[80] i.e., such a weapon system is capable of being employed in compliance with the targeting rules.

It seems clear that the ultimate goal of much contemporary research is complete autonomous decision-making in attack.[81] There is, however, no internationally agreed legal definition of what automated and, respectively, autonomous attack decision-making means.[82]

Current UK doctrine refers to highly automated systems that 'are constrained by algorithms that determine their responses by imposing rules of engagement and setting mission parameters which limit their ability to act independently'.[83] Such a system is not remotely controlled, but functions in a self-contained and independent manner once deployed. It independently verifies or detects a particular type of target and then fires or detonates.[84] Such technologies are not new, having been employed for example in the past in mines and booby-traps.[85]

Autonomous systems differ from automated ones in that they can understand higher-level intent and direction; and 'from this understanding and its perception of its environment, such a system is able to [take] appropriate action to bring about a desired state'.[86] So autonomous systems independently identify and decide to engage targets. They are not pre-programmed to target a specified object or person. It is the software that decides which target to prosecute, how and when.

For the foreseeable future, autonomous attack decision-making is most unlikely to be capable of employment in conformity with targeting law principles and rules, particularly the distinction, discrimination, and proportionality principles and the rules as to precautions in attack.[87] At a Chatham House Conference on autonomous weapons systems, there was broad agreement that 'except in very unique battle spaces (where the likelihood of civilians was nonexistent), deployment of autonomous weapon systems today would not be consistent with the requirements of International Humanitarian Law'.[88] Accordingly, a legal review will generally reject the entirely autonomous use of weapon systems employing such technology.

Human Rights Watch has called for a legally binding instrument and national laws banning fully autonomous attack technologies, a notion which it describes as 'robots that are capable of selecting targets and delivering force without any human input or interaction' or that, although they operate under the oversight of a human operator who can override the robot's actions, are subject to such limited supervision that there is no effective human input or interaction.[89]

A UNIDIR discussion paper argues that there might be a difference in the acceptability of an autonomous but static system that is a 'last line of defence' to counter an incoming attack versus a system that employs superhuman decision-making speed to carry out an attack.[90] Such a last line of defence system, similar perhaps to Israel's Iron Dome or the Phalanx system, would be programmed to engage only inbound threats that are by definition military objectives in a defensive operational context in which collateral damage is unlikely to be a prohibitive consideration. Such systems are already in operational use and are likely to be more discriminating than human decisions made in the stressful, high speed, and potentially overwhelming conditions that necessitate the employment of the autonomous/highly automated system.

There has been criticism of the Human Rights Watch report[91] and its definition of 'fully autonomous weapons' which seems to include automated as well as autonomous systems. For the reasons discussed in the previous paragraph, states are likely to consider a prohibition of these technologies premature or inappropriate; so highly automated decision-making, informed certainly by human input before mission commencement, may increasingly become the norm.[92]

Improvement of artificial intelligence enabling a weapon system to learn and base its decisions on what it has learned seems likely to be the basis on which autonomy will emerge. Such learning systems might apply lessons learned in the battlespace to develop their own criteria against which to recognise a target; or they may observe and record the pattern of life in the target area and subsequently apply its observations and pre-learned lessons to decide what to attack. Perhaps a future ALI system could detect that a planned attack would no longer comply with the discrimination rule, for example because it detects that hostages have entered the target area.

More realistically, a weapon system might be able to detect whether the previous pattern of life in the target area that it has been observing has changed materially, such that the search for targets by the automated system should not proceed.

States are likely also to consider the ethical issues arising from the acquisition of autonomous weapons, such as whether the decision to initiate the use of lethal force can be legitimately delegated to an automated process.[93] As noted previously, however, it is existing law that should be applied to decide the legal acceptability of such technology in warfare[94] and this involves applying the normal weapons law criteria. Thereafter, the weapon review should assess whether the targeting law rules can be complied with despite the absence of a person from target decision-making.

We will start, therefore, by considering the normal weapons law criteria. The superfluous injury and unnecessary suffering principle is likely to be irrelevant, concerned as it is with the nature of the injury caused by the weapon as opposed to the automated or autonomous nature of its targeting decision-making. For similar reasons, the environmental protection rules are unlikely to be relevant to the automated or autonomous aspect of the weapon system.

When applying the indiscriminate weapons rule, the performance of the autonomous or automated target recognition technology, whether during tests or in the course of previous actual hostilities, should be evaluated with care. A weapon system only breaches the indiscriminate weapons rule, however, if it cannot be directed at a specific military objective or if its effects cannot be limited as required by international law and if the result in either case is that the nature of the weapon is to strike military objectives and civilians or civilian objects without distinction. So if attack technology is designed to recognise the particular characteristics of, say, a tank, and if the recognition software performs satisfactorily in tests that realistically reflect the intended circumstances of use the indiscriminate weapons rule is likely to be satisfied. If, however, such software does not perform satisfactorily in such tests and thus the weapon strikes civilian objects and military objectives without distinction, the system will breach the rule. In some cases it may be necessary in the text of the weapon review to draw attention to restricted circumstances in which the weapon system will comply with the indiscriminate weapons principle,[95] and to explain what action is required in order to ensure that use of the weapon system does not result in indiscriminate attacks. There are no specific prohibitions or restrictions on the use of autonomous or automated attack technology in customary or treaty law.[96]

The weapon reviewer should then consider whether the automated or autonomous weapon system is capable of being used in accordance with the targeting rules. This will involve consideration of all targeting rules that seem to be relevant to the particular weapon system. The precautions required of an attacker by article 57 of API illustrate clearly some of the relevant challenges. Article 57(1), to which the reviewer should draw attention, requires that constant care be taken to spare the civilian population, civilians and civilian objects, and this provides the context for what follows.

Those who 'plan or decide upon an attack' and thus have obligations under Article 57(2) would seem to include those who prepared the mission, programmed the automated or autonomous software, reviewed available information, prescribed the areas to be searched and the times of such searches, set the target identification criteria for the weapon control software and so on.

The weapon reviewer will need to be satisfied that the characteristics of the weapon system and the arrangements that are being made for its employment are such that the decisions to attack made by the automated or autonomous weapon system apply these provisions, whether by virtue of action taken by the equipment itself or because of what personnel operating the weapon systems, supervising them, or planning the sortie are enabled to do in advance of, or during, the sortie.

Everything 'feasible' must be done to fulfil the obligations in Article 57(2)(a)(i) and (ii). If a manned mission would be capable of fulfilling an Article 57 obligation which the automated or autonomous mission cannot fulfil, then the manned mission should be employed, or some other method should be found of achieving the desired military purpose. This may become a difficult issue if an autonomous or automated weapon system cannot be programmed to recognise when its employment would preclude the taking of sub-paragraph (i) or (ii) precautions that, in contrast, could be taken if some other weapon system, such as a manned one, were to be employed. The mere fact that an autonomous or automated system cannot fulfil an obligation does not render the obligation non-feasible if it can be fulfilled using an alternative weapon system traditionally used for the relevant purpose.

The Article 57(2)(a)(i) requirements to do everything feasible to verify that the object of the attack is a military objective and that it is not entitled to special protection[97] are likely to be complied with by using algorithm-based technologies, for example, that are found in tests satisfactorily to differentiate between the objects they are programmed to identify and those they are not, and thus between military objects and civilian objects.[98]

Shifting the focus of attention from targeting objects to targeting persons, the challenge for autonomous or automated target recognition technology under the first element of article 57(2)(a)(i) would be to show that it can satisfactorily distinguish between lawful targets, namely able-bodied combatants and able-bodied civilians directly participating in the hostilities, and persons whom the law protects, such as combatants who are *hors de combat*, non-combatants, and civilians who do not directly participate.

Article 57(2)(a)(i) also requires that attackers do everything feasible to verify that it is not prohibited by the Protocol to attack the intended targets. Prohibited attacks include those which would breach Article 51(4) (discrimination principle), 51(5)(a) (separate and distinct military objectives treated as one), 51(5)(b) (proportionality), 53 (cultural objects), 54 (objects indispensable to the survival of the civilian population), 35(3) and 55 (protection of the natural environment), 56 (works and installations containing dangerous forces and military objectives in their vicinity), 41 (safeguarding of persons *hors de combat*), 12 and 15 (protection of medical units and personnel), and 21 to 28 (protection of medical transports).

To the extent that these rules simply prohibit attacks directed at specified objects or persons, the weapon reviewer will be concerned to establish that the automated or autonomous weapon system, in the manner in which it is intended to be used, will detect that a person or object comes within one of these protected categories and will accordingly refrain from attacking it. It remains to be seen whether, for example, software can be developed that distinguishes between an able-bodied combatant and one who comes within article 41 as being *hors de combat*.[99]

The precautionary requirements of Article 57, however, go beyond target recognition so an autonomous or automated weapon system must also be able to comply, for example, with the evaluative judgements involved in Article 51(5)(a), in the proportionality assessment referred to in Articles 51(5)(b) and 57, and in the tests in Article 57(2)(a)(ii) and 57(3) of API. Taking Article 57(2)(a)(ii) as an example, the weapon system will have to be able to decide whether an attack should be undertaken using an operator-controlled, automated, or autonomous platform with a view to minimising incidental civilian injury and damage. The weapon reviewer will need to be satisfied that the available technology facilitates the making of each of these evaluative assessments.

The difficulty may, however, be overcome if human planners or operators are enabled to make the necessary evaluations and thus take the required precautions[100]; but, if there is no such human involvement, the weapon reviewer will need to be satisfied that the weapon system itself can discharge the complex decision-making that has been discussed.

For the foreseeable future, therefore, autonomous or automated attack capabilities can only be used lawfully if required precautions are taken by personnel, probably in advance of the sortie. Appreciating, however, that technology will develop with time and that autonomous or automated weapon systems will tend to be used in conjunction with other support systems, the reviewer's task will be to determine whether the method of warfare as a whole can comply with the legal rules we have discussed. Current technology requires a person to be in a position to cancel autonomous and some automated attack operations if the need should arise. That person will need to remain sufficiently engaged, suitably located, and appropriately tasked to know what is taking place and, if necessary, to override the system's attack decisions.

Notes

1 *Concise Oxford English Dictionary*, 11th edn, 2006.
2 J. M. McClelland, 'The Review of Weapons in Accordance with Article 36 of Additional Protocol I', *International Review of the Red Cross*, No. 850, 2003, p. 397.
3 ICRC Guide at p. 8, Note 17 referring to a presentation by W. Hays Parks to the Expert Meeting on Legal Reviews of Weapons and the SIrUS Project, Jongny sur Vervey, 29–31 January 2001.
4 AP I Commentary, paragraph 1402. *The Manual on International Law Applicable to Air and Missile Warfare* (Program on Humanitarian Policy and Conflict Research, Harvard University, 15 May 2009) (hereafter AMW Manual) refers to 'methods of warfare' as the various general categories of operations, rule 1v on p. 43, and 'means of warfare' as weapons, weapon systems or platforms employed for the purposes of attack; rule 1(t) on p. 41.
5 'Instructions for the Government of Armies of the United States in the Field', 24 April 1863.
6 Lieber Code, Article 16.
7 Regulations Respecting the Laws and Customs of Wars on Land, Annexed to Hague Convention IV, 1907.
8 'St Petersburg Declaration renouncing the use, in time of war, of explosive projectiles under 400 grammes weight', 1868.
9 *The Manual of the Law of Armed Conflict* (Oxford: Oxford University Press, 2004) (hereafter UK Manual), para 6.10.1. While the prohibition is likely based on the customary superfluous injury and unnecessary suffering principle, discussed above, as opposed to the application of the treaty, the prohibition on the use of exploding ammunition against personnel is nevertheless now a customary rule in its own right.
10 1899 Hague Declaration II, para 2.
11 1899 Hague Declaration III, para 2.
12 UK Manual (see note 9 above), para 6.9 and Note 32. A resolution adopted on 10 June 2010 by the First Review Conference for the Rome Statute of the International Criminal Court amended article 8 of the Statute by adding to the list of war crimes in armed conflicts not of an international character, *inter alia*, the offence of 'employing bullets which expand or flatten easily in the human body, such as bullets with a hard envelope which does not entirely cover the core or is pierced with incisions'; RC/Res 5 adopted at the 12th Plenary Meeting. A preambular paragraph to the Resolution and one of the Elements of the crime make it clear that the crime is only committed in connection with a non-international armed conflict 'if the perpetrator employs the bullets to uselessly aggravate suffering or the wounding effect upon the target of such bullets, as reflected in customary international law'; preambular paragraph 9. It is therefore not a crime under the Statute to use such bullets in connection with such conflicts if there is a good military reason for doing so, such as may arise in, for example, the kinds of circumstance discussed in the main text.
13 Preamble to the St Petersburg Declaration, 1868 (see note 8 above), tirets 2 to 6.
14 Project of an International Declaration concerning the Laws and Customs of War, Brussels, 27 August 1874, article 12, and The Laws of War on Land, Oxford, 9 September 1880, article 9; US Field Manual 27–10, paragraph 34 on p. 18 explains that weapons breaching the rule include 'lances with barbed heads, irregular shaped bullets, projectiles filled with glass, substances on bullets tending unnecessarily to inflame the wound, scoring the surface or filing off the ends of the hard cases of bullets'.
15 API, Article 35(2).
16 'A weapon causes unnecessary suffering when in practice it inevitably causes injury or suffering disproportionate to its military effectiveness. In determining the military effectiveness of a weapon, one looks at the primary purpose for which it was designed', W. J. Fenrick, 'The Conventional Weapons Convention: A modest but useful treaty', *International Review of the Red Cross*, Vol. 30, Issue 279 (1990), p. 500.
17 W. H. Boothby (2016) *Weapons and the Law of Armed Conflict*, Oxford: OUP, p. 63, where the criteria prepared by E. R. Cummings, W. A. Solf and H. Almond as the basis for the original United States Department of Defense weapons review directive are discussed. The UK

Manual (see note 9 above) explains that the current practice is to regard the principle as 'a guiding principle upon which specific prohibitions or restrictions [in the law of weaponry] can be built'; UK Manual, paragraph 6.1.5.

18 F. Kalshoven, 'Arms, Armaments and International Law', Extract from (1985-II) 191 *Hague Recueil des Cours*, p. 236.

19 API, Article 51(4).

20 J. M. Spaight, *Air Power and War Rights,* 3rd edn (Gale, 1947), p. 215; Spaight notes that such a weapon was not 'banned in terms by any international convention'.

21 Geneva Protocol for the Prohibition of the Use in War of Asphyxiating, Poisonous or Other Gases, and of Bacteriological Methods of Warfare, 1925.

22 States have in practice withdrawn their 'no first use' reservations as they have become party to the Chemical Weapons Convention of 1993 and to the Biological Weapons Convention of 1972.

23 United Nations Convention on the Prohibition of Military or any other Hostile Use of Environmental Modification Techniques (ENMOD), adopted on 2 September 1976.

24 ENMOD, Article 1(1).

25 The Convention was adopted on 10 October 1980 in Geneva.

26 Convention on Prohibitions or Restrictions on the Use of Certain Conventional Weapons, Protocol I, adopted in Geneva on 10 October 1980.

27 Ibid.

28 Adopted in Geneva on 3 May 1996.

29 'Mine means any munition placed under, on or near the ground or other surface area and designed to be detonated or exploded by the presence, proximity or contact of a person or vehicle.' Protocol II, article 2(1). The equivalent definition in Amended Protocol II, article 2(1) is broadly similar.

30 'Other devices means manually-emplaced munitions and devices designed to kill, injure or damage and which are actuated by remote control or automatically after a lapse of time.' Protocol II, article 1(3). Under Amended Protocol II, 'Other devices means manually-emplaced munitions and devices including improvised explosive devices designed to kill, injure or damage and which are actuated manually, by remote control or automatically after a lapse of time.' Amended Protocol II, article 2(5).

31 Protocol II, article 1(2).

32 Protocol II, article 6(1)(a).

33 Protocol II, Article 6(1), APII article 7(1).

34 APII, article 3(5).

35 APII, article 3(6).

36 APII, article 4 and Technical Annex, paragraph 2(a), but note that some technical requirements depend on the date of construction, see paragraph 2(b).

37 APII, article 6(2).

38 APII, Technical Annex, para 3(a) taken with (b).

39 APII, article 6(3).

40 APII, article 7(2).

41 CCW, Protocol IV, Article 1. Blinding as an incidental or collateral effect of the legitimate use of laser systems is not prohibited; article 3. Equally, laser systems which are not specifically designed to cause permanent blindness are also not prohibited by this provision. Permanent blindness means irreversible and uncorrectable loss of vision which is seriously disabling with no prospect of recovery, and serious disability is equivalent to visual acuity of less than 20/200 Snellen measured using both eyes. Protocol IV, article 4. Under article 2, when using laser weapons not prohibited by the Protocol, States party must take 'all feasible precautions to avoid the incidence of permanent blindness to unenhanced vision'.

42 Convention on the Prohibition of the Development, Production and Stockpiling of Bacteriological (Biological) and Toxin Weapons and on their Destruction 1972, opened for signature on 10 April 1972 (Biological Weapons Convention).

43 UK Manual (see note 9 above), p. 104, Note 8.

44 ICRC Customary Humanitarian Law Study, Vol. 1, Rule 73, p. 256.

45 Convention on the Prohibition of the Development, Production, Stockpiling and Use of Chemical Weapons and on Their Destruction, Paris, 13 January 1993 (Chemical Weapons Convention).

46 Chemical Weapons Convention, Article I(1).
47 Chemical Weapons Convention, Article II(1).
48 Chemical Weapons Convention, Article II(2).
49 Chemical Weapons Convention, Article II(3).
50 Chemical Weapons Convention, Article II(9).
51 ICRC Customary Humanitarian Law Study Report, Rule 74; Boothby, *Weapons and the Law*, p. 137 (see note 17 above), and note that there are now 190 States party to the Chemical Weapons Convention, www.opcw.org/about-opcw/ (accessed 22 May 2015).
52 Chemical Weapons Convention, 1993, Article II(7).
53 CCW, Article 8(2)(b).
54 Convention on the Prohibition of the Use, Stockpiling, Production and Transfer of Anti-Personnel Mines and on their Destruction, Oslo, adopted on 18 September 1997.
55 Ottawa Convention, article 2(1).
56 Cluster Munitions Convention, Article 1(1). See also the discussion by Brian Rappert in Chapter 5.
57 Cluster Munitions Convention, Article 2(2).
58 Cluster Munitions Convention, Article 2(3).
59 CCW Review Conference, December 2001.
60 CCW, Article 1(2) as amended.
61 CCW, APII, Article 1(2).
62 Consider article 2.
63 Protocol on Prohibitions or Restrictions on the Use of Incendiary Weapons (Protocol III) adopted in Geneva 10 October 1980.
64 Protocol III article 1(1). The definition excludes from the Protocol munitions with incidental incendiary effects, such as tracers or illuminants, and combined effects munitions in which the incendiary effect is designed to be used against objects not persons.
65 A 'concentration of civilians' may be permanent or temporary, and can include inhabited parts of cities, towns, villages, camps, columns of refugees or groups of nomads; Protocol III article 1(2).
66 Protocol III, article 2(2).
67 Protocol III, article 2(3).
68 Protocol III, article 2(4).
69 NATO Policy on Non-Lethal Weapons dated 27 September 1999.
70 At the time of writing there were 174 states party to API. Source: https://ihl-databases.icrc.org/applic/ihl/ihl.nsf/INTRO/470 (accessed 20 July 2018).
71 API, article 36; for a discussion of weapon reviews, see Geoffrey S. Corn et al., *The Law of Armed Conflict: An Operational Approach* (New York: Wolters Kluwer Law and Business, 2012), pp. 203–4.
72 Boothby, *Weapons and the Law*, p. 345.
73 Consider I. Daoust, R. Coupland and R. Ishoey, 'New Wars, New Weapons? The Obligation of States to Assess the Legality of Means and Methods of Warfare', *International Review of the Red Cross*, No. 846, pp. 345, 348, 2002.
74 AMW Manual (see note 4 above), Rule 9.
75 Schmitt, Michael N. (Gen. ed.) (2013). *Tallinn Manual on the International Law Applicable to Cyber Warfare*, New York: Cambridge University Press, Rule 48(a). The Commentary to this Rule also cites article 1 common to the Geneva Conventions of 1949. References in military manuals include the UK Manual, paragraphs 6.20–6.20.1; the United States Naval Commanders' Handbook on the Conduct of Naval Operations, NWP 1–14, paragraph 5.3.4; the Canadian Manual paragraph 530; and the German Manual paragraph 405. See also W. Hays Parks, 'Conventional Weapons and Weapons Reviews', *Yearbook of International Humanitarian Law*, Vol 8 (2005), pp. 55–7.
76 K. Lawand, *A Guide to the Legal Review of New Weapons, Means and Methods of Warfare: Measures to Implement Article 36 of Additional Protocol I of 1977*, ICRC, 2006, p. 4.
77 Ibid., p. 5.
78 *Tallinn Manual* (see note 75 above), commentary accompanying Rule 48, paragraph 3.
79 See N. Melzer, 'Human Rights Implications of the Usage of Drones and Unmanned Systems in Warfare', Geneva Papers 11–2013, 8.

80 Report of ICRC Meeting on Autonomous Weapon Systems: technical, military, legal and humanitarian aspects, Geneva, 26–28 March 2014, pp. 4–5 (ICRC Report).
81 For systems with autonomous features already in use, consider the maritime Phalanx system in service with the Royal Navy and described at www.royalnavy.mod.uk/The-Fleet/Ships/Weapons-Systems/Phalanx; the United States Navy MK 15 – Phalanx Close-In Weapons System, described at www.navy.mil/navydata/fact_display.asp?cid=2100&tid=487&ct=2; the Russian Arena-E Active Protection System; the Mutual Active Protection System; the Diehl BGT Mutual Active Protection System described at www.defense-update.com/20110112_maps.html, South Korean border security arrangements discussed in 'South Korea deploys robot capable of killing intruders along border with north', *Daily Telegraph*, 13 July 2010 available at www.telegraph.co.uk/news/worldnews/asia/southkorea/7887217/South-Korea-deploys-robot-capable-of-killing-intruders-along-border-with-North.html and the Israel Aircraft Industries Harpy autonomous anti-radar SEAD system as to which see www.iai.co.il/2013/16143-16153-en/IAI.aspx.
82 ICRC Report, p. 1 (see note 80 above).
83 UK Ministry of Defence, UK Air and Space Doctrine, JDN 0.30, para 215. Automated systems 'do not involve a human operator during the actual deployment but rather the necessary data is fed into the system prior to deployment of the system', and WW2 V-1 and V-2 rockets, automated sentry guns and sensor-fused ammunition are examples.
84 J. Kellenberger, 'International humanitarian law and new weapon technologies', 34th Round Table on Current Issues of International Humanitarian Law, San Remo, 8–10 September 2011, p. 5.
85 A. Backstrom and I. Henderson, 'New capabilities in warfare: an overview of contemporary technological developments and the associated legal and engineering issues in Article 36 weapons reviews', *International Review of the Red Cross*, No. 886, 2012, pp. 488–90.
86 JDN 0–30 Lexicon-5; note the UK view that autonomous systems are self-governing and set their own rules, that this is neither welcome nor useful in the military context and that the UK is committed to maintaining human oversight over weapons release decisions; ibid, para. 215.
87 E. Quintana, 'The Ethics and Legal Implications of Military Unmanned Vehicles', (2008) RUSI available at www.rusi.org/downloads/assets//RUSI_ethics.pdf.
88 See ICRC Report (note 80 above), p. 5. Current UK policy 'is that the operation of weapon systems will always be under human control'; ibid, pp. 10–11.
89 Human Rights Watch, *Losing Humanity: The Case Against Killer Robots* (2012) available at www.hrw.org/sites/default/files/reports/arms1112ForUpload_0_0.pdf, 2, 5.
90 UNIDIR, Framing Discussions on the Weaponisation of Increasingly Autonomous Technologies (2014), available at www.unidir.org/files/publications/pdfs/framing-discussions-on-the-weaponization-of-increasingly-autonomous-technologies-en-606.pdf, 6.
91 M. N. Schmitt and J. S. Thurnher, '"Out of the Loop": Autonomous Weapon Systems and the Law of Armed Conflict', *Harvard National Security Journal*, Vol. 4, 2013, pp. 231–81.
92 K. Anderson and M. Waxman, *Law and ethics for autonomous weapon systems, why a ban won't work and how the law of war can*, Hoover Institution, Stanford University (2013) available at www.hoover.org/taskforces/national-security, 5 and footnote 16.
93 P. Asaro, 'On banning autonomous weapon systems: human rights, automation, and the dehumanization of lethal decision-making', *International Review of the Red Cross* Vol. 94, Issue 886, p. 689 and see for example *Losing Humanity* (see note 89 above).
94 See Anderson and Waxman, *Law and ethics* (note 92 above), M. N. Schmitt, 'Autonomous Weapon Systems and International Humanitarian Law: A Reply to the Critics', Harvard National Security Journal Feature, 2013, available at: http://ssrn.com/abstract=2184826 or http://dx.doi.org/10.2139/ssrn.2184826. See also Schmitt and Thurnher, '"Out of the Loop"' (see note 91 above).
95 See Schmitt and Thurnher, '"Out of the Loop", pp. 245–50, 271–6 (note 91 above).
96 S. S. 'Lotus' (Fr. v. Turk.), 1927 P.C.I.J. (ser. A) No. 10 (Sep. 7), 18; the absence of specific reference to such technologies implies that they are not the subject of a prohibition.
97 See AMW Manual (see note 4 above), Rule 32(a).
98 The reference here to 'military objects' is intentional; the technology is likely to be configured so as to distinguish, for example, between an artillery piece or a tank on the one hand

and a civilian vehicle of comparable size on the other; see M. Lewis et al., 'Scaling up Wide-area-search Munition Teams', *IEEE Intelligent Systems*, Vol. 24, Issue 3 (May-June 2009), p. 10.

99 As to relevant engineering challenges, see Backstrom and Henderson, 'New capabilities in warfare' (see note 85 above), pp. 510–13.

100 Performance of the system cannot be determined in advance with certainty. These matters must however be considered; and appropriate limitations on circumstances of planned use must be developed to ensure that the discrimination and precautions rules will likely be complied with.

4

A DEFENCE TECHNOLOGIST'S VIEW OF INTERNATIONAL HUMANITARIAN LAW

Tony Gillespie

The 1949 Geneva Conventions are supplemented by two Additional Protocols. The first (API) concerns the conduct of international conflicts. This includes Article 36, a requirement for reviews of new means and methods of warfare.[1] Formal reviews are usually carried out at major milestones – points which mark the end of a major stage in a programme for the procurement of a new weapon. The definition of a major stage will vary depending on the nature of the programme and its technologies. This chapter discusses Article 36 reviews from a technologist's point of view, identifies important milestones, and proposes the types of technical evidence required at key points.

Modern weapons employ an increasing level of automation. This is often accompanied by increased levels of automated decision-support tools used in the military command and control (C2) chain. This chapter takes the view that the use of a weapon cannot be separated from the surrounding system. However, there must be clear definitions of its interfaces to the surrounding system. Otherwise, a review of a new weapon, or modifications to it, would lead to continual reviews of complete sections of operational infrastructure.

Platform lifetimes

The time taken to develop large military platforms such as aircraft and ships has increased over the last century. It is not always realised that modern platforms also have a greatly increased lifetime. Table 4.1 shows the timescales for four UK/European aircraft. Note that this does not include the time for the research programmes which developed many of the key technologies in the aircraft.

Table 4.1 Lifetimes for four major aircraft

	Sopwith Camel[2]	*Hawker Hunter*	*Panavia Tornado*	*Eurofighter Typhoon*
Start design	1916	1948	1970	1986
In-service	1917	1953	1980	2003
Out of service	1919	1970	2019	2050?
Lifetime (years)	2	17	39	>47

It is necessary to consider the reasons for these timescales and the accompanying increased costs, as they affect the scope and complexity of any legal reviews. The aircraft examples given here are typical of all three domains (air, sea, and land).

One reason for the high costs and long timescales is the sophistication of the technologies used in their manufacture. This is because military systems must operate in all climatic conditions, anywhere in the world, and have a tangible, even if small, military superiority over the opponent's systems. This is commonly known by engineers as designing to the corners of the specification. As an environmental example, in the Second World War Russian antifreeze in their fuel allowed their armoured vehicles to operate at a lower temperature than their German counterparts – a battle-winning advantage.[3] A performance example is the use of a communication link from an aircraft to the missile it has just launched. This enables the pilot to fire at a target whilst at a long range and guide the missile to a point where the missile's own seeker can ensure a hit. An aircraft without this capability has to come closer before firing, a definite disadvantage.

Long lifetimes bring three problems:

1 The design must allow upgrades with technologies which may not have been developed when the platform entered service. One example is that military forces are becoming more integrated across all three domains with increasingly capable data-links and C2 networks, giving requirements for high-bandwidth digital radios and processors. These will be in addition to, and not replace, analogue systems as these are needed to communicate with other aircraft in a coalition operation.
2 The shape of warfare is evolving. Typhoon was designed as a third generation Cold War air-to-air fighter. However, its first operational use of weapons was firing anti-tank missiles (Brimstone) against armoured vehicles in Libya.
3 Long-term support costs begin to dominate budgets. Maintaining decades-old technology in a reliable manner brings many expensive problems. As an example, virtually all electronic circuits manufactured before 2000 used solder containing lead. Environmental legislation mandates that modern electronics must use lead-free solder, so now most companies find it uneconomic to repair circuits with lead-based solder even if they can charge high prices and obtain local exceptions to workplace legislation.

Management and engineering solutions to problems arising from extended lifetimes are found using the Concept, Assessment, Development, Manufacture, In-service, Disposal (CADMID) Cycle and its accompanying Defence Lines Of Development (DLODS). Transition from one step in the CADMID cycle to the next is also often a major step in the release of funding and so makes an appropriate legal review point.[4] Unfortunately, the 'I' part of the CADMID cycle can now be much longer than the timescales for procurement of new technologies, as discussed below, and the platform's capabilities will change significantly over this time.

The consequent problem that arises for the IHL lawyer is deciding when a platform or weapon upgrade is a new or novel means of warfare or merely a variation on current methods not requiring an Article 36 review.

Counter-examples to the extending development timescales can be found, such as the development of unmanned air vehicles (UAVs), also known as remotely piloted aircraft (RPA) or drones. The Reaper UAV was developed in the late 1980s and entered service in the late 1990s in Kosovo, with widespread use after 2000. However, it is a very specialised surveillance aircraft with virtually no defensive capabilities, unlike other major platforms. It has been weaponised, as an upgrade, but with limited flexibility. Unmanned aircraft, however, are not a new development and have a history dating back to the 1900s, with extensive development around the world since the 1920s. It is amusing to note that Samuel Cody mentioned the legal problems of using unmanned aircraft rather than manned ones in his address to the Aeronautical Society on 8 December 1908.

Military capabilities and engineering

The major procurement problem faced by a military planner is how to deliver military effects in a range of predicted scenarios with limited budgets and a range of views about the feasibility of technological solutions. This is addressed by a combination of Operational Analysis (OA) and engineering methodologies (Capability Engineering, Systems Engineering and Product Design). These do not have exact definitions, but they are generally understood to be the following: *OA* is the translation of government policy into possible methods to implement it in a range of future scenarios. It is known as Operational Research (OR) in civilian applications. *Capability engineering* takes the OA results and specifies, in general terms, the military capabilities required to deliver the desired effect. *Systems engineering* identifies the combinations of equipment, services, and personnel which can deliver the specified capability. *Product design* is the design and development of the various components of the required system or systems.

OA is generally dominated by government input and product design is dominated by industry, with considerable overlap of all aspects in the other stages. Transition from one stage to the next does, however, create a potential legal review point. The programme manager should define these points in the project planning phase. Research and military judgement underpin all stages. Their precise roles will depend on the nature of the capability and technologies under consideration.

Product design has been the basis of engineering for centuries. It now embraces software, hardware, and firmware design as an integrated product. It is essential that the product has a definitive specification and defined interfaces and is able to be tested thoroughly. This is easy to state, but can be difficult to achieve in practice.

In contrast, the identification and separation of the first three stages (OA, capability engineering and systems engineering) are relatively new and reflect the increasing complexity of modern technology and products in civil and military applications. OA came to the fore in the First World War and systems engineering during the late 1980s. Capability engineering may be considered to be a 1990s response to capped or restricted military funding when full-scale war was considered to be a low risk. It addresses the problem of delivering effect in the most cost-effective manner and specifically includes solutions which use as much existing capability as possible.

Capabilities developed for narrow objectives can be very successful in meeting these objectives, but fail when used for other objectives. Two examples from the Second World War are the close coordination between the Luftwaffe and German Army for rapid ground advances in the Blitzkrieg in France[5] which became worse than irrelevant in the Battle of Britain when the Luftwaffe bombers had no radio contact with their escorting fighters. The British Home Chain radar warning system, *used as part of an Integrated Air Defence System (IADS)*, was highly successful in defending the UK, but of very limited value for offensive air operations over Europe.

Once we have systems comprising subsystems with several interactions between them it becomes difficult to describe completely and definitively the behaviour of the system when it interacts with external systems. Complex systems can display 'emergent behaviours'. Even though each subsystem can be specified sufficiently well for it to be designed and produced, the system as a whole will exhibit behaviour which could not be predicted or can only be explained with hindsight. This is particularly true if the systems behave non-linearly, i.e. a change in input values does not produce an output proportional to the input.[6]

An extra level of complexity comes when the system, with or without emergent behaviours, interacts with external systems with non-linear or random outputs. These can be other technical systems or human operators. Producing a complex human/machine system at an affordable cost which has completely predictable properties is beyond the capability of modern engineering. One example is the safe control of large numbers of aircraft, a 'wicked' problem that is managed by modern air traffic management (ATM) organisations and processes. Despite a decades-long process of continuous improvement and mandated complex regulations, emergent properties still arise leading to accidents or near-misses. Increasing aircraft numbers have led to initiatives such as the Single European Sky proposed in 1999, but that is still a long way from implementation due to its technical, legal, and political complications.[7]

There is no reason why complex military systems should be different from civil ones. The UK set up a new Military Airworthiness Authority (MAA) as a result of the Haddon-Cave report[8] into one incident involving the loss of an aircraft. This was in an area where there was a well-established set of regulations and procedures which dominate engineering design and in-service support, but there was still a need for the MAA and revised procedures.

Article 36 reviews are not usually considered to be drivers of engineering design, although the measured performance of engineering designs and methods of use are part of the review.[9] As stated earlier, it is essential that decisions are made at an early stage about what constitutes a new or novel means of warfare and the boundaries of what that means. It should be clear from the description of emergent properties and wicked problems that actual system performance is difficult to report in a complete manner. This makes it essential to understand where the boundaries of the review lie. The question of how to provide evidence of actual performance is discussed later in this chapter.

Technology timescales

It is often stated that technical progress is much faster now than it has ever been. However, developments which rely on the introduction of new results from basic physics and chemistry laboratories still take decades to reach commercial (civil and military) products. The driver is the market need, which dictates the level of funding for product development. The military need has to be urgent for immature technologies to be developed rapidly into military service. One example is the metal Super Bainite, announced in 2011.[10] It allows deformation to absorb shock, but does not allow penetration and is a revolutionary material for protecting armoured fighting vehicles. Bainite was originally developed in the 1920s, but it was not until the late 1990s that interest in it was revived, leading to the new product which was introduced into service rapidly due to the urgent military needs in Iraq and Afghanistan.

Once a new technology has reached maturity, it is exploited in as many applications as there are economic drivers for it. In the 1950s and 1960s, much research and development was funded for military use, but this is no longer the case except where there is a specific need. Now research funding for technologies in civil markets far exceeds that for defence applications and has produced a much wider range of capabilities. Civil-market products sometimes do not meet full military requirements such as environmental conditions in operations or long-term support, and therefore cannot be used as made. (See for example the discussion in the earlier section 'Platform lifetimes'.) Military effect can, however, often be achieved by applying the developments underpinning the civil product to defence problems. This has produced a fundamental change to the military procurement process and policy changes for military procurement exemplified by the UK's 2012 White Paper, *National Security through Technology*. Essentially this proposes that all equipment is bought 'off the shelf' with only limited exceptions. The 'shelf' can have either civilian or military products on it. Designing to the 'corners of the specification' does mean that many civil products have to be modified to meet military requirements. This leads to decisions about the economics of commercial off the shelf (COTS) products versus different ranges of modification, sometimes known as custom off the shelf, also known, confusingly, as COTS, or preferably modified off the shelf (MOTS).

What has changed public perception is that semiconductor technology has matured in the second half of the twentieth century, along with the software engineering techniques to exploit it. Processing power, expressed as the number of transistors per unit area on an integrated circuit, known as Moore's Law, can be regarded as a direct result of the maturing of semiconductor science into an engineering capability. Economic drivers now lead to their exploitation by engineers in many fields, including military ones.

Modern software engineering techniques allow complex programmes to be developed to exploit processor developments. The two are now intertwined.[11] The result is that the provision of a new military capability by the application of mathematical techniques as algorithms now takes only a few years. As discussed earlier, this is much shorter than the lifetime of a modern military platform.

There is one class of soft, hard, and firm-ware that has not developed at the above speed. This is for safety-critical systems. Exact definitions of this term depend on the industry or application, but it is generally agreed that a system or component is safety critical when a failure will lead to catastrophic failure and probable loss of life. A range of techniques has been developed for safety-critical systems, but their widespread adoption is limited by their high cost and slow design cycles. The reason for the latter is that the design process is highly regulated, with intense process verification and testing at every stage. When completed, there can be no modifications to the system without an extensive retesting regime. There is research underway to overcome these limitations and reduce cost. See for example the Advanced Mission Systems Demonstrations & Experimentation to Realise Integrated System Concepts (AMS DERISC) programme,[12] which is introducing new approaches in the certification of safety critical systems for air applications.

It is convenient to describe the progress of a technology from a theoretical concept to its inclusion in a new weapon or weapon system in terms of technology readiness levels (TRLs). These were originally proposed by NASA and are now used in many industries. Table 4.2 gives the definitions adapted for military use.[13]

Table 4.2 Technology readiness levels

TRL	*Description*
1	Basic principles observed and reported
2	Technology concept and/or application formulated
3	Analytical and experimental critical function and/or characteristic proof of concept
4	Component and/or breadboard validation in a laboratory environment
5	Component and/or breadboard validation in a relevant environment
6	System/subsystem model or prototype demonstration in a relevant environment
7	System prototype demonstration in an operational environment
8	Actual system completed and qualified through test and demonstration
9	Actual system proven through successful mission operations

When a specific technology such as a new fuse mechanism is developed, there is a reasonably clear transition point between levels. In this example, TRL 4 is laboratory tests, TRL 5 might be a prototype in the required space, and TRL 6 would be demonstration under the shock conditions of weapon firing. TRLs have proven to be very effective in this type of application, and transitions between TRLs are often taken as procurement milestones for major releases of funding. As a consequence, they are often used as milestones for Article 36 reviews.

Unfortunately TRLs have limited utility when a system or system concept is considered. This is partly due to the difficulty in agreeing on a generic definition of a system and partly because systems have implementation problems due to interactions between the subsystems and the so-called emergent properties. There have been proposals to define system readiness levels (SRL) but these have not gained general acceptance and are not used as procurement milestones.[14] In the opinion of Kujawski, they can be ignored as a consistent progress measurement, but can act as a useful checklist and prompt for the programme's Technical Authority.

Measured performance

Ultimately, any legal process concerning the use of a weapon must consider its performance in operational use. Therefore it is important that the review at the time of release to service should consider measured performance, not just the specifications. With certain types of weapon, such as a warhead, it is realistic for the reviewer to expect a comprehensive data pack including both trial measurements of explosive power and reliability statistics from a tool such as Failure Mode Effects and Criticality Analysis (FMECA). Many of these tests will have been carried out in the normal course of engineering design and testing. It may not be possible to test completely the more complex systems due to cost or complexity issues. The discussion earlier about emergent properties and wicked problems should illustrate the difficulties. However, these are not new problems for engineers to solve; the relevant question is how can we apply these solutions to weapons in a way that allows a satisfactory Article 36 review.

One important aspect of systems and product engineering is the decomposition of a capability requirement down to component level. This produces a hierarchy of requirement and specification documents which are used as the basis of procurement contracts throughout the supply chain. Each one provides the basis for the design of the deliverable at that level in the chain. Similarly, there are matching sets of test requirement and specification documents with associated test plans and test records. The technical evidence for a review of a new warhead would be: the warhead specification; the test plan with evidence of the completeness of the tests; and the test results. It is assumed here that the legal authority has decided whether the new warhead is simply a component change in the ordnance or is a change in capability offering new means of warfare. That decision may well affect the aims of the test plan and the detail required from the test results.

The early stages of introducing a new technology include deciding on a Concept of Use (CONUSE). Although not defined by NATO,[15] a CONUSE is generally understood to describe the way in which specific equipment is to be used in a range of operations or scenarios. This is not to be confused with Concept of Operations (CONOPS) which is defined as a clear and concise statement of the line of action chosen by a commander in order to accomplish his given mission. The CONUSE is the correct OA level for an Article 36 review.[16]

It is almost certain that any military appraisal of research results will assess them against at least an outline CONUSE. This should immediately provide guidance as to whether the technology will lead to a prohibited means of warfare or not and to consequent funding decisions. A common next step after research is to set up a Technology Demonstration Programme (TDP) which raises the TRL of key technologies to TRL 4 or 5. At this stage, an assessment of Article 36 requirements is a clear necessity.

The transition from OA to capability engineering will require CONOPS to be developed for realistic scenarios. Without CONOPS, it is very difficult for capability and systems engineers to specify procurement requirements to an acceptable level of detail. 'Acceptable' here means that the product specifications give measurable parameters that can be used to test the delivered product sufficiently to allow payments to be made that meet government accounting standards.[17]

Modern systems engineering processes are based on testing deliverables from component level up through subsystems to system and system-of-system levels. This allows control of the integration process and restricts the number of emergent properties to those at each level. In principle, the undesirable properties can be dealt with in a controlled and cost-effective way. Testing at systems and system-of-system level does require an operational context which can only come from CONOPS in generic scenarios. Full testing with actual systems is complex and expensive, so extensive use is made of modelling and simulation, with a NATO Standard.[18] Synthetic environments are well-known tools for testing the human–machine interface and verifying performance.

The impossibility of conducting exhaustive system tests under all circumstances and conditions leads to contractual acceptance of a system being a compromise between the ideal and the practical. A complex system is introduced into service over a period of time. These test and evaluation times can be lengthy and are used by the services to develop their CONOPS and Tactics Training and Procedure (TTP) requirements. The Article 36 review is not concerned with the detail of operation, only that the weapon can be used legally – a much easier requirement. This moves the area of concern from detailed performance issues to an assessment of the emergent properties of the system. As these will be detected during system integration and setting-to-work phases, a review must take place after these have been completed, i.e. at Initial Operating Capability (IOC). This review should be able to allow entry into service, but it may well identify areas of concern which will need to be addressed later.

The legal authorities setting rules of engagement (ROE) have to consider detailed trial results against CONOPS for a specific campaign and so may require more detail of specific aspects of weapon system performance.

Review points and types of technical evidence required

The arguments presented in this chapter identify several points in technology development programmes that are suitable for Article 36 reviews. These are given below, with the type of evidence required for each one. Not all of these review points will apply to every new technology during its progress into military service.

Technical evidence should be reviewed to ascertain whether the technology is likely to improve or degrade adherence to the four principles of the Laws of Armed Conflict (LOAC): necessity, distinction, humanity, and proportionality.

End of research programme

The research should produce evidence that the technology can be used legally. It must show that military use will not breach any of the bans on specific methods or materials. The technologists and military sponsors could also give an opinion on whether the new capability is a new means or method of warfare or provides a more efficient support to current means. The legal reviewer can then decide on the need for later Article 36 reviews.

CONUSE development at the end of OA (end of concept phase)

There may be little technical evidence that can be presented here for an Article 36 review. The review can, however, specify the type of technical evidence that would be expected to be presented before introduction into service.

Technology Demonstration Programme

TDPs are aimed at raising the TRL of a technology. Therefore they should lead to improved adherence to one or more of the four principles which should be identified at the release of TDP funding. The review question at the end of the TDP is whether this aim has been met and whether there will be any degradation in adherence to the others. The technologist will probably not be able to comment on whether this is a new or novel means of warfare beyond any opinions expressed after the preceding research programme.

Design contract review stage

The traditional 'Assessment' and 'Demonstration' phases of the CADMID cycle do not necessarily relate directly to modern procurement programmes which will not include building prototypes of the final equipment or service. TDPs or detailed analysis of specific subsystem performance are approximately equivalent to these phases. These may have been produced under contract by suppliers, or may have been carried out within the government. In both cases, all test results will be used in assessing the bids for the design phase contract.

Technical evidence will be the results of modelling and simulation of deliverables that the procurement authority considers can be produced by industry for the available budget. This should be part of the normal process, but, as stated earlier, the modelling and simulation may have been aimed at deriving specifications from CONOPS rather than the legality of the CONUSE. Additional technical analysis may be needed and presented to assess two types of principles: the newness of the means of warfare; and the changes to adherence to the four principles.

Manufacture

This will be at the end of the design phase. The technical evidence should simply be to show whether design changes due to technical issues or to changes in requirements will make any difference in the results of the review at design contract award. The timing of this technical assessment will depend on the overlap of, or gap between, the design and manufacturing stages.

The technical reviewers should be able to give guidance on the technical problems which will be encountered when integrating the design into the military infrastructure. This will provide a key reference point for the programmatic timing of the review or reviews needed during introduction into service. It should also provide guidance for the initial planning of TTP development and introduction into service.

Introduction into service

A complex system must have technical evidence about identified emergent behaviours: how they were detected, assessed, and dealt with; the probability of occurrence of future emergent behaviours; the principles used in specifying the trials and simulations for Article 36 review; and the quantitative performance of the new system and the relationship of that performance to adherence to the four LOAC principles.

Conclusion

The procurement process for new technologies and systems has been discussed from a technology viewpoint. Up to six review points are proposed. The type of technical evidence required at each of them has been presented, although the details will be strongly dependent on the specific technology and procurement programme.

Notes

1 1949 Geneva Conventions, Additional Protocols, Article 36 – *New weapons*: In the study, development, acquisition, or adoption of a new weapon, means, or method of warfare, a High Contracting Party is under an obligation to determine whether its employment would, in some or all circumstances, be prohibited by this Protocol or by any other rule or international law applicable to the High Contracting Party.
2 Sopwith Camel data taken from Wikipedia and references therein.
3 Wikipedia: *Operation Barbarossa*; and anecdotal evidence provided to the author by participants in Barbarossa.
4 J. M. McClelland, 'The Review of Weapons in Accordance with Article 36 of Additional Protocol I', *International Review of the Red Cross*, No. 850, 2003, p. 397.
5 James Holland, *The Battle of Britain* (London: Bantam Press, 2010), pp. 78–9.
6 Distortion in a sound system is an example of a non-linear process.
7 See the IATA factsheet at www.iata.org/pressroom/facts_figures/fact_sheets/Pages/ses.aspx.
8 Charles Haddon-Cave QC, *The Nimrod Review*, 28 October 2009, available at www.gov.uk/government/uploads/system/uploads/attachment_data/file/229037/1025.pdf.
9 See 'A guide to the legal review of weapons', *International Review of the Red Cross*, Vol. 88, Issue 864, December 2006.
10 Available at www.gov.uk/government/news/new-armour-steel-showcased-at-dsei.
11 Microsoft Windows and Intel processors are examples of this.
12 Available at www.amsderisc.com/related-programmes.
13 Taken from Department of Defense, *Technology Readiness Assessment (TRA) Deskbook*, July 2009.
14 An example of the debate can be seen in Edouard Kujawski; 'Analysis and Critique of the System Readiness Level', *IEEE Transactions on Systems, Man, and Cybernetics: Systems*, Vol. 43, Issue 4 (2012), pp. 979–87.
15 NATO publication AAP-06, Edition 2014, *NATO Glossary of Terms and Definitions*.
16 J. M. McClelland, ,The Review of Weapons in Accordance with Article 36 of Additional Protocol I, *International Review of the Red Cross*, No. 850, 2003.
17 The NATO standard for product assurance is *Engineering for System Assurance in NATO Programmes*, AEP-67, Edition 1, February 2010.
18 AMSP-01: *NATO Modelling and Simulation Standards Profile*, January 2012.

5

CAN THE LAW REGULATE THE HUMANITARIAN EFFECTS OF TECHNOLOGIES?

Brian Rappert

Scientific and technological developments are often accompanied by concern over their ethical, legal, social, and political implications. In the context of armed conflict, moral and humanitarian concerns are often raised about the means and methods of war. Autonomous weapons, drone technologies, directed energy weapons, and cyber warfare are but a few of the examples that have garnered such attention in recent years.

Agreeing on prohibitions or control measures for particular types of weapons has often proven difficult, for many reasons. One is disagreement. Within the context of conflict, the questions of which lives should count and for what often prove highly contentious. Another is uncertainty. When controls are relatively easy to introduce in the early stages of the development of a technology, they are often difficult to justify because of the lack of demonstrable harms. Yet, the prospective humanitarian effects of weapons are often difficult to predict. At the time of their introduction artillery, tanks, and chemical weapons were all said to be so effective that wars would come to a quick end, thus ultimately minimising the loss of combatants and non-combatants.[1] However, when the need for controls becomes widely acknowledged, it is often more expensive and difficult to put them in place because the technology has become established within the practices and routines of militaries.[2] Historically, where weapon types have already been developed and widely deployed, it has taken a considerable effort over many years (if not decades) to put in place humanitarian-inspired controls (for instance, in the case of anti-personnel mines, mines other than anti-personnel mines, and cluster munitions). Whenever choices get made about force options, counter-factual questions can be asked about what lives might have been saved, whether wars might have started, progressed, or ended otherwise, etc., if a different set of capabilities had been available.

Today, as at many other times, efforts to place limits on the means of armed conflict are challenged because of the transforming character of conflict as well as the shifting rationales and instruments used to determine what counts as legitimate force. Prominent considerations for the former include how new technical abilities blur what counts as 'armed conflict' and the movement away from state-to-state conflict. Against the prevalence of situations labelled as 'military operations other than war', 'non-obvious warfare', 'peacekeeping', 'not-war', etc., many have sought to advance standards in international human rights law that privilege the right to life. The development of case law at the international level through organs such as the International Court of Justice continues to provide new markers for defining legality.

Against this complex set of developments, this chapter focuses on the central tenets of international humanitarian law (IHL). At its heart, IHL embodies the notion that armed conflict requires that a balance be struck between the need for military necessity and concerns for humanity.[3] The balancing of these two principles is embodied in a number of specific legal rules, such as those regarding superfluous injury and unnecessary suffering, indiscriminate attacks, and proportionality.

IHL is the object of attention in this chapter for a number of reasons: While controls over the means and methods of armed conflict in recent decades have often been couched through IHL, the sufficiency of this framework has been disputed. Additional treaties and conventions have been adopted to correct for perceived deficiencies in the core principles and rules of IHL. In addition, many debates about the legitimacy of weapons have either been framed in the language of IHL or been positioned in contrast to it.

In asking how the possible humanitarian effects of technologies can factor into their governance today, this chapter sets out to contrast the logic of balancing military necessity and humanity central to IHL against the rationales associated with some past and ongoing efforts to restrict certain weapons. As will be argued, evaluations about the legitimacy of weapons and what needs to be done are being offered that go beyond what is deemed as justified under the principle and rules of IHL. In placing harms prominently, these logics can be said to be more 'precautionary' in manner. As part of this, users and stockpilers of weapons are being asked to justify their established practices rather than the onus being on those calling for reform. Through this process, categorical normative evaluations of weapons are being pursued and secured. The implications of such forms of reasoning have already underpinned significant international developments and, potentially, could foster novel further ones in the near future.

In making this contrast, though, the purpose of this chapter is not to contend that a hard and fast distinction exists between IHL and more precautionary approaches. Each can support or challenge the other, and in practice it might be difficult to neatly distinguish which is in operation because the reasoning overlaps or the terms of discussion are ambiguous. Instead of making a binary split, the purpose is to draw out certain distinctions in order to sharpen our awareness of the choices possible in thinking about the humanitarian effects and how account can be taken in the governance of technology. This is done, in part, by surveying and evaluating several recent disarmament initiatives. What they share is the advancing normative evaluations of harms that: go beyond the logic of weighing military utility-civilian consequences for specific scenarios, shift the onus for proof, and (as a result) justify categorical treatments of weapons.

Locating the balance

In any armed conflict, the right of the parties to the conflict to choose methods or means of warfare is not unlimited. In making this declaration, Article 35(1) of 1977 First Additional Protocol to the Geneva Conventions of 1949 expresses a central tenet of the legal regulation of modern warfare.

IHL is the body of law which most directly addresses how the methods or means of warfare in armed conflict are not unlimited. Central to it is a basic 'utility-consequence' obligation: the needs of military necessity must be balanced by concerns for civilian lives and property. This balancing is laid out in a number of legal rules; such as the rules regarding superfluous injury and unnecessary suffering, distinction, indiscriminate attacks, and feasible precautions.[4] The call for a balancing between military necessity and humanity is most straightforwardly evident in the rule of proportionality – this requires military commanders to 'carefully consider' anticipated incidental loss of civilian life and damage to civilian property against the anticipated military advantage in each instance of an attack against a legitimate object.

Given the difficulty of weighing military advantage versus civilian damage on a case-by-case basis, it is challenging to justify outright bans of weapons. Doing so would require judging that the harm to civilians would be excessive across expected use scenarios as well as the variations of any particular weapon.

Perhaps the starkest example of the difficulty of justifying categorical restrictions through IHL and international law more generally is the case of nuclear weapons. In December 1994, the UN General Assembly requested that the International Court of Justice offer an advisory opinion on the question 'Is the threat or use of nuclear weapons in any circumstance permitted under international law?' In coming to their decision the judges had to determine whether the envisioned threats or uses of these weapons across all expected use scenarios would necessarily violate the principles and rules of international humanitarian law. However, to the extent that the judges generalised about the effects of 'nuclear weapons' as a whole category their judgments could be challenged.

On 8 July 1996 the Court delivered its opinion. The decision can be interpreted as exhibiting the tensions associated with offering categorical evaluations while also acknowledging circumstantial contingencies. The judges agreed that the existing rules of international law neither universally prohibited nor authorised the threat or use of nuclear weapons. It was further agreed that the use of nuclear weapons had to comply with the tenets of international law. To the main issue of legality though, by a vote of seven to seven decided through the second vote of the President of the Court, the judges ruled that:

> the threat or use of nuclear weapons would generally be contrary to the rules of international law applicable in armed conflict, and in particular the principles and rules of humanitarian law;
>
> However, in view of the current state of international law, and of the elements of fact at its disposal, the Court cannot conclude definitively whether the threat or use of nuclear weapons would be lawful or unlawful in an extreme circumstance of self-defence, in which the very survival of a State would be at stake …

Responding to the claims forwarded by the UK and other states regarding the potential for low civilian casualties from nuclear weapons in certain settings (e.g., on the high sea) and in low-yield varieties, the judges ruled that while the use of nuclear weapons seemed 'scarcely reconcilable' with respect to international law, they could not 'conclude with certainty that the use of nuclear weapons would necessarily be at variance with the principles and rules of law applicable in armed conflict in any circumstance'.[5] So while the threat or use of nuclear weapons was *generally* held to be against international law, the judges could not determine that it *always* would be. As such, no categorical evaluation could be justified. Just what would constitute 'the very survival of a State' was not defined in the opinion.

Beyond balance – past achievements

Against this example, it is noteworthy that alternative ways of reasoning have been offered to justify categorical prohibitions. For instance, utility-consequence calculations have not been the only or even dominant way of making sense of biological weapons. They stand as one example of how a whole category of weapons has been rendered effectively out of bounds because of international standards about what counts as legitimate force. Their categorical unacceptability in customary law derives from the Geneva Protocol, the Biological and Toxin Weapons Convention (BTWC; also known as the Biological Weapons Convention (BWC)), and official statements, rather than a legal ruling or widespread consensus that they fall foul of IHL principles and rules.[6]

In prohibiting the development, production, acquisition, transfer, retention, and stockpiling of biological and toxin weapons, the BWC represents a major achievement of the international community. The categorical nature of the ban established through its General Purpose Criterion was not only a historical precedent at the time the convention came into force, but it remains unsurpassed today.

At the time of writing, biological weapons are roundly deemed beyond redemption. It does not matter if, in some scenarios, specific form of bioweapons used in certain ways might be judged as proportionate, discriminate, etc. according to the rules of IHL or any other part of international law – they are simply unacceptable within international diplomacy.[7] In contrast to other fields of science, the use of the life sciences for destructive purposes is not advocated or undertaken in a substantial manner by practising researchers.

Arguably this prevalent norm against bioweapons will be ever more important in the future. While at present it is reasonable to conclude that the biological weapon capabilities of sub-state groups, individuals, and even certain states could not be highly effective in terms of causing mass casualties,[8] this may well change due to the continuing development of civilian and commercial science. In the absence of positive and integrated action in the years ahead, a reasonable worry is that many more individuals and groups will have the capabilities required to cause major disruption and harm through biology. The source of concern is not simply about the proliferation of laboratory agents and equipment, but how the information and techniques generated through life science research are enabling new capabilities.

Given the breadth of challenges and potential sources of concern, a strong and widely shared normative stigma that fosters community self-policing is vital in preventing the hostile use of biological agents and toxins. It offers a flexibility in being able to guard against the biology being put to malign use, whatever form that takes.

As part of becoming aware of the challenges associated with decisions about the governance of warfare, it is important to acknowledge the potential dangers associated with the categorical prohibition of biological weapons as well. Several have been proposed:

1 *Exceptionalism*: As often taken for granted standards about what is right or wrong, norms can reinforce beliefs about what is acceptable in ways that are open to question. For instance, what have traditionally been labelled as 'unconventional' weapons – nuclear, biological, and chemical weapons – have been subject to much diplomatic attention in recent decades. That they kill and injure in ways unlike traditional kinetic force weapons is one of the reasons cited as to why unconventional weapons should be deemed insidious and inhumane. Yet, a danger with this is that conventional weapons become more and more normalised as 'conventional' – even if they can kill and maim to degrees comparable to nuclear, biological, or chemical means. For instance, fuel-air explosives might have the destructive power of small nuclear weapons, but they have not been subject to anything like the same level of scrutiny.[9]

2 *Geopolitical Realpolitik*: As some commentators have maintained, the categorical attention to certain weapons has not only been selective, but also self-serving. Against the illegitimacy of the possession of biological and chemical weapons, the legitimacy accorded to the continuing stockpiling of nuclear weapons by certain states has been taken as emblematic of the uneven distribution of geopolitical power.[10]

3 *Heightening Disruption*: Even if most feasible bioattacks today are not likely to cause mass casualties, they can be highly disruptive and economically costly (as in the 2001 US anthrax letter mailings). Arguably the manner in which biological weapons are held as distinct, special, extraordinary, etc. in the minds of political leaders, publics, and policy commentators works to increase their effectiveness as weapons of mass *disruption*.

4 *Unthinkable Weapons, Unthinkable Responses*: The manner in which biological weapons have been rendered as unthinkable force options within the life science communities also hampers attempts to promote a professional response to emerging concerns. As research undertaken by this author and others has indicated, the destructive application of the information and techniques generated through benign life science research is not something that many science practitioners have considered.[11] For good and for bad, in many meaningful ways, biological weapons have been rendered unthinkable.

These dangers suggest the need carefully to question the nature and impact of the categorical evaluations. The effects of this might not be wholly positive, even if positive overall.

Beyond balance – recent developments

The past orientation towards biological weapons that goes beyond the logic of balancing expected 'harms and utilities' is aligned with recent and ongoing developments in disarmament. This section highlights three such developments: one recent international prohibition achievement and two other ongoing discussions. Just what the latter efforts will lead to by way of a formal agreement (if any) is an open question at this time. Yet in seeking to move on from the status quo, it is clear that some are challenging the sufficiency of IHL to protect civilians and, as part of this, seeking to move beyond the logic of balance central to IHL.

1 The Convention on Cluster Munitions

On 3 December 2008, 94 states signed the Convention on Cluster Munitions (CCM). While allowing for certain exclusions, the CCM prohibits all of those weapons commonly identified under the terminology of 'cluster munitions' that have been documented as causing significant humanitarian harms in the past.[12] On 1 August 2010, the Convention came into effect. As of December 2018, it has 120 signatories and 106 States Parties.

For over 40 years before the signing of the CCM, some governments, NGOs, international organisations, and others had voiced concerns about the consequences from cluster munitions in places such as Laos, Cambodia, Vietnam, Lebanon, Western Sahara, Chechnya, Ethiopia, Eritrea, Afghanistan, and Iraq.

Against such concerns, in the years and decades prior to the agreement on the CCM, major user states frequently argued that cluster munitions had been tested against international law and had been used in accordance with the law.[13] With the basic logic of weighing military advantage versus civilian damage on a case-by-case basis under IHL, user and stockpile states argued that any kind of prohibition on this category of weapons could not be justified since that would require judging that the harm to civilians would be excessive across all expected use scenarios.[14] Against the many different use scenarios and types of cluster munitions, it was always possible to suggest instances in which they would not always lead to civilian harms. Some even went further. For instance, as argued by Christopher Greenwood QC in a legal analysis for the UK government, the 'whole picture' of the effects of different weapons would need to be examined when considering what to do about cluster munitions.

If this did not happen 'it may be that the protection of the civilian population is diminished rather than enhanced'.[15]

As a result, since the time they came to international attention, few detailed legal assessments suggested that cluster munitions per se were unlawful under the terms of IHL.[16] Instead, lawyers and scholars shared with many states the importance of reform measures to get rid of 'legacy munitions' with a high failure rate, to improve targeting practices, and to undertake other modifications intended to get rid of the 'worst of the worst'. Some lawyers and governments also called for a clarification of what factors should go into the balancing of military necessity versus humanity. As part of this, for instance, the proposal was forwarded that the likely long-term effects of sub-munitions should be taken into account.

In response to the year-after-year failure to adequately address humanitarian concerns, in February 2007 a Core Group of governments initiated a series of multilateral conferences – the Oslo Process – that sought to establish a binding treaty that would prohibit 'cluster munitions that cause unacceptable harm to civilians'.

With this normative starting point, the Oslo Process followed a different logic than set out in IHL, one that can be characterised as 'precautionary' in nature. It did so by combining the starting assumption that some cluster munitions 'cause unacceptable harm to civilians' with a definition structure that all weapons falling within the initial, broad understanding of cluster munitions were impermissible *until* the case was made otherwise. So early on in the process Article 1 prohibited states from ever using developing, acquiring, stockpiling, retaining, or transferring cluster munitions. In Article 2, the definition was set out as:

> 'Cluster munition' means a munition that is designed to disperse or release explosive sub-munitions, and includes those explosive sub-munitions. It does not mean the following:
>
> (a) ...
> (b) ...
> (c) ...[17]
>
> Exclusions were argued as part of '(a) ... (b) ... (c) ...'

Thus instead of following the typical pattern in disarmament negotiations of making critics justify why specific weapons should be banned, the Oslo Process started with a wide-ranging ban and then put it to those arguing for exclusions to justify why anything within that definition did *not* cause 'unacceptable harm'. This was not based on a sense that the case had already been made against all cluster munitions in all use scenarios with absolute certainty, but rather that history justified a starting orientation that presumed they were problematic.

In other words, rather than specifying what should be *prohibited*, the definition structure demanded that countries make a case for what exclusions should be *allowed*. Herein, the burden of proof was on those seeking to retain options – exclusions had to be 'argued in', rather than allowances 'argued out'.[18] With the near complete lack of detailed evidence or argument about the humanitarian damage or even the military utility of cluster munitions from users and stockpilers of cluster munitions during the Oslo Process, those wishing to retain cluster munitions were constantly put on the back foot.[19]

2 *Explosive weapons (in populated areas)*

As expressed by those directing attention to 'explosive violence', a stated goal is to achieve a 'reframing of conventional attitudes to weapons and violence'.[20] Explosive weapons include artillery shells, missiles, bombs (including mortar bombs, aircraft bombs, suicide bombs), grenades, landmines, and rockets. What unites this diverse range of technologies is that they function by projecting an area blast from an explosion in order to inflict injury, damage, and death.

An initial move made as part of the recent attention to this category has been to maintain that harms to civilians and damage to infrastructure from explosive weapons – especially when they are used in populated areas – are readily foreseeable. In short, there is a pattern. Air strikes in Georgia, grenades in Nigeria, artillery attacks in Gaza, shelling in Yemen, rockets used as artillery in Bosnia, and improvised explosive devices in Baghdad produce patterns of significant harm to civilians.

Such conclusions about the pattern of effects were echoed in a 2009 report to the UN Security Council. In this the UN Secretary-General wrote of the severe humanitarian concerns associated with explosive weapons with area effects in densely populated areas (as witnessed in Sri Lanka and Gaza).[21] In July 2010, the UN Under-Secretary-General for Humanitarian Affairs and Emergency Relief Coordinator likewise stated:

> The use of 'ordinary' explosive weapons in populated areas also repeatedly causes unacceptably high levels of harm to civilians. From air strikes and artillery attacks in Afghanistan, Somalia, Yemen and Gaza to rockets launched at Israeli civilian areas by Palestinian militants and car bombs and suicide attacks in Pakistan or Iraq, use of explosive weapons and explosives has resulted in severe civilian suffering.[22]

March 2011 saw high-level UN reiteration of concerns about explosive weapons in statements by the Coordinator of UN Emergency Relief regarding the shelling and bombardment of populated civilian areas in Libya[23] and Ivory Coast.[24] In 2012 the UN Secretary-General Ban Ki-Moon condemned the use of 'heavy artillery and the shelling of civilian areas' in Syria.[25] The UN Secretary-General's Reports have repeatedly raised concerns about the humanitarian impact of 'explosive weapons use in densely populated areas'.[26] Contributing to the debate, the International Committee of the Red Cross has stated that 'the use of explosive weapons in densely populated areas exposes the civilian population and infrastructure to heightened – and even extreme – risks of incidental or indiscriminate death, injury or destruction'. The continuing conflict in Syria has become identified as a prime case of the dangers of explosive weapons and the failure of international law to adequately protect civilians.

In such ways, active deliberation is now underway about what restrictions should be placed on explosive weapons, especially when used in populated areas.[27] Representing civil society, the International Network on Explosive Weapons called on states and others to:

- Acknowledge that use of explosive weapons in populated areas tends to cause severe harm to individuals and communities and furthers suffering by damaging vital infrastructure;
- Strive to avoid such harm and suffering in any situation, review and strengthen national policies and practices on use of explosive weapons and gather and make available relevant data;
- Work for full realisation of the rights of victims and survivors;
- Develop stronger international standards, including certain prohibitions and restrictions on the use of explosive weapons in populated areas.

The first expert meeting on explosive weapons attended by governments, intergovernmental agencies, and civil society was convened by the United Nations Office for the Coordination of Humanitarian Affairs and Chatham House on 23–24 September 2013.

In the current focus on explosive weapons, the reframing of violence is not only based on studies of effects from previous conflicts. Is also based on a sense of widely shared standards. Explosive weapons are already treated in practice as a coherent, taken-for-granted category despite the diversity of the technologies covered by the term. So they rarely figure within the context of domestic law enforcement. When this happens – as is the case of the Mexican government's response to drug cartels – it is typically regarded as a breakdown of law and order. Instead the entire category of technologies that fall under the heading of 'explosive weapons' are overwhelmingly used by military forces in the context of external or internal armed conflict. The latter, too, is typically regarded as a failure of state authority. This treatment differs from that of firearms, which also cause death and injury, but are widely accepted as force options for the military and police.

Once this categorical management across different settings is recognised, it is possible to ask whether it is appropriate. 'Who should be put at risk for injury' would seem to turn on political accountability. States rarely use explosive weapons when they are directly accountable to the populations that might be adversely affected by them. With a view to reducing humanitarian harm, the question being asked by those NGOs, intergovernmental organisations, and governments expressing concerns today is this: if explosive weapons are generally regarded as intolerable to use within a national border, when – and why – should they be tolerable elsewhere? Especially in situations that fall short of outright battlefield warfare, why should only some people be doomed to be put at risk? As such, the concerns being raised today extend far beyond the question of whether this or that type of explosive weapon was used in an indiscriminate or disproportional manner.

3 Banning nuclear weapons

Efforts to advance normative assessments of effects are also evident in recent attempts to prohibit nuclear weapons. As mentioned previously, working within the terms of international law the International Court of Justice did not offer a categorical censure of the threat or use of nuclear weapons.

Against this background and the limited achievements within the Non-Proliferation Treaty, many pressed for an alternative forum for debating nuclear weapons. At the 2012 UN First Committee on Disarmament and International Security, some 35 governments insisted that 'All States must intensify their efforts to outlaw nuclear weapons and achieve a world free of nuclear weapons'.[28] In 2013, Norway hosted an international conference to examine the *humanitarian impact* of the detonation of nuclear weapons. This was followed by a second such meeting in early 2014 in Mexico attended by delegates from 146 states. In announcing a follow-on meeting in Austria, the Chair of the latter conference, the deputy Minister of Foreign Affairs of Mexico, argued:

> The broad-based and comprehensive discussions on the humanitarian impact of nuclear weapons should lead to the commitment of States and civil society to reach new international standards and norms, through a legally binding instrument. It is the view of the Chair that the Nayarit Conference has shown that time has come to initiate a diplomatic process conducive to this goal. Our belief is that this process should comprise a specific timeframe, the definition of the most appropriate fora, and a clear and substantive framework, making the humanitarian impact of nuclear weapons the essence of disarmament efforts.[29]

In December 2014, the Vienna Conference on the Humanitarian Impact of Nuclear Weapons concluded with a pledge proposed by the Austrian government. That pledge started from the premise that the risk of nuclear weapons use with their unacceptable consequences can only be avoided when all nuclear weapons have been eliminated.[30] It called for, among other things, 'effective measures to fill the legal gap for the prohibition and elimination of nuclear weapons'. At the time of writing, over 50 governments had aligned themselves with this pledge.

Notably, the framing of nuclear weapons within recent efforts has placed their harms in the forefront. In doing so, at least at times, this has challenged the balancing of military utility and civilian harm in specific scenarios as called for in IHL. In this regard, the 35 countries that spoke at the 2012 UN First Committee on Disarmament and International Security stated, among other things:

> If such weapons were to be used, be it intentionally or accidentally, immense humanitarian consequences would be unavoidable.
>
> Nuclear weapons have the destructive capacity to pose a threat to the survival of humanity and as long as they continue to exist the threat to humanity will remain.
>
> The catastrophic humanitarian consequences of any use of nuclear weapons concern the community of States as a whole.[31]

Accordingly, these states argued that the response must be decisive and categorical:

> It is of utmost importance that nuclear weapons are never used again, under any circumstances. The only way to guarantee this is the total, irreversible and verifiable elimination of nuclear weapons, under effective international control . . .[32]

In a similar manner to how the Oslo Process drew on the past history of cluster munitions to argue for an approach that went beyond treating each case of use in terms of its individual harm versus necessity balancing, so too within these discussions the anticipated catastrophic consequences of nuclear weapons were treated as justifying an evaluation of them as whole – and one that offers a denunciation beyond what is proscribed by IHL. It is this evaluation that underpinned the agreement in 2017 of the Treaty on the Prohibition of Nuclear Weapons (TPNW). The TPNW embodies a categorical prohibition on nuclear weapons, their development, testing, production, stockpiling, stationing, transfer and use, as well as their threat of use. As of early 2019, 70 states have signed the Treaty and 22 have ratified it. It will enter into legal force once 50 nations have signed and ratified it. Beyond the formal signatures, as with the Mine Ban Treaty and Convention on Cluster Munitions, the hope of many supporters of the TPNW is that the instrument will serve as a normative influence on those countries that did not formally sign up to any resulting convention (e.g., nuclear-armed states).

Discussion

By considering both historical and contemporary examples, this chapter has sought to highlight different ways in which weapons can be governed with reference to their humanitarian consequences. A contrast has been set up between the logic of weighing military utility consequences under the principles and rules of IHL and an alternative set of orientations. As argued, the latter is characterised by features such as:

- The centrality of normative evaluations of harms;

- The recognition of the necessity to move beyond 'utility-consequence' or 'cost-benefit' type reasoning (justified because of the severity of harms, their persistence, etc.);
- The need for modes of justification that shift the onus away from what should be *prohibited* to what should be *allowed*; and
- The use of categorical evaluations of weapons, supporting and supported by the previous features.

Within the cases examined, the past and present concerns about biological weapons, cluster munitions, explosive weapons, and nuclear weapons all derived from, but also underpinned, normative standards of what is right and wrong in armed conflict. Indeed, the stigmatising of weapons is meant to have effects well beyond those nations that formally subscribe to international treaties, statements, etc., through setting international standards and thereby customary international law. As a result, consequentialist arguments about the balance of harm are mixing in complex ways with prescriptive normative assessments rooted in the contingencies of history. In the case of explosive weapons, for instance, both consequentialist and normative arguments have been used to advance the category of (conventional) 'explosive weapons' – a category that itself has been offered in response to the manner in which technologies that fall under the category of 'unconventional weapons' have typically garnered ample attention because they are held to be 'unconventional'.[33]

Having made these overall points about the possibilities associated with alternatives to the notion of balancing in IHL, it is important to consider some of the questions that they raise. As a generalising gloss, this author has characterised the features above as more 'precautionary' in their logic – a reference to 'the precautionary principle' popularised through environmental policy. However, just as the cases above exhibited diversity in their formulations, so too do formulations of the precautionary principle. Many variations of it share the premise that the lack of definitive evidence of harm – and so the existence of uncertainties, unknowns, and cases of ignorance – should not prevent deliberation or even action (and, in particular, action designed to prevent harm to the population or the environment).[34] Yet, in practice, this shared kernel gives way to diversity. The lack might be understood as deriving from either the absence of evidence or the inability of evidence to resolve disputes. It might be taken to justify making slight alterations to traditional forms of risk management, offering categorical evaluations of specific technologies, or making sweeping evaluations of whole areas of technology. As such, it does not follow that an appeal to 'precaution' implies a particular course of action. Questions that can be asked include:

- What level (threshold) of threat or potential for harm is sufficient to trigger application of the principle?
- Are the potential threats balanced against other considerations [...] in deciding what precautionary measures to implement?
- Where does the burden of proof rest to show the existence or absence of risk of harm?[35]

These are questions also relevant to the 'precautionary' orientations that have been examined in this chapter, each of which has to account for why and how the features listed above should apply. For each of the cases considered, the momentum that has built up has only been established after: many years of effort, a widespread perceived failure by traditional instruments of IHL, and/or extensive attempts to evidence the severity of past or potential harms.

Additional caveats can be mentioned. One, any attempts to impose restrictions on certain force options have to contend with the question of what will be used in their place. Two, by focusing on categories of technology, the orientations mentioned above are necessarily making generalisations about outcomes that entail bracketing off from consideration many contingencies and complexities. In speaking to both these points, those seeking to restrict weapons have done so under the premise that the weapons under question pose exceptional humanitarian harms vis-à-vis other force options that might be used in their place. While arguments can be given to justify certain determinations, it is not possible to prove such matters beyond all doubt given the counter-factual reasoning entailed.

Three, to the extent that they involve shifting the onus of proof, the orientations set out above do not escape from concerns that can be raised about IHL regarding how to define, measure, and compare the utility of and consequences from the use of force.

Four, all of the initiatives examined in this chapter have been undertaken in response to long-established weapon systems. In that respect they have all been retrospective in character. In line with the manner in which 'precautionary' thinking is intended to be relevant to emerging concerns, it is possible to apply the four features of the alternative set of orientations to up and coming capabilities,[36] though the additional considerations associated with that are beyond the scope of this chapter.

With regard to many of the issues mentioned in the previous several paragraphs, one way to think about the relevance of the alternative logics examined in this chapter is through asking what kinds of conversations they enable. Through considering matters of process, it is possible to appreciate the advantages associated with the alternative orientations. For instance, despite evidence put forward in conflict after conflict about the extra-ordinary civilian harms stemming from cluster munitions, governments rejected calls for significant restrictions on these weapons, often through citing their legality under the principles and rules of IHL. And yet, they did so with little or no reference to their evidence for either the military utility or civilian harms. As a result, the appraisal of the weapons had become deadlocked. In contrast, with its basic argumentative structure, the Oslo Process demanded that states wishing to retain cluster munitions make the case as to why this was justified. In theory this could have led to a robust assessment of variations of cluster munitions and their alternatives in which certain types of these weapons were retained. In practice though, because so little evidence in defence of cluster munitions was offered by governments, this was not the case. What the structure of the dialogue of the Oslo Process enabled was a process of learning how little information states had gathered or circulated about the effects and reliability of the weapons that they advocated as being in accordance with IHL.

In this respect, the alternative approaches identified in the chapter for handling evidence, uncertainty, and onus for proof, offer the prospect of loosening the knots and impasses that frustrate international disarmament today.

The issues at hand in relation to determining the permissibility of the use of force are, in practice, more complicated than it would be possible to present in this chapter, because the consequentialist balancing of military necessity against concerns for civilian lives and property within IHL is complicated by additional forms of reasoning (e.g., the deontological prohibitions on making civilians as such the object of attack) and evaluative distinctions (e.g., the legitimacy of intended versus unintended harms). Nevertheless, the existence of such provisions or other elements of existing international law (such as current requirements for the review of new weapons and the means and methods of warfare under Article 36 of the 1977 Additional Protocol of the Geneva Conventions[37]) has not prevented extensive death and injury to civilian populations in recent conflicts. It is this lamentable state of affairs that ought to be redressed.

Notes

1 For instance, US Air Force Colonel Charles Dunlap quotes John Donne (1621) as stating: 'So by the benefit of this light of reason, they have found out artillery, by which wars come to a quicker end than heretofore, and the great expense of blood is avoided, for the number slain now, since the invention of artillery, are much less than before, when the sword was the executioner.' See Charles Dunlap, 'Technology: Recomplicating moral life for the nation's defenders' *Parameters*, Autumn1999, pp. 24–53.

2 For a general introduction to the dilemmas of controlling technologies, see David Collinridge, *The Social Control of Technology*, New York: St Martin's, 1980.

3 See International Committee of the Red Cross, *Existing Principles and Rules of International Humanitarian Law Applicable to Munitions that May Become Explosive Remnants of War*, Paper Submitted to the Convention on Prohibitions or Restrictions on the Use of Certain Conventional Weapons Which May Be Deemed to Be Excessively Injurious or to Have Indiscriminate Effects, CCW/GGE/XI/WG.1/WP.7, 28 July 2005.

4 International Committee of the Red Cross, *Existing Principles* (see note 3 above).

5 See International Court of Justice, *Legality of the Threat or Use of Nuclear Weapons* – Advisory Opinion of 8 July 1996 (The Hague: ICJ, 1996), Articles 95 and also 94.

6 Jean-Marie Henckaerts and Louise Doswald-Beck (eds.), *Customary International Humanitarian Law*, 2 vols (Cambridge: Cambridge University Press, 2004), Vol. I, Chapter 23.

7 For past examples of such arguments, see Brian Balmer, 'Killing "without the distressing preliminaries": scientists' defence of the British biological warfare programme', *Minerva*, Vol. 40, Issue 1 (2002), pp. 57–75.

8 Jonathan Tucker, *Innovation, Dual-use and Security*, Cambridge, MA: MIT Press, 2012.

9 For a more lengthy discussion of what Nina Tannenwald calls the 'permissive' effects of the nuclear taboo see Nina Tannenwald, *The Nuclear Taboo* (Cambridge: Cambridge University Press, 2007), pp. 317–24.

10 Richard Falk, 'The challenges of biological weaponry', in Susan Wright (ed.), *Biological Warfare and Disarmament* (London: Rowman & Littlefield, 2001), p. 29.

11 See, for instance, Brian Rappert, Experimental Secrets: International Security, Codes, and the Future of Research (New York: University Press of America, 2009) and Brian Rappert (ed.), *Education and Ethics in the Life Sciences,* Canberra: Australian National University E Press, 2010.

12 Article 2 in the final text of the Convention on Cluster Munitions defined a cluster munition as:

> a conventional munition that is designed to disperse or release explosive submunitions each weighing less than 20 kilograms, and includes those explosive submunitions. It does not mean the following:
>
> a A munition or submunition designed to dispense flares, smoke, pyrotechnics or chaff; or a munition designed exclusively for an air defence role;
> b A munition or submunition designed to produce electrical or electronic effects;
> c A munition that, in order to avoid indiscriminate area effects and the risks posed by unexploded submunitions, has all of the following characteristics:
>
> i Each munition contains fewer than ten explosive submunitions;
> ii Each explosive submunition weighs more than four kilograms;
> iii Each explosive submunition is designed to detect and engage a single target object;
> iv Each explosive submunition is equipped with an electronic self-destruction mechanism;
> v Each explosive submunition is equipped with an electronic self-deactivating feature.

See www.icrc.org/applic/ihl/ihl.nsf/ART/620-6?OpenDocument. Also the discussion by Bill Boothby in Chapter 3.

13 Brian Rappert and Richard Moyes, *Failure to Protect*, London: Landmine Action, 2006.

14 See Brian Rappert, *How to Look Good in a War: Justifying and Challenging State Violence*, London: Pluto, 2012.

15 Christopher J. Greenwood QC, *Legal Issues Regarding Explosive Remnants of War*, Group of Government Experts of States Parties to the Convention on Prohibitions or Restrictions on the Use of Certain

Conventional Weapons Which May Be Deemed to Be Excessively Injurious or to Have Indiscriminate Effects, CCW/GGE/I/WP.10, 23 May 2002, p. 8.

16 For a detailed discussion see Rappert and Moyes, *Failure to Protect* (note 13 above).

17 Convention on Cluster Munitions, 'Chair's Discussion Text for Vienna Conference, December 2007. Available at: www.clusterconvention.org/pages/pages_vi/vib_opdoc_chairsvienna.html.

18 Brian Rappert and Richard Moyes, 'The prohibition of cluster munitions: setting international precedents for defining inhumanity', *Non-proliferation Review*, Vol. 16, No. 2 (2009), pp. 237–56.

19 Ibid.

20 Richard Moyes, *Explosive Violence: The Problem of Explosive Weapons* (London: Landmine Action 2009), p. 7. See as well UNIDR, *Explosive Weapons: Framing the Problem* April (Geneva: UNIDIR, 2010).

21 UN Security Council, *Report of the Secretary-General on the Protection of Civilians in Armed Conflict* S/2009/277, 29 May (New York: UNSC, 2009), para. 36.

22 Statement by John Holmes, Under-Secretary-General for Humanitarian Affairs and Emergency Relief Coordinator, during the Security Council Open Debate on the Protection of Civilians in Armed Conflict, 7 July 2010.

23 UN Emergency Relief Coordinator, *United Nations Humanitarian Chief Highlights Humanitarian Consequences of Continued Fighting in Libya*, New York: OCHA, 2011.

24 UN Emergency Relief Coordinator, *United Nations Humanitarian Chief Alarmed at Cote D'Ivoire Violence*, New York: OCHA, 2011.

25 Ban Ki-Moon, *Statement Attributable to the Spokesperson for the Secretary-General on Syria*, 6 February 2012.

26 UN Security Council, *Report of the Secretary-General on the Protection of Civilians in Armed Conflict* S/2009/277, 29 May (New York: UNSC, 2009); UN Security Council, *Report of the Secretary-General on the Protection of Civilians in Armed Conflict* S/2010/579, 11 November (New York: UNSC, 2010); and UN Security Council, *Report of the Secretary-General on the Protection of Civilians in Armed Conflict* S/2012/376, 22 May (New York: UNSC, 2012).

27 See Moyes, *Explosive Violence* (note 20 above).

28 Switzerland, 'Joint Statement on the humanitarian dimension of nuclear disarmament', Delivered at the 67th session of the United Nations General Assembly First Committee, 22 October 2012.

29 Juan Manuel Gomez Robledo, 'Statement by the Deputy Minister of Foreign Affairs of Mexico, Nayarit, Mexico', 14 February 2014.

30 Michael Linhart, *Pledge presented at the Vienna Conference on the Humanitarian Impact of Nuclear Weapons* (Vienna: Austrian Foreign Ministry, 2014).

31 Switzerland, 'Joint Statement on the humanitarian dimension of nuclear disarmament' (note 28 above).

32 Ibid.

33 Richard Moyes, 'Causing problems: classification of humanitarian concerns regarding explosive weapons', in Brian Rappert and Brian Balmer (eds.), *Absence in Science, Security and Policy* (London: Palgrave, 2015), pp. 200–225.

34 David C. Magnus, 'Risk management versus the precautionary principle', in Robert N. Proctor and Londa Schiebinger (eds.), *Agnotology* (Stanford: Stanford University Press, 2008), pp. 250–65.

35 Deborah C. Peterson, 'Precaution: principles and practice in Australian environmental and natural resource management', *Australian Journal of Agricultural Resource Economics*, Vol. 50, Issue 4 (2006), p. 471. doi:10.1111/j.1467-8489.2006.00372.x.

36 Brian Rappert, 'Why has not there been more Research of Concern?' *Frontiers in Public Health*, Vol. 2, July 2014. doi: 10.3389/fpubh.2014.00074.

37 Brian Rappert, Richard Moyes, Anna Crowe and Thomas Nash, 'The Roles of Civil Society in the Development of Standards around New Weapons and other Technologies of Warfare', *International Review of the Red Cross*, Vol. 94, Issue 886, 2012, pp. 765–85.

PART II

Cyber warfare

6

COMPUTER NETWORK ATTACKS UNDER THE *JUS AD BELLUM* AND THE *JUS IN BELLO*

'Armed' – effects and consequences

Elaine Korzak and James Gow

Computer Network Attacks (CNAs) – cyber warfare – present major challenges regarding both the *jus in bello* and the *jus ad bellum* – the two branches of international law concerned with warfare (or, more narrowly and more technically, 'armed conflict').[1] Similar issues arise in and run across the contexts of the *jus ad bellum*, the law on the use of force, governing the rightfulness of a state's use of force, and the *jus in bello*, international humanitarian law (IHL) governing the conduct of armed hostilities.

International law on the use of force, or *jus ad bellum*, governs the resort to force in international relations. It is based on a comprehensive prohibition on the use of force in inter-state relations, which has evolved into a fundamental norm of customary international law. With the creation of the United Nations, the prohibition was codified in Article 2(4) of the Charter. However, the Charter regime acknowledged only two exceptions to this sweeping ban on the use of force, namely a state's 'inherent' right to self-defence as well as the authorisation of forceful measures by the UN Security Council. These exceptions are codified in Article 51 as well as Articles 39 and 42 of the UN Charter, respectively. Taken together, these provisions set out the basic tenets stipulating when states can lawfully use armed force in their international relations.

Although related, international humanitarian law and international law on the use of force are normatively distinct. The norms of international humanitarian law are related to the Charter's framework for the *jus ad bellum*, but both pre-date the Charter in some instances and are distinct from it. While international law on the use of force stipulates *when* states can use armed force in their international relations, international humanitarian law regulates *how* armed force is to be used. And, irrespective of the legality *ad bellum*, international humanitarian law applies equally to all parties in an *armed conflict*, irrespective of the legality or illegality of the specific use of force as judged under the international law on the *use of force*.[2]

Reflecting the experiences of the First and Second World Wars, the *jus ad bellum* regime is premised on the exclusive ability of states to use *armed force* on a scale necessary to engage international law. It is, further, based on a dichotomy between non-forceful measures and forceful actions, with the latter underpinned by a notion of *force* that involves the elements of 'blast, heat and fragmentation'[3] in one form or another. The use of information and communication technologies for military purposes and in particular the emergence of CNAs presents obvious challenges to such an understanding. The characteristics of CNAs, which do not directly cause blast and destruction, bring into question the notion of *force* underpinning the legal paradigm of international law on the use of force. The wide availability of technology also challenges the monopoly of states to use force in international relations, a challenge already seen in connection with changing conflict patterns and the emergence of transnational terrorist actors.[4]

Questions also arise regarding the applicability of IHL. The determination of an *armed conflict* is generally the threshold for application of the *jus in bello*. Do CNAs fall under IHL, given their very different nature, and, if so, in which circumstances? Different aspects of IHL come into play, including the rules pertaining to perfidy, participants in conflict, and measures of special protection, as well as rules forming the law of neutrality.[5] The protection of civilians from the effects of hostilities is a particular concern, given the twenty-first-century reliance on the digital domain and the way this reliance has opened up new vulnerabilities.[6]

IHL is generally regarded as applying in all situations of *armed conflict*, which are fundamentally characterised by the 'resort to armed force' or 'protracted armed violence'. In the case of international armed conflicts, actions have to be conducted by states or be attributable to them under international law. With regard to non-international armed conflicts, any non-state groups involved have to exhibit a certain level of organisation and in the case of Additional Protocol II to the Geneva Conventions also control territory. While international armed conflicts arguably do not have to reach any particular level of intensity or duration to be judged as such (but they will need to be more than isolated shots fired across a border), observations of internal disturbances or acts of violence will not suffice to determine the legal existence of a non-international armed conflict. In all these instances, international humanitarian law applies from the initiation of such armed conflicts and extends beyond the cessation of hostilities until a general conclusion of peace is reached; or, in the case of internal conflicts, a peaceful settlement is achieved. Until that moment, international humanitarian law continues to apply in the whole territory of the warring states, or in the case of internal conflicts, the whole territory under the control of a party, whether or not actual combat takes place.[7]

This chapter explores the challenges that cyber warfare present to both bodies of law relating to warfare. It does so in four sections, recognising the impact and degree of change that CNAs bring. The first section discusses the crucial term 'armed' on which both bodies of law rely and which CNAs significantly complicate, examining the nature of *weaponry* in relation to cyber activity. The second section explores the way in which international lawyers have reacted to the emergence of cyber warfare by shifting the focus of analysis from the means used and the intention of the user to the effects and consequences of such use. The third section examines the notions of *distinction* and *proportionality*, traditional notions within the *jus in bello* tradition, which are significantly compromised by the emergence of cyber weapons, even to the point that the application of these core principle could make certain types of attack more likely, rather than inhibiting them. Finally, we assess the crucial question of *attribution* in relation to both bodies of law, noting that for the right to self-defence to be invoked, an *armed* attack would require *inter alia* identification of a perpetrator, as would any war crimes prosecutions – but that both technical and, ultimately, political constraints might prevent this. Overall, we argue that, while significant progress has been made among international lawyers on finding ways to apply the law to cyber warfare, the unique features of these new types of attacks, particularly their *non-kinetic* mode of operation, their range of possible effects, and their perceived anonymity, create significant difficulties for the application of international law. In very important ways, gaps remain and, indeed, may provide increased, or better, opportunities for hostile action using cyber capabilities.

'Armed': the nature of the weapon

Computer network attacks (CNAs) present fundamental questions for both of the bodies of law relating to war. The prohibition on the use of force appears in Article 2 (4) of the UN Charter. The Charter regime, inevitably, was based on threat categories and perceptions that could not include any type or form of cyber threat, or, indeed, the unique characteristics of such attacks, which were not even imagined at the time. Unsurprisingly, the Geneva Conventions, written when computers were in their infancy, and other relevant instruments of IHL, do not directly address the phenomenon of computer network attacks. This might be taken to mean that they fall outside the regulatory purview of international humanitarian law,[8] making them 'exempt' from it.[9]

Such claims can be countered with relative ease. First, as with the relevant provisions of the UN Charter, the fact that the Geneva Conventions and other instruments do not explicitly refer to computer network attacks does not preclude their applicability. The International Court of Justice addressed this line of reasoning with regard to nuclear weapons.[10] As both computer network attacks and nuclear weapons were developed after relevant humanitarian norms entered into force, the same conclusion could, in principle, be said to apply to the phenomenon of CNAs.[11]

In principle, and specific to IHL, the *Martens Clause* provides in situations which are not covered by international agreements: 'civilians and combatants remain under the protection and authority of the principles of international law derived from established custom, from the principles of humanity and from the dictates of public conscience'.[12] IHL therefore foresees new developments in the evolution of warfare and extends its coverage to those – yet unknown – instances. In the case of Additional Protocol I, Article 36 requires states parties to the Protocol to review new weapons for compliance with IHL.[13] Thus, the fact that CNAs postdate particular instruments of IHL, or that they are not directly addressed therein, does not exclude them from the regulatory purview of those instruments.[14] Rather, the critical question in this context is whether CNAs could amount to an (international or non-international) 'armed conflict' in accordance with relevant provisions of IHL.

Yet, major questions remain. These turn on the interpretation of the crucial term *armed*. This underpins the notions of *armed attack* and *armed conflict* that trigger the application of *ad bellum* and *in bello* law, respectively. IHL becomes applicable in armed conflict, but the emergence of CNAs begs the question of whether these types of attacks can trigger an armed conflict bringing into play the framework of the *jus in bello*. In other words, the fundamental question arises whether CNAs are subject to IHL.

The notion of armed conflict is quintessentially characterised by the use of armed force, as is the notion of armed attack. It is important to note that these notions are similar, but not identical. They are also related to the *use of force* as a concept, which is generally understood to mean a use of *armed* force. In each instance, it is the understanding of 'armed' that is crucial, given that cyber capabilities do not share many of the characteristics commonly associated with conventional warfare – blast and fragmentation, physical destruction and damage. The question of whether CNAs could be considered to be armed, for the purposes of international law, is moot.

Where CNAs are conducted alongside traditional means of warfare, the requirement of armed force does not pose any definitional difficulties. If *normal* weapons are used, the determination of an armed conflict is unproblematic and the relevant CNAs become subject to IHL.[15] The *Tallinn Manual on the International Law Applicable to Cyber Warfare*, developed by leading international lawyers and officials from NATO countries and coordinated by Mike Schmitt, specifies that, where there is already an armed conflict, 'the applicable law of international or non-international armed conflict will govern cyber operations undertaken in relation to that conflict'.[16] In this sense, IHL would cover the attacks on governmental and media websites during the initial phases of the armed conflict between Georgia and the Russian Federation in 2008, as they were part of a wider conventional conflict.[17]

Arguably, CNAs do not resemble familiar types of coercion or force – they do not correspond to previously known categories of armed force or economic and political coercion.[18] For instance, if the information held by the banking systems of a state were to be destroyed, the effects could cripple the financial sector within minutes and without a single immediate casualty.[19] According to Benatar, 'No type of armed, political or economic force can accomplish such a feat, so there simply is no analogy to be made'.[20] This makes it doubtful that CNAs could be regulated by the notion of *force* in Article 2(4) of the UN Charter, as they do not resemble familiar types of coercion.[21] Closely related to this, another line of argumentation posits that CNAs are qualitatively different from previously known forms of economic and political pressure. Traditionally economic or political coercion involved external and gradual pressure; for example, using trade sanctions.[22] In contrast, CNAs enable swift and more devastating economic pressure, while targeting the internal economic structures of a state directly.[23] Crippling the banking sector and closing securities markets represents a qualitatively different phenomenon.[24] This could make debates over whether economic sanctions and other measures should be regarded as *force* nugatory.[25] However, the current interpretative framework may still be applicable – CNAs constitute a new form of hostile activity, which could be viewed as merely another form of economic and political coercion falling outside the scope of force contained in Article 2(4).[26]

The concept of armed force advanced by Benatar requires an act of coercion to be military and physical.[27] However, there is a paucity of legal guidance on what constitutes 'military'.[28] One indicator of the military character of CNAs could be their use by, or inclusion in the arsenals of, armies.[29] It is less likely that they would be accepted as military in character, if they were used by intelligence organisations.[30] The understanding of what constitutes a military technique, or act, can change over time as the character of warfare changes. The United States has incorporated cyber means in its military doctrine.[31] Moreover, in 2009 then Defence Secretary Gates announced the creation of a separate military command dedicated to operations involving cyberspace – US Cyber Command.[32] Other countries have followed suit.[33] The inclusion of cyber capabilities in their military structures clearly gives them a 'military' character.

The requirement for an act of coercion to be 'physical', which, as just noted, Benatar demands (and he is not alone), appears more challenging in the case of CNAs.[34] The need for physical means is often tied to the concept of *weaponry*. Much in the discourse surrounding armed conflict and armed attack depends on the notion of *arms* themselves – the character of the instruments being used, the weapons. This, in turn, has depended on an understanding of weapons as being *kinetic*. The concept of arms for the purposes of international law has typically rested on an understanding of the terms *physical* and *kinetic* involving 'explosive effects with shock waves and heat'.[35] Kinesis is the physical transfer of energy, resulting in change, or movement. *Kinetics* have been aligned with projectile (usually) or detonating weapons that produce an energy transfer that generates blast and fragmentation, producing actual and visible physical damage.[36] In this sense, IHL is 'designed for methods and means that are kinetic in nature'.[37] CNAs, however, are seen as neither physical nor kinetic, by most observers, meaning that such attacks 'fall outside the scope of humanitarian law'.[38] Bit streams of malicious code are seen as intangible and do not correspond to the physical effects of conventional weapons. This conclusion, shared by many authors, rests on the 'invisibility' of CNAs, which makes it 'hard to see [*literally*] how the inherent intangibility of computer attacks can be reconciled with the ... requirement (of) physical/kinetic force'.[39]

Yet, the concept of 'weaponry' is flexible and kinesis, as a process, could, theoretically, apply technically to computer network attacks – if there is no transfer of energy, then there is nothing digital. So, the dominant understanding of both armed conflict and the law to which it relates is not beyond conceptual interrogation. In terms of kinetics, the 'virtual' world of digital action relies on electricity – itself a physical phenomenon inherently involving energy transfer. There is no electricity unless energy is transferred. Therefore, the sense in which 'kinetic' is used by military and legal analysts alike to refer to types of large-scale energy transfer resulting in blast and destruction is more an intellectual convention than a genuine matter of kinesis.

Although certain instruments are expressly created as weapons – knives, swords, pikes, spears, guns, bombs, shells, tanks, and so on – and are designed to inflict physical damage on an opponent, any understanding of weapons restricted to these artefacts would be limited. It is clearly reasonable to see these instruments as weapons – though even then, as John Stone notes, a tank could be used as a large paperweight.[40] As this comment indicates, weapons are defined by purpose and use.[41] A weapon is anything that serves to disrupt an enemy. It is an instrument that gains advantage in armed conflict. Weapons have many forms and characteristics and may be dual-edged – in this sense, irony is classified as a weapon by the *Oxford English Dictionary*. As many a murder inquiry, or murder mystery novel, or parlour game has shown, a candlestick, a telephone, a paperweight, or any other object can be used as a murder weapon. These objects have other social purposes, but they can be adapted to cause violence and to gain a perceived advantage. Weapons are, therefore, social constructions. In some cases, where industrial procedures are clearly intended for use as weapons, their primary purpose is clear (though others might be possible). In others, their purpose may be defined by circumstance. It is the use to which an instrument is put that determines whether or not it is a weapon. Any instrument used to strike blows of any kind against an opponent, attempting to gain an advantage in *armed conflict* is a weapon.[42]

A weapon is not defined by the characteristics of physical or kinetic force – indeed, as just seen, many objects designed to be weapons and cause harm, such as swords, do not rely on the qualities of 'blast and fragmentation' so important to those who socially agree on using the construction 'kinetic force' in relation to armed conflict (or the use of force or armed attack) in relation to international law. The use of spears and pikes, in any sensible realm in the context of armed conflict (clearly, not re-enactment, or a play on the stage), would still be subject to IHL in the context of *armed conflict* – and the armed conflict itself could be determined by their use with sufficient organisation and intensity. A weapon, then, is defined by its use and its ability to gain an advantage over an opponent, which might well involve significant physical damage, but need not necessarily do so. Therefore, the relative, or perceived intangibility of CNAs (noting that, technically, even these have physical qualities) should not preclude classification as weapons. Rather the intent and purpose of a CNA is determinative.[43] Approached this way, we may dispense with the challenging aspect of the intangibility of CNAs by advancing an intent-based definition of 'weapon'. In this understanding, in the present chapter, where many authors use 'kinetic force', we avoid this particular, limiting, social construction and prefer simply to use the broader and more generally descriptive term 'conventional armed force', recognising both the conventions of practice among armed forces and the conventions of use among certain commentators.

Another approach to addressing the problematic aspect of that which is described as 'kinetic', or physical, force in defining 'armed force' is the two-pronged test articulated by Brownlie. He argues that the use of devices that are (1) 'commonly referred to as "weapons" and as forms of "warfare"' and that are (2) 'employed for the destruction of life and property' can be qualified as a use of 'armed force'.[44] While the notion of armed force has conventionally been associated with explosive effects involving shock waves and heat, this approach opens up the social space potentially to encompass cyber capabilities, where they are described as weapons and used for the purposes described. This would be a narrower definition than discussed above, probably, but it would widen the scope for application of the law. Brownlie's first element examines whether CNAs are commonly referred to as *weapons* or forms of *warfare*. A state's use of, or planning to use, CNAs by its military, or armed forces, might be an indicator that it regards CNAs to be part of its weapons inventory.[45] The activities of numerous states and their efforts to incorporate computer network capabilities into their force structures in one way or another shows that CNAs are seen as weapons or forms of warfare. Some commentators have even referred to computer network attacks as 'eWMDs' due to their potential for asymmetric warfare.[46] Likewise, both media and academics have referred to a range of activities as cyber war, CNAs, or information warfare – although whether these would trigger application of the law would depend on factual material relating to the second of Brownlie's tests. Cyber instruments and techniques, such as Trojan horses, viruses, worms (all forms of 'syntactic' attack, where code is affected and changed), or denial of service attacks (where the aim is to overload a system and prevent its working, rather than to change it at the level of code), might then be qualified as forms of armed force, whose use would consequently be proscribed, based on a combination of their social description and context and the outcome of their use. Following Brownlie's test, computer network attacks could, in principle and in practice, be characterised as a form of armed force despite their perceived non-kinetic mode of operation.

Effects and consequences

Computer network attacks conducted on their own without any accompanying traditional or physical force, then, challenge both international law and the notions of weapons and armed action on which prevailing interpretations of that law depend. Do CNAs in isolation qualify as 'armed'? If so, when, under what conditions? Despite the conservative doubts about whether digital attacks constitute armed attacks, the underlying motivation for the application of IHL – namely to limit damage and provide care for those affected – may militate in favour of an expansive interpretation of when IHL should apply.[47] While cyber operations themselves may not rise to the level of violence and destruction that many legal commentators regard as necessary to invoke the law, they can, nonetheless, 'generate violent consequences'.[48] Contemporary societies depend on computers, computer systems, and networks, and it is possible to cause major destruction using means that are 'non-destructive' in conventional terms.[49] To deal with this problem, some scholars have tackled the problematic requirement of armed force by introducing a consequence-based interpretation of the terms 'armed conflict', 'use of force' and 'armed attack'. The analytical focus in this section is, therefore, shifted onto the effects of a CNA and whether these can be considered to be analogous or, even, equivalent, to the results of conventional arms.[50]

Accordingly, CNAs can trigger the application of international humanitarian law if the consequences of their use are equivalent to the damage caused by conventional armed force.[51] In this understanding, damage to the computer programs – even though physical – would not be sufficient to implement international humanitarian law; there would need to be significant physical damage beyond that.[52] Schmitt usefully summarises this position:

> humanitarian law principles apply whenever computer network attacks can be ascribed to a State, are more than merely sporadic and isolated incidents, and are either intended

> to cause injury, death, damage, or destruction (and analogous effects), or such consequences are foreseeable. This is so even though classic *armed* force is not being employed.[53]

Similarly, Schmitt argues that 'A careful reading of Additional Protocol I ... discloses that the concern was not so much with acts which were violent, but rather with those that have harmful consequences (or risk them), in other words, violent consequences'.[54] 'Attacks', as a label, constitutes 'prescriptive shorthand' regarding the protection of persons and property.[55] The object and purpose of Additional Protocol I are to avoid these consequences to the greatest extent possible in light of military necessity.[56] Thus, CNAs that result in, or are expected to result in, death or injury to individuals, or damage or destruction of objects, can be qualified as 'attacks' in accordance with Article 49 (1) of Additional Protocol I,[57] as scholars concur.[58] In short, international law could very likely be applied to most CNAs launched by states or other organised groups in the context of armed conflict, where the attack has a 'physical manifestation' involving characteristics associated with conventional armed force – physical damage and destruction affecting property and life.[59]

The key issue becomes whether CNAs are employed for the destruction of life and property. That is, despite their non-conventional mode of operation, CNAs can be characterised as a form of armed force, if they cause human injury or property damage. The vast majority of legal commentators have taken this consequence-based approach as the basis for classifying CNA as a use of (armed) force, enabling measures that do not clearly fit with traditional interpretations of 'physical force' and the law to be covered by it, as they have the potential to cause considerable physical damage.[60] For Silver, CNA can be considered as 'armed force', when physical injury or property damage arise as a 'direct and foreseeable consequence' of the CNA and 'resemble the injury or damage associated with what, at the time, are generally recognized as military weapons'.[61] In this consequence-based approach, adopted by numerous leading scholars (and following the second element of Brownlie's two-part test, in particular), CNAs constitute a use of force if their consequences resemble the effects of recognised weapons, primarily physical destruction and injury.[62]

However, a focus on the consequences of CNAs results in the need to classify CNAs on a case-by-case basis because of the wide range of consequences that can be effectuated by these attacks,[63] in order to determine whether a particular attack constitutes, or constitutes part of, an 'armed conflict' for IHL to be applied. This line of analysis moves away from the actor-based threshold for application of IHL offered by Pictet's commentary on the Geneva Conventions. Instead, the majority of commentators advance an interpretation of the armed conflict threshold that is contingent on the consequences of a CNA, which have to resemble the use of conventional armed force. Results can vary between the mere inconvenience of not being able to access certain websites to actual physical destruction or injury, meaning that not all attacks would fall under the law.[64] Rather, each CNA has to be analysed *sui generis*. CNAs whose consequences do not involve physically destructive effects would most likely be regarded as falling outside the scope of the prohibition on the 'use of force'. Attacks that directly and foreseeably result in physical destruction or human injury, effects resembling traditional armed force, can clearly be qualified as a prohibited use of force. For example, a CNA used to manipulate the data in the information system controlling a military aircraft that caused the aircraft to crash would most likely be qualified as a use of force.[65] The physical destruction and presumed human injury or death caused by the crash undoubtedly constitute a physical outcome which is a direct and foreseeable result of the CNA. According to Schmitt, CNAs on a large airport's air traffic control system by agents of another state, or attacks intended to destroy oil pipelines by surging oil through them after taking control of computers governing their flow, or causing the meltdown of a nuclear reactor by manipulation of its computerised nerve centre, or using computers to trigger a release of toxic chemicals from production and storage facilities, would all warrant application of IHL. In such a consequence-based analysis, the destructive effects of the CNA in question qualify it as a use of force irrespective of the means used. It is not dispositive that the outcome, a plane crash, was achieved through electronic means as opposed to the use of an anti-aircraft missile or a bomb aboard the aircraft.[66] If attributable to a state, therefore, Stuxnet, for example, might have triggered the application of IHL, if it could be said to have caused material damage to the gas centrifuges at Natanz.

This line of analysis moves away from the actor-based threshold for application of international humanitarian law offered by Pictet's commentary on the Geneva Conventions. Instead, the majority of commentators advance an interpretation of the 'armed conflict' threshold that is contingent on the consequences of a computer network attack, which have to resemble the use of conventional 'armed force'. In light of the wide spectrum of consequences that can arise from them, computer network attacks have to be analysed on a case-by-case basis, as already noted, in order to determine whether a particular attack constitutes, or constitutes part of, an 'armed conflict' international humanitarian law to be applied. Thus, computer network attacks need to be assessed *sui generis* and cannot per se be categorised. Depending on their nature and likely consequences, computer network attacks may be qualified as 'armed' or not. In contrast, IHL would not apply to merely disruptive CNA, such as 'disrupting a university intranet, downloading financial records, shutting down Internet access temporarily, or conducting cyber espionage because … the foreseeable consequences would not include injury, death, damage or destruction'.[67] If a state were to penetrate the information system of another state with the aim of copying data but otherwise leaving the system intact, the unauthorised access to and copying of the data in question constitute theft or espionage. This would not be a use of force because the incident did not cause major physical damage or personal injury and the information system was not manipulated in such a way as to cause malfunction or other intended effects. Attacks that do not result in physical damage, but nevertheless cause loss of functionality, are controversial. Arguably, a CNA that disables an object, requiring its repair, might constitute a reasonable extension of the notion of damage.[68] However, *mere inconvenience* would not be enough.[69]

The consequences of a CNA constitute the linchpin of analysis. Physical damage to property or persons analogous to that of conventional armed force qualifies a computer network attack as a 'resort to armed force', thereby making international humanitarian law applicable. Arguably, the serious disruption of critical infrastructures also qualifies. If CNAs can be characterised as 'armed force' while being used 'between states', the IHL governing international armed conflicts applies. Such attacks have to be attributable to states under the law of state responsibility. In the case of non-international armed conflicts, the additional requirements of organisation and intensity attach.[70] Although the *organisation* criterion is always context-specific, it generally implies that actions in question 'are best understood as those of a group and not its individual members'.[71] Therefore, CNAs conducted by individuals acting alone cannot meet this requirement. In the case of the distributed denial-of-service attacks in Estonia in 2007, the absence of the requisite level of organisation meant that the attacks fell short of the non-international armed conflict threshold, despite the number of individuals involved – attacks in parallel are deemed not to be organised.[72] Last, the degree of intensity required for a non-international armed conflict precludes application of the law to many computer network attacks. Even if attacks proved highly destructive, they would have to occur on a regular basis over a period of time in order to qualify as 'protracted violence'.[73] Thus, unless attributable to a state (in the case of international armed conflicts) or satisfying the requirements of organisation and intensity (for non-international armed conflicts) CNAs would fail to trigger application of international humanitarian law, even if their consequences resembled damage to property or persons. Municipal legal regimes and international human rights law would offer the only legal avenues to cover such attacks.

Conclusion

The emergence of CNAs as a new method or means of warfare poses important questions in the context of the Laws of Armed Conflict and the Use of Force. The non-conventional character of CNAs poses obvious difficulties when assessing *armed attacks*, *armed conflict* and the *use of force*, given that conventional characteristics of armed force, such as blast and fragmentation or significant visible destruction, are absent, at the immediate level, at least. As a result, international lawyers have largely shifted towards the potential physical consequences of CNAs as a way of incorporating them into the legal framework of both *jus in bello* and *jus ad bellum*. If these effects are analogous to those of conventional arms, that is damage to property or human injury, then a CNA will be a prohibited *use of force* and, if the scale and effects are significant enough, an *armed attack*, or part of an *armed conflict*. This, however, means that broad principles are hard to derive, as each situation must be analysed on a case-by-case basis. Computer network attacks are capable of mounting an *armed attack*, or constituting an *armed conflict*, or of manifesting *armed force*, so long as they produce evident damage or destruction of objects and injury to, or loss of, human life.

Notes

1 This chapter is based in part on research for a project funded by the Research Council's UK-DSTL 'Science and Security' *SNT Really Makes Reality: Technological Innovation, Non-Obvious Warfare and Challenges to International Law* ES/K011413/1. Subsequent references to this research are labelled '*SNT*'.

2 Christopher Greenwood, 'Scope of Application of Humanitarian Law', in Dieter Fleck (ed.), *The Handbook of Humanitarian Law in Armed Conflicts* (Oxford: Oxford University Press, 1999), p. 51.

3 Daniel Kuehl, 'Information Operations, Information Warfare, and Computer Network Attack: Their Relationship to National Security in the Information Age', in Michael Schmitt and Brian O'Donnell (eds.), 'Computer Network Attack and International Law', *International Law Studies*, Vol. 76, 2002, p. 35.

4 A discrete aspect of law on the use of force concerns collective security measures authorised by the UN Security Council, which are not discussed in this chapter. For a brief introduction to relevant aspects see Michael Schmitt, 'Cyber Operations in International Law: The Use of Force, Collective Security, Self-Defense, and Armed Conflicts', in Committee on Deterring Cyberattack, *Proceedings of a Workshop on Deterring Cyberattacks. Informing Strategies and Developing Options for U.S. Policy* (Washington, DC: The National Academies Press, 2010), pp. 160–62 as well as Michael Schmitt (ed.), *Tallinn Manual on the International Law Applicable to Cyber Warfare* (Cambridge: Cambridge University Press, 2013), pp. 69–72.

5 For an overview of these issues see generally Marco Roscini, *Cyber Operations and the Use of Force in International Law* (Oxford: Oxford University Press 2014); Heather Harrison Dinniss, *Cyber Warfare and the Laws of War* (Cambridge: Cambridge University Press, 2012); as well as Schmitt, *Tallinn Manual*.

6 Roscini, *Cyber Operations*, p. 165 (see note 5 above).

7 *Judgement, Appeals Chamber Prosecutor v. Dusan 'Disko' Tadic*, ICTY IT-94-1.

8 Michael N. Schmitt, 'Wired Warfare: Computer network attack and jus in bello', *International Review of the Red Cross*, Vol. 84, No. 846 (June 2002), pp. 188–9. Also Robert Hanseman, 'The Realities and Legalities of Information Warfare', *Air Force Law Review*, Vol. 42, 1997, p.183.

9 Schmitt, 'Wired Warfare', p. 188 (see note 8 above).

10 Legality of the Threat or Use of Nuclear Weapons, Advisory Opinion, I.C.J. Reports 1996, 226, para. 86.

11 Schmitt, 'Wired Warfare', p. 189 (see note 8 above). However, Dinniss points out that there is a qualitative difference between nuclear weapons and computer network attacks in as far as computer network attacks cannot be regulated as a category of weapons like nuclear weapons. The damage of a CNA depends entirely on the objective and design of the attack. See Heather Harrison Dinniss, 'The Status and Use of Computer Network Attacks in International Humanitarian Law' (DPhil thesis, 2008), pp. 117–18.

12 See Article 1(2) of Additional Protocol I.
13 Article 36 of Additional Protocol I states: 'In the study, development, acquisition or adoption of a new weapon, means or method of warfare, a High Contracting Party is under an obligation to determine whether its employment would, in some or all circumstances, be prohibited by this Protocol or by any other rule of international law applicable to the High Contracting Party'; see Dinniss, 'The Status and Use of Computer Network Attacks in International Humanitarian Law' (DPhil thesis, 2008), p. 116.
14 See for example Michael Schmitt, 'Computer Network Attack: The Normative Software', *Yearbook of International Humanitarian Law*, Vol. 4, 2001, p. 75; Schmitt, 'Wired Warfare' (see note 8 above), pp. 188–9; Robin Geiß, 'The Conduct of Hostilities in and via Cyberspace', *American Society of International Law Proceedings*, Vol. 104, 2010, p. 371.
15 Louise Doswald-Beck, 'Some Thoughts on Computer Network Attack and the International Law of Armed Conflict', in Schmitt and O'Donnell (eds.), *Computer Network Attack*, p. 165 (see note 3 above).
16 Michael Schmitt (ed.), *Tallinn Manual*, p. 76 (see note 4 above). Rule 20 of the *Tallinn Manual* (Rule 80 in the revised *Tallin Manual 2.0 on the International Law Applicable to Cyber Operations*, with Michael Schmitt, again, as General Editor, Cambridge: Cambridge University Press, 2017, p. 375) states, 'Cyber operations executed in the context of an armed conflict are subject to the law of armed conflict'. The 'term in the context of' is, however, not further defined. For a discussion see Roscini, *Cyber Operations*, pp. 123–5 (see note 5 above).
17 It is important to note that in order for IHL to apply the attacks must be attributable to a state. However, the responsibility of the Russian Federation has never been conclusively established. See Roscini, *Cyber Operations*, p. 123 (see note 5 above). See also Eneken Tikk, Kadri Kaska, and Liis Vihul, *International Cyber Incidents. Legal Considerations* (Tallinn: Cooperative Cyber Defence Centre of Excellence, 2010), p. 74.
18 Marco Benatar, 'The Use of Cyber Force: Need for legal justification?', *Goettingen Journal of International Law*, No. 3, Vol. 1, 2009, p. 391.
19 Ibid.
20 Ibid.
21 Duncan Hollis, 'Why States Need an International Law for Information Operations', *Lewis & Clark Law Review*, Vol. 11, 2007, p. 1042.
22 Daniel B. Silver, 'Computer Network Attacks as a Use of Force under Article 2(4) of the United Nations Charter', *International Law Studies*, Vol. 76, 2002, p. 82.
23 Ibid.
24 Ibid.
25 Ibid.
26 Ibid.
27 Benatar, 'The Use of Cyber Force' (note 18 above), pp. 387–8.
28 Ibid., p. 388.
29 Ibid.
30 Silver, 'Computer Network Attacks as a Use of Force' (note 22 above), p. 84.
31 See discussion in Chapter 1.
32 For initial information see for example the US Cyber Command fact sheet at www.stratcom.mil/factsheets/2/Cyber_Command/ (accessed 22 October 2014).
33 See for example the account of national cyber capabilities in James Lewis and Katrina Timlin, *Cybersecurity and Cyberwarfare. Preliminary Assessment of National Doctrine and Organization* (Washington, DC: Center for Strategic and International Studies, 2011).
34 Schmitt, 'Computer Network Attack', 908; 'The Status and Use of Computer Network Attacks in International Humanitarian Law' (DPhil thesis, 2008), p. 58; Christopher C. Joyner and Catherine Lotrionte, 'Information Warfare as International Coercion: Elements of a Legal Framework', *European Journal of International Law*, Vol. 12, Issue 5 (2001), p. 845.
35 *Brownlie's Principles of Public International Law*, 8th edn (ed. James Crawford), Oxford University Press, 2012, p. 362.
36 Doswald-Beck, 'Some Thoughts on Computer Network Attack' (note 15 above), p. 164.
37 Schmitt, 'Wired Warfare', pp. 188–9 (see note 8 above).
38 Ibid.
39 Benatar, 'The Use of Cyber Force' (note 18 above), p. 389.
40 John Stone, *The Tank Debate: Armour and the Anglo-American Military Tradition*, Amsterdam: Harwood Academic, 2000.

41 Milena Michalski and James Gow, *War, Image and Legitimacy: Viewing Contemporary Conflict* (London and New York: Routledge, 2007), p. 7; Keith Grint and Steve Woolgar, 'Computers, Guns and Roses: What's Social About Being Shot?', *Science, Technology and Human Values*, Vol. 17, No. 3 (1992), pp. 366–80; Grint and Woolgar, *The Machine at Work*, Cambridge: Polity Press, 1997; Bob Kling, 'When Gunfire Shatters Bone: Reducing Socioeconomic Systems to Social Relationships', *Science, Technology and Human Values*, Vol. 17, No. 3 (1992), pp. 381–5.
42 Counter to this understanding of definitions and meanings, it should be noted that Dinniss argues 'the semantics of war and weaponry are no longer a useful criterion in determining whether something is a use of force'. Dinniss, 'The Status and Use of Computer Network Attacks in International Humanitarian Law' (DPhil thesis, 2008), p. 60.
43 Ibid., p. 71.
44 Brownlie, *International Law*, p. 362 (see note 35 above).
45 Silver, 'Computer Network Attacks', p. 84 (note 22 above).
46 John Kelly III and Lauri Almann, 'eWMDs', *Policy Review*, Vol. 152, 2008–2009, pp. 39–50.
47 Doswald-Beck, 'Some Thoughts on Computer Network Attack', p. 164 (see note 15 above).
48 Michael N. Schmitt, 'Classification of Cyber Conflict', *Journal of Conflict & Security Law*, Vol. 17, Issue 2 (2012), p. 51.
49 Roscini, *Cyber Operations*, p. 132 (see note 5 above).
50 Hanseman, 'The Realities and Legalities' (see note 8 above), p. 184.
51 Ibid. It is important to note that Hanseman's analysis and findings are somewhat limited as he conflates notions of *jus ad bellum* and *jus in bello*. For instance, the notions of *armed conflict* and *aggression* are equated. However, the rationale of equivalent outcomes still underpins his analysis.
52 Doswald-Beck, 'Some Thoughts on Computer Network Attack', p. 165 (see note 15 above).
53 Schmitt, 'Wired Warfare', p. 192 (see note 8 above). Emphasis in original.
54 Michael N. Schmitt, '"Attack"' as a Term of Art in International Law: the cyber operations context', in Christian Czosseck, Rain Ottis and Katharina Ziolkowski (eds.), *Proceedings of the 4th International Conference on Cyber Conflict* (Newport, Rhode Island: Naval War College, 2012), p. 290. [Hereafter, 'Term of Art'].
55 Schmitt, 'Wired Warfare', p. 194 (see note 8 above).
56 Schmitt, 'Term of Art' (see note 54 above), p. 290.
57 Ibid., p. 291.
58 Dinniss, 'The Status and Use of Computer Network Attacks in International Humanitarian Law' (DPhil thesis, 2008), p. 172; Roscini, *Cyber Operations*, p. 179 (see note 5 above).
59 Dinniss, 'The Status and Use of Computer Network Attacks in International Humanitarian Law' (DPhil thesis, 2008), p. 124, also similarly pp. 118–19.
60 Benatar, 'The Use of Cyber Force', p. 390 (note 18 above).
61 Silver, 'Computer Network Attacks', pp. 92–3 (note 22 above).
62 Yoram Dinstein, 'Computer Network Attacks and Self-Defense', in Schmitt and O'Donnell (eds.), *Computer Network Attack*, p. 103, and Dinniss, 'The Status and Use of Computer Network Attacks in International Humanitarian Law' (DPhil thesis, 2008), p. 75.
63 Silver, 'Computer Network Attacks', pp. 84–5 (note 22 above).
64 Ibid.
65 This example is adapted from James Bond, *Peacetime Foreign Data Manipulation as one Aspect of Offensive Information Warfare: Questions of Legality under the United Nations Charter Article 2(4)* (Newport, Rhode Island; Naval War College, 1996), pp. 84–5.
66 Ibid., p. 84.
67 Schmitt, 'Wired Warfare', p. 192 (see note 8 above).
68 Schmitt, 'Classification', p. 252 (see note 48 above).
69 Roscini, *Cyber Operations*, p. 135 (see note 5 above). Emphasis in original. Schmitt argues in a somewhat similar vein that 'a *de minimis* threshold should attach. In much the same way that a soldier throwing a rock across the border does not propel the States concerned into international armed conflict, it would not suffice, for instance, to merely disable a single computer that performs non-essential functions.' See Schmitt, 'Classification', p. 252 (note 48 above).
70 Additionally, in the case of Additional Protocol II the requirements of territorial control and responsible command in order to implement the Protocol. See Article 1(1) of Additional Protocol II.
71 Schmitt, 'Classification', pp. 255–6 (see note 48 above).
72 Ibid., p. 256.
73 Ibid., p. 258.

7

COMPUTER NETWORK ATTACKS UNDER THE *JUS AD BELLUM* AND THE *JUS IN BELLO*

Distinction, proportionality, ambiguity and attribution

Elaine Korzak and James Gow

When international humanitarian law applies, questions arise regarding the law of targeting – who and what can be attacked, and how?[1] New methods and means of warfare, even if deemed lawful per se, will still have to be used in accordance with these fundamental principles in order to qualify as lawful.[2] The emergence of computer network attacks, therefore, raises a number of issues in the application of two fundamental principles of IHL: *distinction* and *proportionality*. These are discussed below. As emerges in the discussion, traditional concerns for these important principles may be rendered moot by questions attribution cloaked in problems of ambiguity, prompting concern that the legal protections might not be available, in practice.

Distinction and proportionality

Under the principle of distinction, attacks may only be directed against military objectives (defined as objects making an effective contribution to military action) and which offer a definite military advantage through their destruction. Applying this principle in the context of computer network attacks does not change these requirements. But it raises a number of issues regarding their implementation, the most important of which is the possibility that an increasing number of objects will qualify as 'military objectives' and will thus become targets liable for attack.

At first glance, the non-conventional nature of CNAs does not impact on the application of the principle of distinction. As Geiß and Lahmann point out, 'Whether a given military objective is attacked via cyberspace or via the air by a drone or fighter place, for the purposes of IHL, essentially makes no difference'.[3] Thus, if an object is to be subjected to CNAs, a state would first have to establish whether the object in question qualifies as a legitimate military objective in accordance with Article 52 (2) of Additional Protocol I.[4]

The potential use of information and communication technologies for military purposes has not only given rise to a novel, non-conventional way of creating physical consequences, but has also increased the importance of the information and communication infrastructure. In light of societies' increased reliance on this infrastructure, its components will become targets in and of themselves.[5] With this, determining which components qualify as a legitimate *military objective* becomes ever more important. As discussed above, the first requirement holds that an object must make an effective contribution to military action by its nature, location, purpose, or use. Objects that would effectively contribute to military action due to their nature would include computers or systems specifically to be used as components of weapon systems or to facilitate logistics operations.[6] Obviously, military command, control and communication networks, as well as military air defence networks would equally qualify.[7] The premises of US CYBERCOM would also qualify as a *military objective* due to their military nature.[8] As for effective contribution by use, an example would include the use of a server by the military that is normally reserved for civilian purposes.[9]

However, the requirement for an effective contribution to mean 'affect the use of an object' potentially introduces difficulties in the context of CNAs because information networks are interconnected. The use of an object by the military potentially renders a network or system a *military objective*. The problem lies in the fact that most computer technology, hardware and software, has become dual-use – that is, technology that is used by civilians and the military at the same time.[10] Servers, routers, cables, satellites and software, used to make effective contributions to military action, but also linked to civilian purposes, would qualify as 'legitimate military targets'.[11] Further, given intertwined military and civilian spheres, 'significant parts of the civilian cyber infrastructure will be used to make an effective contribution to military action'.[12] Ninety-eight percent of US government communications, for example, travel through civilian-owned, or civilian-operated, networks.[13] In addition, some systems, initially designed for military purposes, have become integral parts of civilian life. For instance, disruption of the Global Positioning System (GPS), which has been integrated into civilian applications, such as air traffic control, would cause serious civilian effects.[14]

As a result, the interconnectedness of systems and the dual-use character of technology render an increasing number of objects potentially liable to attack. This is even more the case with the expansive interpretation of *military objective* advanced by the United States. Effective contribution is not ascertained in terms of 'military action' but the broader notion of 'war-sustaining' or 'war-supporting' capability. Under this interpretation activities that are not directly connected to hostilities, in particular economic operations and facilities, would qualify as well. Highly developed information societies would offer a plethora of potential targets whose destruction or incapacitation would significantly impair a country's political and economic activities.[15] Roscini argues that this view would, for instance, legitimise the use of the Shamoon virus that targeted and disabled the computers of Saudi Arabia's national oil company Saudi Aramco. If perpetrated during an armed conflict the malware would have considerably impaired the country's ability to pump oil, damaging its overall economy and capacity to fuel war machines, and thereby reducing its 'war-supporting' capacity.[16]

Secondly, in determining whether an object can be attacked as a *military objective*, its destruction, capture or neutralisation would have to provide a 'definite military advantage' in the circumstances prevailing at the time. In the context of CNAs, this determination may be hampered by the difficulties of measuring the effects of an attack.[17] Indeed, at the time of writing, it had still not been conclusively established whether the Stuxnet malware actually led to the physical destruction of centrifuges and if so of how many centrifuges.

In the end, the application of the principle of distinction and the definition of *military objective* in the context of CNAs may expose an increasing number of objects to attack. Due to the pervasive interconnectedness of information networks and the inherently dual-use character of information technology an increasing number of objects may meet the criteria set out in Article 52 (2) of Additional Protocol I. In theory, a country's entire cyber infrastructure could potentially be qualified as a *military objective*, once it engages in an *armed conflict*.[18] Yet, the civilian component of dual-use objects would still need to be taken into account under the principle of *proportionality*.

As with *distinction*, the abstract test provided by *proportionality* finds equal application in the case of CNAs, despite their non-conventional mode of operation.[19] However, the implementation of this principle has already proven controversial in the context of conventional attacks. CNAs promise to compound further the problems of assessing the legality of an attack under *proportionality*.

Before addressing these aspects it is important to point out that Article 57(2) (iii) of Additional Protocol I stipulates that *proportionality* applies to an *attack* on a *military objective*. CNAs with no violent consequences would fall outside the scope of this provision (notwithstanding the analysis above, outside the mainstream, that deleting digits is, indeed, kinetic and an act of physical violence). A CNA that alters, deletes or otherwise corrupts information without causing any physical consequences would not qualify as an *attack* according to Article 49(1) of Additional Protocol I.[20] Thus, the question of *proportionality* would not arise with regard to these types of attack.

Conversely, if a CNA crosses the threshold of *attack*, it is subject to *proportionality*. Consequently, its effects have to be assessed as collateral damage and put into relation to the military advantage anticipated from the CNA in question. As with kinetic attacks, any damage to military objectives, or injury and death of combatants, would not count towards incidental loss of life or damage to objects – only damage to civilians or civilian objects is considered. However, the assessment of civilian damage or incidental loss of life is complicated in the context of computer network attacks because the uncertainties involved are greater than those usually associated with conventional attacks,[21] where certain effects are resolutely predictable – *x* amount of explosive will produce *y* effect against *z* material. Thus, a major difficulty in applying *proportionality* to CNAs concerns the assessment of an attack's effects. This influences both sides of the proportionality equation, i.e. the expected incidental damage, as well as the concrete and direct military advantage anticipated. It is unclear to what extent knock-on, or reverberating, effects would be taken into account under the principle of proportionality. Although this problem is well known in the context of conventional attacks – as controversies surrounding attacks on electrical power plants and electricity grids illustrate – it is exacerbated in the context of CNAs where systems and networks are pervasively interconnected.[22] The civilian harm ensuing from knock-on effects may be more significant than the direct effects of the attack,[23] meaning that it is uncertain how many levels of 'cascading' effect would need to be considered by planners.[24] One standard supported by the use of the adjective 'expected' in describing collateral damage in Article 57 (2) (iii) of Additional Protocol I is the test of whether effects were reasonably foreseeable. Both direct and indirect effects that 'should' be expected by those planning, approving, or executing operations should be included in consideration of cyber attacks.[25]

Another standard for the assessment of knock-on effects sees effects that 'would not have occurred "but for" the attack' included.[26] Others argue that effects on infrastructure operated by attacked computer systems as well as the effects on persons relying on the functioning of these attacked computer systems and infrastructures should fall under the notion of incidental damage.[27] For one US State Department Legal Advisor, *proportionality* requires parties to a conflict to assess: (1) the effects of cyber weapons on both military and civilian infrastructure and users, including shared physical infrastructure (such as a dam or a power grid) that would affect civilians; (2) the potential physical damage that a cyber attack may cause, such as death or injury that may result from effects on critical infrastructure; and (3) the potential effects of a cyber attack on civilian objects that are not military objectives, such as private, civilian computers that hold no military significance, but may be networked to computers that are military objectives.[28]

The last point, the effects on civilian computers networked to military objectives, could arguably include the negative repercussions arising from the possibility that malware used in a CNA could also infect and spread to other computer systems[29] (as happened with Stuxnet). Overall, the potentially significant reverberating effects of CNAs resulting from the interconnectedness add to existing controversies over the inclusion of such effects in the *proportionality* equation.

The second difficulty encountered in applying *proportionality* concerns the definitional boundaries of the concept of *damage* in the context of computer network attacks. The question arises whether incidental damage to civilian objects goes beyond physical damage to include the loss of functionality. Whereas the definition of military objective in Article 52 (2) of Additional Protocol I refers to the 'destruction, capture or neutralization' of an object, the provisions of the principle of proportionality apply to the broader notion of incidental *damage* to civilian objects.[30] Roscini and others have thus argued that collateral damage should not only cover the destruction of networked infrastructure, but also its incapacitation or loss of functionality, which might affect the civilian population more (for example, disconnection from communication services, including internet financial transactions, or energy supplies – even though the latter would need to be factored into a conventional armed attack).[31] Roscini also argued that certain disruptive cyber activities are *attacks* under IHL by expanding the concept of *violence* to include 'not only material damage to objects, but also incapacitation of infrastructures without destruction'.[32] Schmitt similarly contends that destruction includes operations that, while not causing physical damage, nevertheless *break* an object, rendering it inoperable, as in the case of a cyber operation that causes a computer reliant system to no longer function unless repaired.[33]

However, he notes that only cyber operations going beyond 'mere inconvenience' and causing 'functional harm to structures or systems' would qualify as '*attacks* in the sense of Article 49 (1)'.[34] *Damage*, for the purposes of *proportionality*, does not include 'inconvenience, irritation, stress, or fear' generated by loss of functionality.[35] Effects such as these do not rise to the standard of 'loss of civilian life, injury to civilians, damage to civilian objects' expressed in Article 57 (2) (iii) of Additional Protocol I.

A significant implication of any interpretation of *damage* going beyond physical destruction to include inoperability, or incapacitation, is that CNAs could actually provide for physically less destructive ways of attacking dual-use objects. Incidental injury and collateral damage might be minimised, if infrastructure were to be incapacitated, rather than physically destroyed. Thus, CNAs could significantly alter the *proportionality* equation. Where previously physical destruction might have been required, in the twenty-first century, cyber attacks can 'turn off' an opponent's airports, air traffic control, or power production and distribution.[36] With this, certain attacks against *military objectives*, which would be unlawful using conventional weapons because they could reasonably be anticipated to cause excessive incidental civilian damage, could be lawful, if conducted by disruptive cyber operations.[37]

Generally, an increased number of civilian objects may become liable to attack, because of dual-use. On the other hand, the interconnectedness of systems may result in complex, reverberating, effects, whose assessment in terms of *proportionality* could equally counter this development. At a minimum, given the interconnectedness of computer networks and the dual-use character of information systems, *proportionality* plays an even more significant role in the protection of civilians than the principle of *distinction*.[38]

Some authors have tried a different approach, focusing on the distinction made between *military operations* and *attacks* by most authors. Nils Melzer argues that *distinction* and *proportionality* apply not only to *attacks* but to the broader category of 'hostilities'. As such, the restraints imposed by the law pertain to 'whether they constitute part of the "hostilities" ', not on whether the operations are classified as *attacks*. A CNA disrupting a radar system would not qualify as an *attack* if there are no physical consequences. Yet, it would still be subject to the restrictions of IHL because it can be categorised as an act of hostilities.[39]

Importantly, the International Committee of the Red Cross (ICRC) has also challenged the conclusion that CNAs do not qualify as *attacks* as long as they do not result in personal injury or physical damage. In its view, *attacks* may only be directed at *military objectives*. Objects not falling within that definition are civilian and may not be attacked. The definition of *military objectives* is not dependent on the method of warfare used and must be applied to both conventional and non-conventional means. The fact that a cyber operation does not lead to the destruction of an attacked object is irrelevant. Pursuant to Article 52 (2) of Additional Protocol I, only objects that make an effective contribution to military action and whose total, or partial, destruction, capture, or neutralisation, offers a definite military advantage, may be attacked. By referring not only to destruction, or capture, of the object, but also to its neutralisation, the definition implies that it is immaterial whether an object is disabled through destruction, or in any other way.[40]

Thus, the ICRC argues that CNAs that do not result in physical damage, or human injury, likewise fall under the notion of *attack* and need to comply with the principle of *distinction*. However, as Roscini and Schmitt both point out, the ICRC's reasoning goes against the majority interpretation of the provisions in Additional Protocol I. This holds that *distinction* applies only to *attacks*.[41] Further, it could be noted that 'neutralisation' was included in Article 52 (2) of Additional Protocol I with regard to the effects of a conventional attack,[42] making the ICRC's approach no more than arguable – even if the argument has some logic, given the nature of cyber weapons.

In the end, the different interpretative approaches advanced by the ICRC and others illustrate that the scope of the term *operations* in Article 48 of Additional Protocol I becomes pivotal in the context of CNAs whereas different interpretations of *operations* and *attacks* have not had significant implications in the context of conventional warfare. CNAs have heightened controversy over the scope of *distinction* and whether it attaches to *attacks*, or the broader notion of *operations*. According to the majority of analysts, *distinction* attaches to actions characterised as *attacks*, so CNAs not resulting in physical damage or personal injury would not be covered by *distinction* and *proportionality*. A disturbing consequence of this, as already indicated, is that states could target, but not *attack*, civilian objects via CNAs that do not produce evident (or visible) physical consequences. The qualification of effects that do not clearly fall outside, or inside, the remit of physical consequences – such as disabling computer systems so that they need to be repaired – remains debatable. However, even if CNAs that do not result in obvious physical consequences are subjected to *distinction* and *proportionality*, as noted, significant ramifications for the implementation of these principles remain.

Attribution

Perhaps the most significant issue arising from the advent of cyber warfare is that of *attribution*. In many senses, the discussions and debates over *distinction* and *proportionality*, or the nature of an *attack* covered by the law, or, indeed, the very notion of what *armed* means, are rendered nugatory and irrelevant by the issue of *attribution*. CNAs, as complicated as the preceding discussion reveals them to be, become even more complex when added to discussion of the right to self-defence, itself an realm characterised partly by deep doctrinal divides. On one level, the issue of whether a CNA could trigger a state's right to self-defence, under Article 51 of the UN Charter, turns on the issues of definition, interpretation, and effects and consequences, as discussed in Chapter 6. In addition to the issues discussed already, the perceived anonymity of CNAs give rise to significant difficulties regarding the *attribution* of a CNA – that is, tracing the attack and establishing the identity of the attacker, a necessary prerequisite to taking action in self-defence. If this were settled, then, on a second level, questions arise of how the principles of *necessity*, *immediacy* and *proportionality* that govern the exercise of this fundamental right would apply in the domain of CNAs. These considerations are further complicated if the availability of cyber capabilities to groups and even individuals is taken into account, invoking questions concerning states' right to self-defence against non-state actors. Whether related to state or non-state activity, and to attacks as events or attacks as consequences of action, the problems surrounding *attribution* – the identification of an attacker (both technical and, most of all, political) for application of the legal right to self-defence (as well as other purposes) means that, often, other aspects on the legality of CNAs fall aside, as we show in this last section of the chapter.

Clearly, as discussed above, the intangibility of CNAs raises the questions whether and how such an attack could be qualified as an *armed attack*. While other approaches could be argued, as noted above, the dominant view among scholars is that a consequence-based approach can be applied to determine whether *force* has been used, or whether an *armed conflict* exists, or whether an *attack* has occurred. The determination of a CNA as an *armed attack* hinges upon physical consequences such as property damage or injury to human life. Therefore, CNAs may trigger a state's right to self-defence if they result in physical consequences of sufficient gravity, with each instance having to be addressed, once again, on a case-by-case basis.[43]

Once a CNA can be characterised as an *armed attack*, questions arise about the exercise of the right to self-defence. How are the principles of *necessity*, *proportionality*, and *immediacy* implemented in response to a CNA? Since the modalities governing the exercise of self-defence comprise general principles that are context-specific, their application to CNAs does not pose a problem, as such. It is only in their case-specific application to the unique characteristics of CNAs that significant difficulties are revealed. Most notably, the perceived anonymity of attacks resulting in the so-called 'attribution problem' raises significant issues.

With regard to *proportionality*, the first question was whether it would ever be proportionate for a state to use conventional armed force in response to a CNA. *Proportionality* in the exercise of self-defence does not require that a victim state resorts to the same weapons or the same number of armed forces as the attacking state. Therefore, a state may respond to a CNA with conventional force, so long as its response is proportionate to the overall threat posed.[44] None the less, although a forceful response may be permitted under the law of self-defence, it might not always represent the best, or most appropriate, response.[45]

In comparison, *necessity* and *immediacy* prove more challenging to apply in response to a CNA that otherwise triggers the right to self-defence. *Necessity* requires that any *use of force* taken in self-defence must have been *necessary* for that purpose. Applying this principle to CNAs gives rise to several obligations. The victim state must verify that the computer network attack in question was not accidental[46] – although the scale of physical destruction required for an *armed attack* would be unlikely to be the result of technical malfunction or failure. Most important, *necessity* implies that the victim state has to identify the responsible state (or non-state actor) behind the CNA order to be able to respond in self-defence.[47] In some instances, the source of an attack might be identifiable with relative ease, either because the attacker reveals themselves or because the CNA forms part of a *kinetic attack* that is more easily *attributable*.[48] However, CNAs will more often present significant difficulties in this regard. Identifying the source of an attack as well as establishing and verifying the identity of the attacker might prove an intractable problem in the case of CNAs.

These difficulties also impact the closely related requirement of *immediacy*. Forceful measures taken in self-defence must generally be taken without undue delay. If the perpetrator of a CNA can be revealed shortly after the attack, then subsequent measures taken in self-defence would meet the requirement of *immediacy*. However, in instances where the source of the attack and the identity of the attacker are unknown, the time needed to trace and collect evidence in order to establish the identity of the perpetrator might render the forceful response a prohibited armed reprisal rather than legitimate action in self-defence.[49] Of course, *immediacy* should not be interpreted too strictly – even where there is no doubt about the perpetrator of an act, it may take many months to marshal forces to respond, as with the US-led response to the Iraqi invasion of Kuwait, in 1990, which needed six months' preparation.[50] Flexibility is particularly warranted in the case of CNAs, where there are several obvious time delays before any response: cyber actions can reverberate around the world in the blink of an eye;[51] the initial strike might incapacitate military computer capabilities, or other capabilities, that would require time to restore before reaction in self-defence were feasible; or an aggressor used logic or time bombs, where the actual damage could occur, or be registered, well after the attack itself.[52]

Nonetheless, even if *immediacy* is construed broadly, identifying the source of an attack as well as establishing the identity of the attacker present major challenges. This involves *attribution* on a technical level, meaning the tracing of a CNA and determining which machine (or machines) initiated or controlled the attack, as well as *attribution* at a human level to identify the person or organisation responsible for the attack. In addition, for purposes of legal analysis, it needs to be established whether the individual or organisation responsible can be linked to a state in order to *attribute* the attack in question to a state under the law of state responsibility. Legal *attribution* and possible state sponsorship of attacks are particularly relevant for the exercise of the right to self-defence against a non-state actor.

Significant doctrinal controversies exist over the status of non-state actors launching armed attacks, as well as the degree of state involvement necessary to justify a forceful response by the victim state. Computer network attacks augment these debates. First, CNAs are potentially available to a significantly larger pool of actors than conventional armed force ever has been. Whereas traditional warfare has been shaped largely by states broadly exercising a monopoly on the use of force in international relations (or by would-be states, or state-like formations),[53] computer network attacks can be carried out not only by states, but also by groups, or even individuals. In theory, almost anyone anywhere, with the right knowledge and access to information could bring about a devastating attack.[54] Although other elements such as intelligence on the target system might also be required, a critical aspect of CNAs is their potentially low financial and technological entry requirements, which significantly increase the number of potential actors.[55] CNAs are often far less expensive than the major weapons systems characteristic of conventional warfare.[56] And unlike traditional warfare, the knowledge required to conduct a CNA is available outside governmental structures – even if governments and their agencies tend still to dominate. As Boebert points out, 'brilliance in software development can be found anywhere' and only a laptop computer and an internet connection are needed.[57] Therefore, in comparison with conventional warfare, CNAs produce a significantly expanded pool of players, both among states and non-state actors.[58]

The increased prominence of non-state actors is compounded by the perceived anonymity of CNAs. As Shackelford points out, the internet itself provides the 'perfect platform for plausible deniability'.[59] This anonymity significantly complicates the process of legal *attribution* – that is, imputing an attack to a state under international law. Of course, the *attribution* of attacks has also been problematic in cases involving conventional armed means, as *armed attacks* are often carried out anonymously. Victim states have to identify the perpetrators and in the case of non-state actors establish possible connections to a state before responding in self-defence. Although responsibility is, sometimes, claimed by non-state actors, states are generally reluctant to come forward and claim, or even take, responsibility for an attack, particularly in the case of state-sponsored terrorism. The already challenging problems associated with establishing a link between non-state actors and a state – as in the Nicaragua and Tadić cases – are significantly compounded in the case of CNAs and might require completely new thinking, or law.[60] While showing links can be hard, at least conventional physical attacks involve staging activity and forensic evidence at the attack site that might be detected,[61] in the case of CNAs, only 'virtual' signals, bits and bytes, similar to other civilian and government data, exist, which bear 'neither state insignia nor other markers of military allegiance or intent'.[62] Proving state involvement has been difficult enough in the context of conventional armed attacks, which inevitably involve some 'physical' elements. The intangibility of CNAs, coupled with their perceived anonymity, significantly compounds existing difficulties of attribution.

The technical and socio-political issues surrounding *attribution* are, ultimately, linked because those capable of technical attribution, when they have such evidence, often find it hard to reveal the information they have and how they obtained it. At a domestic level, this can be seen in the 1994 criminal court case of Richard Pryce, where, although he was ultimately convicted for intrusions into the US Air Force's networks, his conviction imposed only a minor fine because the case that could be proven against him beyond reasonable doubt was limited. It was limited, in the main, not because the US authorities and the UK police had been unable to trace and attribute actions to Pryce, but because they refused, *inter alia*, to hand over Pryce's hard disk, as it was said to have three security sensitive files on it, or to reveal the US Air Force's original logs, let alone prove that he had tampered with them.[63] In the end, there was greater sensitivity about revealing further information by confirming the evidence for attribution than about securing a conviction.[64]

This sensitivity in municipal criminal proceedings is shared at the international level, where attribution would reveal more about a prosecutor's capabilities and how it came to be able to attribute responsibility for an unclaimed attack with certainty than that actor was prepared to disclose. Therefore, in the most serious cases, *attribution*, technically difficult enough to achieve, possibly, in the first place, would be unlikely to be offered as a basis for self-defence in international law (or, alternatively as part of an international criminal prosecution), lest it give too much away in the process.

Short of this, the standards that have generally been applied regarding accusations and attribution of cyber attacks have fallen well short of the 'beyond reasonable doubt' test applied to criminal cases, although they may constitute a fair balance of probabilities (the legal test in civil cases, or when presenting *prima facie* criminal indictments). Six criteria invoked for *attribution* were identified in the research for the SNT project, although each of them was found to be potentially misleading, if not false.[65] The first criterion is the geopolitical context. Many attacks are relatively straightforwardly attributed on the basis of the context in which they occur, even if they may be technically difficult to attribute. However, it is important to note that this can be wholly misleading, as the 'Solar Sunrise' case in 1998 demonstrates – attribution to Iraq because of the geo-political situation was misplaced, as three teenagers were eventually found to be responsible. The second criterion, the political nature of the victim, is linked directly to the third, *cuid bono*, or 'whom does it benefit'.[66] Again, the 'Solar Sunrise' case illustrates both the basis on which this works – the US, as victim, must be a target of Iraq, which will benefit from the attack – and yet an otherwise insignificant third-party group of teenagers was, in the end, responsible. The apparent origin of an attack is the fourth criterion, irrespective of other factors, such as geopolitical context. This is true even though attackers can mimic IP addresses, and they have ways to divert their activity via different proxies located in different countries, as happened in the Pryce case mentioned above and in the cyber attack on Estonia in 2007. It can be difficult to know if an address is genuine or a false proxy. In any case, the uncertainties this creates make certain attribution problematic. In this case, the best circumstantial corroborator is a lack of cooperation – indeed, Richard Clarke, former chief counter-terrorism adviser during the US Presidency of George W. Bush recommended that the United States should 'judge a lack of serious cooperation in investigations of attacks as the equivalent of participation in the attack'.[67] However, this approach has clear limitations – in terms of international law and state responsibility this does not amount to 'instruction', 'direction', or 'control' and, more importantly, states simply might not have the technical information to share, for one reason or another. The fifth criterion used to assert responsibility is sophistication. This 'capability-centred approach' relies on the assumptions that capabilities differ from one group to another and that this makes it is possible to identify an attacker – Stuxnet is an example here, but it may be a false understanding that 'sophisticated' attacks can only be carried out by certain actors and that the term 'sophisticated' has a clear and agreed meaning in the context of cyber attacks, putting it beyond manipulation by those attributing responsibility. The final and often-used sixth criterion is the scale of the attack. However, measuring the scale of an attack is not easy and may involve assessment of intangible qualities, such as reputation. Yet, this criterion is also the most used and most twisted, with evidence adduced to fit hypotheses of state sponsorship, not deriving the attribution from the evidence. Yet, attribution is rarely contested and the evidence on which it is based is either not examined or beyond examination.

In the end, *attribution* is both a difficult technical process that might not always be successfully completed, and it is a phenomenon where, irrespective of technical completeness, assertions of responsibility will rarely, if ever, be contested. On both sides of this equation, there is a reluctance to reveal capabilities publicly. The attacker wishes to retain anonymity, or, at least, ambiguity, so as not to legitimise a response. The victim may accuse without a firm evidence base, so as not to reveal that they do not know for sure. Or, if they do have the technical means to be sure, they might well simply say nothing, so as not to reveal their own capabilities and what they know about the attackers' capabilities. In the end, application of the right to self-defence, or international criminal law, or other aspects of the law are unlikely to occur in response to cyber attacks, as this would acquire attribution that is either elusive, or retained.

Conclusion

Consideration of *distinction* and *proportionality*, pivotal to the application of IHL, reveals interesting and potentially concerning findings. The all-encompassing interconnectedness of civilian and military networks, as well as the inherent dual-use character of information and communication technologies, generates an increasing number of potential *military objectives*, liable to attack, in accordance with the principle of *distinction*. Apparently civilian objects could become legitimate military targets. Similarly, hitherto controversial attacks on dual-use objects that would have been impossible with conventional arms because of the risks of collateral damage might pass the *proportionality* test if computer network attacks are used, as these can be a physically less destructive way of achieving the same objective. More important, because *distinction* and *proportionality* attach only to *attacks* in the sense of Article 49 (1) of Additional Protocol I, a range of CNAs could be made, potentially without any regulation, lawful in the realm of conventional armed conflict or beyond the protection of IHL. Computer network attacks that do not result in violent consequences could be employed regardless of *distinction* and *proportionality* – or, perhaps better, because the prevailing interpretations of these principles would render such attacks legitimate.

With the advent of computer network attacks, the applicability of the law, generally, is brought into question in relation to the issue of *attribution*. Assuming that international law can apply, in principle, to computer network attacks, the implementation of the right to self-defence is complicated by the perceived anonymity of CNAs and the major problems that arise with *attribution* of responsibility for them. Attributing an attack with certainty is challenging. Yet, establishing the identity of the perpetrator is crucial to satisfying the legal requirements governing the exercise of self-defence. A CNA needs to be imputable to a state, under the law of state responsibility, or to some other non-state armed organisation, possibly. In this respect, significant controversies exist with regard to actions of non-state actors and levels of state involvement necessary in the context of the right to self-defence, irrespective of the cyber sphere. When the latter is introduced, the picture becomes all the more complex and problematic. While significant progress has been made among international lawyers on finding ways to apply the law to cyber warfare, its features create significant difficulties for the application of international law. In very important ways, gaps remain and, indeed, may provide increased, or better, opportunities for the hostile action using cyber capabilities.

However, in the final analysis, the crucial question of *attribution* and the realities surrounding it make much of the argument about applying the laws of war and international humanitarian law to cyber warfare moot. In relation to both bodies of law, the identification of an attacker is essential for the law to apply. For the right to self-defence to be invoked, an *armed attack* requires the identification of an aggressor. The prosecution of alleged war crimes, equally, would require identification of those responsible. However, in practice, both technical and, ultimately, political constraints might prevent any such detection. Without the means to attribute responsibility under the law, the law itself could, therefore, be largely irrelevant.

Notes

1 Marco Roscini, *Cyber Operations and the Use of Force in International Law* (Oxford: Oxford University Press 2014), p. 176.
2 Methods of warfare generally describe operational modes used by parties to an armed conflict whereas means of warfare refer to weapons, weapon systems, and material. While IHL prohibits certain means and methods of warfare it also seeks to regulate the use of lawful ones. Once it is established that a means or method is not inherently unlawful, it becomes necessary to examine the legality of the modalities of its use. See Yoram Dinstein, *The Conduct of Hostilities under the Law of International Armed Conflict*, 2nd edn (Cambridge: Cambridge University Press, 2010), p. 1. Also Roscini, *Cyber Operations*, p. 168 (see note 1 above).
3 Robin Geiß and Henning Lahmann, 'Cyber Warfare: Applying the Principle of Distinction in an Interconnected Space', *Israel Law Review*, Vol. 45, Issue 3 (2012), p. 384.
4 Ibid.
5 Ibid.
6 Roscini, *Cyber Operations*, p. 184 (see note 1 above).
7 Ibid.
8 Ibid.
9 Roscini, *Cyber Operations*, pp. 184–5 (see note 1 above).
10 Dinniss, 'The Status and Use of Computer Network Attacks in International Humanitarian Law' (DPhil thesis, 2008), pp. 168–9; Roscini, *Cyber Operations*, 185 (see note 1 above).
11 Geiß and Lahmann, 'Cyber Warfare', pp. 385–6 (see note 3 above).
12 Ibid., p. 386.
13 Roscini, *Cyber Operations*, p. 185 (see note 1 above).
14 Dinniss, 'The Status and Use of Computer Network Attacks in International Humanitarian Law' (DPhil thesis, 2008), p. 169.
15 Ibid., p. 164.
16 Roscini, *Cyber Operations*, p. 186 (see note 1 above).
17 Roscini, *Cyber Operations*, pp. 187–8 (see note 1 above).
18 Geiß and Lahmann, 'Cyber Warfare', p. 390 (see note 3 above).

19 Ibid., p. 395.
20 See for example Roscini, *Cyber Operations*, p. 222 (see note 1 above).
21 William A. Owens, Kenneth W. Dam, and Herbert S. Lin, *Technology, Policy, Law, and Ethics Regarding U.S. Acquisition and Use of Cyberattack Capabilities* (Washington, DC: The National Academies Press, 2009), p. 262.
22 Geiß and Lahmann, 'Cyber Warfare', p. 396 (see note 3 above).
23 Dinniss, 'The Status and Use of Computer Network Attacks in International Humanitarian Law' (DPhil thesis, 2008), p. 186.
24 Ibid., p. 187.
25 Schmitt (ed.), *Tallinn Manual*, p. 160.
26 Michael Schmitt, Heather Harrison Dinniss and Thomas C. Wingfield, *Computers and War: The Legal Battlespace*, Background Paper prepared for Informal High-Level Expert Meeting on Current Challenges to International Humanitarian Law, Cambridge, 25–27 June 2004, p. 9, available at www.hpcrresearch.org/sites/default/files/publications/schmittetal.pdf (accessed 21 October 2014).
27 Roscini, *Cyber Operations*, pp. 220–21 (see note 1 above).
28 Harold Hongju Koh, *International Law in Cyberspace*, Remarks at USCYBERCOM Inter-Agency Legal Conference, Ft. Meade, MD, 18 September 2012, available at www.state.gov/s/l/releases/remarks/197924.htm (accessed 21 October 2014).
29 Roscini, *Cyber Operations*, pp. 223–4 (see note 1 above).
30 Articles 52(2) and 57(2) (iii) of Additional Protocol I.
31 Roscini, *Cyber Operations*, pp. 222–3 (see note 1 above); Geiß and Lahmann, 'Cyber Warfare', p. 397 (see note 3 above); Schmitt, *Tallinn Manual*, 108 (see note 25 above).
32 Roscini, *Cyber Operations*, p. 181 (see note 1 above).
33 Michael N. Schmitt, '"Attack"' as a Term of Art in International Law: the cyber operations context', in Christian Czosseck, Rain Ottis and Katharina Ziolkowski (eds.), *Proceedings of the 4th International Conference on Cyber Conflict* (Newport, Rhode Island: Naval War College, 2012).
34 Ibid.
35 Roscini, *Cyber Operations*, p. 181 (see note 1 above).
36 Michael N. Schmitt, 'Wired Warfare: Computer network attack and jus in bello', *International Review of the Red Cross*, Vol. 84, No. 846 (June 2002), p. 204.
37 Roscini, *Cyber Operations*, p. 223 (see note 1 above).
38 Ibid., p. 220.
39 Nils Melzer, *Cyberwarfare and International Law* (Geneva: UNIDIR, 2011), 27–8; available at http://unidir.org/files/publications/pdfs/cyberwarfare-and-international-law-382.pdf (accessed 21 October 2014).
40 International Committee of the Red Cross, 31st International Conference of the Red Cross and Red Crescent. International Humanitarian Law and the Challenges of Contemporary Armed Conflicts, Report, 31IC/11/5.1.2, 37.
41 Roscini, *Cyber Operations*, p. 181 (see note 1 above); Schmitt, 'Term of Art', pp. 292–3 (see note 33 above).
42 Dinniss, 'The Status and Use of Computer Network Attacks in International Humanitarian Law' (DPhil thesis, 2008), pp. 172–3.
43 See also Dinniss arguing that the spectrum of consequences makes 'classification based on the type of computer network attack impossible'. 'The Status and Use of Computer Network Attacks in International Humanitarian Law' (DPhil thesis, 2008), p.82.
44 Ibid., p. 100.
45 Ibid.
46 See Dinstein, 'Computer Network Attacks and Self-Defense', in Michael Schmitt and Brian O'Donnell (eds.), 'Computer Network Attack and International Law', *International Law Studies*, Vol. 76, 2002, p. 109 and Marco Roscini, 'World Wide Warfare – "Jus Ad Bellum" and the Use of Cyber Force', *Max Planck Yearbook of United Nations Law*, Vol. 14, 2010, p. 119.
47 Dinstein, 'Computer Network Attacks and Self Defense', 109 and Roscini, 'World Wide Warfare', p. 119 (see note 46 above).
48 Dinniss, 'The Status and Use of Computer Network Attacks in International Humanitarian Law' (DPhil thesis, 2008), p. 95. For example, the electronic attack on Syria's air defence systems was followed by a kinetic strike clearly attributable to Israel.

49 Dinniss, 'The Status and Use of Computer Network Attacks in International Humanitarian Law' (DPhil thesis, 2008), pp. 98–9.
50 See the discussion of the right to self-defence in James Gow, *Defending the West*, Cambridge: Polity, 2005.
51 Dinstein, 'Computer Network Attacks', p. 110 (see note 46 above).
52 Roscini, 'World Wide Warfare', p. 120 (see note 46 above).
53 See Ernst Dijxhoorn, *Quasi-states, critical legitimacy and international criminal justice*, Abingdon: Routledge, 2017.
54 Roger Barnett, 'A Different Kettle of Fish: Computer Network Attack', in Michael Schmitt and Brian O'Donnell (eds.), 'Computer Network Attack and International Law', *International Law Studies*, Vol. 76, 2002, p. 22.
55 Sean Watts, 'Low-Intensity Computer Network Attack and Self-Defense', *International Law Studies*, 2010, p. 77.
56 Ibid., p. 73.
57 Earl Boebert, 'A Survey of Challenges in Attribution, in Committee on Deterring Cyberattack', *Proceedings of a Workshop on Deterring Cyber Attacks: Informing Strategies and Developing Options for U.S. Policy* (Washington, DC: The National Academies Press, 2010), p. 43.
58 Watts, 'Low-Intensity Computer Network Attack', p. 61 (note 55 above).
59 Scott Shackelford, 'From Nuclear War to Net War: Analogizing Cyber Attacks in International Law', *Berkeley Journal of International Law*, Vol. 27, 2009, p. 208.
60 Dinniss, 'The Status and Use of Computer Network Attacks in International Humanitarian Law' (DPhil thesis, 2008), pp. 96–7.
61 Susan Brenner, '"At Light Speed": Attribution and response to cybercrime/terrorism/warfare' *Journal of Criminal Law & Criminology*, Vol. 97 (2006–2007), p. 424.
62 Ibid., p. 425.
63 Peter Sommer, 'Intrusion detection systems as evidence', *Computer Networks: The International Journal of Computer and Telecommunications Networking*, Vol. 31, No. 23–24 (1999).
64 *SNT* Unpublished Research Paper, 2014.
65 This analysis is based on *SNT*, Unpublished Research Paper, 2014.
66 This logic can be found for instance in Dmitri Alperovitch, *Revealed: Operation Shady RAT* (Santa Clara: McAfee, 2011).
67 Richard A. Clarke and Robert K. Knake, *Cyber War: the next threat to national security and what to do about it* (New York: Ecco, 2010), p. 178. Thanks to Gordon Burck for observing that this same approach can be adopted in more conventional circumstances, such as Russian non-cooperation over the shooting down of civilian airliner MH17 in 2013, during the conflict in eastern Ukraine.

8

PROPORTIONALITY IN CYBER TARGETING

Marco Roscini

According to Article 51(5)(b) of Additional Protocol I (AP 1) to the 1949 Geneva Conventions on the Protection of Victims of War,[1] an attack in an international armed conflict would be indiscriminate, and thus prohibited, if it 'may be expected to cause incidental loss of civilian life, injury to civilians, damage to civilian objects, or a combination thereof, which would be excessive in relation to the concrete and direct military advantage anticipated'.[2] This provision, commonly referred to as the principle of proportionality,[3] has been incorporated in several military manuals, including the British Military Manual,[4] the French *Manuel de droit des conflits armés*[5] and the US *Joint Doctrine for Targeting*.[6] It reflects customary international law and, although it does not appear expressly in Additional Protocol II, it is generally accepted that it also applies to attacks in non-international armed conflicts.[7]

The Final Report by the Committee established to Review the NATO Bombing Campaign Against the Federal Republic of Yugoslavia noted that '[t]he main problem with the principle of proportionality is not whether or not it exists but what it means and how it is to be applied'.[8] Indeed, '[t]he intellectual process of balancing the various elements is so complicated, needs to take into account such a huge amount of data and so many factors, that any attempt to design a formula which is both comprehensive and precise would be ridiculous'.[9] These complexities are even more evident in the cyber context, where 'uncertainties in outcome [...] are significantly greater than those usually associated with kinetic attacks in the sense that there may not be an analytic or experiential basis for estimating uncertainties at all'.[10]

It is not disputed, however, that the principle of proportionality extends to cyber operations committed during an armed conflict and having a belligerent nexus with it.[11] The question, then, is not if, but *how* the principle of proportionality applies to them. The principle entails comparing two parameters – incidental damage to civilians and civilian property on the one hand, and the attacker's concrete and direct military advantage on the other – of different natures but of equal standing in specific attacks.[12] But what do *damage* and *military advantage* mean in the cyber context? This chapter first addresses the former and then moves to an analysis of the latter, before finally discussing their relationship to each other. But before examining the application of the principle of proportionality, the type of cyber operations in armed conflict that would fall under its scope needs to be identified: this is the object of the following section.

What cyber operations amount to 'attack' under the law of armed conflict?

Article 51(5)(b) of AP I, containing the principle of proportionality, only applies to indiscriminate *attacks*. What cyber operations qualify as such? *Attack* is defined in Article 49(1) of AP I as 'acts of violence against the adversary, whether in offence or in defence'. In other words, attacks are only those acts of hostilities characterised by *violence*: unlike other acts of hostilities, such as military espionage, non-violent military harm is not sufficient. It is not the author, the target or the intention that define an *act of violence*. Rather, a cyber operation amounts to an *attack* in the sense of Article 49(1) of AP I when it employs means or methods of warfare that result or are reasonably likely to result in violent effects. If a cyber operation causes or is likely to cause loss of life or injury to persons or more than minimal material damage to property, then it is an *attack* and the law of targeting fully applies, including the proportionality rule.[13] Had it been conducted in the context of an armed conflict between Iran and those states allegedly responsible for the cyber operation, for instance, the Stuxnet cyber operation would have been an example of such an 'attack' because of the damage it caused to the centrifuges of Iran's Natanz uranium enrichment facility.[14] The relevant violent effects of a cyber attack include 'any reasonably foreseeable consequential damage, destruction, injury, or death', whether or not the computer system is damaged or data corrupted.[15] If the attack is intercepted and the reasonably expected violent effects do not occur, or occur to a lesser degree, the operation would still qualify as an *attack* for the purposes of Article 49(1).[16]

The problem is more complicated with regard to cyber operations that merely disrupt the functionality of infrastructures without causing material damage. Rule 92 of the *Tallinn Manual 2.0 on the International Law Applicable to Cyber Operations* appears to exclude that such operations qualify as *attacks*. The majority of the experts that drafted the Manual maintained that disruptive cyber operations may be *attacks* only 'if restoration of functionality requires replacement of physical components'.[17] The problem with this view, which still relies on the occurrence of physical damage, is that the attacker may not be able to know in advance whether the restoration of functionality will require replacement of physical components or mere reinstallation of the operating system: the attacker could claim, therefore, that it was not aware that it was conducting an *attack* and thus avoid the application of the law of targeting.

The limits of the doctrine of kinetic equivalence, which requires the occurrence of physical consequences for a cyber operation to be an *attack*, become evident if one considers that, under the *Tallinn Manual 2.0*'s approach, a cyber attack that shuts down the national grid or erases the data of the entire banking system of a state would not be an *attack*, while the physical destruction of one server would. Some commentators have therefore tried to extend the notion of *attack* to include at least some disruptive cyber operations. Dörmann, for instance, recalls that the definition of *military objective* in Article 52(2) of API mentions not only destruction but also 'neutralization' of the object and concludes that, when the object (person or property) is civilian, '[i]t is irrelevant whether [it] is disabled through destruction or in any other way'.[18] Therefore, the incapacitation of an object, such as a civilian power station, without destroying it would still qualify as an *attack*. Melzer adopts a different approach to reach the same conclusion and argues that the principles of distinction, proportionality, and precautions apply not to *attacks*, but rather to the broader notion of *hostilities*: therefore,

> the applicability of the restraints imposed by IHL [international humanitarian law] on the conduct of hostilities to cyber operations depends not on whether the operations in question qualify as 'attacks' (that is, the predominant form of conducting hostilities), but on whether they constitute part of the 'hostilities' within the meaning of IHL.[19]

According to this view, cyber operations disrupting the enemy radar system would not amount to an *attack* because of the lack of violent consequences, but, as an act of hostilities, they would still be subject to the restrictions on the choice and use of methods and means of warfare.[20] This position, however, is inconsistent with the prevailing view according to which the rules contained in Part IV, Section I of AP I essentially apply to *attacks* and not to *hostilities* or *military operations*,[21] and with the letter of Article 51(5)(b), which expressly refers to *attacks*.

It is submitted in this chapter that a better way of including at least certain disruptive cyber operations in the definition of *attack* under Article 49(1) of API is to interpret the provision evolutively, taking into account recent technological developments, and to expand the concept of *violence* to include not only material damage to objects, but also severe incapacitation of physical infrastructures without destruction.[22] This is suggested by Panama in its views on cyber security submitted to the UN Secretary-General, where it qualifies cyber operations as a 'new form of violence'.[23] Indeed, the dependency of modern societies on computers, computer systems, and networks has made it possible to cause significant harm through non-destructive means: cyber technologies can produce results comparable to those of kinetic weapons without the need for physical damage. After all, if the use of graphite bombs, which spread a cloud of extremely fine carbon filaments over electrical components, thus causing a short-circuit and a disruption of the electrical supply, would arguably be considered an *attack*, even though they do not cause more than nominal physical damage to the infrastructure, one cannot see why the same conclusion should not apply to the use of viruses and other malware that achieve the same effect. It is, however, only those cyber operations that go beyond transient effects and mere inconvenience and cause significant functional harm to infrastructures that can qualify as 'attacks' in the sense of Article 49(1). During the crisis between Ukraine and Russia over Crimea for instance, a limited disruption of Ukrainian mobile communications through Distributed Denial of Service (DDoS) attacks and the defacement of certain state-run news websites and social media (the content of which was replaced with pro-Russian propaganda) were reported: because of their limited disruptive effects, such operations would not be *attacks* for the purposes of the law of targeting.[24]

Incidental damage to civilians and civilian property caused by cyber operations

It is therefore those cyber operations that result or are reasonably likely to result either in physical damage to property or persons or in significant incapacitation of infrastructures, with consequent disruption of services, that are subject to the principle of proportionality when conducted in armed conflict. The first element to balance in the proportionality equation is that of the incidental damage caused by the cyber operation. As in kinetic attacks, it is only the incidental damage to civilians and civilian property that is relevant in the application of the principle of proportionality: damage to military objectives and injury/death of combatants and civilians taking direct part in hostilities do not count in this context.[25] What damage to civilians and civilian objects is then 'incidentally' caused by a cyber operation? The effects of a cyber operation can be distinguished in primary effects, i.e., those on the attacked data, computer system, or network; secondary effects, i.e., those on the infrastructure operated by the attacked system or network (if any)[26]; and tertiary effects, i.e., those on the users of the attacked system or infrastructure. Primary, secondary, and tertiary effects all fall within the notion of incidental damage to civilians and civilian objects when they are an intended or at least foreseeable consequence of the attack.[27] The fact that secondary and tertiary effects, at least those that are intended or foreseeable, should also be included in the proportionality calculation results clearly from the US *Counterinsurgency Field Manual*, which explicitly states, 'Leaders must consider not only the first-order, desired effects of a munition or action but also possible second- and third-order effects – including undesired ones'.[28] According to the US *Commander's Handbook on the Law of Naval Operations*, in case of a non-kinetic computer network attacks (CNAs),

> factors involved in weighing anticipated incidental injury/death to protected persons can include, depending on the target, indirect effects (for example, the anticipated incidental injury/death that may occur from disrupting an electricity-generating plant that supplies power to a military headquarters and to a hospital).[29]

The former US Department of State's Legal Advisor confirmed that proportionality

> requires parties to a conflict to assess: (1) the effects of cyber weapons on both military and civilian infrastructure and users, including shared physical infrastructure (such as a dam or a power grid) that would affect civilians; (2) the potential physical damage that a cyber attack may cause, such as death or injury that may result from effects on critical infrastructure; and (3) the potential effects of a cyber attack on civilian objects that are not military objectives, such as private, civilian computers that hold no military significance, but may be networked to computers that are military objectives.[30]

The Commentary to Rule 113 of the *Tallinn Manual 2.0* endorses the inclusion of both direct and indirect effects that 'should be expected by those individuals planning, approving, or executing a cyber attack'.[31] With regard to long-term effects, some commentators have suggested that they should be included in the proportionality equation when such effects 'would not have occurred "but for" the attack'.[32] Others have suggested that a weaponised code, once it becomes available, could be misused by malicious third parties and thus should be factored in the proportionality calculation.[33] This view, however, goes too far. As the use of the adjective 'expected' in relation to incidental damage suggests, the crux of the matter is whether the effect is a reasonably likely or foreseeable consequence of the operation on the basis of the information available at the time of the attack: 'remote effects will generally be beyond the attacking commander's ability to reliably predict and are probably within the defenders' control'.[34]

With regard to the primary effects of cyber operations, mere corruption of software or data without physical consequences would amount to *damage* for the purposes of the proportionality calculation only in exceptional cases, e.g., data corruptible in tangible objects, such as bank account records, so that if the data are destroyed, so are the tangible objects, or data that have an intrinsic value, as in the case of digital art.[35] In all other cases, a cyber operation that only modifies or deletes information without further violent consequences in the analogue world would not be an 'attack' in the sense of Article 49 (1) of AP I and therefore issues of proportionality would not arise.

As to the secondary effects of cyber operations, the question arises whether incidental *damage* to protected objects includes not only physical damage, but also the loss of functionality.[36] It has been noted that, while Article 52(2) of API distinguishes between destruction and neutralisation, Article 51(5)(b) only refers to *damage*, which is broad enough to include the loss of functionality without physical destruction.[37] According to these commentators, 'It would appear counter-intuitive that only the physical destruction of a civilian object should be taken into consideration, whereas functionality loss – even if it affects the civilian population much more severely – should be irrelevant'.[38] This view is correct. Indeed, whereas incidental damage for the purposes of proportionality clearly does not include 'inconvenience, irritation, stress, or fear', as they cannot be compared to 'loss of civilian life, injury to civilians, damage to civilian objects',[39] the incapacitation of networked critical infrastructure has potentially as severe effects on protected persons as those of kinetic attacks. All in all, what matters is that the infrastructure is rendered inoperable, whether by destroying or incapacitating it. Like the notion of *attack*, then, the principle of proportionality, and in particular the notion of *incidental damage*, should be evolutively interpreted to take into account the digitalisation of essential services in modern information societies: as the Israeli Supreme Court stated, 'new reality at times requires new interpretation'.[40] In their proportionality calculation, then, targeteers will need to take into account the consequences arising from the loss of functionality of dual-use infrastructures caused by the cyber operation amounting to an *attack*.[41] This is spelt out in the 2010 *US Joint Terminology for Cyberspace Operations*: according to this document, the 'collateral effect' of cyber operations in the context of targeting includes the

> unintentional or incidental [...] effects on civilian or dual-use computers, networks, information, or infrastructure [when] there is a reasonable probability of loss of life, serious injury, *or* serious adverse effect on the affected nation's national security, economic security, public safety, or any combination of such effects.[42]

It has been claimed that, as cyber operations can be used to incapacitate, instead of destroying, an object, they might expand the scope of what is targetable. A commentator, for instance, has maintained, 'The potentially nonlethal nature of cyber weapons may cloud the assessment of an attack's legality, leading to more frequent violations of the principle of distinction in this new form of warfare than in conventional warfare'.[43] According to this view, international humanitarian law protects civilian objects because of the severe effects that conventional attacks have on them[44]: as disruptive cyber operations leave the object intact, they could be carried out against objects that only indirectly contribute to military action or whose neutralisation offers a non-definite military advantage.[45] This view cannot be accepted, as it does not take into account that, in today's information societies, shutting down the computer systems controlling national critical infrastructures could have far more serious effects on protected persons than certain kinetic attacks. Objects not fulfilling the definition of military objective under Article 52(2) of AP I, therefore, are and remain civilian objects and cannot be attacked. However, certain attacks *against military objectives*, which would be unlawful if executed with kinetic weapons because they may be expected to cause excessive incidental civilian damage, may be lawful if conducted by way of disruptive cyber operations.[46]

Unlike in kinetic attacks and similarly to biological weapons, the incidental damage caused by a cyber operation is not only that to objects and persons located within or near the attacked military objective or to the civilian function performed by the attacked dual-use installation, but also that caused to computer systems (and the infrastructures they might operate) to which the weaponised code may spread as a consequence of the interconnectivity of networks, if this is a foreseeable consequence. The proportionality calculation in a cyber operation that shuts down a dual-use power station, for instance, will have to factor in both the loss of the civilian function performed by the installation, with consequent negative repercussions on its civilian users, and the fact that the malware might infect other computer systems. The only inherently discriminatory cyber operations from this perspective are those on systems that are part of a closed military network or DDoS attacks, which do not use malware and only affect the system overloaded by the multiple requests.

In the *Kupreškić* judgment, the International Criminal Tribunal for the Former Yugoslavia (ICTY) held that

> in case of repeated attacks, all or most of them falling within the grey area between indisputable legality and unlawfulness, it might be warranted to conclude that the cumulative effect of such acts entails that they may not be in keeping with international law. Indeed, this pattern of military conduct may turn out to jeopardise excessively the lives and assets of civilians, contrary to the demands of humanity.[47]

This is particularly relevant for multiple low-intensity cyber attacks (the 'death by a thousand cuts' scenario). According to the ICTY Final Report, however, the *Kupreškić* statement must be interpreted as referring to 'an *overall* assessment of the totality of civilian victims as against the goals of the military campaign', since 'the mere *cumulation* of such instances, all of which are deemed to have been lawful, cannot *ipso facto* be said to amount to a crime'.[48]

'Concrete and direct military advantage' in the cyber context

Let us now move to the other element of the proportionality equation, against which the incidental damage to civilians and civilian objects must be balanced, namely the *concrete and direct military advantage* anticipated from the attack. If, in the context of the definition of *military objective* (Article 52(2) of AP I), the military advantage has to be 'definite', Article 51(5)(b) requires it to be 'concrete and direct', which is a stronger standard and imposes stricter limits on the attacker when incidental damage is expected. As a consequence, 'the advantage concerned should be substantial and relatively close, and [...] advantages which are hardly perceptible and those which would only appear in the long term should be disregarded'.[49] If 'concrete and direct' means 'a real and quantifiable benefit',[50] however, the problem with cyber operations is that measurement of their effects can be difficult: it is still not clear, for instance, if Stuxnet did destroy any centrifuges at Natanz and, if so, with what consequences on the Iranian nuclear programme (while Iran denied that the incident caused significant damage, the International Atomic Energy Agency (IAEA) reported that Iran stopped feeding uranium into thousands of centrifuges at Natanz[51]: it is however unclear whether this was due to Stuxnet or to technical malfunctions inherent to the equipment used).[52]

In any case, according to the ICRC, in the context of the principle of proportionality '[a] military advantage can only consist in ground gained and in annihilating or weakening the enemy armed forces'.[53] In contrast to this narrow interpretation, some states have claimed that military advantage should also include the protection of the attacking forces.[54] Canada's *Joint Doctrine Manual*, for instance, recalls that '[m]ilitary advantage may include a variety of considerations including the security of the attacking forces'.[55] Australia and New Zealand also issued declarations at the ratification of AP I emphasising that *military advantage* includes the 'security of attacking forces'.[56] Israel's document on the 2009 operation in Gaza also claims that military advantage 'may legitimately include not only the need to neutralise the adversary's weapons and ammunition and dismantle military or terrorist infrastructure, but also – as a relevant but not overriding consideration – protecting the security of the commander's own forces'.[57] In the Targeted Killings Judgment, Israel's Supreme Court also supported the inclusion of force protection in the proportionality calculation, holding that 'The state's duty to protect the lives of its soldiers and civilians must be balanced against its duty to protect the lives of innocent civilians harmed during attacks on terrorists'.[58] If this view is correct, the remote character of cyber operations, and thus the enhanced security for the attacking forces, increases the military advantage they provide and would therefore justify a higher amount of incidental damage on civilians and civilian objects.

The number of states that support a broad notion of *military advantage* that includes force protection, however, is relatively limited and other states have adopted different positions.[59] The broad interpretation also has the major disadvantage of introducing a further subjective element in the calculation of proportionality: for instance, what is the value of the safety of military personnel with respect to civilian lives? Does it depend on the rank or the specialisation of the member of the armed forces in question? Is the evaluation different when it is the safety of military *matériel* that is at stake? It should not be forgotten that only a *concrete and direct* military advantage is relevant in the calculation of the proportionality of the attack, and not *abstract protectiveness of the means and methods used to attack*: a belligerent 'cannot justify higher numbers of civilian casualties for the sole reason that it has opted for a more secure [...] operation instead of a less secure [...] operation'.[60] If force protection cannot be a determinant factor in the calculation of proportionality, it could however be relevant in the context of the duty to take precautions.[61] Article 57(2)(a)(ii) of AP I requires the belligerents to 'take all feasible precautions in the choice of means and methods of attack with a view to avoiding, and in any event to minimizing, incidental loss of civilian life, injury to civilians and damage to civilian objects'. It is clear that the determination of what is a 'feasible' precaution will consider the additional risks encountered by the attacking forces, which are under no obligation to sacrifice themselves. That this is the correct relevance of force protection in the framework of the law of targeting is confirmed in the UK Military Manual, which includes the 'risks to his own troops' among the factors that a commander needs to evaluate in the context of the duty to take precautions when choosing means or methods of attack.[62] A targeter, then, *may* go for a certain means or method of warfare, including cyber operations, that minimises the risk for the attacking forces (even if it is at the cost of reducing the military advantage he anticipates) but providing that this does not increase the expected incidental damage to civilians or civilian objects.

As with incidental damage, when the attack is composed of multiple hostile acts the military advantage is what results from the attack considered as a whole. Several NATO states added interpretive declarations in relation to Article 51(5)(b), stating that the attack has to be considered in its totality, not in its specific parts. Italy, for instance, declared that 'the military advantage anticipated from an attack is intended to refer to the advantage anticipated from the attack considered as a whole and not only from isolated or particular parts of the attack'.[63] Read in the light of these declarations, the concrete and direct military advantage of the Israeli cyber operation that allegedly switched off the Syrian radar system with the aim to facilitate the bombing of a nuclear reactor in 2007 has to be evaluated jointly with the airstrike that followed it. This is also relevant for several coordinated low-intensity cyber attacks. In such cases,

> if a CNA [computer network attack] is mounted systematically against a whole array of enemy computers, the military advantage accruing from the destruction of – or intrusion into – any particular target computer may be of little consequence by itself. Only an examination of the larger picture would divulge what is at stake.[64]

The Eritrea-Ethiopia Claims Commission (EECC), however, has gone further than the NATO reservation and opined that

> the term 'military advantage' can only properly be understood in the context of the military operations between the Parties taken as a whole, not simply in the context of specific attack [and that] a definite military advantage must be considered in the context of its relation to the armed conflict as a whole at the time of the attack.[65]

These views have been rightly criticised by the EECC President van Houtte in his Separate Opinion[66] and by several commentators[67] for excessively conflating incompatible notions of *military advantage*: if military advantage is defined in relation to the armed conflict as a whole and the ultimate objective of defeating the adversary, it would justify virtually any level of incidental damage to protected persons and objects.

Balancing the two factors of the equation

Having established what *incidental damage* and *concrete and direct military advantage* mean in the cyber context, the two parameters must now be balanced against each other. In particular, the expected incidental damage must not be *excessive* with respect to the anticipated military advantage. *Excessive* should not be confused with *extensive*:[68] the principle of proportionality permits even massive incidental damage to civilians and civilian property if this is matched by a correspondently significant military advantage. On the other hand,

> if – while disrupting some military electronic systems in a minor way – [a cyber operation] causes irreparable damage to the civilian infrastructure (eg water management, research centres, banking systems, stock exchanges), this should be adjudged 'excessive'.[69]

It should be noted that, under Article 51(5)(b), it is not the incidental damage that actually occurs or the military advantage that is effectively gained from the attack that count, but rather the *expected* damage and the *anticipated* military advantage. The difficulty of calculating the *expected* incidental damage and the 'anticipated' military advantage are already well known in relation to traditional warfare, but the problems are exacerbated in the cyber context, where the interconnectivity of networks and the reverberating effects of cyber operations often make the *ex ante* evaluation an esoteric prediction.

The difficulties of measuring each term of the equation must be added to those of balancing the terms against each other, which is necessarily a subjective process, depending on social and historical factors as well as on the background of the specific targeteers involved. As the ICTY Report states,

> It is unlikely that a human rights lawyer and an experienced combat commander would assign the same relative values to military advantage and to injury to non-combatants. Further, it is unlikely that military commanders with different doctrinal backgrounds and differing degrees of combat experience or national military histories would always agree in close cases.[70]

In an attempt to objectivise the test, the ICTY Trial Chamber held that, to assess the proportionality of an attack, it is necessary to determine 'whether a reasonably well-informed person in the circumstances of the actual perpetrator, making reasonable use of the information available to him or her, could have expected excessive civilian casualties to result from the attack'.[71] The ICTY Final Report refers to this person as a *reasonable military commander*.[72] Because of the technicalities of cyber warfare, the *reasonable military commander* will almost inevitably have to be assisted by cyber engineers when making any decision with regard to the proportionality of the cyber attack, unless he is a trained cyber expert himself.[73] Collecting information about the architecture of the attacked network (network mapping) or operating system (footprinting) will also be of decisive importance, as the damaging effects of a cyber operation greatly depend on the characteristics of the targeted systems. All in all, the issue is one of degree: if the effects of a cyber operation are *entirely* unclear and unforeseeable, then the attack would be indiscriminate and must not be carried out.[74]

Conclusion

Cyber operations present both opportunities and dangers for the principle of proportionality in attack. On the one hand, their potentially less damaging character might offer a better means to minimise incidental damage to civilians and civilian property, which can be seen in the context of the trend towards effects-based warfare.[75] Cyber operations also present advantages for the attacking state, as they entail virtually no risk for its forces thanks to their remote character and the difficulties with regard to identification and attribution.[76] On the other hand, the interconnectivity of military and civilian networks raises the question of the uncontrolled spreading of malware to other computers and networks, which might be difficult to predict and therefore to avoid or minimise.[77] As required by Article 57(2) of AP I, then, all feasible precautions must be adopted to ensure that the attack is consistent with the principle of proportionality.[78] It appears for instance that, even though international humanitarian law was not applicable, those that developed and deployed Stuxnet went a long way to prevent or at least minimise incidental damage on targets other than the Natanz uranium enrichment facility: the worm was activated only when it found the presence of the specific Siemens software used at Natanz; each infected computer could spread the worm only to three other computers; even when a computer was infected, the worm did not sufficiently self-replicate to inhibit computer functions and therefore only caused annoyance; and it contained a command that deactivated the worm on 24 June 2012.[79]

Notes

1 This chapter is based, with some amendments and updates, on the author's book *Cyber Operations and the Use of Force in International Law* (Oxford: OUP, 2014), pp. 219–29.

2 The same language appears in Art. 57(2)(a)(iii) and Art. 57(2)(b) of Additional Protocol I in relation to the duty to take precautions in attack.

3 The expression is explicitly used in the British Military Manual (UK Ministry of Defence, *The Manual of the Law of Armed Conflict* (Oxford: OUP, 2004), p. 68 (hereinafter 'UK Military Manual'). Proportionality operates differently in *jus in bello* and *jus ad bellum*: while in the latter case it is a requirement for the legality of a self-defence reaction, applies to the operation as a whole and balances the armed reaction against the purpose of defeating an armed attack, in the former it is a limitation that applies to each individual attack and balances the concrete and direct military advantage anticipated from the attack against the expected incidental damage to protected property and persons. Furthermore, unlike in the *jus ad bellum* notion of proportionality, where the interest of the attacked state is given superior standing with respect to that of the attacker, *jus in bello* proportionality is a normative technique that aims to reconcile two values of equal rank (E. Cannizzaro, 'Contextualizing proportionality: *jus ad bellum* and *jus in bello* in the Lebanese War', *International Review of the Red Cross*, Vol. 88, 2006, pp. 786–7).

4 UK Military Manual, p. 86 (see note 3 above).

5 Ministère de la Défense, *Manuel de droit des conflicts armés*, 4, 58, at www.defense.gouv.fr/sga/le-sga-en-action/droit-et-defense/droit-des-conflits-armes/droit-des-conflits-armes.

6 *Joint Doctrine for Targeting*, Joint Publication 3–60, 17 January 2002, Appendix A, A-1, at www.dtic.mil/doctrine/jel/new_pubs/jp3_60.pdf.

7 See Rule 14 of the ICRC Study on Customary International Humanitarian Law (J.-M. Henckaerts and L. Doswald-Beck (eds.), *Customary International Humanitarian Law* (Cambridge: Cambridge University Press, 2005) (first published 2004), Vol. I, p. 46. Indeed, it follows from Art. 13(1) of Additional Protocol II, according to which civilians 'enjoy general protection against the dangers arising from military operations', that incidental civilian losses must be avoided or at least minimised (W. H. Boothby, *The Law of Targeting* (Oxford: OUP, 2012), p. 436); see also Israel's Supreme Court, *The Public Committee against Torture in Israel v. The Government of Israel*, Judgment of 11 December 2005, para. 42 (Barak); and M. N. Schmitt, C. H. B. Garraway and Y. Dinstein, *The Manual of the Law of Non-International Armed Conflict* (Sanremo, 2006), para. 2.1.1.4.

8 *Final Report by the Committee Established to Review the NATO Bombing Campaign against the Federal Republic of Yugoslavia*, 8 June 2000, para. 48, at www.icty.org/x/file/Press/nato061300.pdf.
9 S. Oeter, 'Methods and means of combat', in D. Fleck, *The Handbook of International Humanitarian Law*, 3rd edn (Oxford: OUP, 2013), p. 191.
10 W. A. Owens, K. W. Dam and H. S. Lin (eds.), *Technology, Policy, Law, and Ethics Regarding U.S. Acquisition and Use of Cyberattack Capabilities* (Washington, DC: The National Academies Press, 2009), p. 262.
11 H. Koh, 'International law in cyberspace', Speech at the USCYBERCOM Inter-Agency Legal Conference, 18 September 2012, at http://opiniojuris.org/2012/09/19/harold-koh-on-international-law-in-cyberspace; see Rule 113 of the *Tallinn Manual 2.0 on the International Law Applicable to Cyber Operations* (Cambridge: Cambridge University Press, 2018), p. 470 [hereinafter *Tallinn Manual 2.0*]; and the French White Paper on defence and national security, which states, with regard to hostile cyber attacks, that 'Les capacités d'identification et d'action offensive sont essentielles pour une riposte éventuelle et proportionnée à l'attaque' (*Livre blanc, Défense et sécurité nationale*, 2013, p. 73, at www.elysee.fr/assets/pdf/Livre-Blanc.pdf).
12 According to the ICTY *Final Report*, 'It is much easier to formulate the principle of proportionality in general terms than it is to apply it to a particular set of circumstances because the comparison is often between unlike quantities and values' (ICTY *Final Report*, para. 38).
13 Rule 92 of the *Tallinn Manual 2.0*, for instance, defines a cyber attack as 'a cyber operation, whether offensive or defensive, that is reasonably expected to cause injury or death to persons or damage or destruction to objects' (*Tallinn Manual 2.0*, 470 (see note 11 above); the Manual includes 'serious illness and severe mental suffering' in the notion of 'injury' (p. 417).
14 On the Stuxnet attack, see M. Roscini, 'Cyber operations as nuclear counterproliferation measures', *Journal of Conflict and Security Law*, Vol. 19, 2014, pp. 133–57.
15 *Tallinn Manual 2.0*, p. 416 (see note 11 above).
16 Commentary to Rule 92, in *Tallinn Manual 2.0*, p. 419 (see note 11 above).
17 *Tallinn Manual 2.0*, p. 417 (see note 11 above).
18 K. Dörmann, *Applicability of the Additional Protocols to Computer Network Attacks*, at 6, www.icrc.org/eng/assets/files/other/applicabilityofihltocna.pdf.
19 N. Melzer, *Cyberwarfare and International Law* (UNIDIR, 2011) p. 27, at www.isn.ethz.ch/Digital-Library/Publications/Detail/?lng=en&id=134218.
20 Ibid., pp. 27–8.
21 Y. Sandoz, Ch. Swinarski, and B. Zimmermann (eds.), *Commentary on the Additional Protocols of 8 June 1977 to the Geneva Conventions of 12 August 1949* (Dordrecht: Nijhoff, 1987), para. 1875 (hereinafter 'ICRC Commentary'); see M. N. Schmitt, 'Wired warfare: computer network attack and *Jus in Bello*', in M. N. Schmitt and B. T. O'Donnell (eds.) *Computer Network Attack and International Law, International Law Studies*, No. 76 (2002), pp. 193–4; D. Turns, 'Cyber war and the concept of "attack" in International Humanitarian Law', in D. Saxon (ed.), *International Humanitarian Law and the Changing Technology of War* (Leiden: Brill, 2013), p. 217; and Robin Geiß and Henning Lahmann, 'Cyber Warfare: Applying the Principle of Distinction in an Interconnected Space', *Israel Law Review*, Vol. 45, Issue 3 (2012).
22 On evolutive interpretation in the cyber context, see Roscini, *Cyber Operations*, pp. 20–24, 280–81 (see note 14 above).
23 UN DOC A/57/166/ add. 1, 29 August 2002, p.5.
24 M. Roscini, 'Is there a "cyber war" between Ukraine and Russia?', OUPBlog, 31 March 2014, at http://blog.oup.com/2014/03/is-there-a-cyber-war-between-ukraine-and-russia-pil; to clarify the distinctions: denial of service (DoS) attacks, of which 'flood attacks' are an example, aim to inundate the targeted system with excessive calls, messages, enquiries or requests in order to overload it and force its shut down; permanent DoS attacks are particularly serious attacks that damage the system and cause its replacement or reinstallation of hardware; and when the DoS attack is carried out by a large number of computers organised in botnets, it is referred to as a 'distributed denial of service' (DDoS) attack.
25 In the *Nuclear Weapons* Advisory Opinion, the ICJ also held that 'States must take environmental considerations into account when assessing what is necessary and proportionate in the pursuit of legitimate military objectives' (*Legality of the threat and use of nuclear weapons*, Advisory Opinion of 8 July 1996, *ICJ Reports* (1996), para. 30).
26 Supervisory Control and Data Acquisition (SCADA) systems, for instance, are computer-controlled industrial control systems that monitor and control industrial processes or physical infrastructures.

27 For instance, if a power station used by both the military and civilians is destroyed or incapacitated, hospitals and civilian infrastructures like water purification plants might be deprived of electricity for a certain amount of time (H. Shue and D. Wippman, 'Limiting attacks on dual-use facilities performing indispensable civilian functions', *Cornell International Law Journal*, Vol. 35, No. 3 (2002), p. 564; see also Boothby, *The Law of Targeting*, pp. 384–5 (note 7 above); J. W. Crawford, III, 'The law of noncombatant immunity and the targeting of national electrical power systems', *Fletcher Forum of World Affairs*, Vol. 21(1997), p. 114; *Technology, Policy, Law, and Ethics*, p. 264 (see note 10 above).

28 The U.S. Army – Marine Corps, *Counterinsurgency Field Manual*, U.S. Army Field Manual No. 3–24-Marine Corps Warfighting Publication No. 3–33.5 (Chicago, London, 2007), pp. 7–36.

29 *The Commander's Handbook on the Law of Naval Operations*, July 2007, para. 8.11.4, at www.usnwc.edu/getattachment/a9b8e92d-2c8d-4779-9925-0defea93325c/.

30 Koh, 'International law' (see note 11 above).

31 *Tallinn Manual 2.0*, p. 472 (see note 11 above); the relevance of secondary and tertiary effects when assessing the legality of the conduct of hostilities is also confirmed by provisions such as Articles 54 and 56 of API, which prohibit attacks on certain objects because they are indispensable to the survival of the civilian population or because they might cause the release of dangerous forces with obvious consequences for the population; see also Switzerland's declaration in relation to explosive remnants of war (ERW): 'The military commander's proportionality assessment with regard to the choice and use of a particular means or method of warfare must also take into account the foreseeable incidental long-term effects of an attack such as the humanitarian costs caused by duds becoming ERW' (CCW/GGE/XI/WG.1/WP.13, 3 August 2005, para. 15) [similar declarations were made by Austria (CCW/GGE/XI/WG.1/WP.14, 4 August 2005, para. 9) and Norway (CCW/GGE/XI/WG.1/WP.7, 28 July 2005, para. 21)].

32 M. N. Schmitt, H. A. Harrison Dinniss and T. C. Wingfield, 'Computers and war: the legal battlespace', 2004, p. 9, at www.hpcrresearch.org/sites/default/files/publications/schmittetal.pdf.

33 J. Richmond, 'Evolving battlefields: does Stuxnet demonstrate a need for modifications to the law of armed conflict?' *Fordham International Law Journal*, Vol. 35, 2011–2012, p. 893.

34 J. Holland, 'Military objectives and collateral damage: their relationship and dynamics', *Yearbook of International Humanitarian Law*, Vol. 7, 2004, p. 62; see also ICTY, *Prosecutor v. Galić*, Judgment, 5 December 2003, para. 58.

35 M. N. Schmitt, 'Cyber operations and the *Jus in Bello*: key issues', in R.A. Pedrozo and D. P. Wollschlaeger (eds.), 'International Law and the Changing Character of War', *International Law Studies*, Vol. 87, 2011, p. 96; this is so especially when the work of art exists only in its digital version.

36 E. T. Jensen, 'Cyber attacks: proportionality and precautions in attack' *International Law Studies*, Vol. 89, 2013, pp. 206–7.

37 Geiß and Lahmann, 'Cyber warfare', p. 397 (see note 21 above); see also the Commentary to Rule 51 of the *Tallinn Manual 2.0*, p. 472 (see note 11 above).

38 Geiß and Lahmann, 'Cyber warfare', p. 397 (see note 21 above).

39 *Tallinn Manual 2.0*, p. 472 (see note 11 above); similarly, mere 'passing through' civilian computers without causing damage would not count as incidental damage; on this see J. Richardson, 'Stuxnet as cyberwarfare: applying the law of war to the virtual battlefield', *Journal of Computer and Information Law*, Vol. 29, 2011, p. 26.

40 *Targeted Killings* Judgment, para. 28 (Barak).

41 *Tallinn Manual 2.0*, p. 472 (see note 11 above); Shue and Wippman, 'Limiting attacks', 570 (see note 26 above).

42 *US Joint Terminology for Cyberspace Operations*, 3, at www.nsci-va.org/CyberReferenceLib/2010-11-joint%20Terminology%20for%20Cyberspace%20Operations.pdf (emphasis added).

43 J.T.G. Kelsey, 'Hacking into International Humanitarian Law: the principles of distinction and neutrality in the age of cyber warfare', *Michigan Law Review*, Vol. 106, 2008, p. 1439.

44 Kelsey, 'Hacking', p. 1440 (see note 42 above); M. R. Shulman, 'Discrimination in the laws of information warfare', *Columbia Journal of Transnational Law*, Vol. 37, 1998–1999, p. 964.

45 Kelsey, 'Hacking', p. 1448 (see note 42 above); see also L. T. Greenberg, S. E. Goodman, and K. J. Soo Hoo, *Information Warfare and International Law* (National Defense University Press, 1998), p. 12; and a similar argument seems to be implicit in Wedgwood's suggestion that proportionality could be conceived dynamically so to tolerate greater collateral damage to civilian objects to eliminate a security threats 'so long as the damage is reversible or, indeed, aid is given in its restoration', see R. Wedgwood, 'Proportionality, cyberwar, and the law of war', in Michael Schmitt and Brian O'Donnell (eds.), 'Computer Network Attack and International Law', *International Law Studies*, Vol. 76, 2002, p. 228.

46 Dörmann, *Applicability*, p. 6 (see note 18 above).

47 ICTY, *Kupreškić*, Trial Chamber, 14 January 2000, para. 526; according to the Tribunal, this interpretation follows from the application of the Martens clause codified in Art. 1(2) of APl I.

48 ICTY Final Report, para. 52; see also comments by N. Ronzitti, 'Is the non liquet of the Final Report by the Committee Established to Review the NATO Bombing Campaign against the Federal Republic of Yugoslavia acceptable?' (2000) 82 *International Review of the Red Cross*, p. 1017.

49 ICRC Commentary, para. 2209.

50 Commentary to Rule 113, *Tallinn Manual 2.0*, p. 473 (see note 11 above).

51 W. J. Broad, 'Report suggests problems with Iran's nuclear effort', *The New York Times*, 23 November 2010, at www.nytimes.com/2010/11/24/world/middleeast/24nuke.html.

52 K Ziolkowski, 'Stuxnet – legal considerations', NATO Cooperative Cyber Defence Centre of Excellence, 2012, p. 5.

53 Sandoz, Swinarski, and Zimmermann (eds.), *Commentary on the Additional Protocols of 8 June 1977 to the Geneva Conventions of 12 August 1949* para. 2218; this interpretation has been criticised for being too narrow: according to the Commentary to the Harvard Manual on International Law Applicable to Air and Missile Warfare (*HPCR Manual on International Law Applicable to Air and Missile Warfare* (Cambridge: Cambridge University Press, 2013), p. 36)

> [a] better approach is to understand military advantage as any consequence of an attack which directly enhances friendly military operations or hinders those of the enemy. This could, e.g., be an attack that reduces the mobility of the enemy forces without actually weakening them, such as the blocking of an important line of communication.

54 In literature, see R. Geiß, 'The principle of proportionality: "force protection" as a military advantage', *Israel Law Review*, Vol. 45 (2012), p. 77; see also Y. Dinstein, *The Conduct of Hostilities under the Law of International Armed Conflict* (Cambridge: Cambridge University Press, 2010), pp. 141–2, and I. Henderson, *The Contemporary Law of Targeting* (Leiden: Nijhoff, 2009), p. 205, but *contra*, see W. J. Fenrick, 'Attacking the enemy civilian as a punishable offence', *Duke Journal of Comparative and International Law*, Vol. 7, (1996–1997), p. 549, and G. D. Solis, *The Law of Armed Conflict* (Cambridge: Cambridge University Press, 2010), p. 285.

55 *Joint Doctrine Manual, Law of Armed Conflict at the Operational and Tactical Levels*, B-GJ-005 104/FP-021, 2003, 4–4; note however that in the cited sentence, 'military advantage' is not preceded by 'concrete and direct'.

56 A. Roberts and R. Guelff, *Documents on the Laws of War* (Oxford: OUP, 2000), pp. 500, 508; force protection is also mentioned in France's declaration upon ratification of AP I, but in relation to the second sentence of Art. 50(1) (see declaration of 11 April 2011, para. 9, at www.icrc.org/applic/ihl/ihl.nsf/Notification.xsp?action=openDocument&documentId=D8041036B40EBC44C1256A34004897B2).

57 *The Operation in Gaza (27 December 2008–18 January 2009). Factual and Legal Aspects*, July 2009, para. 126, at www.mfa.gov.il/MFA_Graphics/MFA%20Gallery/Documents/GazaOperation%20w%20Links.pdf.

58 *Targeted Killings* Judgment, para. 46 (Barak).

59 The US Counterinsurgency Field Manual, for instance, states that the principle of proportionality and discrimination require combatants to '[a]ssume additional risk to minimize potential harm [to noncombatants]' (US Counterinsurgency Field Manual, pp. 7–30).

60 Geiß, 'The principle', p. 87 (see note 53 above).

61 Oeter, 'Methods', p. 210 (see note 9 above).

62 UK Military Manual, p. 84 (see note 3 above); further, The Final Report to Congress on the Conduct of the Persian Gulf War also recalls that coalition forces took the 'risk to aircraft and aircrews' into account when choosing means of warfare for attacks on targets in populated areas (US Department of Defense, *Conduct of the Persian Gulf War: Final Report to Congress* (1992) 31 *International Legal Materials*, p. 622).

63 Roberts and Guelff, *Documents*, p. 507 (see note 55 above); see also the UK declaration (*Documents*, p. 511). The ICRC Commentary considers such statements 'redundant' as 'it goes without saying that an attack carried out in a concerted manner in numerous places can only be judged in its entirety' (ICRC Commentary, para. 2218).

64 Y. Dinstein, 'The principle of distinction and cyber war in international armed conflicts', *Journal of Conflict and Security Law*, Vol. 17, 2012, p. 271.

65 Eritrea-Ethiopia Claims Commission (EECC), Partial Award, *Western Front, Aerial Bombardment and Related Claims Eritrea's Claims 1, 3, 5, 9–13, 14, 21, 25 & 26*, 19 December 2005, para. 113; the same questionable position appears in certain US documents, including the Final Report to Congress on the

Conduct of the Persian Gulf War, p. 622; see also Iran's position in *Yearbook of International Humanitarian Law*, Vol. 6, 2003, p. 496, according to which '"military advantage" will be the advantage expected from an invasion in its entirety and not part of it'.
66 *Western Front*, Separate Opinion of President van Houtte, paras. 8, 10 (see note 64 above).
67 G. Venturini, 'International Law and the Conduct of Military Operations', in A. de Guttry, H. H. G. Post and G. Venturini (eds.), *The 1998–2000 War between Eritrea and Ethiopia* (Asser Press: The Hague, 2009), p. 301; L. Vierucci, 'Sulla nozione di obiettivo militare nella guerra aerea: recenti sviluppi della giurisprudenza internazionale' *Rivista di diritto internazionale*, Vol. 89 (2006), pp. 704–6; and Y. Dinstein, 'Air warfare', in *Max Planck Encyclopedia of Public International Law* (OUP, 2012), Vol. I, p. 255.
68 Dinstein, 'The Principle', p. 272 (see note 63 above).
69 Dinstein, 'The Principle' (see note 63 above).
70 ICTY Final Report, para. 50.
71 *Prosecutor v. Galić*, para. 58.
72 ICTY Final Report, para. 50; see also Israel's Supreme Court, *Beit Sourik Village v. The Government of Israel*, HCJ 2056/04, para. 46, and the declarations issued by Germany, Belgium, Italy, the Netherlands and Spain in relation to Art. 51 of AP I, in Roberts and Guelff, *Documents*, 505, 501, 507, 508, 509, respectively.
73 H. Harrison Dinniss, *Cyber Warfare and the Laws of War* (Cambridge: Cambridge University Press, 2012), pp. 206–7.
74 Louise Doswald-Beck, 'Some Thoughts on Computer Network Attack and the International Law of Armed Conflict', in Michael Schmitt and Brian O'Donnell (eds.), 'Computer Network Attack and International Law', *International Law Studies*, Vol. 76, 2002, p. 170.
75 The EECC has emphasised the 'increased emphasis on avoiding unnecessary injury and suffering by civilians resulting from armed conflict' that characterises modern effect-based warfare (*Western Front*, para. 104) (see note 64 above).
76 On the identification and attribution problems of cyber operations, see Roscini, *Cyber Operations*, pp. 33–40 (see note 1 above).
77 Schmitt, 'Wired warfare', p. 204 (see note 21 above).
78 Additional Protocol I, Art. 57(2)(a).
79 Richmond, 'Evolving battlefields', p. 856 (see note 32 above).

9

DIGITAL INTELLIGENCE AND ARMED CONFLICT AFTER SNOWDEN

Sir David Omand

The use of intelligence to support military operations dates back to antiquity. That military intelligence lineage can still be seen in much of the present organisation of secret intelligence, for example in the popular names MI5 and MI6 for the British Security Service and Secret Intelligence Service respectively, and in the funding within the US defence budget of the National Security Agency (NSA) with its Director holding senior military rank. In the twentieth century the dominant source for all branches of defence activity was signals intelligence based on radio interception.[1] Today, it is digital communications and data that are relied upon to provide vital information needed by military commanders and planning staffs as well as for cyber defence and underpinning the development of offensive options for the use of cyberspace as an adjunct to conventional operations.[2]

This chapter puts forward a model for examining how digital support for the UK armed forces relates to the use of the same technologies for national security missions, such as countering terrorism, subversion and proliferation and for counter-intelligence as well as for law enforcement, including frustrating cyber and other serious criminal groups. These are the same technologies that were alleged by Edward Snowden and civil rights activists drawing on the material that Snowden stole from the NSA and the UK's Government Communications Headquarters (GCHQ), to be used by these organisations for mass surveillance and invasion of privacy. These allegations of the misuse of digital intelligence have generated a global controversy over the legitimacy of digital intelligence gathering by the US and UK (and by their 'Five Eyes' partners Canada, Australia, and New Zealand). There are still international calls for such activity to be regulated and at least severely curtailed, as well as welcome recognition that in the wrong hands without adequate regulation and oversight developments in digital intelligence technology are also capable of providing authoritarian states with unprecedented levels of surveillance and control of their citizens and repression of their dissidents. This chapter examines whether such understandable concerns for personal privacy could nevertheless have the unintended effect of threatening the legitimate intelligence missions of democratic states such as the UK, including support for military operations.

Broadly speaking, the demands for intelligence to support the defence mission can be seen as falling into three categories, which are discussed in the following three sections.

The requirement for information on the military capabilities and intentions of potential adversaries

An obvious first area of interest to the defence establishment is the continuing requirement for information on the military capabilities of potential adversaries, including their order of battle, organisation, military doctrine, plans, tactics, techniques and procedures, and weapons systems and their intentions (both strategic and tactical). Much of this demand is met by the traditional 'INTs' including OSINT (open sources), HUMINT (human intelligence), IMINT (imagery), ELINT (electronic intelligence from radar and other emissions) and MASINT (measurement and signature intelligence such as monitoring telemetry from the testing of new weapons systems) and of course SIGINT (signals intelligence). Such traditional military intelligence activity continues but now looks limited in the present context: in the past, much SIGINT, or signals intelligence, would have come from discovering and then monitoring the specific radio frequencies used by defence forces such as the command net of a military formation.[3] Today, relevant communications are likely to be found on virtual private networks carried over the internet or in encrypted messages using the many systems available that work on the internet protocol (IP). Additionally, in theatres of recent military operations, such as Afghanistan, Iraq, and Libya, armed forces have become accustomed to their adversaries using the same mobile internet devices – mobile phones, tablets and laptops, – and the same social media platforms as everyone else. These communications are all carried over the global packet-switched networks of the internet. Media coverage[4] of the social media usage of the paramilitaries operating in Eastern Ukraine shows the same pattern of usage of modern digital communications as did the communications of British and other European jihadists that joined ISIL in Syria and Iraq to fight for so-called Islamic State.[5]

So the legitimate demands for intelligence to support military operations, to locate enemy combatants, and to anticipate their next moves will for most operations today and into the future involve an element of signals intelligence to be derived from digital internet communications, as well as the more traditional radio transmissions. The equivalent of the traffic analysis that was used to great effect to monitor the order of battle of the Group of Soviet Forces in Germany during the Cold War, largely from the pattern rather than the content of their wireless communications, is paralleled today in the analysis of communications data from the mobile devices used by insurgents or terrorists – who called or emailed or texted whom, when, and where. And the equivalent of the direction finding of radio signals that pinpointed the location of a transmitting warship is the geolocation of the mobile device from the GPS signal used by apps on the device or by the triangulation of signals on the mobile telephone network.

The requirement for information about the individuals who pose a threat

A second area of interest for modern military intelligence is obtaining information about the adversary as individuals rather than in terms of military structures, deployments, and weaponry. It will be evident that in many peacekeeping, peace enforcement, and other situations of asymmetric warfare intelligence about people will be critical to successful operations. On active duty, J2 intelligence officers need to be able to provide situational awareness to their commander of the threats they face and the risks to their intended plans, a reasonable explanation of why the adversary appears to be deployed and behaving as he does, and at least an estimate of what might be likely to develop next – all matters that may be largely determined by the personalities and experience of the leadership of the opponents. At a tactical level, pre-emptive intelligence will be sought to allow the deployment of friendly forces to frustrate the adversary's move or next attack, be it interdicting an arms supply, preventing the mortaring of a friendly position, or uncovering the placing of an improvised explosive device (IED) to explode under a supply convoy. So the demand would be for intelligence on, for example, the identities, locations, movements, communications, financing, and intentions of members of an insurgent group who are trying to blend with the community inside the tactical area of operations. Interdicting narcotics or arms smuggling, countering piracy, and enforcing United Nations embargoes are further examples.

An obvious source of such intelligence on individuals of interest is from their use of mobile digital communications. Mobile networks in the developing world have expanded very rapidly indeed (by comparison with the Western experience of the slow and expensive historical experience of fixed-line telephony). The infrastructure is cheap to install and at least simple handsets are affordable by most of the population. From a world population of around 7.3 billion, there are already well over 5.5 billion mobile telephone users and around three billion internet users worldwide, increasing currently at the rate of 50 million per year. Another source of relevant intelligence is 'data at rest', the digital traces that individuals leave in any reasonably developed economy and are stored in databases either in the private sector (airline bookings for example) or by government (such as identity cards).

A recent development is the derivation of intelligence from the adversary's social media use, called SOCMINT.[6] This has become a very important source of intelligence both for tactical and operational purposes, and for strategic analysis. Around 1.2 billion people now use a social media platform at least once a month. Facebook is the largest, with over one billion regular users, but the Russian language VK network has 190 million and the Chinese QQ network 700 million users. Individuals post detailed information about themselves, their friends and their associates, their movements and daily doings, and their transactions, not just using networking sites but also specialised applications and the blogs associated with thousands of apps. Had this technology been invented and in general use by the population in, say, 1970 and thus potentially available to be drawn upon then, the support provided to the civil power by the British Army in Northern Ireland would have been much more effective.

Unsurprisingly, SOCMINT is now accepted by the Home Office as a form of intelligence on social disorder, as recommended by Her Majesty's Inspectorate of Constabulary after the experience of the use of social media by rioters in English cities in 2012. A powerful 'all sources hub' which includes open-source social media collection was established by the Metropolitan Police and was used to calibrate the policing of protests and demonstrations at the time of the 2012 London Olympics. Military intelligence has similarly already seized on SOCMINT as a potential source of information on the attitude or sentiment of local populations towards the security forces and the local government, for example during the recent period of operations in Afghanistan. It must be expected to be part of the framework of intelligence needed for any likely future operation.

On operations overseas, the legal authority for the interception of global communications (including by GCHQ acting in support of the armed forces) would rest on a warrant signed by the Foreign Secretary.[7] Provided that the activity was necessary and proportionate within the context of the operational mission, such digital intelligence activity would be unlikely to attract controversy. Digital interception is now standard practice internationally in support of military operations. The journalists reporting on some of Edward Snowden's material did nevertheless misinterpret an NSA programme, Boundless Informant, as meaning that the NSA was intercepting large numbers of mobile telephone calls in countries including France, Germany, Spain, Norway, the Netherlands, and Sweden, causing considerable political controversy in those countries. It has now been persuasively argued[8] that the Snowden material was misunderstood and that these were telephone intercepts carried out by those countries' own intelligence services in support of a NATO military operation (assumed to be Afghanistan) and shared with the US.

For the UK, the legal basis – when need arises – for access to domestic communications to derive intelligence is now laid down in the Regulation of Investigatory Powers Act 2016 (RIPA 2016).[9] Should the armed forces be deployed in the future in support of the United Kingdom civil power (as previously during Operation Banner, the Northern Ireland campaign that preceded the peace process), then in such a case the intelligence lead would rest with the civil authorities, supported by the armed forces, as necessary. There would be a demand in such circumstances for intelligence derived from the communications of individuals in the British Islands, for which warrants would be signed at the level of a Secretary of State, such as the Home Secretary or the Northern Ireland Secretary, or a senior Scottish minister in Edinburgh, all of which would require to be judicially reviewed by a Judicial Commissioner before entering into force. It would normally be the civil power that would apply for and execute such warrants. For SOCMINT there is a spectrum, ranging from unclassified tweets at one end – 'imagine opening the digital door and shouting your message' – through to closed social media groups or encrypted messaging services. In peacetime of course the armed forces would need training in the use of such techniques for their potential overseas operations, and that too might require specific legal authority if conducted within the United Kingdom.

The requirement for intelligence to support defensive and offensive cyber operations

The third area of defence interest in digital intelligence is the support needed for defensive cyber security and for offensive cyber operations. Snowden has given glimpses of what may be possible for the US and UK intelligence communities in that respect.

The defence interest in cyberspace has to be broad. The Ministry of Defence has to protect itself and the armed forces from a range of threats that could be termed (as an acronym) the 'cesspit' of modernity: crime, espionage, sabotage and subversion perverting internet technology.

There is an 80/20 rule for security today, as during the Cold War. Good security hygiene and personnel security and looking after staff will help tackle the lowest 80 percent of the threat. The Snowden case, for example, revealed significant weaknesses in personnel security by the commercial companies Dell and Booz Allen that were employing him as a contractor for the US intelligence community while failing to vet properly an employee given special access rights. But although it is a necessary step, improving passive defences, such as personal vetting, is not sufficient to stop advanced persistent threats from State actors. To reduce the likelihood of advanced exploits, intelligence-led active defence is needed that uses the techniques of intelligence access and analysis to anticipate the types and direction of attack and to be able to respond fast to neutralise the danger. Simply erecting barrier defences such as firewalls around key networks will not be sufficient. An active approach to defending military networks will involve, among other techniques, monitoring activity to spot anomalies and unexpected data exfiltration, detecting and removing malware from incoming traffic, and protecting users of the network from being spoofed into accessing bad websites on which the adversary has positioned malware.[10]

Reducing the likelihood of advanced attacks on defence networks and combat and logistic information and communications technology (ICT) systems will therefore depend on a combination of good security practices and security education, on one hand, and on intelligence-led defences on the other hand. Offensive responses against an attacker may also need to be deployed, proportionately to an attack (but not necessarily symmetrically and not necessarily in cyberspace). Coupled with the implied threat of using all elements of national power should there be a devastating attack, this means deterrence by denial coupled with deterrence by threatened punishment. In addition to good defensive security, this will require the ability to uncover the cyber attackers' networks, understand their methods, and counter their stratagems while avoiding discovery of the intrusion.

Government and the private sector companies supporting military activity and the wider civil infrastructure will also have to establish highly trusting relationships[11] to allow the sharing of sensitive information about cyber attacks, the cyber vectors used, and information relevant to the identities or national affiliation of cyber attackers to assist in attributing attacks. Any information about attacks and anticipated attacks has to be shared at network speed between the machines patrolling the digital frontiers of the defence and national critical infrastructure cyber frontier, government intelligence agencies, law enforcement, and other actors such as commercial enterprises that need to be forewarned before they suffer the same attack. As discussed later in this chapter, the allegations flowing from the Snowden material initially cast a shadow over many of the relationships that are needed between government and the major internet players.

The intelligence required for effective advanced cyber defence, including insights into the latest techniques that adversaries are using to try to defeat the defence (and will also involve research into the next generation of possible attack vectors), can inform decisions on the offensive use of cyber attacks. Cabinet Minister Phillip Hammond[12] told the Conservative Party conference that the UK was the first advanced country to admit to investing in offensive cyber attack capabilities. Although no details have been released about the Ministry of Defence's work on this subject (assumed to be in close cooperation with GCHQ), it would be a reasonable reading of the UK Government's position that it expects UK Armed Forces in any future major operations to be supported by offensive cyber capabilities.[13] *The Washington Post* first leaked the top secret US Presidential Policy Directive 20[14] calling on America's national security leaders to develop OCEO (Offensive Cyber Effect Operations); that is, destructive cyber warfare capabilities that 'can offer unique and unconventional opportunities to advance U.S. national objectives around the world, with potential effects ranging from the subtle to the severely damaging'. Since then published US Cyber Strategy has incorporated the concept of offensive cyber as a key capability for modern warfare.[15] Globally, cyber weapons will undoubtedly form part of the armoury for future conflicts, as we have already seen with the use of cyber means to disable air defence systems before the invasion of Iraq and, it is claimed, before the Israeli attack on Syrian nuclear facilities.[16] The existence of such capability is also likely to have some deterrent effect on an adversary. To cite the words of the US statement on space deterrence, what is needed is 'Being prepared to respond to an attack on U.S. or allied space systems proportionally, but not necessarily symmetrically and not necessarily in space, using all elements of national power'.

Cyber operations to support military activity take careful preparation. Thomas Rid[17] has written of the relationship between the amount of damage that is intended and the extent to which the attack can be very specifically targeted, and that in turn must relate to the texture and detail of the intelligence available on the target. This general statement is likely to apply whether the intended target is an adversary's air defence system, battlefield communications, or air to ground communications, or is the dislocation of some rear area logistic system supporting the frontline or function that is crucial to mobilisation. The intelligence support must be sufficiently precise, for example, to allow assessment that if used the cyber weapon will comply with the requirements of international law including discrimination and assessment of civilian casualties.[18]

Now it is possible that the intelligence requirement could be obtained by conventional means, for example the defection of a key technician involved in the design or programming of the system to be attacked, or with the cooperative support of a company that exported the system to an adversary, or by human intelligence methods including the theft of handbooks and schematics. But even with such support it is very unlikely that digital intelligence attained through attack on the networked communications of the target will not be needed.

Another, rather different, area of difficulty is the exploitation of previously unknown flaws in software. Pressure from industry, from internet gurus, and from the media means that there will have to be a much tougher process of 'equities' balancing the value of having a library of exploits to be used when the need arises and the wider risks to internet security from others' less well-intentioned discovery of the same flaws.[19] It would be inconsistent with the primacy of UK cyber security strategy to condone intelligence actions that deliberately weakened the systems on which essential internet activities rest or that, as acts of omission, allowed weaknesses to remain unreported, especially relating to systems carrying financial transactions, although this is an allegation that has followed the Snowden revelations.[20]

The US Administration has addressed this issue specifically,[21] as has the UK,[22] to stress the importance of having rigorous processes in connection with such a vulnerability disclosure that is able to answer questions such as:

- How much is the vulnerable system used in the core internet infrastructure, in other critical infrastructure systems, in the economy and/or in national security systems?
- Does the vulnerability, if left unpatched, impose a significant risk?
- How much harm could an adversary nation or criminal group do with knowledge of this vulnerability?
- How likely is it that we would know if someone else was exploiting it?
- How badly do we need the intelligence we think we can get from exploiting the vulnerability?
- Are there other ways we can get it?
- Could we utilise the vulnerability for a short period of time before we disclose it?
- How likely is it that someone else will discover the vulnerability?
- Can the vulnerability be patched or otherwise mitigated?

To use a rough historical parallel, the UK Chiefs of Staff and their Joint Intelligence Committee during the Second World War had to evolve processes to examine the 'equities' involved in resolving comparable conflicts, but with earlier technology, between the value of allowing targets to continue to broadcast in order to gain valuable signals intelligence and the direct military value of destruction of the targets by RAF bombing. The essence of the issue over so-called zero day vulnerabilities and whether they should be exploited or revealed to be fixed is not therefore new to students of intelligence and military studies.

A deliberate process is certainly required that is biased towards responsibly disclosing cyber vulnerabilities when discovered. When organisations such as the Ministry of Defence and GCHQ have to weigh up a vulnerability in terms of keeping it for future offensive use versus the defensive value of getting it fixed, if it is a close call, then defensive should always tend to win. That is not an ethical judgement but a practical military one. A defence being breached with certainty is much more serious than losing the future hypothetical value of an offensive tool. The priority will be getting the defence fixed so that confidence in the internet is retained.

Another consideration that defence planners in the MOD will need to weigh is that every attack actually launched raises the risk that the latest techniques will be revealed to the cyber analytic world[23] and is an invitation to others to copy them and for their relevant defences to be tightened. So even a successful attack against target A in country B may well lead to the more important potential need in the event of major conflict for attack on target X in country Y being negated. That consideration may well argue for great restraint in the Western use of offensive cyber so that the techniques – and the intelligence access on which they may depend – can remain undetected until really needed for strategic effect.

The privacy critique of bulk access to digital data

As described above, the demands in all three areas of interest to military intelligence require that access be sought to digital communications and data in order to help provide the intelligence required on the capabilities of potential or actual adversaries; intelligence about the individuals who may pose the threat; and intelligence needed for the development of digital cyber defences and cyber weapons. It will be apparent that the techniques that can be employed to obtain such digital intelligence from communications – data in motion – and from analysis of digital databases – data at rest – are in their fundamentals the same as those that have to be employed by national intelligence agencies to support other aspects of national security such as domestic counter-terrorism, counter-subversion, counter-proliferation, and serious criminal investigations by law enforcement.

It is the potential for these powerful digital tools to access information about the general public – what has wrongly been categorised as 'mass surveillance' – that was highlighted by the coverage given to the documents stolen by Edward Snowden from the US National Security Agency and the UK's GCHQ.

For the US the most controversial revelation[24] was the exposure of a huge NSA collection and storage programme of all Americans' phone call data, secretly authorised under s.215 of the Patriot Act. For the Federal Government to amass such information on US citizens struck Snowden as unconstitutional (the matter remained contested). What made the issue more toxic was that the full details of the programme were allegedly kept from the Congressional Oversight Committee members who heard an apparently misleading answer in testimony by Director of National Intelligence Jim Clapper, which subsequently had to be corrected.[25] President Obama and senior officials were obliged to release much new information about such activities and to restrict their application.[26]

Snowden certainly therefore lifted a big lid on many aspects of US (and UK) digital capabilities to collect and analyse communications data. Snowden of course stole and gave to journalists a mass of information taken from the NSA (a figure of 1.7 million documents has been quoted) and 58,000 documents from a GCHQ internal wiki that had been mirrored to the NSA. The journalists who exploited the material highlighted many sensational aspects of digital sources and methods of gathering and processing intelligence as well as active methods for 'close access' to the devices and computers of targets or tracking their activity across the internet. In the case of the UK they did not always understand the legal framework[27] under which such work is carried out, nor take into sufficient account that many of the same techniques are used for intelligence purposes that have nothing to do with domestic surveillance, such as support for military operations, and thus that publicising intelligence methods can do a lot of inadvertent damage.

In order to examine more closely how these adverse consequences for defence arise, it is helpful to consider in turn three levels or layers of human activity on the internet: the need for everyday security, the needs of law enforcement, and the work of national intelligence including in support of the armed forces. The impact of the Snowden relations themselves and of restrictions aimed at safeguarding everyday privacy at the top level will be seen potentially to impact both law enforcement and counter-terrorism and national intelligence and thus the support available to defence.

Layer 1: everyday activity on the internet

The first top layer comprises our everyday activity on the internet: communicating, sharing, entertaining, and trading. This is the level at which data protection legislation, both national and European,[28] kicks in to try to protect the personal data of citizens from unauthorised use. Internet applications have privacy settings that the user can activate; and there are encryption options available to secure devices by pin numbers and passcodes or to add additional levels of protection, for example in confirming identity in multiple ways before making financial transactions.

This everyday level of internet activity is under constant attack from cyber criminals. We do, as a consequence, need good cyber security practices and very secure encryption in everyday communications to protect our privacy and intellectual property and defeat the cyber criminals. Retaining confidence in the internet and its financial systems and transactions is fundamental for economic well-being.

The policy priority for successive British governments has been clearly stated to be the economy.[29] We rely on the internet for our everyday lives. Our economy – especially the financial sector – would not function without it. Our future economic prosperity will rely on tapping the creativity inherent in the way the internet is an open-source domain. The development of new digital services, for which good public key cryptography is essential, will power economic growth, which is why such importance is being given to retaining multi-stakeholder governance of the internet, including a strong technical voice for the industry, and not encouraging those who want to Balkanise the internet or to increase state control over its development.

The British government is pursuing in the national interest an active cyber security strategy, with GCHQ's cyber security arm, the National Cyber Security Centre (NCSC), in the technical lead, and is spending an additional £860 million on major cyber security programmes over the next few years.[30]

Also communicating via the internet on that same everyday top level are others who show hostile intent towards UK interests or that seek criminal gains at public expense. These are the dictators, terrorists, insurgents, proliferators, narcotics gangs, criminal groups, and people traffickers that are the targets of digital intelligence, not to mention the Russian paramilitaries in Ukraine and violent ISIL jihadists in North Africa and elsewhere. For the future, as previously noted, to this list can be added the adversaries of all likely military interventions and operations, including service-assisted evacuations of British nationals from trouble spots and hostage rescue.

There is also the dark net, beyond the indexing of Google, and accessible only using anonymisation software such as TOR.[31] On websites in the dark net jihadist beheading videos are circulated, and extremists of all kinds communicate. On the hidden markets that TOR and comparable programs make accessible can be found for sale weapons of all kinds, counterfeit goods, malware, drugs, sex, and slaves. With payments made in the untraceable cyber-currency Bitcoin, there is a very low overall probability of detection by the authorities of an illegal transaction.

All of the everyday internet traffic, legitimate and illegitimate, is passed from sender to recipient via the multiple interconnected networks that use the internet protocol with individual components of any communication (be it data, text, photograph, video, etc.) broken down into small packets each with a header describing the origin and destination and that are directed automatically by servers to take at each stage the less crowded/cheapest route. The packets making up a single communication may indeed travel down different routes before being reassembled at the destination server. It is in the nature of these global interconnected internet networks, therefore, that the major communications channels connecting the networks, such as fibre-optic cables, satellite links, and microwave links, are carrying a mix of packets of data of all types and users. The traffic will also include data passing to and from 'Cloud' services, including users accessing programs (such as language translation) too large to fit on their own devices such as mobile phones and tablets. It is the access that NSA and GCHQ have to these internet bearers that attracted the ire of Edward Snowden and his supporters since that opens up for the critics the spectre of large-scale invasion of the personal privacy of everyday communications and interactions. The access is, as argued in this chapter, needed for the very different purposes of trying to obtain the communications of legitimate targets, not least for law enforcement and military operations.

Layer 2: law enforcement activity on the internet

Underneath, supporting this everyday activity and trying to police the worst abuses on the internet, such as paedophile images, is a law enforcement layer. To protect society, the police have the right to obtain information about the patterns of internet (as well as traditional) communications of suspects, terrorists and criminals of all sorts, under conditions that society legislates and oversees. When necessary and proportionate, the police have the right to seek warrants to be able to access the content of those communications. Whether it is trying to locate a missing schoolgirl, test an alibi, or uncover a terrorist assassination plot, the police say that access to communications data is the most important investigative tool they have. Communications data has been used as evidence in 95 percent of all serious organised crime cases handled by the Crown Prosecution Service. The British Home Secretary has said that it has played a significant role in every MI5 counter-terrorism operation over the last decade. Comparable opinions would be expressed by intelligence officers supporting recent operations in Afghanistan and other theatres.

At the law enforcement level, there is regulated international cooperation, for example through advance passenger information and watch list data exchanges and liaison on suspects through Interpol and Europol. But there are growing and serious problems for the law enforcement level since the rapidity of the growth of crime on the internet is running well ahead of the capabilities of law enforcement, with criminals both using internet technology directly through malware and by using the internet simply as a more efficient way to conduct traditional crimes such as fraud, but at increasing scale.[32] The tools or exploits for cyber crime can be bought from hacking specialists so those conducting cyber crime no longer need to be software hackers themselves; and the most serious cyber criminals are based in jurisdictions overseas where Mutual Legal Assistance requests and European arrest warrants may not be respected.

More fundamentally for the conclusions of this chapter, the advent of digital technology is making the task for the authorities of obtaining communications data and warranted communications much harder.[33] Even where they wish to cooperate, the traditional telecommunications and cable companies are increasingly physically unable to respond to legal warrants and provide the information to which the authorities are legally entitled since they have no business need to collect or retain information about their customers' use of digital services that are free at the point of use and are covered by a flat rate subscription. Many of the modern internet service providers (ISPs) that do have the data are located overseas. The US internet service providers for example apply their own company judgements whether to provide British authorities with communications data (for which they have obtained UK legal authority to seek).

Another area of difficulty that affects both law enforcement and the armed forces on operations is gaining access to encrypted material on suspects' computers and mobile devices on which vital evidence may lie, for example in discovering terrorist networks or uncovering areas that have been surveilled for attack such as security forces bases.

Finally, there is the sheer diversity of means of hiding communications using the multitude of apps, social media, or even online video games at which terrorists and insurgents have become adept as they learn more (not least from the Snowden revelations) about the capabilities that can be brought to bear against them. Again, Snowden did the armed forces as well as law enforcement a disservice through the publicity that has been given to digital intelligence capabilities, even when misunderstood by the media, with the effect of pointing to the limitations as well as the length of the reach of the authorities.

Layer 3: intelligence activity on the internet

Faced with these problems the national intelligence community has increasingly been relied upon for assistance on digital intelligence support to the armed forces and to law enforcement (following a traditional role in support of the armed forces, but in respect of law enforcement was one that was specifically provided for by Parliament in the 1989 Security Service and 1994 Intelligence Services Acts).

It is important, therefore, to recognise that there is a third layer of activity on the internet, until recently largely cloaked in secrecy or at least decent obscurity, which is the work of national intelligence agencies. Their main role has been and remains national security, including supporting the armed forces and diplomacy, countering proliferation and uncovering state-sponsored cyber attacks for which they have developed sophisticated means of electronic monitoring.

In the UK, at least, they do also have by law an important and legitimate role in responding to requests from the police to detect and prevent terrorism and serious crime. For example, the intelligence level is providing invaluable help in countering the threat represented by violent jihadis returning from Syria and Iraq intent upon conducting attacks in the UK and elsewhere in Europe. Another example is the assistance that the intelligence agencies have given in managing the advanced, persistent cyber threat from other states intent upon intellectual property theft; identifying some of the worst criminals operating in the dark net, for example through 'watering hole'[34] attacks on those visiting child abuse sharing websites; as well as supporting law enforcement operations conducted by the armed forces such as naval counter-narcotics patrols, counter-piracy, and embargo monitoring.

Traditionally, this third, intelligence, level was hidden from everyday sight, unavowed and largely unregulated. It was part of the 'secret state'. Remarkably, the UK, in advance of most European partners, decided over 20 years ago to legislate for its intelligence agencies and to impose the same basic regulatory regime for intrusive investigative activity as applied to law enforcement.[35] Thus in the UK the paradigm shifted to 'the Protecting State'[36] in which it was recognised that intelligence is a legitimate function of government and one that has to be regulated by law, despite its unique characteristics and its need for secrecy over sources and methods. It is also worth recognising at this point that not all countries adopted that model. Comparable publicity to that Snowden has generated on the US and UK did not emerge on those countries that hold to the old Secret State model, including Russia and China with their digital intelligence activity around the globe that certainly does not have the legal safeguards, ethical constraints, and oversight exercised in the US or the UK.

A key aspect of the UK approach is that legislation specifies in black letter law that the vital principles of proportionality and necessity must be applied to UK intelligence activity (including relevant military intelligence work). Independent commissioners (very senior retired judges) check on legal compliance with the relevant Acts, especially the IPA 2016, including activity carried out for defence purposes. There is Parliamentary oversight by the Intelligence and

Security Committee of Intelligence Agency policies and operational activity, and the committee can also examine the work of defence intelligence for national purposes. In 2013 the committee was given greater powers of access to the agencies.[37] The former Interception Commissioner, Sir Anthony May, a former Court of Appeal judge, has reported publicly that everything GCHQ was doing to access digital communications was properly authorised and legally justified including under Article 8 of the European Human Rights Convention regarding personal privacy.[38]

UK intelligence activity continues to benefit greatly from the longstanding, close relationship with the US NSA (and equivalent signals intelligence agencies in Canada, Australia, and New Zealand). The Interception Commission also confirmed that there is no jurisdiction hopping by GCHQ to get these agencies to carry out tasks that UK legislation would prohibit, or for which UK analysts have not obtained the necessary legal authority.

The UK intelligence agencies accept all of these restrictions and oversight as the price for public acceptability of their role in a modern democracy and as an inevitable consequence of the use of their advanced digital methods to support national security and the missions of the armed forces while helping to maintain the safety and security of the public.

In this context it is also relevant to consider where there are particular differences in the US and UK legal situations covering access to digital intelligence. The Snowden affair revealed at least three areas of potential difficulty for digital intelligence gathering in circumstances where US and UK forces are operating together.

The first is that the US may take a broader view of the retention of data in bulk for future discrimination and analysis. The UK approach is more tightly defined to necessity for the time it takes to complete a task of identifying relevant material for analysis, and it would not allow unsorted information to be retained just in case one day, for some mission as yet unspecified, it might come in useful.

A second difference relates to the legal distinction in the IPA 2016 between the content of a communication and communications data (the digital equivalent of who called whom, when, and where that allows packets of data to be directed between servers). There is a broader term, 'metadata', not used in UK law but in common media currency, especially in the US, that is used to describe not just communications data but also such information available from digital sources as an individual's browser history. The UK definition in the IPA 2016 is much narrower than this commonly used US metadata expression and this is recognised in requiring a higher level of authority in the UK to examine such matters as internet browsing history or digital address books accessed on mobile devices, both of which examples would count as 'content', not communications data.

A third difference is to be found in the way that territorial authority is expressed. For the US, under the Constitution's 4th Amendment, all US persons have their property protected (wherever they are in the world) from unreasonable search and seizure including their communications. But foreigners in the US have a lesser degree of protection, just as they do outside the US. For the UK the criterion is geographical: when in the British Islands all enjoy the same Article 8 Human Rights privacy protection regardless of nationality. The UK thus treats as domestic communications those between individuals in the British Islands, covering the domestic communications of UK nationals, those with the right to remain in the UK, and foreign visitors on temporary visas as well as EU citizens exercising their rights of free movement. Thus when intercepting external (overseas) communications[39] UK authorities do not have to consider the nationality of those who may be intercepted in the way in which US agencies must have regard as to whether their target overseas is a US person. The IPA 2016 recognises, however, that when intercepting digital packet switched global networks, some domestic traffic may also be intercepted by UK authorities when seeking external communications. But the safeguards in the

Act cover this eventuality and provide a practical and effective way of governing the intrusion into privacy of people in the UK, whether the communications are internal or external.

National and industry responses to the revelations

President Obama set up a Review Group to examine US signals intelligence practice in the light of the Snowden allegations, and to make recommendations, that reported in December 2013.[40] The President's response[41] included, for the first time, making public his Presidential Decision Directive (PDD-28) to govern US signals intelligence activities at home and abroad. The Directive explicitly recognised that the evolution of technology had created a world where those communications important to our national security and the various communications that all of us make as part of our daily lives are transmitted through the same channels. This presents new and diverse opportunities for, and challenges with respect to, the collection of intelligence – and especially signals intelligence. Locating new or emerging threats and other vital national security information is difficult, as such information is often hidden within the large and complex system of modern global communications. The directive is clear that the United States must consequently collect signals intelligence in bulk in certain circumstances in order to identify these threats, including threats to US or allied armed forces or other US or allied personnel.

The media accounts from the Snowden material did create significant diplomatic embarrassment for the United States over alleged spying on friendly nations, including the mobile telephone of German Chancellor Angela Merkel and the communications of the Brazilian President. In his directive referred to above, President Obama when in office made clear to his intelligence community that the US will not monitor the communications of heads of state and government of close friends and allies unless, he added, there was a compelling national security purpose. Although attempts were made by the German government to turn this into a binding agreement with the US administration, this was rebuffed. So at the secret intelligence level there are never going to be binding 'no-spy' agreements.

There is in fact no *international* law regulating the intelligence level; and there never will be since all nations engage in it, there is no agreed definition of what it is, and few will admit fully to it. Every nation of course makes intelligence activity directed against it an offence. Some nations have bilateral or multilateral agreements governing the handling of classified intelligence material among them (such as applies to NATO member states).

There is a growing international recognition that military operations cannot responsibly be conducted without intelligence support, both for the prosecution of the mission and for the self-protection of the forces concerned, a lesson that the United Nations has been learning since the Bosnian war in the 1980s. The UN had traditionally shied away from the term 'intelligence' on the grounds that it carries the implication that one member state must have spied on another, something that the UN institutionally has had difficulty accepting. Nevertheless, after 9/11, the UN Security Council passed Resolution 1373 accepting the value of intelligence in combating terrorism. It is also of some significance for the future that the UN has organised the flying of remotely piloted aircraft for reconnaissance purposes in support of its mission in the eastern part of the Democratic Republic of the Congo.[42]

The dilemma facing legislatures can be summarised thus. If, on the one hand, the regulation of digital intelligence access, especially using bulk powers, fails to provide sufficient public confidence in the necessary restraint being exercised, then the resulting unease on the part of a vocal section of the public will destabilise the very intelligence community whose work is needed to manage twenty-first century risks. If on the other hand, and it is a risk in some European jurisdictions, parliaments are over-zealous in trying to constrain national digital intelligence-gathering capability, then there will be obvious problems for law enforcement in managing the risks from terrorism, cyber crime and other criminality and threats to the public, and for the provision of intelligence on which sound policy decisions can be made. In addition there will be consequences for the provision of support for military operations.

The response of some nations outside the Five Eyes community, including especially Brazil and Germany, to the Snowden-related accusations that their citizens may have had their communications intercepted by bulk access to the local servers, internet bearers such as cables, and satellite links and from the major internet companies (without their knowledge) has been calls to nationalise their internet clouds and force internet companies to localise data about their citizens in servers on their own territory. But further study of the issue has led to greater understanding that such measures will not advance their own economic and social interests in an open internet by erecting the equivalent of tariff barriers, making it more expensive to offer digital services in that country. Nor in practice can localisation be expected to improve security. Such naive thinking also encourages those authoritarian nations that do wish to fragment the internet to facilitate their own censorship and social control. There are, however, also commercial pressures that may persuade some nations to invest more in domestic digital industries rather than relying on the US-dominated industry. Brazil has announced[43] an intention to have a submarine cable laid across the Atlantic that would mean less of her internet traffic would be carried by the bearers that Snowden alleged were sources of bulk data for the NSA and GCHQ.

The damage done to intelligence work by the Snowden revelations

Although this is disputed by the media outlets that were foremost in publicising the Snowden material, there is growing official evidence that the actions of Edward Snowden created serious difficulties for the authorities, amplified by the way that some journalists chose to report stories drawn from the cache of stolen documents. In the words of the Director of NSA, this has already caused 'irreversible damage for the United States'.[44] For the UK, according to former GCHQ Director Iain Lobban, the Snowden leaks have done 'immense damage to Britain's counter-terrorism efforts'.[45] What has not received coverage, of course, is that any increase in difficulty in conducting digital intelligence work will not just affect the US and its close allies but any nation with an advanced intelligence capability or seeking one (which includes most nations). Insofar as greater cyber security awareness spreads, that is welcome for maintaining general confidence in the top layer of everyday internet activity, as described earlier. But if, as seems likely, the effect is also to shut out law enforcement and specialised digital support to military operations then the effect on the national interest will be increasingly negative.

The most obvious casualties, and those that have received most coverage, are the losses of coverage of terrorists and serious criminals as a result of their greater awareness of the reach of the authorities. Since the same methods are in use to provide intelligence on adversaries in military operations, such as recently in Afghanistan, the impact will also increasingly be felt in defence intelligence circles.

Turning to the kinds of damage done, the first and most obvious problem is that generated by the sheer global coverage of the issues of digital intelligence. There has, for example, been a general increase in awareness of the potential for cyber attack for intelligence gathering by using a network attack, and once inside a network by accessing data at rest. Security is being tightened by users all over the internet including, for example, the violent jihadist terrorists. Similarly, there is added pressure on the global ICT community to identify – and then fix – so-called zero day vulnerabilities in commonly used software, including engineering control software.

Digital intelligence gathering is likely to be seriously impacted by the decision by the internet companies to introduce end-to-end hard encryption in all major communications bearers and between the servers of the major internet service providers, all factors that will make it harder to obtain intelligence from communications and harder to achieve the kind of bulk access needed to conduct a successful network attack on a potential adversary.

Damage was also caused by the Snowden-induced pressure on the communications and internet industry not to be seen to cooperate with Western governments for fear that their commercial reputation in other markets will suffer. We can see how the Snowden affair initially made internet companies feel commercially obliged to minimise their exposure to any cooperation with the government, including legitimate demands from law enforcement and intelligence. The passage of time has lessened that effect, as has the switch of focus of critics to their access to and use of personal data for commercial advertising reasons. Where law enforcement has the 'probable cause' material on a criminal or terrorist suspect to serve a traditional warrant on an internet company, then it may be executed if the material sought on a suspect is readily identifiable. But that would not necessarily cover the discovery of new leads for investigation that could be uncovered by data mining and layering.

Similarly, the device manufacturers have reacted to the publicity from the Snowden affair, wanting to assure their customers that the government cannot access metadata or content from their devices. For example, Apple announced that to give more confidence to its customers it had deliberately written its iOS 8 software to be so secure that the company itself cannot unlock mobile devices protected with a full password. But that means that when the police present a legal warrant authorising examination of, say, a terrorist's or kidnapper's laptop, the company will deliberately have put itself in a position of not being able to help. Whether it is possible to require companies to make access technically possible in response to legal warrants without their having to introduce weaknesses into their software that could be exploited by criminals remains to be seen.[46]

Yet a further level of damage arises from the publicity attached to certain specialised techniques, including those of close access to the target and implants in devices used by suspects and other techniques likely to be of value in the support of military operations.

An indirect risk is that of over-reaction by government, both at home and overseas, with in particular increasing emphasis being given to data protection legislation and higher hurdles being erected for obtaining government access. In policy terms this is where the Snowden revelations are causing the most difficulty. The media, and many parliamentarians, still find it hard to distinguish in their comment between their future fears of 'mass surveillance' of the domestic population on the one hand and on the other hand the continuing necessity of digital intelligence work with its 'bulk access' to the internet required to find suspect communications within the vast volume of digital traffic.

'Mass surveillance' means the placing of the population or a substantial part of it under observation by the authorities. The then UK statutory Interception Commissioner, Sir Anthony May (in the report to Parliament cited earlier) was very clear that mass surveillance of the British population does not happen and that is not how the Snowden material must be interpreted. Instead, there is a twofold logic. The first part concerns the authority needed to be able to search for the most likely traffic streams in which the communications being sought might be found and for which there is a valid legal authorisation, in other words access to the haystack. The second part consists of the application of an algorithmic discriminator, such as a suspect IP address, to pull out the needles, the very small amount of communications data or traffic required to be seen by a human being. Computers are not conscious, and to describe the former process as a mass violation of personal privacy is in today's world not a reasonable line to take, provided that there is respect for privacy rights shown in the design of the algorithms and assurance that only a very small proportion can be seen by a human being, the intelligence analyst. With the passage of the IPA 2016, the UK government finally explained what is involved in meeting legitimate demands for digital intelligence for law enforcement, national security, and support of our armed forces, applying the safeguards of the necessary and proportionate test.

The European Court of Human Rights has always recognised in its judgements that the balancing act that the authorities must seek is within the basket of human rights. Personal privacy of the one can be overridden in the interests of the right to protection of the other, *provided* that a set of conditions are in place: powers set down in legislation; the principles of proportionality and necessity followed; right authority, an audit trail, independent oversight, and independent adjudication of claims of abuse all established; and so on. The UK has legislation that incorporates these safeguards, as well as those of the Human Rights Act 1998 that incorporates the provisions of the European Convention on Human Rights, and is therefore a model for other nations to follow.

The category error described between bulk access and mass surveillance continued to be the source of controversy both in the UK and across Europe. One consequence will undoubtedly be calls for more robust oversight by judicial and parliamentary bodies. That of itself is a good thing if it increases public confidence. It need not affect the ability of the intelligence system to support the police and armed forces when engaged in Military Assistance to the Civil Powers (MACP) operations. But it cannot be ruled out that new legal restrictions will be sought, and that from Brussels and Strasbourg we will see pressure to limit data-sharing, time periods for data retention for analysis, and greater protection for domestic communications including for information that could be derived from social media use. Such developments are likely to impede the flow of intelligence to support MACP operations.

Conclusion

This chapter has highlighted the challenges faced by intelligence communities in meeting the increasing *demands* for intelligence to support public safety and national security alike. The chapter has also identified the potential for the *supply* of intelligence, especially about people, from digital sources including information gathered by private internet companies. Snowden revealed much about the dynamic interaction going on between such demand and supply. But media comment under-reported the major efforts by the governments of the democracies to ensure that their intelligence communities can *continue to achieve their mission while behaving ethically in accordance with modern views of human rights, including respect for personal privacy*, in a world where deference to authority and automatic acceptance of the confidentiality of government business no longer holds sway.

The interaction exposed by Snowden between demand and supply for digital information must be *seen* by the public to be regulated by applying safeguards that are recognised and will give assurance of ethical behaviour, in accordance with modern views of human rights, including respect for personal privacy in accordance with the Human Rights Act Article 8.

The UK armed forces have the great advantage that GCHQ does have sophisticated methods of bulk access by computer to the internet. Those allow carefully targeted, highly discriminating selection of the communications of those who intend harm to our armed forces and our society. Thinking about the interactions of the three layers of internet activity – the everyday, the law enforcement, and the intelligence levels – helps dispel much of the confusion caused by the reporting of the material that Snowden stole but also identifies dangers for the digital intelligence support needed for our armed forces.

The essential need for security for our everyday personal data and transactions to keep out the cyber criminals should not be taken as absolute. The law enforcement layer in protecting society from that very criminality has a legitimate need to access information on suspects, and in the digital era it will increasingly have to turn to the intelligence layer to help. The intelligence layer must have the legal authority to respond, but also the ability to continue to pursue its national security mission that is its primary purpose, including the vital support for the armed forces. The techniques involved in obtaining intelligence on people, be they criminals or terrorist insurgents, are to a large degree the same; and over-zealous measures to protect personal data at the everyday level will have the unintended consequence of harming national security.

Notes

1 The world's oldest signal interception station at Irton Moor, Scarborough, was built to monitor the German Fleet in the First World War. Now part of the Government Communications Headquarters, it celebrated 100 years of continuous service in 2014; see www.gchq.gov.uk/press_and_media/press_releases/Pages/HRH-The-Prince-of-Wales-visits-GCHQ-Scarborough.aspx, accessed 30 Oct 2014.

2 Omand, D., *Understanding Digital Intelligence and the Norms that Might Govern It*, London: Chatham House and Ottawa: CIGI, 2015, available at https://www.cigionline.org/publications/understanding-digital-intelligence-and-norms-might-govern-it

3 A point emphasised by the Director of GCHQ in public evidence to the Parliamentary Intelligence and Security Committee, see www.gchq.gov.uk/press_and_media/news_and_features/Pages/Highlights-of-the-ISC-open-evidence-session.aspx, accessed 30 Oct 2014.

4 For example, following the shooting down of the Malaysian airliner over Eastern Ukraine on 17 July 2014.

5 See the analysis by the Centre for the Study of Radicalisation and Political Violence, King's College London, at http://icsr.info/2014/04/icsr-insight-inspires-syrian-foreign-fighters/, accessed 30 Oct 2014.

6 D. Omand, J. Bartlett, and C. Miller, 'Introducing Social Media Intelligence', *Intelligence and National Security*, Vol. 27, No. 6 (December 2012).

7 Under the Regulation of Investigatory Powers Act of 2000 (RIPA 2000), Part I, s.8(4).

8 By Matthew Aid, a knowledgeable author who has written extensively about the NSA; see www.matthewaid.com/post/67998278561/greenwalds-interpretation-of-boundlessinformant-nsa, accessed 30 Oct 2014.

9 Investigatory Powers Act 2016, UK Parliament, available at http://www.legislation.gov.uk/ukpga/2016/25/contents/enacted

10 www.google.co.uk/webhp?sourceid=chrome-instant&ion=1&espv=2&ie=UTF-8#q=US%20statement%20on%20space%20deterrence, accessed 30 Oct 2014.

11 See www.gov.uk/government/news/government-launches-information-sharing-partnership-on-cyber-security, accessed 11 November 2014.

12 As Defence Secretary; he was appointed Foreign Secretary in 2014.

13 UK National Security Capability Review, HMG March 2018, available at https://assets.publishing.service.gov.uk/government/uploads/system/uploads/attachment_data/file/705347/6.4391_CO_National-Security-Review_web.pdf

14 www.washingtonpost.com/world/national-security/secret-cyber-directive-calls-for-ability-to-attack-without-warning/2013/06/07/6a4cc762-cfc0-11e2-9f1a-1a7cdee20287_story.html, accessed 30 Oct 2014.

15 US National Cyber Strategy, Washington DC, September 2018, available at https://www.state.gov/r/pa/prs/ps/2018/09/286093.htm

16 For example by *The Register*, see www.theregister.co.uk/2007/11/22/israel_air_raid_syria_hack_network_vuln_intrusion/, accessed 30 Oct 2014.

17 T. Rid, *Cyberwar Will Not Take Place*, London: Hurst, 2014.

18 NATO guidance in the *Tallinn Manual on the International Law applicable to Cyber Warfare* makes clear that cyber attacks must be compliant with international humanitarian law in the same way as any other weapons system; available at www.ccdcoe.org/tallinn-manual.html, accessed 30 Oct 2014.

19 Statement by Caitlin Hayden, 12 April 2014, see www.nytimes.com/2014/04/13/us/politics/obama-lets-nsa-exploit-some-internet-flaws-officials-say.html?_r=0, accessed 28 Oct 2014.

20 Tim Berners-Lee has made this claim; see www.channel4.com/news/berners-lee-internet-spying-privacy-data-snowden, accessed 30 Oct 2014.
21 www.whitehouse.gov/blog/2014/04/28/heartbleed-understanding-when-we-disclose-cyber-vulner abilities, accessed 30 Oct 2014.
22 The UK Equities Process, Nov 2018, London: NCSC, available at https://www.ncsc.gov.uk/news/gchq-and-ncsc-publish-uk-equities-process
23 The STUXNET attack on the Iranian nuclear centrifuges at Natanz was reverse-engineered and the results published in detail, allowing techniques to be imitated in other attacks, see http://spectrum.ieee.org/telecom/security/the-real-story-of-stuxnet, accessed 30 Oct 2014.
24 www.dni.gov/index.php/newsroom/press-releases/191-press-releases-2013/869-dni-statement-on-activities-authorized-under-section-702-of-fisa, accessed 30 Oct 2014.
25 www.dni.gov/index.php/newsroom/press-releases/191-press-releases-2013/889-dni-clapper-letter-misunderstandings-arising-from-march-12th-appearance-before-the-senate-select-committee-on-intelligence, accessed 11 November 2014.
26 www.intelligence.senate.gov/130926/joint.pdf, accessed 30 Oct 2014.
27 Snowden was a systems administrator, had never been an intelligence analyst, and had no knowledge of the development of UK intelligence and law enforcement practice and regulation.
28 The centrepiece of existing EU legislation on personal data protection, Directive 95/46/EC[1], was adopted in 1995 with two objectives in mind: to protect the fundamental right to data protection and to guarantee the free flow of personal data between Member States.
29 Update on the UK National Cyber Security Strategy, see www.gov.uk/government/publications/national-cyber-security-strategy-2-years-on, accessed 28 Oct 2014.
30 UK Cyber Security Strategy, Nov 2016, London: HMG available at https://www.gov.uk/government/publications/national-cyber-security-strategy-2016-to-2021
31 TOR (the onion router) is a free software program for enabling anonymity online and for evading censorship.
32 Detail can be found at www.europol.europa.eu/content/megamenu/european-cybercrime-centre-ec3-1837, accessed 30 Oct 2014.
33 The issues are well set out by FBI Director Comey in his Brookings Institute speech of 16 October 2014; see www.brookings.edu/blogs/brookings-now/posts/2014/10/fbi-director-james-comey-technology-law-enforcement-going-dark, accessed 30 Oct 2014.
34 For example, where those visiting certain websites have their devices infected with a 'cookie' that enables their subsequent internet history to be tracked.
35 Through the Interceptions of Communications Act 1985 and later RIPA 2000.
36 P. Hennessy, 'From secret state to protective state', in P. Hennessy (ed.), *The New Protective State*, London: Continuum, 2007.
37 Explained at www.legislation.gov.uk/ukpga/2013/18/notes, accessed 30 Oct 2014.
38 Sir Anthony May, Annual Report, 8 April 2014, available at www.iocco-uk.info/ *Liberty and Security in a Changing World* (Washington: White House, 12 December 2013).
39 Defined in RIPA 2000 as a communication sent or received outside the British Islands.
40 *Liberty and Security in a Changing World* (see note 38 above).
41 President Obama, speech at the Deparment of Justice, Washington, 17 Jan 2014, see www.whitehouse.gov/blog/2014/01/17/president-obama-discusses-us-intelligence-programs-department-justice, accessed 23 May 2015.
42 www.un.org/apps/news/story.asp?NewsID=46650, accessed 30 Oct 2014.
43 http://euobserver.com/justice/123260, accessed 30 Oct 2014.
44 Reported at www.theguardian.com/world/2013/jun/23/nsa-director-snowden-hong-kong, accessed 30 Oct 2014.
45 Reported at www.bbc.co.uk/news/uk-25937478, accessed 30 Oct 2014.
46 The suggestion has been floated by the Director of the FBI in his Brookings speech (cited above, see www.brookings.edu/blogs/brookings-now/posts/2014/10/fbi-director-james-comey-technology-law-enforcement-going-dark), but has been subject to heavy criticism.

10

THE AMBIGUITIES OF CYBER SECURITY

Offence and the human factor

James Gow

In the summer of 2018, the insurance industry began to acknowledge a new reality. Perhaps two years before that, at the earliest, it did not recognise cyber terrorism. Prior to the 2017–18 period, when change emerged, if a terrorist organisation blew up a building using a conventional explosive, the building would have been covered by the owner's terrorism insurance. But, if the terrorists had blown up the building by hacking into it and causing it to blow up by cyber means, that was not covered. Each incident would have been a blow struck by an armed political-military movement, but only one of them would be recognised as such. By June 2018, it was evident that insurers had shifted their ground and recognised that cyber attacks had to be covered – even if there were still uncertainties about whether cover should only be restricted to consequences involving physical damage.[1] This is an example of how the whole world is trying to come to terms with this new technology and the issues to which it gives rise. Slowly, understanding was emerging that cyber attacks had to be understood to be a part of conflict spectrum in contemporary warfare.

As seen elsewhere in this volume, even where there was some acceptance of the emergence of cyber weapons, this remained limited. There had been considerable interest and concern about cyber attacks, and a whole body of thought arose around attribution. Much of that discussion concerns the ambiguities that make cyber warfare a disputed notion and obscure the use of digital means in conflict, as well as making the identification of actors and the classification of their actions problematic. The debate surrounding the emergence of the cyber dimension in relation to the laws of war has been largely conducted in traditional terms – the use of force, the right to self-defence and so on.[2] Amid the discussions about how existing law might be applied in the domain of cyber warfare, or where new law might be needed even, there was little sense, however, of two highly important factors, discussed in this chapter. The first of these is the relationship between offence and defence in cyber warfare. The second important aspect concerns the human factor – the reality that, for all the legal and technical discussion that there might be, the key to cyber success or failure largely rests at the level of the individual. These two factors are discussed below. The first two sections discuss cyber attacks and the problem of responding, with the second section considering the importance and problems of sharing information as a response to attacks. The final section considers the human factor.

Cyber attacks

Responding to cyber attack is far from straightforward. For those with the resources and technical knowledge and understanding, it is relatively easy to have cyber weapons and to deploy them gradually. Because of graduated and ambiguous attacks, response is awkward. Even if an attack is an attempt to pull down a whole country's infrastructure, the response is not an easy question to solve. While countries such as the UK have said publicly that a complete infrastructure attack would be considered to be an act of war, what that really means in terms of action in self-defence is far from evident. An attack on a lesser level, which only seeks to try to slow down the system and cause discomfort, not necessarily death, makes the issue of how to deal with the attack that much more complicated. The official position on how to respond for the Americans, the British and most European countries is to say that it will be treated like any other attack. This means that any response will not necessarily be 'in kind' – which is to say that the response to a cyber attack will not necessarily involve a return cyber attack. The response could be enhanced sanctions, or anything up to traditional warfare. The reality is that it is not easy – something that makes Western countries struggle, while those on the front foot to cause disruption, such as Russia, take advantage.

Two major infrastructure attacks in 2017 illustrate the problems in handling an attack: the global WannaCry attack that was noted, in particular, for hitting parts of the British National Health Service in the UK, and the later NotPetya attack.[3] The WannaCry attack was probably the nearest to a Tier 1 cyber attack in the UK classification of attack scheme that there had been. Indeed, although officially, it was not a Tier 1 attack and no attack at that level had, at the time of writing, been judged to have occurred, it is arguable that, in retrospect, WannaCry was probably an attack of sufficient degree that it should have been classified as Tier 1 and should have been treated at that level.[4]

One part of the response had to be to contain the damage caused. Another was to find out who was responsible. Both of these had to come before any other action, including forceful responses of any kind. In the case of WannaCry, the view was that the DPRK was responsible. North Korea launched the attack and it was a ransomware attack, which is an attack where the targeted data is encrypted and the attacker demands an amount of Bitcoins to decrypt the information and restore normal use – in effect, cyber ransom. The official position is never to pay the ransom demanded.

WannaCry exemplifies well why that is a sound position. When the UK forensically analysed the coding, it was impossible to decrypt it. So, even if the ransom had been paid, the North Koreans did not have the capability actually to undo that which had been done and to decrypt the data. In effect, the DPRK tried to blackmail the UK – at least, the NHS – but did not have the capability to accept the rewards and to meet their proposed bargain, had the deal been met. This was perhaps a function of the North Korean polity, where if the leader asks if an attack can be launched, those involved might not be brave enough to point out that it is not fully ready yet – it can be launched, but not in such a way that the goal can be achieved.

The DPRK used a stolen American exploit – that is, a piece of software, a cluster of data, or a coding sequence designed to take advantage of a weakness in a system. This exploit, known as Eternal Blue, was used to spread WannaCry (and also NotPetya). Eternal Blue was part of the leaks that came from the US NSA (National Security Agency). One week, Eternal Blue was a top secret NSA indoctrination tool to spread malware; that is, to spread viruses. A week later, it was available on the internet, as a business proposition.

The DPRK's attempt to use this exploit went out of control. In large part, this is because they also used a different kind of ransomware to encrypt the virus, thereby creating an impossible combination that could not be undone. The first was part was immensely sophisticated, and the second was immensely basic. As they blended the two together in that way, it ended up with an attack – at least an event that was perceived to be an attack, though the North Koreans probably only saw cyber crime as a way of generating funds for their government. The DPRK had already behaved in the same way previously, carrying out the Bank of Bangladesh SWIFT attack, in which they attempted to steal USD 1 billion. They got USD 70 million and transferred it to the Philippines, where they were able to take the cash. But the attack was spotted and the spoils limited.

Identification

Working out who committed a cyber attack can be a major challenge, given the possible ambiguities involved (as discussed in chapters 6, 7 and 8). Identification and attribution in relation to an attack take time. While the first priority is to stop it and prevent the effects from getting worse, very quickly there is a need to try to identify who is responsible. This can be intensely frustrating and certainly can complicate any response – to begin with, responding to an unknown attacker is hard. It is possible to use inference to make initial progress. Part of this can be done because there are commonalities. So, for example, in banking, if there are similarities in the coding signature between the attack and the North Korean one on Sony pictures, then there are grounds to suppose that the attack emanated from the DPRK. Or, if the Ukrainian power system goes down at a time of heightened tension between Russia and Ukraine, it is unlikely that North Korea will be responsible for the attack and also far more probable that the source was Russian. However, this sense of probabilities and that which is likely is no basis on which to launch an attack – certainly not in terms of international law. Proof would be necessary – at least to the satisfaction of the government authorising the riposte.

There is, of course, an important distinction between attribution and proof. Even if there were a high degree of confidence about the source of an attack, the authority in question would be highly unlikely to release the evidence that it had. This might still leave it sure enough to be ready to act, asserting its certainty on the basis of secret information. However, the inability to put evidence into the public domain might also constrain the scope for action. This issue is illustrated by the Kaspersky case.

Kaspersky was a piece of Russian-origin anti-virus software. It came to be understood as a weapon in the US and the UK. An apparently legitimate anti-virus programme was inferred to have links to the Russian Federation by, first, the US government, and then the British government. In part, however, the hostile nature of this commercial software emerged most clearly when Nghia Hoang Pho, a long-serving contractor at the NSA, took his work home (against Agency regulations and US Federal law).[5] He was working on designing malware code for American attacks. He had Kaspersky installed on his home computer. Once anti-virus software is installed on a computer, the company providing the software can suck up everything on the computer. Although most companies only deal with potential viruses and behave ethically, there is a considerable degree of trust involved when installing and using this kind of software. In the Kaspersky case, however, extracting everything is exactly what happened. The coder was subsequently prosecuted and convicted, receiving a prison sentence.

Kaspersky had sucked out a lot of American codewords and coding, it was evident. But the US did not, and almost certainly never would, give details, or release the proof, although unofficial details appeared in *The Wall Street Journal*.[6] The owner of Kaspersky denied that anything had been shared with anyone else – including Russia's various security agencies – and asked anyone with relevant information that the company's systems might have been exploited to 'responsibly to provide ... verifiable information'. In his version of the events, which was largely consistent with the detail available in the original *Wall Street Journal* account, aside, curiously, from the dates involved, on the crucial matter of NSA capabilities, Eugene Kaspersky said that an analyst in his company had flagged up malware that turned out to be new variants of previously identified NSA hacking tools, which, as possible malware, was uploaded to the Kaspersky lab for analysis and when it was identified and reported to him, 'the decision was made to delete the archive from all the company's systems'.[7] However, from the initial US reporting, it would seem likely that the breach was actually detected, one or two years after it had occurred, because the US had identified Russian activity using its malware with high-level authority.

The loss of technical capabilities, such as the hacking tools involved in the Kaspersky case, can be a big blow. This could be seen more clearly in an example involving Wikileaks. When Wikileaks started, the NSA had a program to do coding in Arabic and other languages. When, in 2015 a French TV station, TV5Monde, was taken down, it was initially attributed to something called the 'Cyber Caliphate' and it looked as though it was a terrorist attack. However, subsequent analysis showed that Russia was responsible, benefiting from information revealed by Wikileaks that had allowed Russia to disguise its attack. The initial ambiguity and misunderstanding about this attack, as with Kaspersky and many other attacks, demonstrated how the use of cyber means clearly lies in the middle of what Clausewitz called the 'fog' of war – the conflict environment in which it is hard to get a clear and definitive picture of the situation while it is happening.[8] It is this fog of obscurity that can make many observers – or, even more, non-observers who are not well informed – sceptical even about the reality of war as it happens around them, because it does not look, or feel, like the familiar idea of combat involving obviously identifiable armed forces using weapons of blast and destruction.

Part of this fog, of course, is the characterisation of activity, as well as its actual nature and intention. The Kaspersky position on the NSA leak, *mutatis mutandis*, was completely plausible. Kaspersky was operating in the way that any large cyber company would – collecting vast amounts of data to enhance the service and products it provides. On the surface – and notwithstanding suggestions that the Russian authorities had benefited from the leak, despite Kaspersky's insistence that nothing was shared – the company was only acting in the same way that other, Western, companies would. From a Russian perspective, this could be seen as similar to the activity of Google in the post-9/11 period, which gathered data and assisted the US authorities – activity that was tempered after the Edward Snowden affair.[9] Equally, from a Russian or Chinese perspective, Google (and other operators) are perceived as the equivalents of Kaspersky, sweeping up information from within the country in question.

A strong capability, such as that of the NSA, cannot provide protection or detection services the whole of the time, but it is essential to security in the age of cyber warfare. The UK may be at the forefront, in this respect, with the work of GCHQ (the Government Communications Headquarters), which has immense expertise and whose staff increased in size significantly in the cyber era, notably after the introduction of the UK Cyber Strategy in 2016.[10] They also have associates who work with private companies and can investigate that which has happened. When large parts of the NHS were closed down by WannaCry, immediately this sort of forensic analysis took place, looking at the kind of signatures on the coding that is on the system and that has caused the problem. This points to the need for partnership and sharing, so far as possible, in handling attacks.

When the National Cyber Security Centre was set up in 2016, it was set up specifically to facilitate the sharing of intelligence between government and the private sector. As the front end of GCHQ, its role was to share intelligence with people who needed it. There can be no point in knowing stuff and not sharing it with those who need to know. If the government discovers that the Chinese have stolen all the plans from Ford Motor Company, then it will have to find a way to advise Ford about the breach and what to do about it.

Internationally, the Five Eyes community (Australia, Canada, New Zealand, the UK and the US) has had a history of sharing. Because of terrorism, the moral imperative to share intelligence is higher and higher. But it is shared in a form where sources are protected. If intelligence is shared, those involved know that they are taking a risk – even for the UK with the US. If someone has political motivation and leaks it, that will make it harder to get such intelligence in the future. Nonetheless, this concern is trumped by the imperative that when it relates to saving lives, it has to be shared. So, a way has to be found to share, while protecting sources at the same time. Most intelligence sharing occurs bilaterally. Law enforcement sharing through EUROPOL is far more multilateral, with everything going into one big pot because everyone needs to know, for instance, that a particular individual represents a threat. The higher-level, geopolitical type of sharing tends to be more bilateral. The UK in Europe is probably the highest calibre, most effective and successful security and intelligence player and has contributed to wider European safety, making it a 'prized partner for other European intelligence agencies'.[11] The security community was very keen for that to continue whatever happened in the wider trade and Brexit negotiations because it was the right thing to do, and believed that this field should not become a political pawn in that wider negotiation.

This also has bearing for the idea of controlling cyber weapons globally. While there are clearly legal dimensions to cyber security and warfare (as discussed in chapters 6–8 and 29–30, where UN-focused initiatives for new international law in the realm of cyber security are discussed), there is good reason to consider the benefits of an arms control and confidence-building approach. Indeed, there is a strong argument that a confidence-building process, involving transparency, sharing, and cooperation, is likely to be more effective than a top-down UN treaty approach. There is evidence of this in the US–China bilateral agreement,[12] with the agreement to try not to steal each other's secrets – with the wording wisely recognising that some of this would happen anyway. The agreement was effectively saying that the scale of cyber espionage had to be cut down from the levels that had seen China embarrassed by having many agents arrested. Not surprisingly, it was felt that this would have no effect. However, in practice, it appeared to change Chinese behaviour and reduce the scale of cyber espionage. That appeared to be a positive effect and a measure that could build confidence.

The human factor and the hierarchy of capabilities

While concern about private – that is, mainly, criminal and terrorist – cyber capabilities is justified, it is at the state level that controls, akin to arms control or the application of law relating to war are really relevant. High-end cyber capabilities remain the preserve of states, as they have the resources and development necessary. This is especially true of the major states that engage in cyber activity globally against each other. But private organisations – criminal ones – are also significant, albeit that they may also be linked to state organisations. This challenges international law and leaves a situation in which municipal criminal law is unlikely to be used.

Certain states have really impressive capabilities in the cyber sphere. The top division, the countries that are recognised as being especially strong in the cyber field – the US, the UK, and Israel – have the most advanced and effective capabilities. The next tier down probably includes the Russians and the Chinese and the Iranians. Following this, the North Korean capability might be classified as close to that of Iran, but it is probably on a par with criminal organisations. Below this tier, it is striking that the next tier down involves organised criminal organisations. However, most of the gangs with capabilities at this level are based in Russia. They operate at scale. They operate like businesses. Their capabilities are almost as good as those of states. The criminal organisations based in Russia are almost certainly working with the acquiescence of the SVR or FSB (Russian State security institutions – Sluzhba Vneshneii Razvedki and Federalnaya Sluzhba Bezopasnosti, respectively) and, quite possibly, even in cooperation with them. Without doubt, these organisations are indulged because the Russian authorities calculate that it does them no harm if criminal cyber operations create problems in the West. From their perspective, this is not a bad thing for Russia. Of course, the attitude would be completely different if the criminal organisations were to start to attack Russian banks. The fact that they do not would seem to indicate both that the gangs understand the situation and that the Russian security apparatus indulges them, at a minimum. The line between state actors and criminal groups has pretty much dissolved, it seems, in Russia and in some other non-Western countries – indeed, countries that are Western adversaries.

China is in a similar position to Russia, in terms of its capability and also in terms of cyber kleptomania – that is, efforts to steal important information, whether state, industrial, or commercial, from Western targets. The Chinese can do this at scale. However, after the US and the UK signed a bilateral Memorandum of Understanding, in 2016,[13] the scale of Chinese stealing of secrets from the industrial sector went down significantly. As the Memorandum was agreed, US Secretary of State, Ash Carter, likened the two countries' joining forces to their cooperation in the Second World War against Nazi Germany (although his suggestion implicitly overplayed the US role in breaking German codes[14]). The allusion cemented the vision of strong joint action, building on the creation of a joint cyber cell the previous year;[15] of a determined allied campaign. The agreement was brief, did not appear to contain a lot on the surface, and could easily be regarded sceptically. However, there was a clear impact on Chinese activity following this (and also the US–China agreement – see above), with the scale of thefts in the industrial sector dropping and Chinese attacks becoming much more targeted.

There is thought to be less criminal fusion in Iran and North Korea. Rather, the Iranians have excellent mathematicians and capabilities. They see cyber attacks and espionage as particularly fair for them to be doing. This is because of the Stuxnet and the Olympic Games programmes that the US launched against Tehran's nuclear programme. Iran quickly retaliated, albeit against Saudi Arabia and not the US, regarding this as a 'just' attack, in effect operating a form of 'just war' reasoning.

The key thing in all of this is the focus is on offensive capabilities. While attention will be paid to blocking attack and protection, the reality is both that offence constitutes the best form of defence and it is offensive use of cyber capabilities that dominates the agenda. It is clear that in cyber security, offence trumps defence. In terms of defending against a cyber attack, the situation is somewhat akin to that of protection against conventional terrorism. Traditionally, in terrorism the task is defensively to stop every single explosion (or other form of physical attack) possible. But that is really difficult to achieve, even if several lines are crossed, or circumvented, concerning privacy and using police-state types of organisational surveillance – no police state appears to have been completely successful in stopping every single attack. The authorities, whether crossing ethical lines or not, have to succeed 100 percent of the time, but, the terrorist only needs to get lucky once to have effect – a member of the IRA (Irish Republican Army – the 'military wing', or terrorist branch, of the Irish nationalist movement) once told a British adversary. In terms of cyber attacks, the same equation applies: defence will not always succeed. However, whereas anything that might be identified as 'offensive' in terms of counter-terrorism would generally be acceptable in a liberal democracy – such as actively targeting terrorist group members with physical violence – the ethical aspect is more permissive of offensive operations in the cyber domain. The chances of party A defending are greatly increased if the capabilities of party B who is seeking to attack A are damaged and so attacks are defrayed (even if absolute success, as with counter-terrorism, can never be guaranteed).

However, while offensive activity is more acceptable in cyber warfare, it is clearly easier for some countries, or actors, than others, both in terms of capability and ethics – as the scale of Russian and Chinese activity shows, relatively free from internal legal (or political) restraints. By contrast, there timidity remains in countries such as the UK, even though the UK openly declares its offensive posture. The British prime minister will always be very concerned by what the lawyers advise about attacks. If an attack were considered to take down an electricity network, the UK lawyers would ask if it could inadvertently cause death as a result, or whether the impact would only be a nuisance to those deprived of internet access. A response causing death would likely be regarded as disproportionate. This is a concern which probably does not inhibit, for example, Vladimir Putin, the Russian President – and certainly not in the way that it would a Western European leader.

The major cyber states discussed here, in particular, have invested more and more into the development of capabilities. In part, this is simply because it is necessary – the means are there, so states need to be prepared. It is also because of a cost–effect balance. Compared with weapons of mass destruction, such as nuclear weapons, a cyber capability is remarkably easy to obtain and the price is coming down. In addition, while the skills and technology required are highly specialised and costly, the technical expertise needed to construct cyber-based weapons is more easily found than that required to produce biological weapons, or nuclear weapons. This makes a cyber capability considerably easier to achieve. So, what is here is very much that which is coming. For the next 50 years, at least, this trend will continue. Major states regard cyber capability as vital – and, indeed, an offensive cyber capability. Every year, more and more states and, as noted, other actors, will have an offensive cyber capability. It is relatively easy to gain one. It has become an ineluctable part of international relations. it will take more and more of a state's budget to maintain, or achieve, high-calibre capability – for example, the initial price for the UK's National Cyber Security Strategy was £1.9 billion.

In addition, capabilities will improve – and fairly rapidly.[16] The speed of change can be seen in Russian attacks on Ukraine. Where earlier attacks by various actors had not only struck targets, but then contingently spread to hit unintended targets, it was increasingly possible to gain precision in attacks and to limit their knock-on effects. Twice power stations were taken down, but when the attacks were analysed it was evident that the hacks were designed not to spread beyond Ukraine. This was in contrast to the US Stuxnet attack on Iran, which, after striking the Iranian nuclear programme, accidentally spread further and caused problems beyond Iran. Thus, learning is a key part of the evolving landscape and those marshalling offensive capabilities are refining their weapons. It does not matter whether you achieve your effect, your end, by computer means, or you do it by physically cutting cables – the response will be in the manner and with the means deemed most appropriate and desirable. Whether or not there are back-up generators, it would be extremely difficult to guarantee that no one would inadvertently die.

The focus here has been mainly on states, with some attention to non-government actors, especially in terms of criminal activity. There is also concern about the risk of potential terrorist uses of cyber attack – and terrorist attacks have occurred, or, as in the case of TV5 Monde, were initially suspected to have taken place. In some instances, the lines between actors are blurred. Linked to this, there is political use of cyber attacks – whether or not these would be deemed to be terrorist might well be a matter of interpretation; in the eye of the beholder. A group such as Hacktivist could fall into this category. But there is a big gap between the first two groups identified above – the top state actors and large organised crime (with state fusion) groups, and either of these last two actors.

This marks the great divide between private organisations and states. As could be seen from the case of Eternal Blue, discussed above, gaps can close very swiftly – in this case, one week from top secrecy to criminal commercial availability and the completely loss of an asset that had resulted from high investment of all kinds. The drift-down of state-level capabilities to criminal groups has got faster and faster. And the key to this shift is the human being – and not the ingenuity involved in catching up, but the banality of weakness and fallibility.

The transformation of Eternal Blue from secret to public domain via Wikileaks was the result of human action – in that case, a conscious decision to steal and publicly share material. In the Kaspersky case, it was the combined errors of the NSA contractor, who both took top-secret work home to complete and also had the Russian-origin virus protection software on his personal computer, against advice (though not against policy, as it was only after this horse had bolted that the US government moved to ban Kaspersky from government systems). These two examples show how the human factor is central to failed cyber security – one a matter of conscious wrongdoing, the other involving simple human errors that almost anyone might make in a given moment of 'not thinking' or careless misjudgement.

The insider threat, therefore, is the common theme in defending against cyber attacks across the full range of actors – states, criminals, terrorists, and the Hacktivist-not-yet-terrorist groups. That is to say, even in cases where the most sophisticated attacks occur, they involve somebody from inside the organisation. That subject might be helping actively because they want to do so (for money, for belief, for ideology – whatever their motivation), or simply because they carelessly click on a malware link that they should not have. In either case, they are the weak link on which protection fails.

The reality is that the majority of attacks that are launched use an attachment, or a link to click, in a seemingly normal email. Clicking on the attachment, or the link, would enable the attacker to get onto the system. Some attacks of this kind have become familiar to those informed and aware. The archetypal example is the Nigerian conman, who would send out, at scale, emails saying 'My father, the prince, has just died and if you send me a few hundred dollars, I'll be able to unlock his fortune and give you some back…' This would be an immediate delete for most informed users. Yet, there are still many people who respond – and only one is enough. Most large financial institutions have expensive, big programmes to try to sensitise their staff not to do this. But, even in this heavily managed situation, actually, an 85–90 percent success rate is considered 'good'. That suggests that one in ten people will, nevertheless, despite all the warnings, click on the malware link.

The key point is that an aggressor, or infiltrator, for example, only needs one person to enable access to an entire system – for example, the King's College London network and IT system. If the attacker targets everyone who has a King's email address, by pretending to be James Gow and sending malware, it would only take one person to click on the bad link for the whole system to be compromised. The official guidance given by the UK government is that recipients should check if a given message is actually coming from the purported author – in this instance, Gow. But, of course, people do not do that. Recipients may be getting hundreds of emails every day, and they simply do not check whether a given email actually comes from the purported email address – that is, that the apparent message from James Gow actually comes from him. Minor human mistakes can have major consequences.

Conclusion

Cyber warfare challenges the conventional idioms of UN-era international law, which generally restricted the use of force to self-defence, or actions authorised by the UN Security Council.

While existing law needed to be applied so far as it could, understandings of law were under pressure and needing new interpretation, or new law, even more than had been the case to date. This was potentially leading lawyers into uncomfortable twists that might invert conventional legal approaches. Traditional approaches to law and force, especially since 1945, created an environment in which force could only be thought of in defensive terms. Yet, as shown in this chapter, in cyber warfare, offence trumps defence, makes more strategic and operational sense – and possibly allows actions that would be clearly unlawful if conventional munitions were to be used. An effective cyber security response and strategy is likely to be offensive. Defence cannot succeed 100 percent of the time. Paradoxically, the chances of self-protection are increased by offensive action. In part, this is also because the chances of impairing opponents' capacity to cause harm are increased by attacks to exploit vulnerabilities.

Those vulnerabilities are crucial to the success and failure of cyber attack and cyber defence – and they are focused on the human factor. The reality is that, for all the legal and technical discussion that might take place, the key to cyber success or failure largely rests at the level of the individual. Most cyber attacks with malware rely on one individual human behaving carelessly, or maliciously, in a way that allows digital corruption to occur. The most obvious way of doing this is the commonplace click on an attachment or link that contains a virus. This will be a factor in international politics and international law, but overwhelmingly this will be a question at the domestic level and, in terms of legal action or protections, the individual in question is more likely to be subject to municipal law than any form of international law (though the latter cannot be wholly ruled out, of course). Given the character of cyber warfare, offensive response and the human factor are far more significant than any other aspect.

Notes

1 JLT Specialty Insurance and Risk Management Services, 1 June 2018, available at www.jltspecialty.com/our-insights/publications/cyber-decoder/insurers-look-to-close-cyber-terrorism-gaps at 18 June 2018.
2 These debates are reflected in chapters 6 and 7 in this handbook.
3 National Cyber Security Centre and National Crime Agency, *The Cyber Threat to UK Business 2017–18 Report*, No Place: National Cyber Security Centre, 2018, pp. 8 and 15.
4 This is a level where COBRA – the Cabinet Office Briefing Room – would be convened, with the prime minister and cabinet ministers meeting to be briefed and to decide on how to respond.
5 Reuters, 1 December 2017.
6 *The Wall Street Journal*, 5 October 2017 available at www.wsj.com/articles/russian-hackers-stole-nsa-data-on-u-s-cyber-defense-1507222108 at 30 May 2018.
7 *The Guardian*, 26 October 2016 available at www.theguardian.com/technology/2017/oct/26/kaspersky-russia-nsa-contractor-leaked-us-hacking-tools-by-mistake-pirating-microsoft-office at 30 May 2018.
8 Carl von Clausewitz, *On War*, trans. J. J. Graham, 'Introduction' and 'Notes' by Colonel F. N. Maude, C. B. (Late R. E.) and 'Introduction to the New Edition', by Jan Willem Honig, New York: Barnes and Noble, 2004, Book 1, Chapter 3.
9 This is discussed in Chapter 9.
10 HM Government *National Cyber Security Strategy 2016–2021*, 1 November 2016 available at https://assets.publishing.service.gov.uk/government/uploads/system/uploads/attachment_data/file/567242/national_cyber_security_strategy_2016.pdf at 18 June 2018.
11 *The Guardian*, 13 May 2018, available at www.theguardian.com/uk-news/2018/may/13/uk-and-european-intelligence-more-vital-than-ever-warns-m15-head at 18 June 2018.
12 This was an agreement between US President Barack Obama and China President Xi Jinping during the latter's state visit to the US in September 2015. It was part of a range of agreements, which included a more substantial Memorandum of Understanding that established a framework for development cooperation. See *FACT SHEET: President Xi Jinping's State Visit to the United States*, available at https://obamawhitehouse.archives.gov/the-press-office/2015/09/25/fact-sheet-president-xi-jinpings-state-visit-united-states at 18 June 2018.

13 *US-UK Cyber Communiqué* available at https://assets.publishing.service.gov.uk/government/uploads/system/uploads/attachment_data/file/62647/CyberCommunique-Final.pdf.

14 *Department of Defense News*, 8 September 2016 available at www.defense.gov/News/Article/Article/937878/us-uk-cybe/ at 23 June 2018; on historic cooperation at Bletchley Park, see Telford Taylor (1993), 'Anglo-American signals intelligence co-operation' in F. H. Hinsley and Alan Stripp eds., *Codebreakers: The inside story of Bletchley Park*, Oxford: Oxford University Press, 1993, pp. 71–3.

15 *FACT SHEET: U.S.-United Kingdom Cybersecurity Cooperation*, 16 January 2015 available at https://obamawhitehouse.archives.gov/the-press-office/2015/01/16/fact-sheet-us-united-kingdom-cyber security-cooperation at 18 June 2018.

16 *Reuters* 1 November 2016, available at www.reuters.com/article/us-britain-cyber-idUSKBN12W39K at 18 June 2018.

PART III

Autonomy, robotics and drones

11

AUTONOMY OF HUMANS AND ROBOTS

Thrishantha Nanayakkara

It is often argued that mobility is an essential ingredient of intelligence.[1] Reasonably advanced robots are beginning to join humans in various fields including the military, with rapid advances in solving basic problems to do with mobility. With these developments, debates on the definition of a robot, its autonomy, how they are related to humans, and how robots should be forced to be safer are gaining public participation like never before. While such debates are healthy for the public's engagement with the field of robotics, and can even contribute to its advancement, if misconstrued they can also delay the benefits that robotics can offer to society. In this discussion, I will briefly mention some of the advancements of robotic mobility in three main modes – legged locomotion, flight, and swimming – to emphasise the important role played by fundamental knowledge about locomotion in the advancement of autonomous robots. I will then raise several questions about the notion of autonomy of human body and mind, followed by a discussion on autonomy in robots. Then I will share some thoughts on how bounded autonomy should be addressed when humans and robots are involved in lethal force, and the question of how developed nations can engage with advances in the developing world in the area of autonomous robots to ensure a safer world. Finally, I will highlight some future challenges that should be addressed by multidisciplinary groups involving roboticists and war studies experts.

Main modes of locomotion

Here, I will briefly discuss recent advances in robotic legged locomotion, flight, and swimming, highlighting their technological advantages and challenges.

A Legged locomotion

Legged locomotion offers many advantages such as the opportunity to come into contact with a rough terrain at selected locations to maintain stability, ability to deform soft soil to aid forward movement, and room for various modes of dynamic locomotion such as scrawling, trotting, and galloping. However, the punctuated forces at each foot collision with the ground and the piecewise non-linear interaction dynamics experienced during each collision makes walking harder than many other modes of locomotion. Technology on robotic legged locomotion has made significant advances in the recent past. Starting from simple passive dynamic walking experiments[2] in the 1980s, our understanding about metastability of walking[3] – the fact that there is a chance that the walker can fall even if they are stable most of the time – and the dominant sources of variability of walking that causes metastability[4] has grown rapidly. Perhaps the Big Dog developed by Boston Dynamics[5] is the most successful recent robotic experiment in legged locomotion on unstructured outdoor terrains. The Big Dog uses laser range finders to estimate the location of obstacles like trees and stereo cameras to estimate the terrain roughness. It then reduces the 3D world to a 2D obstacle avoidance plan (imagine 2D shapes of no-go regions projected onto a floor). It also benefits from clever algorithms of legged locomotion to stay balanced. Unlike many other robots, Big Dog has the remarkable ability to decide when to give up active control to let the body roll down safely.

Following the Big Dog, the US Defense Advanced Research Projects Agency (DARPA) funded Maximum Mobility and Manipulability (M3) programme has led to a new generation of legged robots that can perform dynamic running[6] inspired by the morphology of the cheetah. In addition, the 2013 DARPA grand challenge also showed the worldwide interest in autonomous biped locomotion in uncalibrated environments. This challenge helped to make a number of advances in autonomous humanoid robotic capability to use tools designed for human operators in a disaster response operation.

The rapidly growing soft robotics community has also produced some attractive soft robots inspired by muscular hydrostats like the Octopus.[7] For instance, walkers with continuum legs[8] have the potential to burrow through collapsed buildings and rubble by having multi-functional limbs – limbs that can wrap around objects to pull the body, as well as functioning like stiffened legs.

B Advances in aerial vehicles

Among all semi-autonomous robots, drones dominate public discussion[9] due to their routine military applications in remote theatres. Semi-autonomous flying robots have grown in diversity over the past few years due to their potential in remote sensing and accessibility. Next to fixed-wing drones, a large community of researchers are working on quadrotors[10] due to their ability to hover at a given point in the air and make quick path and orientation changing manoeuvres. However, their biggest challenges are limited endurance and payload carrying capacity. In addition to fixed-wing and quadrotor flight, a number of attempts have been made in the recent past to achieve flapping flight in micro-robots. A prototype of a bio-inspired unmanned aerial vehicle (UAV) imitating the hovering flapping pattern of humming-birds and dragonflies is presented by Fenelon.[11] Senda et al.[12] analyse the periodic flapping flight of a butterfly, highlighting the role of free vortexes in flight stability. Tanaka et al.[13] demonstrate a flying robot mimicking the high load to flapping frequency ratio of a butterfly. Their results show that when the centre of gravity is in the right place, the body achieves passive stable forward flight with a cyclical change in attack angle; furthermore they showed that the change in attack angle during wing flaps increased the net upward aerodynamic force. A small butterfly-style flapping robot with veins has been fabricated[14] by Fujikawa et al. to investigate flight characteristics for different design parameters such as swept-forward wing angle and centre of mass (COM). Experiments on the 'robot' showed that the body pitch angle was controlled by the swept-forward wing angle and the relative positions of the COM and centre of lift.

In terms of mechanisms for micro-aerial flapping robots, a piezoelectric fibre composite driving mechanism has been presented by Ming et al.[15] Baek et al.[16] proposed a resonant drive to reduce average battery power consumption for DC motor-driven flapping-wing robots.

C Swimming

Apart from many conventional underwater robots that use rigid bodies driven by propeller- or thruster-based locomotion, there have been recent developments in soft body robots that take advantage of passive dynamics of swimming inspired by biology. It has been experimentally shown by Beal et al.[17] that the passive degrees of freedom and their correct physical distribution along the body can help to propel the dead body of a fish against the stream. A robotic fish called MARMT (mobile autonomous robot for mechanical testing) has been developed, by Long et al.,[18] which uses a biomimetic tale to generate undulating movements to propel the body. According to Roper et al.,[19] the hidden potential of these soft robots is that their internal stiffness can be controlled to give rise to passive interaction dynamics that can extend the robot's endurance. It would be very interesting to have a swimming robot that can drift down or move up a stream of water just by changing the stiffness distribution of the body, which consumes a relatively insignificant amount of energy compared to that required for locomotion. This may become a reality in the near future given related advancements like the soft ray from Moothred[20] or the flapping swimmer AQUA[21] from Zhang et al.

The notion of autonomy

The above discussion on various modes of locomotion helps us to note that bodily adaptations and computational basis of control of locomotion are essentially inter-dependent. For instance, the flapping controllers in the brain of a butterfly cannot be transplanted in the brain of a hummingbird to give any meaningful behaviour, even if they use flapping as the principal mechanism of locomotion. In addition, the flapping controllers in the brain of a butterfly itself will have to go through adaptation if its own bodily morphology were to change. In fact, there is no sharp boundary between the role of the brain and body in the computation of locomotion. The example of the dead trout swimming against the turbulent water breaks down the conventional notion of computation as one confined to the brain, to suggest that body itself can work in tandem with the environment to maintain computation required for locomotion without any intervention from a regulatory mechanism such as the brain. This raises the question of what regulates autonomy of locomotion and action, and the nature of autonomy itself.

1 The neuroscientific and philosophical basis of autonomy of human body and mind

There has been much debate on what autonomy means. Still there are open questions to be answered. For instance, are we humans as autonomous as we think we are in terms of the ability to control mental states according to objectives and plans? If so, why do we find it difficult to stay focused on one thought for more than a minute or find it difficult to stick to plans we set? Does this wandering nature of mind play an essential role in human-level 'cognition'? How much of our thoughts and actions are guided by spontaneous reactions arising from situation-action patterns conditioned in the past, as opposed to action guided by thoughtful plans? This also proposes that a significant part of autonomy is about the freedom to move from one state to another (mental or physical) based on inputs received from the environment and from within the memory itself.

(a) *What we know from neuroscience.* It is known that the human central nervous system (CNS) is organised as a real-time computer that tries to be embedded in the natural parallelism in the sensory stream it receives, and to be able to make good enough responses within the deadlines imposed by the environment. For instance, when we reach to grab an object over a candle flame, if the heat sensed by the skin exceeds a threshold, it will autonomously activate a spinal reflex feedback to throw the hand away from the flame about 90 milliseconds before the brain gets to know about it. In other words, among the parallel stream of information the CNS received, the intensity of the pain signal decided the path of processing it should take in the CNS in order to meet the deadline imposed by the environment to take the hand away from danger. More cognitive processing would follow at least 240 milliseconds after the initiation of the pain signal. Therefore, when faced with tight deadlines to survive, there is no guarantee that the human CNS would take a conscious decision. Even if the brain is involved in the process, tight deadlines will most likely trigger reactions well conditioned over a long period of training or evolution itself (for example, the run towards or run away reflex). The sense of urgency may also be conditioned by more than one environmental factor. Therefore, a remote operator of a semi-autonomous robot that is involved in lethal force can end up being driven by potentially harmful neural processing streams given unpredictable environmental circumstances.

(b) *Practical implications on operators of semi-autonomous robots.* The above neuroscientific base can be useful for designing training courses for those who operate autonomous robots that can potentially lead to lethal consequences. The obvious solution is to develop advanced training techniques that speed up potentially safe decision-making processes based on ethical intentions through repeated conditioning, so that the probability of them being recruited by the CNS to deal with urgent situations will be increased. At this point, we cannot avoid using technology to augment the sensory space of the operator as well as using advanced techniques to maintain mental stability. Technology can play a major role in drawing attention to the most relevant information in a complex stream of visual information (for example, overlay thermal information with camera feedback, filter noise, etc.). However, the role of technology is not without drawbacks. Reliance on such sensor enhancement makes the whole process vulnerable to sensor spoofing. Including the *Titanic* tragedy, technology advancements in locomotion and manipulation are plagued by incidents where engineers have been overly confident of 'robust' designs until an environmental condition neglected at the design phase causes a catastrophe. A single stream of hacked sensor feedback can potentially cause grave damage to the sensor augmentation process. The scale of damage to human life can be very costly if a proper set of guidelines are not available for 'autonomous' or 'semi-autonomous' robots involved in war.

2 Debates about robotic autonomy and intelligence

When it comes to machines, what is the distinction between 'automated' and 'autonomous' robots? My working definition of autonomy of a robot is that *a robot is autonomous if it has the computational resources to map a situation to an action without any outside interference.* Here, action includes changing the embodiment itself (stiffness control, camouflaging, metamorphism), adding/deleting/changing codes and rules, physical action, and giving up to take further action. A robot is semi-autonomous *if it allows real-time interference from a human agent* in the process of taking action. According to this working definition, a robot's ability to estimate its current state (how it is physically embedded in the environment) is an essential component of autonomy. The notion of goal-oriented perception and action is being studied in my laboratory under the general theme of morphological computation and attention.[22] We go beyond passive perception to active perception by controlling how sensors feel the environment by controlling the stiffness of the body that lies between the sensor and the environment. Any robot is passively constrained to take actions by its own limits of actuators. However, when a special program that cannot be inhibited by other programs of the robot imposes limits on decisions taken by other programs, it can exhibit bounded autonomy.

An 'automated' robot differs from an 'autonomous' robot in the range of actions it can take to change its internal parameters without outside interference. Often an automated robot is 'programmed' to map situations to actions in a predictable manner whereas 'autonomous' robots that can even change their internal parameters and logic with adequate computational resources at their disposal to take an action.

In fact my prediction is that autonomous robotics too will take the same pattern of evolution of species, where giant prehistoric animals were followed by more specialised but simpler and smaller species that can survive in ecological niches. Therefore, future autonomous robots will most likely to be specialised machines. The complex robots too will have concurrent functioning of specialised autonomous 'organs' in them to sense and react to achieve goals.

Is full robotic autonomy dangerous? Former director of the MIT Computer Science and Artificial Intelligence Laboratory Rodney Brooks argued[23] that intelligence in a robot is not in any particular part like its body, programs, or in the environment, but something orchestrated by an interplay of all these parts in the eyes of the observer. For instance, a robot that has a reflex-type motor program to move round an obstacle and another program to be attracted towards

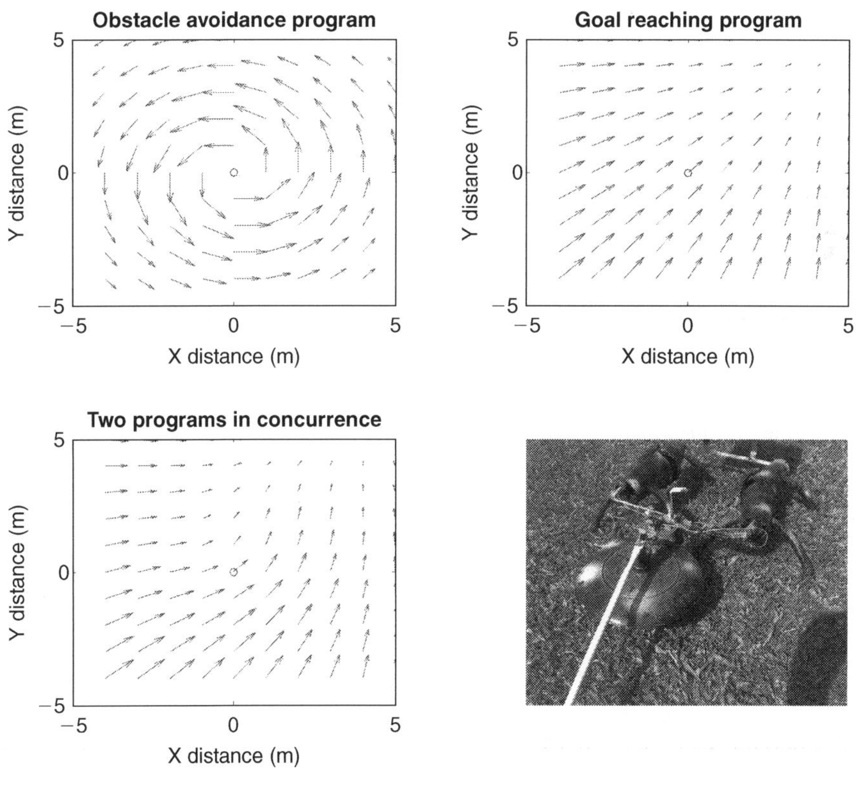

Figure 11.1 How two independent programs – one avoiding an obstacle and the other reaching to a goal – can give rise to complex behaviour when run in concurrence. Arrows represent the direction and magnitude of the force applied on the robot to move at any given point on the X–Y coordinate system.

a goal can give rise to complex goal-seeking behaviours while avoiding different types of obstacles just by running the two programs in concurrence, as shown in Figure 11.1. However, an observer naive as to how the separate programs are run in parallel will come to various conclusions about the 'thinking' and 'planning' ability of the robot, which is not found in the body or in any independent pieces of programs of the robot. Rather, seemingly intelligent or autonomous behaviour emerges as a result of the concurrent processing of independent pieces of programs in a physical robot.

In practice, many do not appreciate that landmines found buried in 60 countries today are autonomous robots according to my above definition. A landmine has a simple program to detect the intended scenario to detonate. The program is encoded in a simple mechanical spring or a more sophisticated mechanical arrangement. Here, I wish to stress that a program does not necessarily have to be a piece of software. It can be a hardware arrangement that is functionally equivalent to what a software code does, as illustrated in Figure 11.2. Hence, in a landmine, the decision to detonate.

The body does computation using hardware. The left side shows a simple pseudocode of the logic used in a landmine, and the right shows that the equivalent hardware arrangement is automated using a mechanical program. Since this program is very primitive in terms of logic, its behaviour too is very generic and does not discriminate between a combatant and a civilian or just a wild animal. Since an autonomous weapon is a mere extension of the freedom to change the code or the logic seen in a landmine, it raises fundamental concerns about agency, legitimacy, and protection embedded in the programs of autonomous weapons.

The transition from a simple landmine to a more autonomous robot that can deliver a warhead to a target simply resides in that part of the program that makes the robot to move and take action. Hence, a drone can be viewed simply as a flying landmine. It is semi-autonomous only in the flying part and the releasing of the warhead section of the operation. Once a warhead is released, the autonomy is transferred to the warhead, which is the most critical part of the function, and the rest of the functionality is not so different from a little more sophisticated version of a landmine. Therefore, it is apparent that unbounded robotic autonomy can of course be dangerous, as implied by early writers such as Isaac Asimov.[24] Strict design constraints have to be imposed on the freedom of the robot to make decisions that may have some implication on human life. One may come to the conclusion that semi-autonomous robots/weapons are less

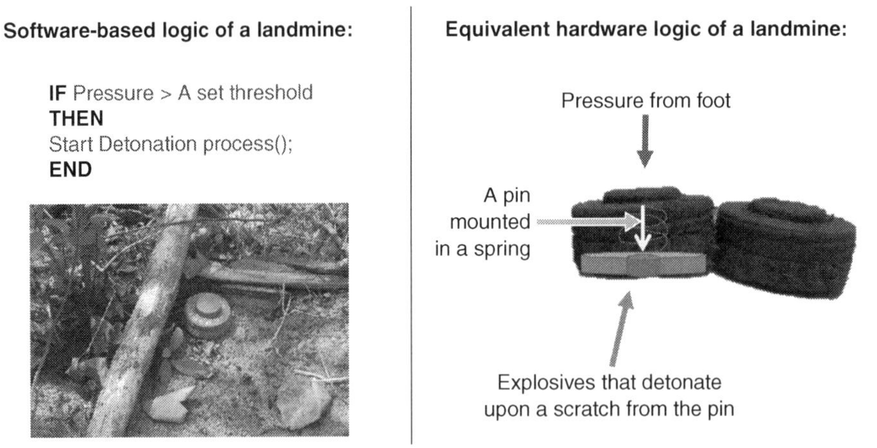

Figure 11.2 A program does not necessarily need to be encoded in software.

dangerous than autonomous counterparts. There are compelling arguments against this stand too. By having a human in the loop, we are opening a channel for adverse human intentions to create flaws in the robot's logic system too. As envisaged by Asimov himself, the challenge is to stop a human from using a robot's autonomous decision-making capability to hurt another human or himself. In that case, what decides whether the robot's actions are within any form of ethics? Should the robot be able to deny human orders if the action leads to a massacre for instance? These interesting questions that lie in the interplay between human intentions and the programmed boundaries in the action space of a robot have to be answered. Therefore, there is a pressing need for an interlocking protocol that maintains checks and balances on all decision-making programs irrespective of whether they are just mechanical (morphological computation), software, or human cognition based.

Are autonomous robots sophisticated and expensive technological masterpieces limited to the developed world? Figure 11.3 shows an experiment carried out in a tropical forest environment in Sri Lanka.[25] A motorised sonar range finder mounted at the front of the robot was swung on a horizontal plane to collect statistical features of obstacle distributions on the left and right side of the robot. Then the robot classified the environment into one of five known classes of environments to decide on one out of five known behaviours. If a planned behaviour resulted in a collision with a tree, a bumper switch was used to skirt around the obstacle. This combination of autonomous recruitment of programmed behaviours and reacting using tactile sensors made

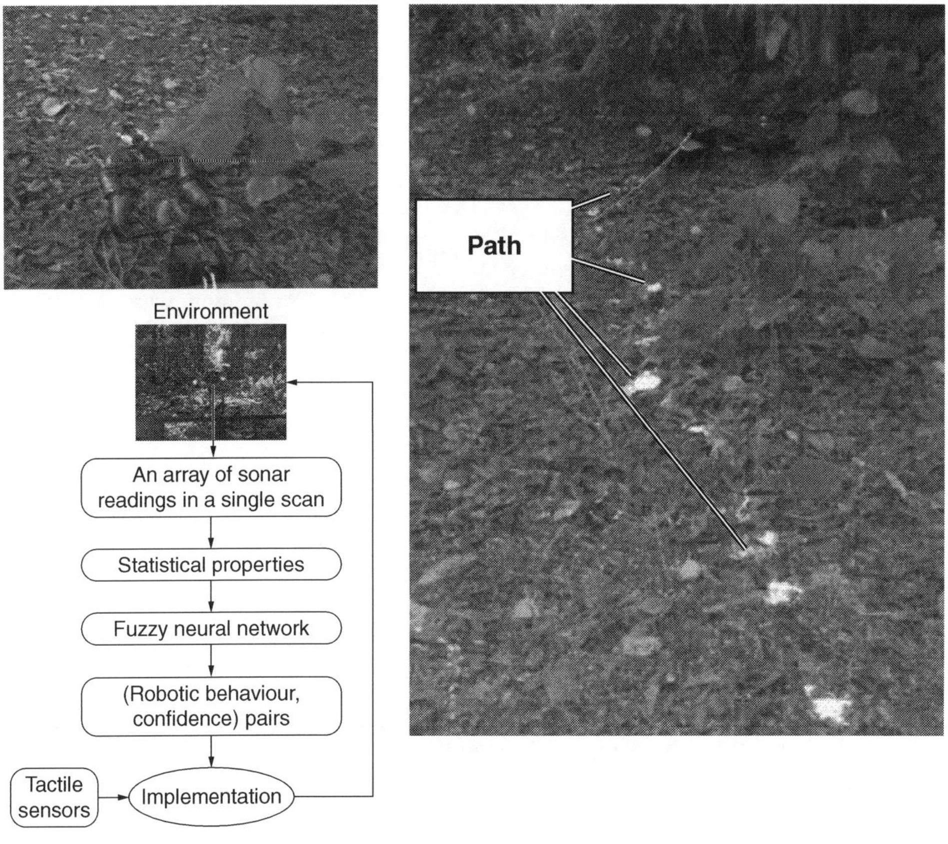

Figure 11.3 Mobility in a tropical forest using multiple independent behaviours.

the robot able to negotiate through trees in a forest till the batteries drained down. Only two motors drove the robot, though it had eight legs in total. Therefore, the robot had many passive degrees of freedom to conform to different geometries of terrains in a forest without burdening a control algorithm to adapt the shape of the body to the terrain. Though the robot was able to demonstrate an impressive level of autonomous behaviour, it was built on a low budget in a developing country like Sri Lanka (less than £200). The only imported components were the embedded microprocessors to build the brain (less than £30), and the two servo motors (around £100) to drive the legs. The rest was salvaged material from junkyards! Therefore, the common notion that autonomous robots are expensive and sophisticated technological pieces is a myth. Reasonable autonomous robots that can go beyond the lab to outdoor environments can be built in any developing country with careful planning to find local niches.

3 What interlocks are needed to impose bounds on autonomy in an international level?

Once lower-level implementation technicalities such as translating ethical limits to programmable equations or sets of rules are solved, the following high-level questions have to be explored by a multidisciplinary community: Who should design the ethical layer of the above constraints for action of an autonomous robot? What kind of supervision should such a designer be subjected to? How do we guarantee that these programs are not hacked and that any shared computational resources they depend on are protected? Should robots be able to argue and negotiate based on ethics? In case of a robot's involvement in human fatalities, what international conventions should be kept in place to scrutinise the design of the programs?

There has been some preliminary work in the direction of human–robot negotiation in my laboratory in a scenario of robotic assistance to firefighters. We first studied the varying confidence level of a blindfolded human participant to follow another human purely based on haptic feedback given via a hard rein. The duo then traced a given path through a process of negotiation. We found out that the person who guides the blindfolded person uses a particular law (a third order predictive relationship between follower's tracking errors and the guider's actions), and that the follower uses another law (a second order reactive relationship between guider's actions and follower's movements).[26,27] We then tested the algorithm in a robotic guider[28] that helped us to understand that robots should account for the asymmetry of perception in humans between leftward and rightward perturbations to provide haptic feedback during guiding. Future challenges remain in the interdisciplinary area of the role of ethics in agency, negotiation, and responsibility sharing in human–'autonomous robot' teams.

Conclusion

Contrary to traditional belief that humans are autonomous beings who can take rational decisions, there are arguments that propose that rationality itself consists of conditioned processes that take uncertain state trajectories. Therefore, conditioning and empowerment through training and technological augmentation of information fed to the process of decision-making is very important for those who operate semi-autonomous robots that provide sensory feedback to humans to take decisions. Since the way on-board sensors of a semi-autonomous robot will perceive and report the environment back to the human inevitably depends on how the human operator probes the environment using the robot, perception and action in tele-operation is a coupled phenomenon. Therefore, the state trajectories of human decision-making in such coupled systems cannot be easily predicted. It becomes clear that novel approaches to develop cross-checking and inter-locking mechanisms are needed to keep collective autonomy of human–robot coupled systems within bounds. This imposes challenges both in the domain of incorporating ethical guidelines in the training process as well as providing flawless technological support. The notion of autonomy in robots too is being debated. Here, I provide reasons to believe that autonomy can emerge in any system regardless of whether there is a software code to control actions or not. Therefore, when imposing bounds on the autonomy of a machine to map a situation to a potentially harmful action, mechanisms should be set in place to limit the entire behaviour as an embodied entity rather than a software code alone. Even in the case of counter-action for hacking, focusing on software alone can lead to potentially dangerous outcomes.

Robots are reducing in cost while rapid advancements are being made in the area of semi-autonomous control of robots. Many recent advancements in technology for locomotion have made robots able to move longer distances with remarkably lower energy consumption. Biological inspirations of how animals exploit passive dynamics to move around by harnessing energy from the environment has been a major factor in these developments. Moreover, performance-driven approaches to build autonomous robots as opposed to traditional connectionist approaches to intelligence and autonomy have led to several low cost advances in the autonomous robotics front not only in developed countries, but also in the developing world.[29]

Notes

1 Brooks, Rodney Allen, *Cambrian intelligence: the early history of the new AI*, Cambridge, MA: MIT Press, 1999.
2 Raibert, Marc H., *Legged robots that balance*, Vol. 3, Cambridge, MA: MIT Press, 1986.
3 R. L. Tedrake, 'Applied optimal control for dynamically stable legged locomotion', PhD thesis, Massachusetts Institute of Technology, 2004.
4 Nanayakkara, Thrishantha, Katie Byl, Hongbin Liu, Xiaojing Song and Tim Villabona, 'Dominant sources of variability in passive walking', in IEEE International Conference on Robotics and Automation (ICRA), pp. 1003–10. IEEE, 2012.
5 Wooden, David, Matthew Malchano, Kevin Blankespoor, Andrew Howardy, Alfred A. Rizzi and Marc Raibert, 'Autonomous navigation for BigDog', in IEEE International Conference on Robotics and Automation (ICRA), pp. 4736–41. IEEE, 2010.
6 Seok, Sangok, Albert Wang, Meng Yee Chuah, David Otten, Jeffrey Lang and Sangbae Kim, 'Design principles for highly efficient quadrupeds and implementation on the MIT Cheetah robot', in IEEE International Conference on Robotics and Automation (ICRA), pp. 3307–12. IEEE, 2013.
7 W. M. Kier and K.K. Smith, 'Tongues, tentacles and trunks: the biomechanics of movement in muscular-hydrostats', *Zoological Journal of the Linnean Society*, Vol. 83, 1985, pp. 307–24.
8 I. S. Godage, T. Nanayakkara, and D. G. Caldwell, 'Locomotion with Continuum Limbs', in IROS, pp. 293–8, October 7–12, 2012 Vilamoura, Algarve, Portugal.
9 Drew, Christopher, 'Military is awash in data from drones', *New York Times,* 11 Jan 2010.
10 Vijay Kumar and Nathan Michael, 'Opportunities and challenges with autonomous micro aerial vehicles', *The International Journal of Robotics Research*, Vol. 11, No. 31 (2012), pp. 1279–91.
11 Michael A. A. Fenelon, 'Biomimetic Flapping Wing Aerial Vehicle', International Conference on Robotics and Biomimetics, Bangkok, 2009, pp. 1053–8.
12 K. Senda, N. Hirai, M. Yokoyama, T. Obara, T. Kobayakawa and K. Yokoi, 'Analysis of flapping flight of butterfly based on experiments and numerical simulations', World Automation Congress, Sept. 2010, pp. 1–5.
13 Tanaka, H., Hoshino, K., Matsumoto, K., Shimoyama, I., 'Flight dynamics of a butterfly-type ornithopter', IEEE/RSJ International Conference on Intelligent Robots and Systems, (IROS 2005), pp. 2706, 2711, 2–6 Aug 2005.
14 T. Fujikawa, Y. Sato, Y. Makata, T. Yamashita and K. Kikuchi, 'Motion analysis of butterfly-style flapping robot for different wing and body design', IEEE International Conference on Robotics and Biomimetics (ROBIO), 2008, pp. 216–21.

15 Aiguo Ming, Luekiatphaisan, N., Shimojo, M., 'Development of flapping robots using piezoelectric fiber composites. Improvement of flapping mechanism inspired from insects with indirect flight muscle', International Conference on Mechatronics and Automation (ICMA), 5–8 Aug 2012, pp. 1880, 1885.
16 S. Baek, K. Ma, R. Fearing, 'Efficient resonant drive of flapping-wing robots', IEEE/RSJ International Conference on Intelligent Robots and Systems (IROS), 2009, pp. 2854–60.
17 D. N. Beal, F. S. Hover, M. S. Triantafyllou, J. C. Liao and G. V. Lauder, 'Passive propulsion in vortex wakes', *J. Fluid Mech.*, Vol. 549, 2006, pp. 385–402.
18 Long, John H., Nicole M. Krenitsky, Sonia F. Roberts, Jonathan Hirokawa, Josh de Leeuw and Marianne E. Porter, 'Testing biomimetic structures in bioinspired robots: how vertebrae control the stiffness of the body and the behavior of fish-like swimmers', *Integrative and comparative biology*, Vol. 51, No. 1 (2011), pp. 158–75.
19 Roper, Daniel, Sanjay Sharma, Robert Sutton, and Philip Culverhouse. 'Energy-Shaping Gait Generation for a Class of Underactuated Robotic Fish', *Marine Technology Society Journal*, Vol. 46, No. 3 (2012), pp. 34–43.
20 Moored, Keith W., Peter A. Dewey, Megan C. Leftwich, Hilary Bart-Smith, and Alexander J. Smits, 'Bioinspired propulsion mechanisms based on manta ray locomotion', *Marine Technology Society Journal*, Vol. 45, No. 4 (2011), pp. 110–18.
21 Zhang, Shiwu, Xu Liang, Lichao Xu and Min Xu, 'Initial Development of a Novel Amphibious Robot with Transformable Fin-Leg Composite Propulsion Mechanisms', *Journal of Bionic Engineering*, Vol. 10, No. 4 (2013), pp. 434–45.
22 Nantachai Sornkarn, Matthew Howard and Thrishantha Nanayakkara, 'Internal Impedance Control helps Information Gain in Embodied Perception', IEEE International Conference on Robotics and Automation (ICRA), 2014 (accepted for publication).
23 Brooks, R. A., 'Intelligence Without Representation', *Artificial Intelligence Journal*, Vol. 47, 1991, pp. 139–59.
24 Anderson, Susan Leigh, 'Asimov's "three laws of robotics" and machine metaethics', *AI & Society*, Vol. 22, No. 4 (2008), pp. 477–93.
25 Nanayakkara, Thrishantha, Lasitha Piyathilaka and Akila Subasinghe, 'A simplified statistical approach to classify vegetated environments for robotic navigation in tropical minefields', in Proceedings of the International Conference on Information and Automation. Colombo, Sri Lanka. 2005.
26 Ranasinghe, Anuradha, Jacques Penders, Prokar Dasgupta, Kaspar Althoefer, and Thrishantha Nanayakkara, 'A two party haptic guidance controller via a hard rein', in *Intelligent Robots and Systems* (IROS), 2013 IEEE/RSJ International Conference on Intelligent Robots and Systems, pp. 116–22. IEEE, 2013.
27 Ranasinghe, A., Dasgupta P., Althoefer K. and Nanayakkara, T., 'Identification of Haptic Based Guiding Using Hard Reins', PLoS ONE 10(7): (2015) e0132020. doi:10.1371/journal.pone.0132020.
28 Anuradha Ranasinghe, Prokar Dasgupta and Thrishantha Nanayakkara, 'Human Behavioral Metrics of a Predictive Model Emerging During Robot Assisted Following Without Visual Feedback', accepted in *IEEE Robotics and Automation Letters (RA-L)*, 2018.
29 For more discussion, see: www.accesstoinsight.org/lib/authors/thanissaro/notself2.html. www.accesstoinsight.org/lib/thai/lee/consciousnesses.html. Giszter, Simon. 'Modelling spinal organization of motor behaviors in the frog' In paper presented at 2nd European Conference on Artificial Life, 1993, pp. 24–26. Dias, M. Bernardine, G. Ayorkor Mills-Tettey and Thrishantha Nanayakkara. 'Robotics, education, and sustainable development', In *Robotics and Automation*, 2005. ICRA 2005. Proceedings of the 2005 IEEE International Conference on Robotics and Automation, pp. 4248–53. IEEE, 2005.

12

AUTONOMOUS AGENTS AND COMMAND RESPONSIBILITY

Jack McDonald

How will the advent of autonomous systems change the military? Stories abound of the encroachment of autonomous machines on the day-to-day lives of citizens in advanced industrial societies. Technology companies plan to market autonomous cars to consumers, while autonomous haulage systems are already in place in the resource extraction industry.[1] In the estimation of some experts, machines will soon be able to replace humans in many professions in the near future, including military professionals. In each industry, we expect significant changes in the workplace, but the prospect of such change in military affairs worries many technical and legal experts.[2] Thus, just over 25 years since Manuel DeLanda highlighted the implications of autonomous weaponry in *War in the Age of Intelligent Machines*,[3] the prospect of states fielding 'killer robots' has sparked an international campaign to pre-emptively ban autonomous weapon systems.[4] This campaign faces numerous challenges, notably a lack of traction with powerful states in the international system.[5] Despite the uncertain prospects for pre-emptive bans or regulation of autonomous weapon systems, the development of autonomous systems, and their use by militaries, raises numerous questions, not least how states can employ lethal autonomous weapon systems (LAWS) and still comply with the law of armed conflict, and the moral requirements of the just war tradition. This is one of the key challenges that autonomous weapons systems highlight: how will the introduction of non-human agents affect the concept of command responsibility?

Command responsibility is only one issue among many. Academics and experts disagree as to whether LAWS can be employed in a lawful manner. Objections to LAWS on moral and legal grounds abound,[6] as do defences that argue these systems can be legitimate means of warfare, conforming to international humanitarian law.[7] Some, like Noel Sharkey, offer foundational objections to allowing machines to make lethal decisions – non-humans shouldn't be given the authority to decide to kill humans.[8] It could be impossible to design a machine that could differentiate between legitimate and illegitimate targets, and besides 'our inability to foresee how they might act in complex operational environments, unanticipated circumstances, and ambiguous situations, there is a further difficulty – how we can test and verify that a newly designed autonomous weapon system meets the requirements imposed by IHL, as required by Article 36 of Additional Protocol I?'[9] This article requires that new weapons systems are tested to make sure that they can be used in accordance with international humanitarian law.[10] General artificial intelligence – cognition on the level that would allow machines to think on par with humans – is a distant prospect, if not a wonderful pipe dream, which means that decisions made by non-human agents are unequal to those made by humans, even if the effects are similar. Indeed, the non-human nature of artificial intelligence is central to the question of whether such systems violate the Martens Clause, which bans means and methods of warfare that 'violate the dictates of the public conscience'.[11]

States and their militaries have responded to this issue by reinforcing the role of the human in armed conflict. For all weapon systems, UK policy emphasises that 'Legal responsibility for any military activity remains with the last person to issue the command authorising a specific activity'.[12] America's first policy on 'Autonomy in Weapon Systems', Directive Number 3000.09, states that 'Autonomous and semi-autonomous weapon systems shall be designed to allow commanders and operators to exercise appropriate levels of human judgment over the use of force'.[13] The states with the advanced industrial and research bases that are a requisite for the effective design and deployment of LAWS appear to want no part in any prospective robot-led apocalypse. Nonetheless activists and smaller states press for a pre-emptive ban, perhaps mirroring the 1868 St Petersburg Declaration that banned fulminating bullets – the earliest example of widespread pre-emptive arms control by international treaty. At the same time, campaigners want humans to exert 'meaningful human control' over the decisions made by autonomous systems, a concept that emerged as a 'major theme' of discussions on the topic at the 2014 United Nations (UN) Convention on Certain Conventional Weapons.[14] There is perhaps good reason for scepticism on the part of campaigners regarding the assertions of powerful states. The adoption and use of technology by militaries is influenced and structured by their culture,[15] but also by interaction with opponents. European militaries relegated the machine gun to a supporting role, or for colonial suppression, until they discovered its distinct advantages on the First World War battlefield.[16] Even if militaries are high-minded now, what is to say that they will stick to their values when faced with opponents that use autonomous weapon systems? The panoply of roles that autonomous machines can potentially fulfil means that the pressure to integrate autonomous systems into militaries – beyond point-defence weapons like the Phalanx CIWS – will only grow over time.

Two competing ideas emerge from the debate on meaningful human control, both centred on the responsibility of non-human agents in armed conflict. The first is a claim by Robert Sparrow that robots cannot be held responsible for their actions.[17] Counter to this, Marcus Schulzke has argued that command responsibility allows for the assignation of responsibility for non-human agents.[18] Since these social practices of command responsibility are contingent on human negotiation, the ultimate impact of non-human agents on human command responsibility structures is not assured.[19] In this chapter, I argue that autonomous systems challenge the social structures that ensure militaries comply with the law of armed conflict, and that this problem should take precedence over the more prominent concern of exerting 'meaningful human control' over individual autonomous systems. As a consequence of this, my argument is that we should be more concerned about autonomous computer systems (ACS) that may potentially be employed for the production of intelligence than the deployment of tangible robotics platforms, commonly referred to as lethal autonomous weapon systems (LAWS). This focus inverts the framing of some of the important legal and moral issues raised by the present debate. Here I am more concerned by the prospect of humans making decisions based upon information provided by ACS, than the replacement of humans by LAWS on the battlefield.

The structure of my argument is as follows. First, I will set out the problem of meaningful human control, and the reason that LAWS are perceived to pose such a challenge to compliance with the legal and moral frameworks that justify and legitimate violence in war. I argue that this places undue focus upon 'bottom up' compliance – seeking adherence to the rules of war through individual ground-level decisions. In contrast, I argue that military structures ensure 'top down' compliance. The combination of the two is, in theory, enough to ensure compliance by state forces to the law of armed conflict. Legal and moral analysis of decision-making and responsibility places great emphasis on individual decisions. LAWS pose the primary challenge to 'bottom up' meaningful human control, but professional militaries do not currently seek to employ LAWS in such a manner that they would pose a significant challenge to compliance with the law of armed conflict. By thinking of individual acts of violence as mediated decisions, however, it is clear that this is not enough to ensure compliance with the law of armed conflict. The important elements of structural compliance are that it places constraints upon individual autonomy and enables them to act based upon limited or fragmentary information passed to them by others. That human beings are the agents responsible for such information is an important element of these human structures. Although LAWS can be controlled to ensure no responsibility gaps occur, the same cannot be said of ACS that could potentially be deployed in intelligence production. Since artificial intelligence and ACS are unlikely to be deployed to make top-level decisions, it is the contribution of ACS to violent action that presents the clearest challenge to maintaining compliance to the law of armed conflict. To conclude, I reflect upon the fact that the focus the law of armed conflict places upon direct decisions to use force distracts from these important structural concerns. The consequence of this is that regulating the decision-making process, as opposed to the outcome of decisions, is unlikely to occur.

The importance of responsibility

The concept of responsibility underpins both the law of armed conflict and the just war tradition. In order for either to regulate or influence human conduct, it is necessary that humans can be held responsible for violations – the commission of war crimes. For this reason the UN Special Rapporteur on Extrajudicial, Summary or Arbitrary Executions has argued that developing lethal autonomous robots 'may be unacceptable because no adequate system of legal accountability can be devised'.[20] Command responsibility, the principal system of legal accountability for violence in war, rests on the understanding that lethal decisions can be attributed to human beings; in other words, that a human can potentially be held accountable for lethal actions. The Campaign to Stop Killer Robots (CTSKR) has advanced the standard of 'meaningful human control' because it seeks to ensure that humans remain responsible for all lethal decisions in war, and thereby for all subsequent acts of violence. Furthermore it hopes that states will soon agree to ban autonomous weapons.[21] Not all agree. Kenneth Anderson, Daniel Reisner and Matthew Waxman pointed out that many of the implied or required elements of this standard have no precise basis in the law of armed conflict.[22] The CTSKR wants humans to make the decisions that result in acts of violence, not machines. Two reasons for this present themselves. One is that autonomous systems are not thought to be able to make decisions in the same way that human beings do; another is that our social concepts of responsibility fail when it comes to non-human agents: how could a robot be held responsible for a decision to kill a human being? For this reason, the CTSKR wants humans to exercise 'meaningful human control' over the actions of autonomous systems. This, however, leads to a definitional issue: at what point is a given technological system autonomous?

In a sense, this is not a new issue by any means. Chris Jenks frames the question of military use of autonomous systems as the '30/30 problem' because 'The Campaign's current proposed ban would encompass some portion of LAWS which over 30 countries have employed for 30 years. This poses a staggering obstacle', notably missile defence.[23] Yet it is at least arguable that the mode of employment of these systems retains human decision-making and 'on the loop' control at a minimum. Moreover, the design and use of these systems places strict limits upon their possible effects. A Phalanx Close-In-Weapon-System (CIWS) is designed to destroy incoming sea-skimming missiles, therefore the types of targets that it is designed to sense are limited relative to human senses, and its authority to engage a target (and attempt to destroy it) is embedded into wider command and control systems such as the US Navy's AEGIS system. Human beings set the rules for a strictly limited form of conduct, and otherwise autonomous systems such as Phalanx carry them out faster than human beings could physically manage. While the prospect of robots replacing soldiers in the near future worries many, the reality is that autonomous systems provide these kind of capabilities in strictly limited environments.

Yet autonomous systems are continuing to improve. The day when people routinely delegate the authority to drive home after work to an autonomous vehicle is nearer than ever. However, while we are comfortable with autonomous systems destroying incoming missiles, delegating authority to machines to make lethal decisions is less palatable since 'There is no way to punish a robot'.[24] From the perspective of the machine, however, there is no real difference between targets that can be detected by sensors. If it were possible to define a human being in terms of sensor input, then systems such as Phalanx could kill humans as easily as they destroy missiles. The plain fact of the matter is that a fair amount of the hair-splitting in present debates on LAWS exists in order to try and split Phalanx and similar systems from those that are designed to sense and target humans. The reason for this is that any ban that included such systems would fail at the first hurdle of military necessity – without this kind of defensive capability, navies the world over would be incredibly vulnerable to attack by anti-ship missiles. If these systems are essential, then how can they be used in compliance with the law of armed conflict?

All sides to the present debate appear to consider that compliance with the law of armed conflict requires maintaining human responsibility for lethal decisions. The current debate on autonomous weapons focuses on two kinds of decision-making relationships: a human–machine bond and, for want of a better term, 'trading places' – the replacement of a human agent with a non-human agent, expected to make the same decisions that a human would. This focus is rooted in a 'bottom up' concept of compliance with the law of armed conflict. In essence, all decisions are required to be directed by humans, or at least supervised by them. This idea of legal compliance is, however, at odds with contemporary military practice. In 2001 Thomas Adams argued that 'military systems (including weapons) now on the horizon will be too fast, too small, too numerous, and will create an environment too complex for humans to direct'.[25] In the last case, military operations are already too complex for single individuals to control. For this reason, in many situations the control that individual service personnel have over the ultimate use of force is decidedly limited. Aerial warfare provides many examples of this. Consider, for example, the position of a fast jet pilot being instructed on a target by a tactical air controller on the ground. Unless the pilot is in possession of information that the use of force would be unlawful, there is every expectation that the pilot would use force against the target described to them by the tactical controller. In this instance, the very nature of military operations reduce the pilot's autonomy. Although we like to think about such choices as being made by rational individuals in full possession of all facts, the practice of warfare means that such decision-making is often distributed between participants. For this reason, it is also necessary to consider the structural features of compliance; in other words, not only a command theory (following orders) but also the design of military command structures and operational methods to ensure compliance in limited information scenarios, even when operations reduce individual autonomy to an extreme degree. The reason that 'bottom up' compliance takes precedence in the current debate is largely due to the way we usually consider decision-making in warfare. In an ideal world, this would be sufficient; however, it is necessary to consider how these decisions work in the context of military practice in order to demonstrate why this form of compliance is generally insufficient, and therefore why the effects of autonomous systems on structures of compliance should be a key concern.

Meaningful human control

The primary focus of the CTSKR is on the replacement of human decisions by those made by machines. This is the 'bottom up' challenge that LAWS pose to the just war tradition and LOAC. The CTSKR argues for changes to international law, and much of the discussion and debate about the use of autonomous weapons takes place in ethical terms.[26] Given that the three interpretations of human control are framed in terms of human decision-making, it is first necessary to contextualise the discussion in terms of military ethics, primarily the just war tradition.

Legal and moral decisions and choices are the fundamental unit of analysis in military law and ethics. Choice can be described and analysed in many ways, but primarily this is judged in terms of state decision-making (the choice to use armed force or engage in war) or at the individual level (whether or not a soldier's given course of action is morally justifiable). Military ethics is divided into two sets of criteria that are commonly required to be satisfied in order to define an action as justified, *jus ad bellum* (regarding the resort to war) and *jus in bello* (regarding the use of violence in war). Although they are traditionally considered separate, recent works have challenged this division, notably by Jeff McMahan, who insists that moral conduct in an unjust war (e.g. a soldier whose conduct satisfies *in bello* criteria on behalf of a state that fails the *ad bellum* criteria) still results in unjustifiable actions.[27] McMahan's requirement to consider the justice of war and that one side will always be unjust has faced criticism from some, such as Nigel Biggar.[28] For present purposes, however, we should consider the limiting case, which is the use of autonomous weapons in a conflict that is somehow justified, since if this is immoral, it would fail the standards asserted by both authors.

The three standards for human–LAWS interaction defined by the CTSKR represent an attempt to describe human relationships to non-human decision processes. To this end, most agree with the idea that the need to exercise 'meaningful human control' is necessary in order to ensure compliance with our existing normative frameworks that justify the use of violence and force in war. The exact nature of this standard is, however, contested. Part of the problem is that this standard has been advanced as part of an attempt to ban an amorphous class of military hardware (and software), so the reticence of states is at least understandable. Each of the three relationships presents different challenges to the notion of humans exerting meaningful control.

The least challenging form of human–LAWS interaction is where humans exert direct control over targeting decisions made by an LAWS. In effect, humans must decide in each and every case whether the use of force is justifiable. This decision is, however, revered in the case of human 'on the loop' interactions, where human beings supervise the operation of LAWS, and the function of human decision-making in this case is to interrupt the lethal decisions made by an LAWS. On being informed that the LAWS plans to perform an action that the human considers wrong, the human being supervising the machine makes a decision and communicates this decision to the machine, which then aborts this course of action. The last case is where such a machine would operate unsupervised, making decisions in an autonomous manner. How, critics ask, could meaningful human control be established over such a system?

One argument is that meaningful human control could be established by ensuring that LAWS conform with the law of armed conflict (and in a similar sense, to the just war tradition).

Meaningful human control would arise from the fact that these machines would, in effect, be perfect soldiers. Unburdened by flaws of human decision-making, and irrational reactions to the stresses of combat, these machines, when given lawful orders, would perform their lethal tasks better than humans could. If an LAWS could even interpret the destruction of its compatriot machines in a way that approaches the human sense of loss, it would still not be inclined to form feelings of malice or hatred towards the perpetrators. With good design, so the understanding goes, we don't have to worry about revenge and reprisals.

Unsurprisingly, this prospect has been attacked on numerous fronts. How, one asks, could machines be programmed to follow the laws of war, when they contain so many grey areas? More to the point, given current limits to machine sensors, processing and so on, how could we design machines to distinguish between civilians and legitimate targets on the battlefield? Humans, it is argued, will always need to be on hand to perform the kind of abstract and symbolic reasoning that lies beyond the capabilities of current systems. More to the point, even as artificial intelligence and probabilistic methods of machine learning increase in scope, speed and accuracy, they still make copious errors. The question of whether machines should replace humans once they beat the (error-prone) human baseline lies beyond the scope of this chapter. More important, I think, is the fact that these kind of atomic decisions, or simple paired relationships, do not accurately portray contemporary warfare.

Two principal challenges present themselves when thinking about, and describing, the moral challenges of war in such granular terms. First is that information and descriptions are taken to be perfect. A soldier firing upon another rests upon the fact that these identities are both knowable, and known – something that should never be taken for granted in war, since, as Clausewitz pointed out, uncertainty is inherent to war itself.[29] Still, for modelling individual decisions, most try to steer clear of the inherent uncertainties of warfare and combat. The second key problem is that warfare is inherently a social activity. Soldiers do not retain full autonomy themselves, but exert carefully circumscribed authority on behalf of their state or political leadership. At a certain level, all moral decisions are made in a social context, but many moral decisions are individual in character. This is key to understanding Schulzke's argument that command responsibility can override the problems presented by non-human agents on an individual level. In Schulzke's argument, command responsibility allows for the social transfer of responsibility, as well as the assignation of responsibility to humans for the actions of a non-human agent.[30] This is not 'fair' – command responsibility is arbitrary by design, but it is designed this way so that in theory there are no 'responsibility gaps' when states wage war.[31]

The kind of analysis made by Jeff McMahan and other revisionists, rooted in the liberal moral theory of John Rawls, erases this social element except as a system of command or transferred responsibility. What this fails to recognise is two things: first that many individuals in military organisations have very limited autonomy as parts of a military organisation, and, more important, that they will routinely lack the information required to make the kind of holistic analysis that revisionists require of them. These two issues – distributed generation/possession of information and the limits placed on individuals – are a function of the fact that most people are essentially mediating the decisions of others. War is primarily mediated violence, not thousands of parallel decisions to use force autonomous from one another.

Violence as a mediated decision

Autonomous systems will not necessarily change the mediation of decision-making. Thinking of violence as a mediated decision does not require reinventing the wheel. The question of what does, and does not, constitute violence is beyond the scope of this chapter, but it allows us to place a couple of limits on its content. One is that the term 'violence' used herein refers to active, rather than latent, forms of violence. In the present day, and for good reason, some argue that social structures can be violent, and that non-physical activities can constitute violence or be equated to legal standards of armed attack in international law.[32] For present purposes, we need not consider the thresholds; instead the most obvious case suffices for present discussion.

Violence is always mediated – for one human to do physical damage to another, some form of physical action has to take place. The differences between the mediation of violence matter: firearms differ from fists. But when we consider the connection between decisions and violence it is also clear that it is necessary to consider mediation in wider terms than tools. If an officer orders a subordinate to fire their weapon at a target, then the decision that originates the act of violence does not belong to the subordinate, although their decision to obey the order is certainly an important factor. Even from such a simple scenario it is clear that a number of academic disciplines can inform our inquiry. Psychology and neuroscience can provide useful perspectives on the actual decision itself. The type of measured and reflective decision-making that we use to discuss the rights and wrongs of violence is what Daniel Kahneman would label 'Type II' thinking, as opposed to the 'Type I' near-automatic decision processes that form the majority of human decisions and consciousness.[33] Mediating decisions via other persons requires that they obey orders and perform the required actions. Perfect obedience is rare in human beings; hence the principal–agent problem is a large area of academic enquiry.[34] One of the important features of the literature on the principal–agent problem is that social concepts and attitudes have a profound effect on human engagement with these kinds of scenarios.[35] Indeed, by characterising this as a military decision, I am presupposing that this exists within an accepted social context, one in which superior officers have the legitimate right to order their subordinates to engage in acts of violence, and every expectation that they will follow socially acceptable orders.

The social character of military violence is important. Bureaucracies tend to be stable mechanisms for transmitting and mediating decisions that result in acts of violence. Probably the prime example of this currently is the US practice of targeted killings. Decisions made in the White House are transmitted via the US military and intelligence agencies to individuals, piloting drones, who ultimately press the buttons, which results in missiles being fired.[36] This is not a new phenomenon, but demonstrates the degree to which lethal decisions are separated from consequent acts of violence by the obedience of subordinates. Although many immediate acts of violence are individual decisions, often soldiers firing weapons in self-defence, most advanced weapons platforms are integrated in systems of decision-making that mean single acts of violence might be constituted by dozens of individual decisions. This is the ultimate aim of many contemporary warfighting concepts, such as Admiral Owens' 'System of Systems' in the 1990s or General McChrystal's attempt to transform JSOC's operational methods because 'it takes a network to kill a network'.[37]

War is a social and political activity, and the social context and political aims inherent in these two elements serve to constrain the individual autonomy of those involved. Warfare, above all, always involves some form of restraint, however small.[38] The concept of the fully autonomous human soldier (or other service personnel) is a fiction, ever more so in today's professional state militaries. While individual soldiers are generally ascribed high degrees of freedom and autonomy (for instance, to kill another human being on the basis that they believe them to be a threat, or a member of the opposing forces), this autonomy is limited and structured by the character of the military that they belong to. This constraint could be very weak – marauding bands of fighters with no care for international law might only be constrained by group membership and identity. Alternatively, this constraint could be exceedingly strong. The history of post-Cold War peacekeeping abounds with stories of soldiers complaining about the inability to use force at times when they perceived that violence could improve a situation, but were denied the ability to do so by rules of engagement, superior orders and the general rule of peacekeeping: the non-use of force except in self-defence.[39] In theory, soldiers are entirely free to adhere to the law of armed conflict, or obey its restrictions. In practice, most Western militaries keep their forces on a tight leash. Rules of engagement that are more restrictive than the law of armed conflict is one example of this. Strict systems of command permission requiring senior authorisation for even planning to engage enemy forces are another, due to the political nature of uses of force.[40] Military cultures that emphasise after-action reviews may deter individuals from taking action in grey-area use of force situations due to the bureaucratic burden that any act of violence entails. Beyond this, most methods of joint warfare require standardised actions and communication with fellow service personnel. It is somewhat difficult to go 'off the reservation' when half a dozen people hold the ability to abort the use of force.

The converse is also true: the reduction of autonomy may also reduce the ability of individuals to disagree with plans of action. In the mediated decision of an officer ordering their subordinate, the same structure that bureaucratises warfare also serves to constrain the ability of individuals to refuse to use force, except where it is possible to articulate such decisions in terms amenable to the structure itself. The subordinate receiving an order to kill another person would find themselves in trouble if they refused to kill an enemy belligerent in a time of war. An answer of 'I don't feel like pulling the trigger' might warrant official sanction; however, answers explained in terms of the rules of war speak directly to military hierarchies themselves. It would, for example, be hard to sanction the subordinate should they refuse an order that requires the murder of a civilian.

None of the my argument so far should be taken to deny the responsibility of individuals for the actions they take, and the decisions that constitute them. Given the weight of evidence, however, the notion of a soldier or other service person as a *tabula rasa* when it comes to the decision to use force is certainly false. The reduction of autonomy inherent in modern warfighting is also due to the lack of information available to each given participant. We often debate use of force decisions as if individuals are Sherlock Holmes, Arthur Conan Doyle's famous detective, in that the weight of decision is placed upon their shoulders, having assessed the evidence at hand. In this framing, the ability of machines to make decisions on par with humans becomes the focal point of debate, or whether states should be able to swap out human agents for autonomous ones.

Distributed information in military operations

How could autonomous systems challenge the structures of LOAC compliance outlined thus far? Two reasons present themselves: ACS serving in a command function, and ACS that generate information used in military operations. While it is natural to worry about computer systems or AI giving out orders – indeed rule by AI is a key feature of many science fiction films and books such as *2001: A Space Odyssey* and *Colossus: The Forbin Project* – here I argue that the production of information by autonomous systems poses a greater problem for LOAC compliance structures. Both of these scenarios essentially resort to the same issue: humans using violence directed in whole or in part by autonomous systems. This is an inversion of the focal problems that drive the LAWS debate, but in my mind is far likelier to cause problems in the near-future.

AI and other systems promise to revolutionise the decision-making of organisations. Businesses the world over are beginning to take advantage of 'big data' – software and systems that are capable of analysing datasets far beyond human capabilities.[41] The promise of such systems is that they help businesses make better decisions. Computers can increase the efficiency of operations by analysing huge volumes of data, and businesses that inform their decision-making with such systems, it is hoped, will outperform others.[42] Software platforms such as IBM's Watson can now perform complex tasks such as supporting clinical decisions in the treatment of cancer,[43] and recent research argues that the IBM Watson for Oncology platform's 'choices today fall within evidence-based standards, Watson For Oncology has the capacity to provide greater precision through iterative training and development'.[44] The relationship between humans and machines in such circumstances is unclear, but, in theory, humans remain in control.

The degree of control that humans retain over their organisations is, however, relative to the involvement of humans in decision-making. It is here that AI and ACS might pose a problem in future. AI is excellent at interpreting the world which can be read by computers. The steady increase in the use of digital devices also increases the range of material that computer systems can easily read, and therefore outcompete humans in examining. While militaries may never rely upon machines to tell human beings whom to kill, it is highly likely that they will one day rely upon them to process data to identify and locate their opponents. Signals intelligence, and similar data processing activity, can now be used to pick individuals out of the morass. The steady increase in digital intelligence capabilities shows no sign of abating. Intelligence agencies such as the US Geospatial Intelligence Agency are turning towards 'activity based intelligence' – cross-comparing datasets to track individuals.[45] Any security organisation hunting for irregular opponents in 'civilian clutter' (to use General Michael Flynn's phrase) is likely to turn to such methods.[46] The use of signals intelligence by US and UK special forces in Iraq to hunt down al Qa'ida in Iraq (AQI) is an example of this, and the use of similar methods will increase in future.

The question, then, is what happens when a substantial section of intelligence processing is turned over to ACS? If human beings are unable to produce this kind of intelligence and unable to check the full rationale of the computer system producing it, would they ever be able to exert 'meaningful human control' over these systems? If these systems cannot be blamed for the decisions they make, will there be someone who can be held accountable? If the answer to both questions is no, then ACS give rise to a significant problem. Rather than worrying about killer robots stalking the battlefields of the future, we should be far more concerned at the prospect of soldiers raiding a house that is identified by an ACS as a military location. By offloading a large portion of the act of identification to machines, humans mediate the decisions that they make. If, as the previous section outlined, responsibility should lie with the decision-maker in such circumstances, this gives rise to the possibility of non-voluntary killings, and situations that, by design, it is impossible to hold humans to account for acts of violence.[47]

The challenge autonomy poses to military structures

If we temporarily set aside the prospect of autonomous systems making any initial or overarching decision to use violence, then our consideration of autonomous machines has to extend to their role in the mediation of that overarching decision, and their role in the human decisions that eventually result in an act of violence. Responsibility for these decisions is managed by the social structure of command responsibility. Schulzke's argument that command responsibility is the answer to Sparrow's critique of robot responsibility rests upon the issue raised in this chapter: the assignation of responsibility to persons for the actions of non-human agents.

Sparrow identified two classes of person to whom responsibility could be assigned – systems developers and commanders – arguing that neither could be held responsible.[48] Schulzke, alongside Asaro, disagrees, arguing that developers could be held responsible for the 'misdeeds' of LAWS and that any reticence on the behalf of developers to build such systems is a good thing as it will improve the design of such machines. The problem with this vision, especially when considering the role of ACS in intelligence analysis, is that failure is a built-in element of such systems. From a statistical standpoint, there will never be a computer system that performs large dataset analysis that eliminates the possibility of either false positives or false negatives. While a false negative in a system employed to identify a potential member of a terrorist network implies that militaries will miss the opportunity to target a person, a false positive implies that someone will be identified as a target when they are not in fact one.

The importance of the probabilistic nature of ACS is that it poses a challenge to the nature of command responsibility structures. Human beings are fallible entities, but it is difficult to determine the precise level of human fallibility, and it is not possible to set a defined standard of fallibility to which all humans must pass. Training regimes for professional militaries seek to guarantee minimum standards of competence, but it is rare that such competence can be measured in an objective fashion, beyond basic cognitive or physical tests. Schulzke argues that commanders should be held 'responsible for the actions of AWS to roughly the same extent that they are now, as they have similar powers to constrain the autonomy of AWS as they have over human soldiers'. While this may be true for LAWS that are physical weapon systems, it is difficult to conceive of how commanders could be expected to compare and contrast non-human agents in the form of decision-support machines with human cognition. Simply put, expert systems and other forms of AI are likely far to surpass human capability in certain respects, but will still remain fallible. This leads to two propositions: one, that such machines should not be employed, and two, that they could, like some AWS, become a more ethical choice if they are far better than humans at the same take. Given that the focus of discussion is the employment of such machines, I will focus upon this second proposition.

What would it mean for commanders to employ an expert ACS that is better than human beings, but known to be fallible? In the context of this chapter, it creates a significant problem when thinking about the assignation of responsibility. Developers could hardly be held to account for a system that is designed to the best possible standards, particularly when they are likely to be unable to test it using real intelligence data. Service personnel that act upon the information provided by such an ACS should not be held to account, since otherwise we are challenging the structure that permits distributed military activity. Two further groups then present themselves: the commanders that authorise the use of such systems, and the analysts that work in tandem with them.

Should an analyst be held responsible for the failure of what is essentially software? I argue that this would be inherently unfair. Imagine, for a second, being asked to assess the output of a computer system in this scenario. In order to challenge the decision that an ACS arrives at, even if it is relatively mundane (identifying a person as a potential node in an insurgent network, for example), would involve a colossal feat of human cognition. First, the analyst would have to understand the computer algorithm governing the behaviour of the system. If this ACS has used machine learning to arrive at its classification or decision tree mechanism, then the analyst would, in the event of any positive result, require knowledge of this training set. Moreover, the analyst would then be required to understand the operation of the algorithm on a dataset that, in the case of big datasets, might even be a stream of data. The human would, in this instance, be too slow. The ACS would effectively be a black box. It is difficult to see how the analyst could be held responsible for producing intelligence products based upon its output. The system would necessarily have to be trusted, representing, in the words of Roger Brownsword, an 'autonomy eroding technology'.[49]

What, then, of commanders? Consider the prospect here: the commander would be held responsible for the functioning of an ACS that is, by definition, unsupervisable. After all, whereas a human 'supervising' the targeting loop of a LAWS can check to match the target to what they can perceive, here a commander is being asked, like the analyst, to second-guess the decisions and classifications of a computer system whose function they cannot possibly comprehend. It is here that Schulzke's argument breaks down, because it is difficult to see who would wish to be placed in such a situation. For this reason, responsibility for the decisions of ACS, and the actions that take place upon them, are difficult to place upon human agents. This is the ultimate challenge of ACS: the possibility for non-voluntary uses of force, in an Aristotelean sense, where responsibility cannot meaningfully be assigned to human agents.

Conclusion

This chapter does not seek to argue that command responsibility is defunct, or that LAWS and ACS pose an existential threat to it. Human beings are likely to adapt to the integration of autonomous agents into our existing legal and moral structures. This chapter does not propose a way forward in this regard, particularly since this is such a nascent issue. Instead, it points to a way forward for thinking about the role of legal and moral structures that ensure compliance with the law of armed conflict. Instead of placing specific focus on the problems raised by LAWS, it is necessary to think in a wider sense about the computer systems that are now part and parcel of military operations. I argue that the rise of ACS has the potential significantly to enable militaries in the future, but it will require rethinking the role of responsibility for the use of force, as well as the design of intelligence production and communication so as to increase human culpability for actions in other ways. For example, and this is just a thought experiment, it may require that humans working upon ACS-produced intelligence may need to take extra precautions when using force subsequent to the role of such systems. As good as AI and expert systems promise to be, we may need to maintain a level of precaution regarding their use, at least until these questions are studied in more depth.

Notes

1 Lata Sundar, 'Komatsu's Autonomous Haulage System passes 330 million tonne landmark', *Australian Mining*, 21 May 2015. Online at: www.australianmining.com.au/news/komatsus-autonomous-haulage-system-passes-330-million-tonne-landmark/ (accessed 6–5–2018).

2 Carpenter, Charlie, 'Beware the killer robots: Inside the debate over autonomous weapons', *Foreign Affairs*, Vol. 3, 2013.

3 DeLanda, Manuel, *War in the Age of Intelligent Machines*, New York: Swerve, 1991.

4 The civil society 'Campaign to Stop Killer Robots' formed by a number of leading NGOs began on October 19, 2012.

5 Jenks, Chris, 'False Rubicons, Moral Panic & Conceptual Cul-De-Sacs: Critiquing & Reframing the Call to Ban Lethal Automatic Weapons', *Pepperdine Law Review*, SMU Dedman School of Law Legal Studies Research Paper No. 243, 2016. Available at SSRN: http://ssrn.com/abstract=2736407.

6 Human Rights Watch, *Losing Humanity: The Case Against Killer Robots*, 2012). Available at: www.hrw.org/report/2012/11/19/losing-humanity/case-against-killer-robots (accessed 7 June 2018).

7 Schmitt, Michael N., 'Autonomous Weapon Systems and International Humanitarian Law: A Reply to the Critics' (December 4, 2012). *Harvard National Security Journal Feature* (2013). Available at SSRN:http://ssrn.com/abstract=2184826 or http://dx.doi.org/10.2139/ssrn.2184826.

8 Sharkey, Noel. 'Saying "no!" to lethal autonomous targeting', *Journal of Military Ethics*, Vol. 9, No. 4, 2010, pp. 369–83, Asaro, Peter. 'On banning autonomous weapon systems: human rights, automation, and the dehumanization of lethal decision-making', *International Review of the Red Cross*, No. 886, 2012, pp. 687–709.

9 Asaro, 'On banning autonomous weapon systems: human rights, automation and the dehumanization of lethal decision-making', *International Review of the Red Cross*, No. 886, 2012, p. 693.
10 J. M. McClelland, 'The Review of Weapons in Accordance with Article 36 of Additional Protocol I', *International Review of the Red Cross*, No. 850, 2003, pp. 397–420.
11 Carpenter C., 'How do Americans feel about fully autonomous weapons?', *Duck of Minerva*, 10 June, 2013. Available at: http://duckofminerva.com/2013/06/how-do-americans-feel-about-fully-autonomous-weapons.html (accessed 5 June 2018); Carpenter, C. 'Who's afraid of killer robots? (and why)'. *The Washington Post's Monkey Cage Blog*, 30 May 2014. Available at: www.washingtonpost.com/blogs/monkey-cage/wp/2014/05/30/whos-afraid-of-killer-robots-and-why/ (accessed 5 June 2018); Horowitz, Michael C., 'Public opinion and the politics of the killer robots debate', *Research & Politics*, Vol. 3, Issue 1 (Feb 2016).
12 Ministry of Defence (2011) www.gov.uk/government/uploads/system/uploads/attachment_data/file/33711/20110505JDN_211_UAS_v2U.pdf (accessed 5 June 2018).
13 Department of Defense (2012) www.dtic.mil/whs/directives/corres/pdf/300009p.pdf.
14 Horowitz, Michael C. and Scharre, Paul, 'An Introduction To Autonomy In Weapon Systems', *Project on Ethical Autonomy Working Paper*, Center for a New American Security. (February 2015) p. 4. https://s3.amazonaws.com/files.cnas.org/documents/Ethical-Autonomy-Working-Paper_021015_v02.pdf.
15 Mahnken, Thomas G., *Technology and the American Way of War since 1945*, Columbia University Press, 2010.
16 Ellis, John, *The social history of the machine gun*, JHU Press, 1986.
17 Sparrow, Robert, 'Killer Robots', *Journal of Applied Philosophy*, Vol. 24, No. 1 (2007), pp. 62–77.
18 Schulzke, Marcus, 'Autonomous weapons and distributed responsibility', *Philosophy & Technology*, Vol. 26, No. 2 (2013), pp. 203–19.
19 Noorman, Merel, 'Responsibility practices and unmanned military technologies', *Science and Engineering Ethics*, Vol. 20, No. 3 (2014), pp. 809–26.
20 United Nations General Assembly, *Report of the Special Rapporteur on extrajudicial, summary or arbitrary executions, Christof Heyns*, A/HRC/23/47. www.ohchr.org/Documents/HRBodies/HRCouncil/RegularSession/Session23/A-HRC-23-47_en.pdf (accessed 8 June 2018).
21 'Recognizing the need for human control', Campaign to Stop Killer Robots, 2015 www.stopkillerrobots.org/2015/04/humancontrol/ (accessed 7 June 2018).
22 Anderson, Kenneth, Daniel Reisner and Matthew C. Waxman, 'Adapting the Law of Armed Conflict to Autonomous Weapon Systems', *International Law Studies*, Vol. 90, September 2014, pp. 386–411.
23 Jenks, 'False Rubicons' (see note 5 above).
24 Sharkey, Noel, 'Saying "no!" to lethal autonomous targeting', p. 380 (see note 8 above).
25 Adams, Thomas K.; 'Future warfare and the decline of human decisionmaking', *Parameters*, Vol. 31, No. 4 (2001), p. 57.
26 Asaro, 'On banning autonomous weapon systems' (see note 9 above); Sparrow, 'Killer Robots' (see note 17 above); Arkin, Ronald, *Governing lethal behavior in autonomous robots*, CRC Press, 2009.
27 McMahan, Jeff, *Killing in war*, OUP: Oxford, 2009.
28 Biggar, Nigel, *In defence of war*, OUP: Oxford, 2013.
29 Clausewitz, Carl von, *On War*, trans. J.J. Graham, 'Introduction' and 'Notes' by Colonel F. N. Maude, C. B (Late R. E.) and 'Introduction to the New Edition', by Jan Willem Honig, New York: Barnes and Noble, 2004.
30 Schulzke, *Autonomous weapons and distributed responsibility* (see note 18 above).
31 McDonald, Jack, *Ethics, Law and Justifying Targeted Killings: The Obama Administration at War*, London: Routledge, 2016.
32 Schmitt, Michael N. (ed.), *Tallinn Manual on the International Law Applicable to Cyber Warfare*, Cambridge: Cambridge University Press, 2013.
33 Kahneman, Daniel, *Thinking, Fast and Slow*, London: Macmillan, 2011.
34 Eisenhardt, Kathleen M., 'Agency theory: An assessment and Review', *Academy of Management Review*, Vol. 14, No. 1 (1989), pp. 57–74.
35 Miller, Gary J. 'The Political Evolution Of Principal-Agent Models', *Annual Review of Political Science*, Vol. 8, 2005, pp. 203–25.
36 Klaidman, Daniel, *Kill or Capture: The War on Terror and the Soul of the Obama Presidency*, Houghton Mifflin Harcourt, 2012.
37 Owens, William A., 'The emerging US system-of-systems', No. 63. National Defense University Washington DC Institute For National Strategic Studies, 1996; McChrystal, General Stanley, *My Share of the Task: A Memoir*, New York: Penguin, 2013.

38 Gow, James, *War and War Crimes*, OUP, Oxford, 2013.
39 Bellamy, Alex J., Paul D. Williams and Stuart Griffin, *Understanding Peacekeeping*, Polity, 2010.
40 Sagan, Scott D., 'Rules of engagement', *Security Studies*, Vol. 1, No. 1 (1991), pp. 78–108.
41 Manyika, James, et al., 'Big data: The next frontier for innovation, competition, and productivity', *McKinsey Global Institute*, 2011.
42 Chen, Hsinchun, Roger H.L. Chiang, and Veda C. Storey, 'Business Intelligence and Analytics: From Big Data to Big Impact', *MIS quarterly*, Vol. 36, No. 4 (2012), pp. 1165–88.
43 Bach, Peter et al., 'Beyond Jeopardy!: Harnessing IBM's Watson to improve oncology decision making', *ASCO Annual Meeting Proceedings*, Vol. 31, No. 15_suppl, 2013.
44 Kris, Mark G. et al., 'Assessing the performance of Watson for oncology, a decision support system, using actual contemporary clinical cases', *ASCO Annual Meeting Proceedings*, Vol. 33, No. 15_suppl, 2015.
45 Long, Letitia A., 'Activity based intelligence: Understanding the unknown', *The Intelligencer: Journal of US Intelligence Studies*, Vol. 20, No. 2 (2013), pp. 7–16.
46 Flynn, Michael T., Rich Juergens and Thomas L. Cantrell, 'Employing ISR SOF Best Practices', *National Defense Univ Washington DC Inst For National Strategic Studies*, 2008.
47 McDonald, *Ethics, Law and Justifying Targeted Killings* (see note 31 above).
48 Sparrow, 'Killer Robots' (see note 17 above).
49 Brownsword, Roger, 'Autonomy, delegation, and responsibility', In M. Hildebrandt & Antoinette Rouvroy (eds.), *The Philosophy of Law Meets the Philosophy of Technology: Autonomic Computing and Transformations of Human Agency* (Routledge, 2011), p. 64.

13

LEGAL-POLICY CHALLENGES OF ARMED DRONES AND AUTONOMOUS WEAPON SYSTEMS

Kenneth Anderson and Matthew C. Waxman

Historically, new weapon technologies often generate anxiety about their presumed deleterious effects on the law and ethics of warfare. This is especially true of weapons that reduce one's own warriors' exposure to risk while posing new threats to the enemy, or that might pose greater risks to civilians. Concerns that armed, remotely piloted or unmanned aerial vehicles (UAVs) and autonomous weapon systems will erode constraints on how force is used, for example, echo tropes once associated with the advent of crossbows, submarines, and air bombardment. As with those previous weapon technologies, one immediate reaction is unrealistic calls either to prohibit them or to develop specific rules for them that would tend to nullify the new military advantages of the weapon.

Attempts to prohibit or nullify new weapon technologies with clear military advantages – let alone technologies that can offer greater targeting precision and reduce levels of battlefield force and harm – are not likely to succeed and, to the extent that they would rule out systems that can reduce battlefield harms, should not do so. Such attempts also risk occupying the policy, diplomatic, and negotiating space among states seeking both the benefits of technology and realistic legal constraints, thereby crowding out sensible regulatory evolution. This is not to say that UAVs and autonomous weapon systems pose no novel issues or that concerns about their impact on legal and ethical constraints are not warranted. They do pose challenges, but these challenges are mostly novel as a matter of degree rather than of kind.

Moreover, and like many other significant advances in military technology, they pose challenges of interpretation and compliance for the law of armed conflict, but they may also enhance legal and ethical constraints on military force. Possession of unmanned or autonomous weapons may lower thresholds for resorting to military force in the first place or in specific situations, but they might also lower the quantum of force used in any particular situation. Moreover, many other recent technological advances – high-altitude precision bombing, stand-off weapons, and so on – have the same effect, which can be both dangerous and beneficial. These systems may increase the real or perceived distance between soldiers and their targets, but so have those other developments – artillery, rockets, or missiles, for example – and this may produce greater or lesser care in targeting, depending on many other factors.

One reason why the legal and ethical challenges posed by new weapon technologies may be greater today and tomorrow than in the past is because the pace of technological change is quickening faster than the pace of legal evolution and refinement. The component technologies, such as sensors and analytical capabilities, are evolving very rapidly. They are also being developed primarily in the fast-moving private sector and flowing to the military sector, rather than vice versa as was often the case in the past. Legal processes and interpretive development, by contrast, tend to move slowly. To the extent that negotiating energy is absorbed by idealistic, but mistaken, efforts to ban or nullify the technologies, attempts at sober, realistic regulation grounded in the fundamentals of weapons law but adapted to the concrete evolution of technology might move even more slowly – if at all.

Another reason why the legal and ethical challenges posed by new weapon technologies may be greater today and tomorrow than in the past is because the use of those technologies is less visible. Remotely piloted vehicles, for example, are often designed and operated to be largely undetectable (and certainly have been used by the United States with that intention); even when their use is widely known, the details of how they are used remains shrouded in secrecy. An effect of these technologies is that their remoteness and precision – which may increase their capabilities for discrimination among targets – tend to increase the ability to use them invisibly, without attribution. Moreover, the most important details of how autonomous weapons operate will be buried deeply in computer code, rather than externally discernible behaviours. Previous steps and leaps in weapon technology were much more observable, as were their uses and the ability to attribute responsibility for an operation – and so, therefore, they could be more readily measured and defended or criticised against international legal standards.

This combination of factors – rapid technological evolution and low visibility – is true not only of technology that delivers kinetic payloads. It may be even truer of cyber capabilities, nanotechnology, and other areas where an advantageous feature of the technology is its relative undetectability.

Even if the law of armed conflict has proven effective and adaptable in regulating past evolutions – and even revolutions – in military technology, this combination of factors may inhibit the forward development of law. The pace with which weapon technology changes means that there is less time than in the past for responsible states to develop common understandings of how the law of armed conflict applies to them, and less time to convince sceptical audiences of the effectiveness of those understandings. The low visibility of these technologies and their use makes it difficult for key actors in the international systems – states, international organisations, non-governmental organisations (NGOs), and others – to appraise them. One need not embrace sweeping bans on such weapon technologies (or think they would help more than hurt) to understand how these uncertainties generate anxiety and scepticism about their legal regulation.

The first section of this chapter addresses legal-policy challenges associated with armed UAVs, and the second section addresses such challenges associated with autonomous weapon systems.[1] The next section argues that the United States and its closest allies can and should adapt the law of armed conflict to deal with these emergent technologies, but doing so effectively will probably require higher levels of public transparency about weapon systems than they are accustomed to.

Armed UAVs: legal-policy issues

UAVs piloted from afar are already a significant component of the United States' arsenal, and they are proliferating rapidly worldwide.[2] The vast majority are not armed today, however, but are instead tools of surveillance – which for a long time to come will remain their most important operational contribution. Moreover, the growth in UAVs in US and other states' arsenals is as significant for the wide and increasing variety of UAVs – from small model-aeroplane types thrown into the air by soldiers seeking to look over the next hill to very large, high-altitude drones intended to maintain constant surveillance over weeks or more.

As remote-piloting and automation technologies improve, the military roles in which drones can play a part will ramify, although armed drones will be only one part of a transformation of military aviation. Further, the military uses of drones in any role will simply mirror, and ultimately be merely a tiny strand of, the broader transformation of aviation over the next few decades, as UAVs in many different sizes and forms and levels of automation enter commercial and domestic life around the world.

There are important caveats to this narrative of relentless expansion and ramification in the military sphere, however. Current generations of UAVs have not been designed for use against militarily advanced powers or, realistically, any state with a functional air defence system. Current-generation UAVs are vulnerable both to kinetic attack and hacks of the communication links; their successes have been in dealing with non-state adversaries and terrorists lacking the most sophisticated means of defence.

The United States and its partners have used armed UAVs in a variety of contexts and types of conflicts. The most significant, and controversial, use of armed UAVs, however, is as a raiding weapon in the transnational armed conflict with al Qa'ida and its affiliates – a weapon platform that can be used to combine ubiquitous surveillance with a remote, precise, unpredictable strike on limited targets in remote places.

Some of the most publicised and debated uses of armed drones by the United States have been in Pakistan and Yemen. Many in the international community would dispute both that the US can legally be engaged in a transnational armed conflict with a non-state actor and that such counter-terrorism raids by armed drones are carried out in a place of armed conflict to which the law of armed conflict applies. Armed UAV technology, on this view, specially enables the extension of uses of force lawful only in so-called active zones of hostilities to many other places far from the 'battlefield.' The use of these weapon systems – especially in counter-terrorism operations – has given rise to concerns among some states, NGOs, and influential international officials that such systems pose special problems for the law and ethics of military operations.

The US government position is that legal questions about armed UAVs are analytically distinct from legal questions about killing al Qa'ida and other terrorism suspects outside of traditional combat zones. It has argued that UAVs should be evaluated legally like any other weapon platform.[3] In 2010, then-Legal Adviser to the US State Department Harold Koh put it this way:

> [S]ome have challenged *the very use of advanced weapons systems*, such as unmanned aerial vehicles, for lethal operations. But the rules that govern targeting do not turn on the type of weapon system used, and there is no prohibition under the laws of war on the use of technologically advanced weapons systems in armed conflict – such as pilotless aircraft or so-called smart bombs – so long as they are employed in conformity with applicable laws of war. Indeed, using such advanced technologies can ensure both that the best intelligence is available for planning operations, and that civilian casualties are minimized in carrying out such operations.[4]

Among the important points implicit in that paragraph are that, from a legal standpoint, armed drones are not much different than manned aircraft, cruise missiles, and other technologies that have long existed on the battlefield – and that this is true of places that critics also might not regard as legitimate 'battlefields', either. Likewise, the point that UAVs can improve compliance with humanitarian law (for example, by improving real-time assessment of possible targets or collateral damage and because the requirements of self-defence for manned aircraft may produce less accurate targeting). Despite some claims from critics that armed drones as a weapon technology do not produce such humanitarian benefits, some major advocacy groups have gradually moved to distinguish the weapon platform as such from their criticism of the strategic uses to which it is put – namely, targeted killing.[5]

Treating armed UAVs like any other weapon platform is a valid position, as far as it goes – with respect to purely legal questions. Once the discussion expands to legal-policy questions, such as how armed UAVs might affect international actors' willingness to use military force or the extent to which they are likely to cause or avert collateral damage, discussion very quickly moves to UAVs' special attributes – particularly, the remoteness with which they are controlled. That is partly because legal and policy questions are not neatly separable, and partly because American use of UAVs is so closely associated in public and diplomatic debates with military actions against non-state terrorist organisations, particularly al Qa'ida and its offshoots or affiliates.[6]

Indeed, it is almost impossible to find a significant piece of commentary on US armed UAV policy – advocacy reports, government speeches or testimony, or academic scholarship – that does not very quickly become a commentary on US counter-terrorist targeted killing policy. That is true of commentary in both directions – supportive and critical.

Figure 13.1 lists the main criticisms that have been levelled against US use of UAVs, divided into two major categories: those concerns that are primarily about the legality of ethical legitimacy of UAV use and those that are primarily claims about their dangerous or counterproductive effects.

As to the first category, the most prominent criticisms include:

- Many lethal UAV operations against terrorism suspects constitute extrajudicial killing, assassination, or deprivations of rights without the basic legal processes required by international law.
- These operations lack sufficient oversight, or those involved in individual targeting decisions are not sufficiently accountable.
- Remote control of targeting decisions is dehumanising and diminishes ethical constraints on killing.
- Remote targeting with UAVs tends to be indiscriminate, resulting in relatively high collateral damage.
- States are more likely to violate other states' sovereignty with UAVs than with manned aircraft or other modes of military attack.

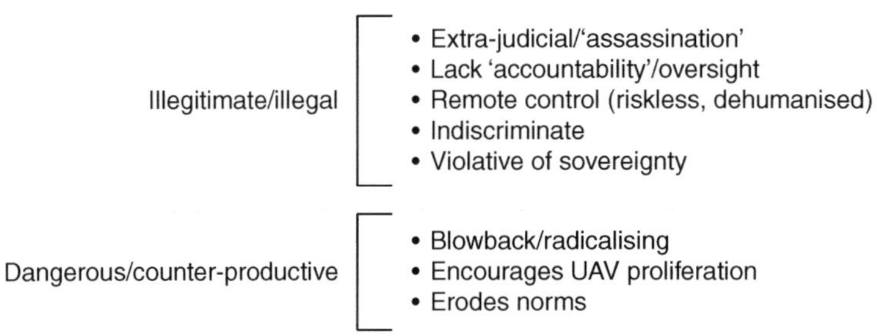

Figure 13.1 Main criticisms of US drones use.

As to the second category, the most prominent criticisms include:

- Lethal targeting operations are counterproductive, because armed strikes and especially collateral damage causes blowback, including violent radicalisation of populations in whose locale strikes take place or among sympathetic communities worldwide.
- They encourage proliferation of armed UAVs among other states and non-state actors.
- They erode norms against resort to armed force outside of traditional battlefields or very narrowly confined categories of cases, particularly because of the difficulties of definitively attributing an operation and holding the actor accountable.

It is important to recognise that although all of these criticisms are frequently levelled in discussions on armed 'drones', only two of them – that remote-control targeting dehumanises military operations and that it encourages UAV proliferation – are really about the technology itself or inherent in the weapon platform. It is sometimes asserted that reliance on UAVs is particularly likely to cause violent resentment among populations local to targeting operations or else sympathetic global populations, but that argument is rarely supported with even the barest empirical evidence.[7]

The vast bulk of these criticisms are actually better aimed at the policy of targeted killing beyond traditional battlefields. They are not really arguments about the means and methods of hostilities, but instead about the 'legal geography' of the battlefield and armed conflict. The same arguments could and would be made if lethal targeting operations against al Qa'ida and affiliated terrorism suspects in places like Pakistan and Yemen were conducted by manned aircraft or ground forces. The extent to which they are problems depends significantly on when, where, and according to what standards and procedures lethal force is used.

That is not to say that the choice of platform – specifically, reliance on armed UAVs – is irrelevant to those concerns. In some cases, reliance on UAVs exacerbates problems. In other cases, however, it may mitigate them. Either way, it is difficult to prise apart questions about armed UAVs from questions about the way they have been used. Conversely, it is not hard to imagine that debates about UAVs would look different today if they had made their prominent debut in uncontroversial inter-state conflict (for example, coalition military operations to dislodge Iraqi forces from Kuwait in 1991).

Today's UAVs – including the Predator and Reaper drones used often by the United States in counter-terrorism strikes – are hardly unique in allowing the United States to conduct military operations against targets in very inhospitable environments and without putting its own personnel at much risk. That relatively riskless distance from combat is at the root of claims that armed UAVs contribute to a dehumanisation of killing or to a willingness by political leaders to resort more quickly and easily to military force in addressing threats. Yet this has long been the situation for the many military technologies, new and old, that permit military forces to attack from a remote, stand-off position. It is not new and not special. Consider artillery bombardment of a city in nineteenth-century warfare; area bombardment of vast urban tracts by high-altitude bombers in the Second World War; Tomahawk cruise missiles fired from long distances against pre-programmed coordinates in the first Gulf War or by the Clinton administration in 1998 against al Qa'ida targets following the East Africa embassy bombings; or NATO aircraft bombing Serbian targets from altitudes above the capabilities of Serb anti-aircraft fire. None of these (among many other examples) was less remote or 'dehumanised' than remotely piloted UAVs today.

The real difference, however, between those technologies and the new technologies of the twenty-first century is that today's technologies are increasingly able to use remoteness to promote precision, not simply piling on kinetic force. In the past, as a general rule, the more remote the operator, the less precise and less discriminating the weapon. This often meant – as in so much Second World War aerial bombardment – using vastly greater, vastly more indiscriminate, explosive tonnage merely in an attempt to hit a discrete target, often with no success. Today's remoteness is not only adjunct to greater precision – but the very fact that soldiers are not at risk, on account of remote piloting and control, allows precision in a different way, the operational luxury under many (more) circumstances not to have to take a shot at a given moment in time. That is, remoteness can aid precision in time as well as space.

The issues of target discrimination and accountability really go to the way UAVs are used and the policies and protocols governing targeting decisions, rather than to the technologies themselves. In some cases, UAVs can be more precise and less likely to cause collateral damage than alternatives, because of their long loiter times and because controllers do not have to worry about self-preservation.[8] Their precision and discrimination in actual attack, however, depends not only on the technologies involved, but often depends heavily on how well UAVs are integrated with various sources of intelligence.[9] Discrimination and accountability are also heavily determined by the administrative and institutional architecture of military forces or government agencies employing UAVs for particular purposes.[10]

The US government has had great difficulty answering objections to UAVs and addressing the concerns just outlined, whatever their merit. That is in part because a considerable swathe of its critics would not be satisfied except by ceasing targeted killing operations altogether; calls for greater transparency are merely a stalking horse for demanding an end to the activity. Yet the lack of transparency has greatly undermined US efforts to give both armed UAVs and targeted killing operations such legitimacy as they might have among the international community. Lack of acknowledgment (up until President Obama's May 2013 speech at the National Defense University even as to the existence of these programmes[11]) has made it difficult for the US government to defend the legitimacy of these tools of counter-terrorism, let alone make the case that they are, in fact, what they are – more precise weapons for counter-terrorism attacks, by comparison to realistic alternatives.

Calls for greater transparency of US lethal UAV targeting have come not only from abroad or from the human rights community, but also from a wide range of former national security officials within the American foreign policy elite. A 2014 task force report on 'US Drone Policy' produced by the Stimson Center in Washington, DC, for example, attracted quite broad support among former US government officials. Among other things, it called for '[i]mproving transparency through the release of a detailed report from the administration explaining the legal basis for US conduct of targeted killings; the approximate number, location and organizational affiliations of those killed by drone strikes; the identities of civilians killed as well as the number of strikes carried out by the military versus the CIA'.[12]

This is almost certainly excessive in terms of either what is realistic or what could prudently be done consistent with operational security in both special operations and intelligence agency activities. Nevertheless, the general call for greater transparency from a broad political spectrum – including not only ardent critics often hostile to the exercise of military power but also those with close ties to the US defence establishment – reveals intense anxiety about how technological development and deployment in this area can be squared with long-term national interests through normative constraints.

Autonomous weapon systems: legal-policy issues

Meanwhile, weapon systems are becoming increasingly automated.[13] In large part, this is occurring as military systems of all kinds become increasingly automated. UAVs, for example, are becoming increasingly automated in their flight and control functions – partly for reasons of efficiency, so that a single pilot can run several craft at once, at least in surveillance mode, and partly because current-generation, large UAVs present problems of control on landing and takeoff for which automation is a useful solution.

Some military systems have become so fully automated that they might be called autonomous; some of these, such as certain missile defence and shipboard defence systems, are weapon systems that have been deployed for years. *Fully autonomous weapon systems* means, for these purposes, systems 'that, once activated, can select and engage targets without further intervention by a human operator'. This definition comes from a 2012 US Department of Defense policy directive,[14] which remains the most extensive public pronouncement by any state on how it intends to proceed with regard to research, development, and deployment of autonomous weapon systems.

Recent technological advances and the possibilities they portend have generated interest – and anxiety – within some governments and their militaries and have spawned a movement of non-governmental activists seeking to ban fully autonomous weapons. In May 2014, the High Contracting Parties of the UN Convention on Certain Conventional Weapons (CCW) convened a wide-ranging discussion of the legal and ethical issues that autonomous weapons raise.

Autonomous weapon systems are slowly becoming incorporated into warfare, as technology advances and capabilities advance, and as systems are increasingly automated one step at a time. Increasing automation in weapons technology results from advances in sensor and analytical capabilities and their integration, and in response to the increasing tempo of military operations. It also results from political pressures to protect not just one's own personnel but also civilians and civilian property.[15] Yet, although automation will be a general feature across battlefield environments and weapon systems, high levels of autonomy or even what many observers would regard as complete autonomy in weapons will probably remain rare for the foreseeable future and will likely be driven by special factors such as decision-making speed required of particular kinds of operations, such as anti-missile defence.[16]

Automation describes a continuum, and there are various ways to define places along it.[17] Terms such as *semi-autonomous*, *man in the loop*, and *man on the loop* are used to describe different configurations of machine–human interaction and degrees of independent machine decision-making. Rather than engaging in definitional debates, a better analytical starting point is the recognition that new autonomous systems will develop incrementally, as more functions, not just of the weapon, but also of the platform (e.g., the vehicle or aircraft), are automated. As all other flight functions are automated for an aircraft, for example, to the point of it being *autonomous* in the (engineering) sense of being able to send it on a mission and see it go and return, some weapon functions will tend to become automated as well, to the point of autonomy, in order for the weapons to be integrated with, and consistent with the speed of, the rest of the system.

For example, intermediate degrees of automation of weapon systems could include a robot that is pre-programmed to look for certain enemy weapon signatures and to alert a human operator of the threat, who then decides whether or not to pull the trigger. In a further degree of automation, the system might be set with the human operator not being required to give an affirmative command – but instead merely deciding whether to override and veto a machine-initiated attack. Perhaps next the system would be designed generally to target and fire autonomously except to wait and call for higher-level authorisation when it assesses the possibility or likelihood of collateral damage, or collateral damage above a certain level.

Weapon systems that would be able to assess civilian status or estimate harm as part of their own independent targeting decisions, as in one of the above examples, do not exist today and research toward such capabilities currently remains in the realm of theory. Still, some modern highly automated – and some would call them autonomous – weapon systems already exist. They are generally for use in battlefield environments, such as naval encounters at sea, where risks to civilians are small – and limited to defensive contexts against other machines, in which human operators activate and monitor the system and can override its operation.[18] Examples include missile systems like the US Patriot and Phalanx and Israel's Iron Dome.[19] Many more could lie ahead, for a variety of battlefields and environments and military missions, in a future that is becoming less and less distant. Increasing automation in weapons technology is also an unsurprising response to the increasing tempo of some military operations and engagements. Increasing automation – and eventually autonomy – of weapon systems also grows from ever-continuing advances in sensor and analytic technologies, machine learning, and their fusion.

Importantly, development of many of the enabling technologies of autonomous weapons systems – artificial intelligence and robotics, for example – is being driven by private industry for many ordinary commercial, civilian purposes (consider self-driving cars, surgical robots, and so on). They are developing and proliferating rapidly, independent of military demand and investment.[20] Such civilian automated systems, such as aircraft landing systems, are already making daily decisions that have potential life-or-death consequences. While many people are generally aware that such systems are highly automated, and have become comfortable with their use, relatively little public discourse has addressed the increasing decision-making role of highly automated or autonomous systems in situations that could threaten lives. As automation and robotics technologies come to be widely understood to be more effective, safe, and reliable than human judgment in many non-military arenas, their use will very likely migrate into military ones.

No doubt, the use and especially the proliferation of autonomous weapons will pose significant risks and difficult challenges for law and regulation. As with any technologically advanced weapon system, dangers include machine malfunction, machines whose design underperforms a legally essential task, or unpredictable effects (including when autonomous weapon systems engage each other or the system decisions generated through probabilistic programming or machine learning).[21] Beyond issues of the autonomous weapon system itself, political and strategic issues include concerns already noted about a state armed with autonomous weapons perhaps being too willing to resort to military force, because these weapons might reduce perceived risks to a side's soldiers or to civilians.[22] Autonomous weapon systems and UAVs raise the same basic issue here: whether the features of the weapon system that limit risk to soldiers and make it more discriminating with respect to civilians are, ironically, the very features that, in the view of some critics, make it not just easier for a state to resort to force, but 'too easy' to do so. Moreover, these systems might be thought to undermine individual discipline and accountability systems in the law of armed conflict.[23]

Note again that many of these of these concerns – machine malfunction, diminution of some political constraints on using force, and abuse or misuse of systems – are not unique to autonomous weapons. As discussed above in the context of UAVs, they are true of many military technologies and targeting practices, including artillery, stand-off manned aircraft, missiles, rockets, and other over-the-horizon weaponry. Moreover, autonomous weapons also offer important potential benefits not only with respect to military effectiveness, but also in terms of humanitarian protection. Existing missile-defence systems mentioned earlier, for example, help protect friendly forces and populations from modes of attack that are too fast or complex for human reaction and decision-making. Like remotely piloted UAVs, some autonomous weapon systems can operate without exposing personnel to enemy fire. This, in turn, reduces pressures on those personnel to employ greater force to neutralise or suppress those threats, with in turn the possibility of greater harms resulting to civilians and civilian objects.

Autonomous weapons systems may also reduce risks to civilians by making targeting more precise and firing decisions more controlled – especially compared to human-soldier failings that are so often exacerbated by emotions such as panic or vengeance, as well as by the limits of human senses and cognition. One of Human Rights Watch's significant, and peculiar, claims in calling for an international ban – that a fundamental objection to autonomous weapon systems is that they take these emotions out of battlefield targeting and firing decisions – flies in the face of how much of the structure of the law of armed conflict actually exists to address, imperfectly, the effects of human soldiers' battlefield emotions, starting with fear, anger, and vengeance, exacerbated under conditions of hunger, exposure, uncertainty, and so on.[24] The ICRC, in contrast to Human Rights Watch, has so far taken a sensibly cautious approach to this question, observing in a 2011 report that 'emotion, the loss of colleagues and personal self-interest is not an issue for a robot, and the record of respect for IHL by human soldiers is far from perfect, to say the least'.[25] Weapon systems with greater and greater levels of automation could, at least in some battlefield contexts, reduce target misidentification, better detect or calculate possible collateral damage, or allow for using smaller amounts of force compared to alternatives.

Some critics argue that legal and ethical deficiencies of autonomous weapons – or the dangerous tendencies that such systems would unleash – warrant an international ban on their use. The assumptions behind such calls to prohibit autonomous weapon systems and the form of proposed bans vary. Some critics doubt that technology can ever be good enough to make sufficiently precise decisions to meet the legal and ethical requirements of targeting, especially distinction or proportionality.[26] Others believe that these legal or ethical principles inherently require human judgement, and that lethal targeting is or should always be guided directly by a moral agent who can be held accountable for culpable failures.[27] Still others acknowledge that though autonomous weapon systems may not be unlawful under existing law, given that technological developments and the costs and benefits autonomy are uncertain, such weapons nevertheless should be banned (at least for the foreseeable future, pending possible development of a special legal framework) as a prophylactic precautionary measure.[28]

One expression of such a ban would define a maximum level of autonomy for any weapon. A variant approach is to define a minimum legal level of human control. Human Rights Watch, for example, has called for a pre-emptive 'ban on fully autonomous weapons', which 'should apply to robotic weapons that can make the choice to use lethal force without human input or supervision'.[29] The International Committee for Robot Arms Control, an NGO dedicated to reducing threats from military robotics, calls for the 'prohibition of the development, deployment and use of armed autonomous unmanned systems'.[30]

Short of a complete and permanent ban, Christof Heyns, the UN Special Rapporteur on Extrajudicial, Summary or Arbitrary Executions, has instead proposed a moratorium, calling for 'all States to declare and implement national moratoria on the testing, production, assembly, transfer, acquisition, deployment and use of [autonomous weapon systems] until such time as an internationally agreed upon framework [...] has been established'.[31] A British non-governmental organisation dedicated to weapons regulation argues that lethal decision-making by automated or autonomous weapons should require 'meaningful human control'.[32] This idea of requiring a minimum level of 'meaningful human control' emerged as a major theme in discussions among states and advocacy groups at the 2014 UN CCW meeting.[33]

A general problem with all of these proposed, categorical formulas is that they do not contain a bright line that would be useful for promoting adherence – though initially they might appear to do so. Each of these seemingly clear-cut definitions leaves many open questions as to what systems would be banned under any particular formulation. Even something as seemingly plain as 'lethal decision-making' by a machine does not address, among other things, the lawfulness of targeting, for example, a tank, ship or aircraft that is ultimately the source of the threat, but inside of which is a human combatant.

In any case, it is also critically important to understand that, before an autonomous weapon system – like any weapon system – is used in a military operation, human commanders and those employing the weapon will generally continue to be expected to exercise judgement about the likely presence of civilians and the likelihood that they may be harmed; expected military advantage; particular environmental features or conditions; the weapon's capabilities, limitations, and safety features; and many other factors. In many cases, even though a weapon system might be autonomous, much of the required legal analysis would be conducted by human decision-makers who elect whether or not to use it in a specific situation. Whether legal requirements are satisfied in a given situation will still depend not simply on the machine's own programming and technical capabilities, but on human judgements as well. Moreover, although there is a common assumption that law of armed conflict requirements have to be met in the heat of battle during an actual engagement, for virtually all operations of organised, professional militaries, the crucial moment of compliance occurs beforehand, in the planning for the operation. Compliance with the law of armed conflict is one element both of the elaboration of an operation's rules of engagement and also of the planning for the use of advanced weapon systems and the rule-set guiding their use, sometimes with machine parameters specific to a particular operation.

Furthermore, whether a highly automated system – say, one with a human supervisor who can override proposed firing decisions – is in practice operating under human oversight or instead autonomously depends on how it is being manned, how operators are trained and how effectively oversight is exercised. It is a point emphasised by the 2012 DOD Directive, which saw the need to ensure that systems thought to be merely automated and subject to human control were not inadvertently 'autonomous' in a functional sense, in any case when humans could not perform the roles anticipated by the design of the weapon. Autonomy also depends on operational context and conditions, in other words, which may limit the degree to which the human role is in any way meaningful.

For these and other reasons, a fully autonomous system and a merely highly automated system will be virtually indistinguishable to an observer unless they know a lot about how that system is authorised to be used in particular operational conditions. And the difference might not matter very much in practice, given the variable performance of human operators. In any event, these systems will be easily transitioned from one to the other. An upshot of the blurriness of these lines is that it will be very difficult to draw and enforce prohibitions on 'fully' autonomous systems or mandates for minimum levels of human decision-making.

In a recent article co-authored with Daniel Reisner, the former head of the international law branch of the Israel Defense Forces, we argued that, for these and other reasons, efforts to prohibit autonomous weapons systems are misguided.[34] We instead argued that the existing law of armed conflict is sufficiently robust to regulate these systems effectively, provided that responsible states collaborate and communicate to develop shared best practices and legal interpretations over time, as technology develops. Gradual adaptation of existing law of armed conflict, including legal requirements for weapons review and targeting law, can more effectively balance military and humanitarian imperatives than would any effort to ban outright autonomous weapons as a category.

In any event, the great practical difficulty of distinguishing between autonomous and highly automated systems means that a legal ban on autonomous systems would be difficult to enforce. Furthermore, imposing a general ban on autonomous systems could carry some highly unfavourable consequences – and possibly dangers. These could include providing a clear advantage in autonomous weapon technology to those states that generally would not join, or in practice likely not comply with, such a ban. Bans on autonomous weapons as such, particularly if coupled with bans on their development, could also mean losing out on the potential advantages of autonomous systems for improving decision-making on the battlefield and possibly limiting human loss of life on both sides and among civilians, including through avoiding emotion-based response and improving system accuracy, thereby perhaps minimising collateral injuries. A persuasive moral argument requires that systems that do not meet the requirements of international weapons law, or do not meet them for one or another battlefield environment, be disallowed – but that there is an affirmative obligation to seek to develop such technologies as might reduce the harms of battle.

And, after all, there is no reason, in principle, why a highly automated or autonomous system could not satisfy the requirements of targeting law.[35] Like any otherwise lawful weapon, it depends on the use and the environment. Uninhabited deserts are different from urban environments. Destroying rockets in the air is different from rural counter-insurgency in which entering civilian villages is necessary, not just for bare security reasons, but to consolidate political relationships. In practical terms, autonomous systems might be better able to satisfy the law in some uses and environments than others, but that is not a matter of principle. Rather, it is a matter of whether and how far technological capability advances relative to the legal standard.

We and Reisner believe that:

> a fundamental principle underlying the gradual development of these standards and rules alongside the evolution of automation technologies ... should be that what matters is ever greater compliance with the core obligations of the law of armed conflict: necessity, distinction, proportionality and humanity. [Furthermore,] [w]hether the actor on the battlefield is a 'who' or a 'what' is not truly the issue, but rather how well that actor performs according to the law of armed conflict.[36]

In that regard, solutions to the challenges posed by autonomous weapons likely lie in a process of gradual international legal development, within the existing law of armed conflict framework – one that evolves in tandem with technology as it advances.

Promoting the ability of the law of armed conflict to effectively regulate new weapon technologies

Beyond arguments over the fundamental morality or lawfulness of these weapons, the United States and its close allies face a dual challenge with regard to advanced weapon technologies of all kinds and, in practical terms, efforts to regulate them through the law of armed conflict. First, it is in their interest to promote international legal interpretations with regard to these systems that help appropriately constrain the actions of states and non-state groups that might otherwise seek to abuse them. Second, it is in their interest to demonstrate the effectiveness of those interpretations in order to help hold off movements to impose unworkable new rules on these technologies – rules that will tend largely to empower bad actors.

As a general matter, the law of armed conflict has historically served the United States and its allies well, and there is good reason to believe that it *could* continue to do so with respect to new weapon technologies. That body of law is premised on reasonable balances of military necessity and humanitarian imperatives. Although always subject to interpretative debates and disagreements, the basic rules are almost universally embraced among states as minimum normative standards to govern military operations. As a result, those rules assist in establishing common standards among the United States and its allies to promote cooperation and permit joint operations; they help earn and sustain necessary buy-in from the officers and lawyers who would actually use or authorise such systems in the field; and they raise the legal, political, and diplomatic costs to adversaries of developing, selling, or using weapon systems that run afoul of them.

A virtue of the law of armed conflict is its adaptability to new technologies – adaptable in the sense of being able to cabin new technologies within its normative structure and fundamental principles. With a few exceptions, generally codified in specific treaties (some of which were mentioned earlier), the general rules governing the use of weapons are applicable to any new weapon system. However, a flip side of this virtue – that the law of armed conflict is adaptable to new technologies – is that the viability of the law of armed conflict as applied to novel weapons depends heavily on its demonstrated effectiveness in practice.

To take the example of manned air power, experience in the first half of the twentieth century led many observers to conclude that capabilities for aerial bombardment would eviscerate traditional legal constraints on military targeting. The frequent use of advancing air power technology during the latter half of that century, however, led to a rough consensus about reasonable military judgements as measured against traditional law of armed conflict standards. States and other international actors advanced arguments and counter-arguments in specific circumstances, of course, yet – though heated controversies remained – the law of armed conflict proved adaptable to air power.[37]

This process of consensus-building – or at least wide agreement – through evolving practice and practical, informal norms usually takes time. It develops most obviously through competitive evaluation of legal claims and counter-claims; it evolves casuistically. As stated at the outset, however, these factors are significantly lacking in the case of some new technologies such as UAVs and autonomous weapons. Technology is developing rapidly, and it is being used – or may be used in the future – in ways that are relatively opaque compared to previous advances in weapon technology. The pace of technological change means that there is little time for incremental legal consensus-building through international practice and assessment. The opacity of new technology's use means that there is not necessarily a clear factual basis evident in practice for favouring one competing legal interpretation over another. An upshot is that greater transparency in the way that responsible states use new weapon systems may be needed if responsible states are to maintain the vitality of the law of armed conflict regime with regard to weapons, especially under conditions of rapid technological change.

It is very difficult to get this right, unfortunately, because even if government leadership buys into this idea of shaping international law as a strategic interest, when it comes to day-to-day operational decisions, that goal almost always seems distant, diffuse, and uncertain – while the benefits of secrecy seem immediate, tangible, and certain. Yet secrecy breeds distrust among observers, as well as fostering perceptions that there exists a vacuum of effective legal regulation, because where no weapon system or its performance is visible, 'law' loses its specificity and grip upon concrete facts and becomes merely vacuous abstractions. This tends to create a perceived normative vacuum that makes radical proposals to prohibit outright certain weapon systems all the more appealing.

The United States has been down this path before, and not to its benefit. Consider, for example, how much more the United States could have been doing during the past decade to declare and explain publicly the legal standards and additional policy limits it imposes on its targeted killing operations with UAVs. This should have included, to start with, the US view of legal categories of lawful targets; information about how it reviews collateral damage; the advantages that UAVs have over other weapon platforms with respect not only to operational necessities but to reducing collateral harm with greater precision; and a host of other 'framework' issues that would not require it to enter into discussion of specific operations. Because the United States government has been slow and vague in its public discussion, however, it has largely lost control of public and international debate about drones – a debate that often equates drone use with lawlessness and portrays the law of armed conflict as a weak framework for dealing with concerns raised by this technology. This is not to say that sceptics opposed to armed UAVs or targeted killing or counter-terrorism use of force operations would be persuaded – or that for every policy made public, more demands for more transparency would not be made; irredentist critics will never be satisfied, or course, and yet legitimacy can be deepened if there is a willingness to say something concrete and then publicly and unapologetically defend it.

The United States and its close friends and allies should take lessons from that often quite negative experience in building and maintaining legitimacy for controversial policies. They should consider, and urge other states to consider, adopting and disclosing national policies with regard to the development and perhaps fielding of autonomous weapons, making as clear as possible the ways that the law of armed conflict governs. The United States has already done so in the form of a Department of Defense Policy Directive. That document emphasises strict internal review processes, including multiple legal reviews, that would be part of any effort to develop and use autonomous weapon systems. It could be supplemented with more detailed discussion of the substantive standards that would govern their use, including not only standards that are legally compelled under the law of armed conflict, but also proposed best practices based on policy and ethical concerns unique to autonomous weapon systems. The Department of Defense could emphasise how this policy directive is directly tied to the extraordinarily detailed and thorough – and publicly available – policy directives and legal orders within the military services for weapons review.

The United States has taken a similar approach in the recent past to other controversial technologies, most notably cluster munitions and landmines, by declaring commitment to specific standards that balance operational necessities with humanitarian imperatives. These policy pronouncements establish parameters internal to the US government and they serve as vehicles for explaining reasoning to outside audiences at home and abroad; they can also be adapted by other states through consultative processes.

Of course, there are limits and strong counter-pressures to transparency here, on account of both special secrecy concerns that surround new weapon technologies and tactics and the practical limits of persuading sceptical audiences about the internal and undisclosed decision-making capacities of rapidly evolving weapon systems. Part of the answer is therefore to emphasise the internal processes by which states consider, test and evaluate autonomous weapon systems. Even when states cannot disclose publicly the details of their automated systems and their internal programming, they should be as open as they can about their vetting procedures, both at the R&D stage and at the deployment stage, including the standards and metrics they use in their evaluations. And, even when states cannot be very open publicly with the results of their tests, for fear of disclosing details of their capabilities to adversaries, they should consider sharing them with military allies as part of an effort to establish common standards.

Conclusion

The law of armed conflict can generally become a meaningful regulatory constraint upon new weapons technologies when it adapts through the gradual evolution of commonly held understandings, interpretations and applications of general principles of law that become widely shared among the parties that might develop, deploy, and use such weapons. The international legal community has a deep bias toward the instant creation and codification of formal law, though. Its first instinct is to create new law, often with little or no attention as to whether it will stick or not. With regard to armed UAVs and autonomous weapon systems, however, formalised international law popping too quickly out of the box will fail, on several counts.

The practical regulatory purchase of norms regarding rapidly evolving, and often subtly evolving, weapon systems will largely depend on the incubation of informal norms over a long period of cooperative discussion. It will depend on leading players, such as the United States, being willing to see that its strategic interests benefit from greater levels of transparency than it would otherwise prefer.

Notes

1 Many of the observations and arguments on autonomous weapon systems are contained also in Kenneth Anderson, Daniel Reisner and Matthew Waxman, 'Adapting the Law of Armed Conflict to Autonomous Weapon Systems', *International Law Studies*, Vol. 90, September 2014.
2 Sarah Kreps and Micah Zenko, 'The next drone wars', *Foreign Affairs* (March/April 2014); David Schaefer, 'Chinese combat drones: ready to go global?', *The National Interest*, 31 October 2014, available at http://nationalinterest.org/blog/the-buzz/chinese-combat-drones-ready-go-global-11583?page=show; and Joshi Shashank and Aaron Stein, 'Emerging drone nations', *Survival*, Vol. 55, No. 5 (1 October 2013), pp. 53–78.
3 William H. Boothby, *Weapons and the Law of Armed Conflict* (Oxford University Press, 2009), p. 231; William H. Boothby, *The Law of Targeting* (Oxford University Press, 2012), generally pp. 275–81; Michael N. Schmitt, 'Extraterritorial lethal targeting: deconstructing the logic of international law', *Columbia Journal of Transnational Law*, Vol. 52, No. 1 (2013), pp. 79–114.
4 State Department Legal Adviser Harold Hongju Koh, 'The Obama administration and international law', address to the Annual Meeting of the American Society of International Law, Washington, DC, 25 March 2010, available at www.state.gov/s/l/releases/remarks/139119.htm.
5 See also some agreement with this: http://justsecurity.org/15521/global-debate-human-rights-council-takes-drones/#more-15521.
6 Anthony Dworkin, 'Drones and targeted killing: defining a European position', European Council on Foreign Relations Policy Brief, July 2013.
7 C. Christine Fair, 'Ethical and methodological issues in assessing drones: civilian impacts in Pakistan', The Monkey Cage (blog), *Washington Post*, 6 October 2014, available atwww.washingtonpost.com/blogs/monkey-cage/wp/2014/10/06/ethical-and-methodological-issues-in-assessing-drones-civilian-impacts-in-pakistan/; 'Drop the pilot: a surprising number of Pakistanis are in favour of drone strikes', *The Economist*, 19 October 2013, available at www.economist.com/news/asia/21588142-surprising-number-pakistanis-are-favour-drone-strikes-drop-pilot.
8 Scott Shane, 'The moral case for drones', *New York Times*, 14 July 2012, available at www.nytimes.com/2012/07/15/sunday-review/the-moral-case-for-drones.html?_r=0.
9 Kate Brannen, 'US options limited by lack of drones over Syria', *Foreign Policy,* 8 October 2014, available at www.foreignpolicy.com/articles/2014/10/08/us_options_limited_by_lack_of_drones_over_syria.
10 Gregory S. McNeal, (Targeted killing and accountability' *Georgetown Law Journal*, Vol. 102, 2014, pp. 681–794.
11 Available at www.whitehouse.gov/the-press-office/2013/05/23/remarks-president-national-defense-university.
12 Available at www.stimson.org/spotlight/recommendations-and-report-of-the-stimson-task-force-on-us-drone-policy/.
13 John Markoff, 'Fearing bombs that can pick whom to kill', *New York Times*, 11 November 2014, available at www.nytimes.com/2014/11/12/science/weapons-directed-by-robots-not-humans-raise-ethical-questions.html?_r=0.
14 Department of Defense, Directive No. 3000.09, 'Autonomy in weapon systems', at 13 (21 November 2012) [Hereinafter Directive No. 3000.09].
15 Kenneth Anderson and Matthew Waxman, 'Killer robots and the laws of war', *Wall Street Journal*, 4 November 2013, available at http://online.wsj.com/news/articles/SB10001424052702304655104579163361884479576; Markoff, 'Fearing bombs'.

16 Peter Singer, *Wired for War: The Robotics Revolution and Conflict in the 21st Century*, Penguin Books, 2009; for a dissenting view, that removing humans from military targeting is unlikely for the foreseeable future, see Werner J. A. Dahm, Op-Ed, 'Killer drones are science fiction', *Wall Street Journal*, 15 February 2012, available at http://online.wsj.com/news/articles/SB10001424052970204883304577221590015475180.

17 Department of Defense, Defense Science Board, *The Role of Autonomy in DoD Programs*, July 2012, pp. 3–5, 23–4.

18 See International Committee of the Red Cross, Report of the ICRC Expert Meeting on Autonomous weapon systems: technical, military, legal and humanitarian aspects, May 14, 2014, available at www.icrc.org/eng/assets/files/2014/expert-meeting-autonomous-weapons-icrc-report-2014-05-09.pdf [hereinafter Report of ICRC Expert Meeting].

19 US Navy Fact File: MK-15 Phalanx Close-In Weapons System (CIWS), available at www.navy.mil/navydata/fact_display.asp?cid=2100&tid=487&ct=2; *MK 15 Phalanx Close-In Weapons System (CIWS)*, Federation of American Scientists, 9 January 2003, available at www.fas.org/man/dod-101/sys/ship/weaps/mk-15.htm; Michael N. Schmitt and Jeffrey S. Thurnher, '"Out of the Loop": Autonomous Weapon Systems and the Law of Armed Conflict', *Harvard National Security Journal*, Vol. 4, 2013, p. 231.

20 See Robert O. Work and Shawn Brimley, 20YY*: Preparing for War in the Robotic Age* (Washington, DC: Center for a New American Security, 2014), pp. 23–7.

21 Noel E. Sharkey, 'The evitability of autonomous robot warfare', *International Review of the Red Cross*, No. 886 (30 June 2012), available at www.icrc.org/eng/resources/documents/article/review-2012/irrc-886-sharkey.htm; Peter Asaro, 'On banning autonomous weapon systems: human rights, automation, and the dehumanization of lethal decision-making', *International Review of the Red Cross*, No. 886, 30 June 2012, at (www.icrc.org/eng/resources/documents/article/review-2012/irrc-886-asaro.htm; and Wendell Wallach and Colin Allen, *Moral Machines: Teaching Robots Right from Wrong* (Oxford: Oxford University Press, 2010), pp. 47–8.

22 See, e.g., UK Ministry of Defence, *Joint Doctrine Note 2/11 The UK Approach to Unmanned Aircraft Systems* (London: MOD Development, Concepts and Doctrine Centre [DCDC], 30 March 2011), Ch. 5; Medea Benjamin, *Drone Warfare: Killing by Remote Contro*l, rev. edn (Verso, 2013), where the author writes, 'the biggest ethical problem with drones is that it makes killing too easy'.

23 See, e.g., Human Rights Watch, 'Sec. VI: Problems of Accountability for Fully Autonomous Weapons', *Losing Humanity: The Case Against Killer Robots*, (New York and London: Human Rights Watch, 19 November 2012), available at www.hrw.org/reports/2012/11/19/losing-humanity-0.

24 Human Rights Watch, *Losing Humanity*, pp. 37–8 (see note 23 above).

25 International Committee of the Red Cross, 'International Humanitarian Law and the challenges of contemporary armed conflicts: report of the 31st Conference of the Red Cross and Red Crescent', (2011) 31 IC/11/5.1.2, 40 [Hereinafter ICRC 31st Conference Report].

26 See Human Rights Watch, *Losing Humanity*, pp. 30–35 (see note 23 above); Noel Sharkey, Op-Ed, 'America's mindless killer robots must be stopped', *Guardian*, 3 December 2012.

27 See Human Rights Watch, *Losing Humanity*, pp. 35–6 (see note 23 above); Asaro, 'On banning autonomous weapon systems', p. 687 (see note 21 above).

28 See Christof Heyns, *Report of the Special Rapporteur on Extrajudicial, Summary or Arbitrary Executions*, UN Human Rights Council, A/HRC/23/47. 9 April 2013, pp. 20–21.

29 Human Rights Watch, *Losing Humanity*, p. 46 (see note 23 above).

30 Available at http://icrac.net/statements/.

31 Heyns, *Report of the Special Rapporteur*, p. 21 (see note 28 above).

32 Article 36, Structuring Debate on Autonomous Weapon Systems, November 2013, available at www.article36.org/wp-content/uploads/2013/11/Autonomous-weapons-memo-for-CCW.pdf.

33 See UN Convention on Certain Conventional Weapons, 'Report of the 2014 Informal Meeting of Experts on Lethal Autonomous Weapons Systems (LAWS)', 10 June 2014, para. 20, available at www.unog.ch/80256EDD006B8954/%28httpAssets%29/350D9ABED1AFA515C1257CF30047A8C7/$file/Report_AdvancedVersion_10June.pdf.

34 Anderson, Reisner and Waxman, 'Adapting the law of armed conflict' (see note 1 above).

35 Schmitt and Thurnher, 'Out of the Loop', pp. 279–80 (see note 19 above).

36 Anderson, Reisner and Waxman, 'Adapting the law of armed conflict' (see note 1 above).

37 This process is discussed briefly in the 'Final Report to the Prosecutor by the Committee Established to Review the NATO Bombing Campaign against the Federal Republic of Yugoslavia', available at www.icty.org/sid/10052.

14

THE 'ROBOTS DON'T RAPE' CONTROVERSY

Maziar Homayounnejad and Richard E. Overill

The idea that 'robots don't rape', whereas human soldiers might, has been a formally acknowledged by a Special Rapporteur in the United Nations (UN) Human Rights Council[1] and adduced by various experts at the UN's Meetings of Experts on LAWS.[2] The argument is distinctly pro-autonomous weapons systems (AWS), and should be seen as part of the broader narrative on the apparent virtues of robotic warfare. These focus on the absence of the human frailties and imperfections that often lead to erroneous targeting, or even the commission of war crimes;[3] hunger, tiredness, hatred, fear, anger, frustration, resentment and the instinct for revenge are all included in this.[4] It is therefore argued that *removal* of such distinctly human traits from the battlefield can enhance respect for international humanitarian law (IHL) and, especially, for the principle of civilian immunity.[5]

In this chapter, the controversy is explored and technical and legal aspects are discussed. The first section covers the detail of the debate. The subsequent sections tackle three limitations that make the application of conventional wartime rape and torture law to the use of autonomous weapons systems problematic: the question of distance; design-led safeguards; and the introduction of 'kill switch' technology.

The debate

In the case of the prohibition of rape, clearly such a rule cannot be violated by wide-area loitering systems, or any similar AWS being developed in the near term, as these will attack from a distance and will generally lack the capability for 'close quarters combat' (CQC) or any sustained arm's reach interaction with humans.[6] The argument is also put more generally that machines are simply not predisposed to any sexual impulse, or any desire to degrade the object of their actions, these being distinctly human traits. Accordingly, '[r]obots do not rape' and this should offer hope for greater IHL compliance.[7] Yet, despite its positive and factual tone, this argument has triggered visceral rebuttals and counter-arguments in other quarters.[8]

On the one hand, the pro-AWS view can be challenged for making a number of incorrect assumptions regarding *who* decides to rape and whether it is solely an *impulse*, sexual or otherwise. Consider, for example, the more recent and broad ICC definition of rape in the *Elements of Crimes*,[9] which holds that a rape occurs when:

> The perpetrator invaded the body of a person by conduct resulting in penetration, *however slight*, of any part of the body of the victim or of the perpetrator with a sexual organ, *or of the anal or genital opening of the victim with any object* or any other part of the body.[10]

Arguably, any sensible application of the above wording would conclude that it is at least possible for a future bipedal/humanoid robot, designed with a degree of 'manual dexterity', to be *programmed* and used as a *deliberate* instrument of rape.[11] Indeed, sexual abuses are not limited to being by-products of war, or *opportunistic* criminal acts of the male sexual impulse, but have been sanctioned at the very highest levels of a command structure.[12] In this connection, rape and other sexual violence can often serve larger *strategic objectives* in times of armed conflict or occupation,[13] such as to:

> terrorise and conquer, expel or control women and their communities; or
> to intimidate, humiliate and punish them in order to break their resolve, gain submission and extract information.[14]

Frequently, this occurs as part of a broader campaign of torture, which by definition serves larger strategic and tactical goals.[15] In that regard, rape and torture may both be part of a strategy of *asymmetric conflict*, to achieve these objectives in the absence of superior firepower (against enemies) or legitimacy (with domestic populations).[16] Perhaps one of the most infamous cases of 'strategic rape' in recent history occurred during the Bosnian War during 1992–1995, which involved the mass rape and torture of an estimated 25,000–50,000 women[17] for ethnic cleansing purposes[18] and, by extension, genocide.[19]

Accordingly, future humanoid robots designed with 'manual dexterity' can potentially be *programmed* to rape and torture, with historical precedent suggesting that some political and military leaders do indeed see the ends as justifying the means; especially where asymmetry is seen as inevitable. In addition, empirical data suggest that such crimes are far from rare: for example, Cohen's study of 86 major civil wars during the period 1980–2009 showed that 71 (83%) involved reports of rape in at least one of the conflict years;[20] and 53 of these (62% of the total) reportedly involved 'significant' rape.[21] Furthermore, in almost all cases (93%, or 66 of the 71 civil wars involving rape) state actors perpetrated the crime,[22] which may seem particularly concerning, given that AWS is likely to be acquired and fielded by states first.[23] This has led some feminists such as Carpenter[24] and Sandvik and Lohne[25] to dismiss the 'robots don't rape' argument as being both factually inaccurate (on the legal and technical possibility of robots being used to commit rape) and based on misleading assumptions (that rape is always a spontaneous, unplanned breach of discipline by individual soldiers).[26] Indeed, as the title of Carpenter's piece claims, these contradictions highlight the entire 'myth of the humanitarian robot'. The resulting logic is that such ill-informed views dangerously mask the true nature and extent of the civilian risk associated with AWS. Accordingly, in their piece entitled 'Killing the 'Robots-don't-Rape' Argument', Sandvik and Lohne argue that such myth-driven views run the risk of:

> undermining hard-fought gender battles, reducing wartime rape to an issue of uncontrolled/uncontrollable male sexuality … instead of recognizing it primarily as an act of violence … which may or may not be deliberate, intentional, and programmed.[27]

In short, these authors argue that the pro-AWS narrative rests on a logical paradox: humans are more likely than robots to break the law, but it is untrustworthy and nefarious humans who would be the principals issuing instructions, which highly efficient robots would execute without question.[28] Accordingly, the feminist narrative is to argue for a pre-emptive ban on the development, production, deployment and use of AWS, not because such machines will be indiscriminate, but because they will be highly *effective* at performing their programmed duties. They will be held back by neither the moral/ethical brakes that regulate human soldiers,[29] nor the physical limitations that would prevent a terrorising rape from continuing indefinitely.[30]

This reasoning follows the narrative adopted by Human Rights Watch in their seminal report, *Losing Humanity*.[31] There, the non-governmental organisation raised similar concerns but in relation to broader issues of tyranny,[32] to argue for a comprehensive and pre-emptive ban on the development, testing, production, deployment, and use of AWS.[33]

Attack from a distance *versus* close quarters combat

Firstly, as already mentioned above, the kind of AWS likely to be fielded in the near-term will not be designed for CQC, or indeed any arm's reach interaction with humans; much less with the manual dexterity required to commit an act of rape.[34] Furthermore, while bipedal and humanoid robots *are* conceivable in the longer term, it would arguably be regulatory overreach to prohibit an entire class of weapon systems because the more advanced and dextrous variety, which is not even in existence yet, may be misused.[35] In a twist of paradox, it could even be argued that such robots, because of their 'strength' and manual dexterity, will *a priori* have the capacity to capture and arrest targeted persons, thereby offering a non-lethal option that may comply with International Human Rights Law (IHRL).[36] Accordingly, the preferred approach is to home in on the *actual* problem – namely, the unlawful *use* of a weapon system – and avoid slippery-slope arguments that may simply deprive militaries of an otherwise lawful, responsive, and tactically superior weapon system.

Design-led safeguards

Secondly, and following this logic, there are possible technical and design-led means to restrict the purposes to which an AWS can be put. As Wagner argues, in the broader debate on AWS being used for repression and acts of torture:

> Whether dictators will find a pliable force to quash dissent is less important than AWS being programmed to replicate emotions that are desirable within the IHL framework – such as compassion when the situation warrants it – while leaving out those emotions that lead to disastrous consequences.[37]

Of course, it is not clear how a distinctly human-like emotion such as compassion can *directly* be programmed into software. However, as Arkin points out, the rules and principles of IHL (and, by extension, ICL and IHRL) are strongly motivated by human compassion, thus programming a robot to follow these rules and incorporating software suppressors that prevent violations of them should establish the requisite level of compassion by proxy.[38] In this regard, two proposals aimed at embedding ethical considerations into software are worth considering.

Arkin's 'ethical governor'

The first of these is Arkin's 'ethical governor' (EG) and 'ethical behaviour controls' (EBC), which are bolted onto/integrated into an AWS software architecture. While these are intended to govern *lethal* autonomous systems deployed by 'good faith' commanders, they can arguably be applied to impose a technical prohibition on robots being used to intentionally violate the law.

Turning to the EG first,[39] this bolt-on software component mediates between a) the main AWS software architecture and b) the actuators,[40] using both *negative* and *positive* criteria. Namely, it evaluates the ethical appropriateness of any lethal action determined by the main software, by assessing that action against the *forbidding* and *obligating* constraints provided by both legal norms and mission rules of engagement (RoE).[41] In turn, this leads to a binary response to either validate or withhold a permission-to-fire (PTF) variable, which is needed for the system to lethally engage its target. Importantly, for PTF to be validated, all forbidding constraints must be upheld, *and* at least one obligating constraint must be true. Where these cumulative criteria are not satisfied, PTF is withheld; in which case, the EG will either constrain lethal action by the actuators, or (if 'manual override' is enabled) it will alert a human operator who has the option to override the EG's determination.

By contrast, the EBC design option[42] is integrated into the main AWS software architecture, and aims to ensure that the *individual* behaviours generated by the system comply with both IHL norms and mission RoE, before they get to the stage of deliberate monitoring by the EG. Thus, rather than being 'externally' imposed, as in the case of the EG, the ethicality of individual behaviours in the EBC model is 'internally' generated. Accordingly, where a single response is selected as output for lethal action, the result is intuitively ethical; but where lethal action consists of a number of discrete responses that are *individually* permissible, the *interaction between them* may well fall short of the required normative and ethical standards, in which case the EG will withhold PTF.

As noted above, Arkin's proof-of-concept concerns *lethal* autonomous action that a commander *may not have foreseen* on deployment; thus it assumes good faith efforts in the first instance.[43] However, far from limiting its relevance to the current context, this arguably makes it *easier* to adapt and apply the model to non-lethal, but *well-defined* and *intentional* violations. For example, any action that conforms to the ICC definition of rape – itself very specifically and objectively defined[44] – can alone be a *forbidding constraint*, which automatically triggers the EBC to veto any such command.[45] Even separate actions that may individually be lawful, but which aggregate into an act of torture,[46] can be denied a PTF; or, more precisely, a PTA ('permission-to-act').

Admittedly, this latter violation will pose a greater programming challenge, because the threshold for torture – '*severe* pain or suffering'[47] – can be highly subjective, and its assessment has been shown in empirical research to be subject to an 'empathy gap'.[48] That said, empathy gaps are arguably a *human* shortcoming,[49] which may be addressed through well-informed programming that takes into account all relevant indicators of pain. Thus, if there are *objective* indicators of pain and its 'severity', such as those that may be detected in facial expressions,[50] voice frequencies,[51] and other emotional responses observed in a given (torture) context,[52] these may be programmed into the robot's control software and into its suppressor (namely, the EG). Of course, no single indicator is determinative, but by combining a number of cues, a robot can minimise error rates. It may also be that specific physical (in)actions, such as an already restrained human subject not exhibiting any resistance,[53] are identified as objective (proxy) indicators that any *further* physical force by the robot is likely to be unnecessary and unlawful, if not torture.[54] If so, this can also be programmed into the suppressor (in this case, the EBC) as a forbidding constraint, which will veto any individual command to engage as soon as these objective indicators are detected.[55] Beyond these difficulties, no complex or contextual reasoning on derogations is required on the part of the robot, as the prohibitions on rape and torture are norms of *jus cogens*. Accordingly, the usual 'manual override' function of the EG can be disapplied in these instances; as can the ability to reassign certain relevant actions from forbidding to obligating constraints.

Arkin's model has attracted a number of principled objections,[56] mainly based on the fact that both the EG and EBC are devoid of *human* emotion[57] and 'hot' empathy.[58] However, and in line with the author's own response,[59] it is arguable that at the point of kinetic action, neither IHL nor International Criminal Law (ICL) requires human self-awareness, empathy or the like. Rather, the sensory hardware and control software must simply recognise certain *factual* criteria that guide AWS actions and behaviours in ways that *objectively* conform to IHL/ICL norms and mission RoE.[60] As argued above, this is highly likely to be true in the case of such well-defined and deliberately intrusive acts as rape and torture, thereby facilitating the use of Arkin's model as a technical prohibition on these international crimes.

Briggs and Scheutz's 'felicity conditions'

More recently, Briggs and Scheutz demonstrated proof-of-concept for a robot rejecting human instructions, based on hard-coded 'felicity conditions'.[61] These consist of five *cumulative* criteria that inform whether a particular action *can* and *should* be done. Hence, they must *all* be satisfied for a human proposal, or a 'call to action', to be explicitly accepted by a robot. The felicity conditions are:

1. Knowledge: Do I know how to do X?
2. Capacity: Am I, both currently and normally, physically able to do X?
3. Goal priority and timing: Am I able to do X right now?
4. Social role and obligation: Am I obligated based on my social role to do X?
5. Normative permissibility: Does it violate any normative principle to do X?[62]

To be sure, the model is designed for general application, to embed meta-reasoning based on ethics and appropriateness in all human–robot interactions, not just those involving AWS. For example, the felicity conditions will ensure that a child will not be able to get a home robot to write his/her homework, while an elder-care robot will reject an instruction by a forgetful owner to 'wash the dirty clothes', if that task has already been done.[63]

More specifically, felicity conditions 1) to 3) prevent inadvertently dangerous situations where robots would otherwise be assigned tasks for which they are not designed, or for which they are ill-suited in the prevailing circumstances. This clearly has potential military utility, in assisting commanders who act in good faith, and who wish to avoid inappropriate deployments that may lead to unacceptable civilian harm. However, in the case of *intentionally unlawful* commands, it is a combination of 4) and, especially, 5) that arguably provides a sufficient basis to reject an instruction to rape or torture a human subject.

While this model is still a work-in-progress,[64] the proof-of-concept has been demonstrated to work at a basic level in a domestic setting.[65] Arguably, with continued research and testing, full realisation of the project goals is not unrealistic by the time bipedal/humanoid robots enter the battlespace, if indeed they ever do.

For now, it is worth noting that, contrary to the feminist charge, the authors do not naively see robots as a panacea: they remain well aware of the potential for technology to be misused,[66] as do many legal commentators.[67] However, Briggs and Scheutz also recognise the potential to restore the teleology with the technology, such that the latter *serves* rather than undermines humanity; indeed, this is exactly what motivates their research.

Can international law mandate the inclusion of software suppressors?

Arguably, mandating the inclusion of either of the above suppressors in the AWS design and manufacturing process is unlikely to pose an insurmountable problem.[68] Indeed, treaty law already makes technical design stipulations for certain weapons, with the aim of mitigating civilian harm. For example, under Article 6(2) of the Amended Mines Protocol,[69] remotely delivered anti-personnel mines must be equipped with an automatic self-destruction mechanism and a self-deactivating feature,[70] 'in compliance with the provisions ... in the Technical Annex'.[71] In turn, Paragraph 3(a) of the Annex, which has the same legal status as the Protocol, requires that such mines are 'designed and constructed' with certain effects-based features[72] that effectively 'prohibit all use of *long-lived* anti-personnel mines outside of marked, monitored and protected areas'.[73] As Boothby points out, the linking of 'design' and 'construction' to self-destruction and self-deactivation makes clear a dual obligation: that the stipulated reliability rates must be both *built into the design* of the weapons and must *actually be achieved in manufacture.*[74] Applied to AWS, such drafting would compel states to require defence contractors to include software suppressors, not as an 'optional extra', but as an *integral* part of early-stage research and development, during manufacture, verification and validation, and also testing and evaluation.

In addition, Article 4 of the Amended Mines Protocol, effectively mandates a 'detectability' component or feature; again, as specified in the Technical Annex,[75] with the effect that the latter also enjoys legally binding status. In turn, Paragraph 2(a) of the Annex stipulates that anti-personnel mines must 'incorporate in their construction' a material or device that provides a response signal that is detectable by commonly available mine detection equipment.[76] Again, this linking back to the 'construction' of the weapon highlights that detectability must be part of the original manufacturing process and is not an 'optional extra', as there are no circumstances in which deployment without detectable properties can be countenanced.[77] Given the propensity for rape and torture as a weapon of war by repressive regimes,[78] it is equally arguable that the development, fielding and (especially) export sales of more advanced and dextrous AWS also cannot be reasonably countenanced, unless these fully incorporate well-designed software suppressors.

In both of the above instances, the technical design stipulations concern relatively simple devices or features,[79] and this may raise questions on their transplantability to a requirement for advanced software suppressors in an AWS-based instrument. Note, however, that the Convention on Cluster Munitions[80] (CCM), which contains five technical design criteria,[81] requires that each explosive submunition be 'designed to detect and engage a single target object'.[82] This seemingly simple requirement actually entails the inclusion of sophisticated automatic target recognition capabilities,[83] thus suggesting that the development and inclusion of advanced technical features, such as software suppressors, can potentially be mandated by treaty law. Ideally, state parties would then create national implementing legislation to transpose the obligation into national law, and to require especially that export licences for AWS are conditional upon the inclusion of such capabilities.

Tamper-proofing the ethical safeguards

However, once states agree to be bound by such measures, there remains the more serious challenge of preventing nefarious leaders and militaries from uninstalling these safeguards. Yet, even in this regard, there are various options aimed at defeating attempts to uninstall legally required software suppressors, which can also be stipulated in administrable international rules[84] that – once ratified – private defence contractors will be legally bound to follow. For example, the core operating system (OS) and the suppressor can both be written onto the same one-time programmable read-only memory (ROM) chip,[85] such that removal or destruction of one would do the same to the other by default, thereby leaving an inoperable weapon system. Necessary software updates by the manufacturer would have to be physically delivered in the form of a whole new ROM chip. Accordingly, while this option offers relatively greater security and a high level of tamper-proofing, it is also highly inconvenient compared with using re-writable memory that can be updated online.

An alternative option might be to keep the two programs physically separate, but for the core OS to be made to depend on certain critical instructions contained within the suppressor to be able to operate.[86] Again, physical removal of the suppressor would result in an inoperable robot by leaving it with an incomplete OS. This option is more convenient in that it does not necessarily require the OS to be written onto a ROM chip, and thus it can be conveniently updated online. However, it is less secure than the first option, as it leaves open the possibility that a sophisticated analyst-programmer, who is able to reverse-engineer the suppressor, may then discover which parts are required to permit the core OS to work, and may then design and install that code separately.[87] Yet, even this may be pre-empted with the use of strong dynamic code encryption that prevents the reverse-engineering of the suppressor code or the missing OS code.[88]

An additional option that might be entertained would be to make the code inter-dependencies between the suppressor and the core OS so dispersed that reverse-engineering and reintegrating them would be somewhat akin to 'unscrambling the eggs'.[89] The intended effect would be to create such a complex software (re)programming task as to be too difficult to execute within a feasible timeframe.[90] However, this option, which involves 'security through obscurity',[91] has two disadvantages: firstly, it is extremely difficult to create such a complex software artefact whilst preserving correctness[92]; and secondly, determined hackers have historically almost always succeeded in penetrating the deliberately introduced obscurity.[93]

Arguably, the most prudent approach to tamper-proofing is to combine all the above methods. None of them are individually fool-proof, but each is likely to take some effort and expertise to circumvent; the *combination* of all, even more so. Thus, combining: a) highly dispersed code inter-dependencies, with b) strong dynamic code encryption, and c) at least the suppressor being written onto ROM, should present a formidable programming challenge that most entities outside the original equipment manufacturer will find extremely difficult to overcome.

The 'kill switch'

Thirdly, should a nefarious dictator or military commander find a way to overcome all the above technical safeguards, their attempts to use a dextrous AWS for rape or torture may still be defeated through the idea of a 'kill switch'.[94] This is an extreme form of 'manual override'[95] that, combined with a real-time feedback loop, enables an accountable human operator to detect and shut down an AWS that may have 'gone rogue'; accordingly, it is a very useful way to integrate human control of a weapon system *within the same military hierarchy*. However, in the context of a weapon system sold to a *foreign* customer, a feedback loop going to the original defence contractor or vendor state would clearly be unacceptable to that foreign state purchasing the AWS. Depending on the nature of the information being fed back, it may well violate military protocols and national security classification[96]; it would almost certainly make the AWS unsaleable to all but internal customers. Conversely, to have a kill switch *without* a feedback loop may be both legally permissible and administrable, even if this omission would tend to slow down the response to any violations of IHL, ICL, or IHRL.

Accordingly, activation of the more 'politically acceptable' kill switch would depend on *independent* evidence gleaned by intelligence agencies (of the vendor State, and beyond),[97] which proves the AWS is being used for torture or sexual violence; or indeed, if it is being used for deliberate civilian targeting or indiscriminate attack,[98] as was the case recently with Saudi targeting in Yemen.[99] Such evidence would be fed back to the relevant vendor state, which would seek to clarify matters with the alleged violating state, and demand a ceasing of all violations. Should these efforts fail to bring about a satisfactory outcome, the vendor state may then activate the kill switch, or legally require the relevant defence contractor to do so.

The device itself usually works through an inexpensive SIM card embedded into the system,[100] which communicates with nearby mobile-network towers,[101] and this allows for remote disabling of the weapon once activated.

On a practical note, however, two important and potentially limiting points should be made. Firstly, SIM-operated kill switches rely on civilian infrastructure for military control of a weapon system and, as such, may fail to work in the event that mobile services are down.[102] Secondly, as the Israeli Air Force demonstrated in 2007 when it reportedly deactivated a Syrian radar system to bomb a suspected nuclear site,[103] kill switches may indeed be vulnerable to enemy hacking; accordingly, they may also be utilised to prevent the *legitimate* use of an AWS, and this could well act as a barrier to their adoption.[104]

Conclusion

State-sanctioned (or leadership authorised or condoned) strategic rape as a weapon of war has been an indisputable feature of atrocity-high armed conflicts since the end of the Cold War – even if the extent to which it occurs can be debated.[105] Inevitably, perhaps, with so many fears surrounding the advent of new technologies such as AWS and calls for them to be banned, concerns about rape have been transferred from men to machines. However, this only draws a very limited picture and is largely misguided. It is important to recall Newton's comment that where the *teleology* and the *technology* of AWS risk diverging, they can indeed be deliberately realigned. Thus, the application of wartime rape and torture standard to AWS, as has been argued by some critics, is problematic and incomplete. This has been shown in this chapter, by reference to spatial issues, safeguards, and kill switches. Indeed, while no one would wish to be in such a situation, if robots were to rape, the prosecution of a crime would be easier than it often is with human violators. This is because it could be said that robots would only 'rape' when in the hand of someone very clearly wishing to do harm that human beings would normally find difficult and uncomfortable, and who had programmed autonomous systems precisely to cause such harm. In that situation, while it would not prevent or relieve the suffering of victims, there would be little ambiguity about intention and responsibility.

Notes

1 Special Rapporteur on Extrajudicial, Summary or Arbitrary Executions (Heyns, C.), *Report of the Special Rapporteur on Extrajudicial, Summary or Arbitrary Executions*, Human Rights Council, U.N. Doc. A/HRC/23/47, 9 April 2013, ¶ 54, at: www.un.org/en/ga/search/view_doc.asp?symbol=A/HRC/23/47.

2 This was a series of informal meetings held between 2013 and 2016 under the auspices of the Convention on Certain Conventional Weapons at the United Nations, Geneva. Specifically on the 'robots don't rape' argument, see (hear) the various audio presentations of the 2014 Meeting of Experts on LAWS, available at: www.unog.ch/__80256ee600585943.nsf/(httpPages)/a038dea1da906f9dc1257dd90042e261?OpenDocument&ExpandSection=1#_Section1.

3 See Offices of the Surgeon General, Multinational Force – Iraq and US Army Medical Command, *Mental Health Advisory Team (MHAT) IV: Operation Iraqi Freedom 05–07, Final Report*, 17 November 2006.
4 See ibid. for the effects that these human imperfections have had on human combatants and their respect for the principle of civilian immunity, as well as their respect for IHL more broadly.
5 Arkin, R. C. (2013) 'Lethal Autonomous Systems and the Plight of the Non-Combatant', *AISB Quarterly*, No. 137, p. 4.
6 Conversely, wide-area loitering that is *loud, persistent*, relatively *close* to a human subject in custody, and repeated over an *extended period of time*, may cause extreme fear, and may contribute towards 'mental torture'.
7 Sassóli, M (2014) 'Autonomous Weapons and International Humanitarian Law: Advantages, Open Technical questions and Legal Issues to be Clarified', *International Law Studies*, Vol. 90, p. 310.
8 See notes 24–30 below.
9 International Criminal Court (ICC) (2011) *Elements of Crimes*, at: www.icc-cpi.int/nr/rdonlyres/336923d8-a6ad-40ec-ad7b-45bf9de73d56/0/elementsofcrimeseng.pdf.
10 Ibid., Element 1 of Article 7(1)(g)-1 (crime against humanity of rape); and Element 1 of Article 8(2)(b) (xxii)-1 and (e)(vi)-1 (war crime of rape) (emphasis added).
11 Note that to make this an international crime, three subsequent elements must be cumulatively fulfilled, which make violation only possible with *human* action and intention. See ibid., Elements 2–4 of each of the three criminal prohibitions.
12 For an account of the interdisciplinary scholarship concerned with sexual violence in armed conflict, see Branche, R. and Virgili, F. (eds.) (2012) *Rape in Wartime*, Basingstoke: Palgrave Macmillan. See also the 2014 symposium, 'Sexual Violence in Armed Conflict', *International Review of the Red Cross*, No. 894, available at: www.icrc.org/en/international-review/sexual-violence-armed-conflict.
13 Fitzpatrick, B., *Tactical Rape in War and Conflict: International Recognition and Response*, Bristol: Policy Press, 2016, distinguishing rape as a *by-product* of war from rape as a *weapon* of war; and recounting the author's own experience of official resistance to accepting the reality of the latter, during the early 1990s. See also Diken, B and Laustsen, C. B. (2005) 'Becoming Abject: Rape as a Weapon of War', *Body and Society*, Vol. 11, No. 1, p. 111; and Smith-Spark, L., 'How Did Rape Become a Weapon of War?', BBC News, 8 December 2014, at: http://news.bbc.co.uk/1/hi/4078677.stm;.
14 Amnesty International (2004) *Lives Blown Apart: Crimes Against Women in Times of Conflict*, Oxford: Alden Press, pp. 20–27, citing case examples of Rwanda, Nepal, Darfur, Chechnya, Iraq and Bangladesh; available at: www.amnesty.org/en/documents/ACT77/075/2004/en/. But note, the exact objectives pursued, as well as the precise causes and consequences of sexual violence, are highly context-specific and vary from one armed conflict (and armed group) to the next. See Meger, S. *Rape Loot Pillage: The Political Economy of Sexual Violence in Armed Conflict*, Oxford: OUP, 2016, expressing scepticism with the conventional 'rape as a weapon of war' paradigm, and aiming to de-homogenise conflict-related sexual violence, to enable more appropriate policy responses for a given conflict.
15 See ICRC 'Prohibition and Punishment of Torture and Other Forms of Ill-Treatment', *ICRC Advisory Service on International Humanitarian Law*, 25 June 2014, available at: www.icrc.org/en/document/prohibition-and-punishment-torture-and-other-forms-ill-treatment, p. 1, which distils three elements common to most definitions of torture. Namely: 1) severe pain or suffering, whether physical or mental; 2) intentionally inflicted; and 3) instrumental, for such purposes as obtaining information or a confession, punishment, intimidation, coercion, or discrimination.
16 Diken and Laustsen (2005) (note 12 above). See also Kaldor, M. (2012) *New & Old Wars: Organised Violence in a Global Era*, 3rd edn, Cambridge: Polity Press, noting the tendency by some parties in 'new wars' to avoid confrontation with a standing army, but instead to aim to control territory by intimidation; sowing fear and hatred; and by creating a climate of insecurity and suspicion.
17 Snyder, C. S., Gabbard, W. J., May, J. D. and Zulcic, N., 'On the Battleground of Women's Bodies: Mass Rape in Bosnia-Herzegovina', *Journal of Women and Social Work*, Vol. 21, No. 2 (2006), p. 189.
18 As the UN General Assembly (UNGA) asserted at the time, 'this heinous practice [of mass rape] constitutes a weapon of war in fulfilling the policy of ethnic cleansing carried out by Serbian forces in Bosnia and Herzegovina'. See UNGA 'Rape and Abuse of Women in the Areas of Armed Conflict in the Former Yugoslavia', UNGA Res. 49/205, 94th Plenary Meeting, A/RES/49/205, 23 December 1994 (8th preambular clause). See also Salzman, T.A., 'Rape Camps as a Means of Ethnic Cleansing: Religious, Cultural, and Ethical Responses to Rape Victims in the Former Yugoslavia', *Human Rights Quarterly*, Vol. 20, No. 2 (1998), p. 348.

19 The UNGA had earlier categorised 'ethnic cleansing' in Bosnia as a form of genocide. See UNGA 'The situation in Bosnia and Herzegovina', UNGA Res. 47/121, 91st Plenary Meeting, A/RES/47/121, 18 December 1992 (9th preambular clause).
20 Cohen, D. K., 'Explaining Rape During Civil War: Cross-National Evidence (1980–2009)', *American Political Science Review*, Vol. 107, No. 3 (2013), p. 467.
21 Ibid.
22 Ibid.
23 This is due to the high cost of early AWS models, which will very likely command state-level resources.
24 Carpenter, C. '"Robot Soldiers Would Never Rape": Un-Packing the Myth of the Humanitarian Robot', *Duck of Minerva*, 14 May 2014, at: http://duckofminerva.com/2014/05/robot-soldiers-would-never-rape-un-packing-the-myth-of-the-humanitarian-war-bot.html.
25 Sandvik, K.B. and Lohne, K., 'Lethal Autonomous Weapons: Killing the "Robots-don't-Rape" Argument', *IntLawGrrls*, 5 August 2015, at: https://ilg2.org/2015/08/05/lethal-autonomous-weapons-killing-the-robots-dont-rape-argument/.
26 Ibid.; Carpenter (2014) (see note 24 above).
27 Sandvik and Lohne (2015) (see note 25 above).
28 Carpenter (2014) (see note 24 above).
29 Ibid.
30 Krupiy, T., 'Of Souls, Spirits and Ghosts: Transposing the Application of the Rules of Targeting to Lethal Autonomous Robots', *Melbourne Journal of International Law*, Vol. 16, No. 1 (2015), p. 199.
31 Human Rights Watch and the IHRL Clinic, Harvard Law School (2012) *Losing Humanity: The Case Against Killer Robots*, at: www.hrw.org/sites/default/files/reports/arms1112ForUpload_0_0.pdf.
32 'Emotionless robots could … serve as tools of repressive dictators seeking to crack down on their own people without fear their troops would turn on them': ibid., p. 4.
33 Ibid., pp. 46 and 47.
34 See notes 6–7, above.
35 Indeed, as Dinstein points out, there is a distinction between the 'very nature and design' of a weapon and its *use* in any given engagement. The latter may occur illegally in an isolated case, but this does not 'stain [the weapon] with an indelible imprint of illegality, since in other operations [it] may be employed within the framework of [IHL]. See Dinstein, Y., *The Conduct of Hostilities Under the Law of International Armed Conflict* (3rd edn, Cambridge: Cambridge University Press, 2016) p. 72.
36 This is not to diminish the seriousness of rape as a weapon of war, but to acknowledge that new technologies can be used for both lawful and unlawful conduct, and that careful thought is needed to curb the latter while retaining the capacity for (and even promoting) the former.
37 Wagner, M., 'The Dehumanization of International Humanitarian Law: Legal, Ethical, and Political Implications of Autonomous Weapon Systems', *Vanderbilt Journal of Transnational Law*, Vol. 47, 2014, p. 1416.
38 Arkin, R. C. *Governing Lethal Behavior in Autonomous Robotics* (Boca Raton: Chapman & Hall/CRC, 2009), p. 143.
39 The following is summarised from Arkin, ibid., pp. 127–33.
40 Actuators are the internal motors that enable robots to move, like muscles in the human body.
41 *Forbidding* constraints are provided by both legal norms and RoE; for example, 'no targeting of any human not positively identified as an enemy combatants' (IHL), or 'no targeting of any positively identified combatant until after being fired upon' (RoE). By contrast, *obligating* constraints are only provided by RoE; for example, 'attack bridges within GPS coordinates XYZ'.
42 The following is summarised from Arkin (2009) (see note 38 above), pp. 133–8.
43 Moreover, for undesirable consequences that even the EBC and EG did not constrain, Arkin also proposes an 'Ethical Adapter' (EA) design option. This enables human reasoning after a battle damage assessment to update the EBC and EG, thereby avoiding the same unintended consequences on subsequent deployments. Accordingly, the EA is *necessarily* dependent on the commander's good faith, thus not relevant for preventing intentional rape or torture. On the Ethical Adapter, see Arkin (2009) (note 38 above), pp. 138–43.
44 See Element 1 in ICC (2011) (see note 9 above).
45 Recall that the definition of rape is binary ('conduct resulting in penetration, *however slight* … of the *anal or genital opening* of the victim with *any object*'); thus, it should be relatively easy to code into a suppressor.

46 Arguably, this is more likely with broader acts of torture than with rape. In the case of the latter, assigning 'forbidden' status to any individual act of 'penetrating any part of the human body' would arguably make 'workarounds' practically impossible.
47 See the three common elements of torture definitions, at note 15, above.
48 Nordgren, L. F., McDonnell, M. H. M. and Loewenstein, G., 'What Constitutes Torture? Psychological Impediments to an Objective Evaluation of Enhanced Interrogation Tactics', *Psychological Science*, Vol. 22, No. 5 (2011), p. 689, describing how human subjects judged the severity of pain from interrogation – both physical and mental – differently, depending on whether they themselves had experienced (a small dose) of that same kind of pain.
49 Namely, they reflect the fact that humans bring to bear very different personal experiences, which are only partly addressed through training, thus giving rise to inconsistent assessments of pain severity.
50 See, for example, Bellantonio, M., Haque, M. A., Rodriguez, P., Nasrollahi, K., Telve, T., Escalera, S., Gonzales, J., Moeslund, T.B., Rasti, P. and Anbarjafari, G., 'Spatio-Temporal Pain Recognition in CNN-Based Super-Resolved Facial Images', *International Conference on Pattern Recognition (ICPR): Workshop on Face and Facial Recognition*, Springer, 2017, at: http://vbn.aau.dk/files/245132512/FFER_Pain_Super_Resolution.pdf.
51 See, for example, developments in post-polygraph technologies, such as computer voice stress analysis (CVSA). Miller, S. 'When Everybody Lies: Voice-Stress Analysis Tackles Lie Detection', *GCN*, 18 March 2014, at: https://gcn.com/articles/2014/03/18/voice-risk-analysis.aspx.
52 In this connection, significant steps have recently been taken towards the goal of developing emotionally sophisticated robots, with the 'emotion chatting machine' (ECM). See Zhoa, H., Huang, M., Zhang, T., Zhu, X. and Liu, B., 'Emotional Chatting Machine: Emotional Conversation Generation with Internal and External Memory', 2017, available at: https://arxiv.org/abs/1704.01074.
53 Presently, surgical robots are designed with effective feedback mechanisms to enable detection and measurement of muscle tone and muscle tension, for pre-surgical checks before invasive surgery. However, these usually involve attaching electrodes to the patient, which is practical in the case of planned and consensual activities like surgery, but rather impractical in an arrest and capture, or interrogation scenario. That said, current research in tactile sensors suggests that the same feedback mechanisms can be included in robotic manipulators. For recent theoretical contributions, see Yakovenko, A., Goryacheva, I. and Dosaev, M. 'Estimating Characteristics of a Contact Between Sensing Element of Medical Robot and Soft Tissue', in Wenger, P. and Flores, P. (eds.), *New Trends in Mechanisms and Machine Science: Theory and Industrial Applications* (Switzerland: Springer International Switzerland, 2016), p. 561; and Cirollo, A., Cirollo, P., De Maria, G., Natale, C. and Pirozzi, S., 'A Distributed Tactile Sensor for Intuitive Human-Robot Interface', *Journal of Sensors*, Vol. 2017, at: www.hindawi.com/journals/js/2017/1357061/. Intuitively, similar mechanisms would enable an AWS to detect whether a human subject is physically resisting or complying, the latter scenario triggering the technical prohibition against any further physical restraint, lest this becomes torture. We are grateful to Dr Hongbin Liu of the Centre for Robotics Research (CORE), King's College London, for alerting us to this possibility.
54 Clearly, any pain or suffering resulting from physical contact in such a scenario cannot be precluded from the definition of torture under Article 7(2)(e), Rome Statute, for being 'inherent in or incidental to, lawful sanctions'.
55 In this sense, assessing the existence of torture and suppressing any robot actions that would lead to it will arguably become more objective and more consistently applied than where human interrogators are involved.
56 See Wagner, 'The dehumanization' (note 37 above), pp. 1414–16, for a brief summary.
57 Ibid., p. 1415.
58 'Hot' empathy refers to the brain's emotion centre – the amygdala – and the tendency to mirror the emotions of another person, as distinct from 'cold' empathy, which is merely cognitively recognising the emotional state of another person. See Dutton, K., *The Wisdom of Psychopaths* (London: Arrow Books, 2013), pp. 16–18. Clearly, any attempt by robots to read human emotions will be 'cold'.
59 Arkin (2009) (see note 38 above).
60 Sassóli (2014) (note 7 above), pp. 318 and 332–4.
61 Briggs, G. and Scheutz, M., '"Sorry I Can't Do That": Developing Mechanisms to Appropriately Reject Directives in Human-Robot Interactions', *AAAI Fall Symposium Series*, 2015, pp. 32–6, at: www.aaai.org/ocs/index.php/FSS/FSS15/paper/view/11709/11522.
62 Ibid., p. 33; detailed reasoning at p. 34.

63 Scheutz, M., 'Why Robots Need to Be Able to Say "No"', *The Conversation*, 8 April 2016, at: https://theconversation.com/why-robots-need-to-be-able-to-say-no-55799.
64 Briggs and Scheutz '"Sorry I Can't Do That"' (see note 61 above). In particular, programming the software to reason about, and address the felicity conditions to the same degree as a human will remain an open challenge for the foreseeable future.
65 Ibid., p. 35, detailing some very simple examples of human–robot interactions, where commands were successfully rejected.
66 Indeed, as they warned in a recent article: 'Don't worry about defiant machines. Devious human masters ... are a bigger threat.' See Briggs, G. and Scheutz, M., 'The Case for Robot Disobedience', *Scientific American*, Vol. 316, 2017, p. 44.
67 For example, this was apparent in Heyns, *Report of the Special Rapporteur* (see note 1 above), who did not rule out the possibility that a robot may rape, but instead pointed out, at ¶ 54, that '... *unless specifically programmed to do so* ... [r]obots do not rape' (emphasis added).
68 The bigger challenge will be to verify their genuinely effective installation, which will require rigorous and exhaustive simulation testing or formal proof of correct behaviour.
69 Protocol on Prohibitions or Restrictions on the Use of Mines, Booby-Traps and Other Devices, as amended on 3 May 1996, 2048 UNTS 93, entered into force 3 December 1998 (Protocol II, Convention on Certain Conventional Weapons).
70 Article 5(2), Amended Mines Protocol, mandates the same in relation to manually-emplaced anti-personnel mines that are left outside marked, monitored and protected areas.
71 Article 6(2), Amended Mines Protocol. This drafting clearly accords to the Technical Annex and its more detailed requirements, the same legal status enjoyed by the Protocol.
72 Technical Annex, ¶ 3(a). Namely, that no more than 10% of activated mines will fail to self-destruct within 30 days of emplacement; and, together with self-deactivation, no more than 1/1,000 (0.01%) of activated mines will function as a mine 120 days after emplacement.
73 Office of the Under-Secretary of Defense for Acquisition, Technology, and Logistics, *Treaty Compliance: CCW: Article by Article Analysis of the Protocol on Use of Mines, Booby-Traps and Other Devices*, at: www.acq.osd.mil/tc/treaties/ccwapl/artbyart_pro2.htm (emphasis added).
74 Boothby, W. H. *Weapons and the Law of Armed Conflict* (Oxford: OUP, 2016), p. 166.
75 Article 4, Amended Mines Protocol.
76 Technical Annex, ¶ 2(a). That is, it must provide 'a response signal equivalent to a signal from 8 grammes or more of iron in a single coherent mass'.
77 Often, detectability is vital for speeding up post-conflict mine clearance, hence civilian risk mitigation; and for post-conflict economic development.
78 As pointed out in notes 11–26, above.
79 Namely, a *self-destruction mechanism* usually consists of an additional initiating mechanism that detonates the mine in the event that the primary mechanism fails; similar to the 'redundant subsystems' concept in reliability engineering. A *Self-deactivation feature* can be as simple as having reliance on an electrical power supply, with a limited-life battery ensuring that the weapon deactivates after a set period. Finally, detectability can be as simple as incorporating a small concentration of iron within the munition.
80 Convention on Cluster Munitions, 30 May 2008, 2688 UNTS 190, entered into force 1 August 2010.
81 Article 2(2)(c), CCM, which requires that (i) each *munition* contains fewer than 10 explosive submunitions; and, in turn, each explosive *submunition* (ii) weighs more than four kilograms; (iii) is designed to detect and engage a single target object; (iv) is equipped with an electronic self-destruction mechanism; and (v) is equipped with an electronic self-deactivation feature. When cumulatively satisfied, these technical characteristics presume the lawfulness of a (sensor-fused) weapon, by formally excluding it from the treaty's definition of the prohibited 'cluster munition'.
82 Article 2(2)(c)(iii), CCM.
83 See, for example, the (CCM-compliant) 155 BONUS sensor-fused weapon, '155 BONUS Datasheet', at: www.baesystems.com/en/download-en/20151124120132/1434555557732.pdf, noting that its 'Dual Mode Sensor' detects and identifies targets by a) processing images received from infrared sensors, then b) combining the results with data received from the profile sensor. This allows 'combat-worthy targets [to] be separated from false targets' (p. 2).
84 Similar to Article 2(2)(c), CCM.
85 See Griffin, J., Matas, B. and de Suberbasaux, C. (1996) *Memory 1996*, Scottsdale: ICE Corp., Chapter 9 on 'ROM, EPROM, and EEPROM Technology', available at: http://smithsonianchips.si.edu/ice/cd/MEM96/SEC09.pdf.

86 For example, by using coroutines. See Knuth, D.E. (1997) *The Art of Computer Programming, Vol. 1: Fundamental Algorithms*, 3rd edn, Boston: Addison-Wesley, Section 1.4.2: Coroutines, pp. 193–200.
87 Ibid.
88 Delfs, H. and Knebl, H., *Introduction to Cryptography: Principles and Applications*, 3rd edn, Berlin/Heidelberg: Springer, 2017.
89 This term is used in merger control to describe the difficulty (often verging on near-impossibility) for a merged firm to comply with a dissolution order, after that firm has fully integrated the operational, financial and administrative functions of the previously independent undertakings. Such dissolution orders are issued by, for example, the European Commission under Article 8(4), *EU Merger Regulation*.
90 Schneier, B., 'A Plea for Simplicity: You Can't Secure What You Don't Understand', *Schneier on Security*, 19 November 1999, at: www.schneier.com/essays/archives/1999/11/a_plea_for_simplicit.html.
91 Schneier, B., 'Secrecy, Security, and Obscurity', *Schneier on Security*, 15 May 2002, at: www.schneier.com/crypto-gram/archives/2002/0515.html#1.
92 Schneier, 'A Plea for Simplicity' (see note 90 above).
93 Schneier, B., 'The Nonsecurity of Secrecy', *Communications of the ACM*, Vol. 47, No. 10, October 2004, p. 120, at: https://cacm.acm.org/magazines/2004/10/6404-the-nonsecurity-of-secrecy/fulltext.
94 Zittrain, J., 'The Case for Kill Switches in Military Weaponry', *Scientific American*, 3 September 2014, at: www.scientificamerican.com/article/the-case-for-kill-switches-in-military-weaponry/.
95 Extreme in the sense that a 'kill switch' destroys the system or shuts it down, whereas 'manual override' may involve any degree of intervention, from as little as human operators simply vetoing individual robotic actions.
96 For example, in a US context, Executive Order 13526 (Obama, 29 December 2009), includes, at § 1.4, 'military plans, weapons systems, or *operations*' as a category of information that is potentially subject to classification (emphasis added). Thus, the battlefield actions of a AWS cannot be subject to an ongoing feedback loop to any entity outside of authorised military command or civilian control.
97 Or independent insights from international organisations and/or non-government organisations.
98 Zittrain (2014) (see note 94 above), points out that a kill switch can – in a similar way to switches found in mobile phones – also be used to destroy a weapon system that has been seized through hacking or theft. This is particularly important to prevent unauthorised proliferation of AWS capabilities to known repressive regimes, or to states that might reverse-engineer the software and proliferate it. Of course, this assumes the hackers do not take full control of the system, and the communication and kill switch functions remain intact, which may be possible with stronger levels of encryption in these more critical areas.
99 See 'Yemen Conflict: Saudi-Led Coalition Targeting Civilians, UN Says', BBC News, 27 January 2016, at: www.bbc.co.uk/news/world-middle-east-35423282; and Human Rights Watch (2016) *Bombing Business: Saudi Coalition Airstrikes on Yemen's Civilian Infrastructure*, at: www.hrw.org/sites/default/files/report_pdf/yemen0716web.pdf.
100 'Smart Weapons: Kill Switches and Safety Catches', *The Economist Technology Quarterly: Q4 2013*, 30 November 2013, at: www.economist.com/news/technology-quarterly/21590764-arms-control-new-technologies-make-it-easier-track-small-arms-and-stop-them.
101 Ibid., noting that this also provides an independent approximation of the machine's location.
102 While modern mobile technologies are generally reliable, civilian infrastructure may nonetheless be collaterally damaged during an armed conflict, and mobile towers in particular may be targeted if they become military objectives by 'use' or by 'purpose'. See Article 52(2), Protocol Additional to the Geneva Conventions of 12 August 1949, and Relating to the Protection of Victims of International Armed Conflicts, 8 June 1977, 1125 UNTS 3, entered into force 7 December 1978, on the definition of 'military objective'.
103 Adee, S., 'The Hunt for the Kill Switch', *IEEE Spectrum*, 1 May 2008, at: http://spectrum.ieee.org/semiconductors/design/the-hunt-for-the-kill-switch.
104 Davidian, D., 'The Programmable Diplomatic Kill Switch', *Modern Diplomacy*, 22 January 2017: http://moderndiplomacy.eu/index.php?option=com_k2&view=item&id=2168:the-programmable-diplomatic-kill-switch&Itemid=156.
105 For example, Wood has presented evidence that widespread rape is often neither opportunistic nor strategic, but more accurately seen as a 'practice'. See Wood, E. J., 'Conflict-Related Sexual Violence and the Policy Implications of Recent Research', *International Review of the Red Cross*, No. 894, 2015, p. 457.

15

HUMANITY AND LETHAL ROBOTS

An engineering perspective

Tony Gillespie

There are concerns about the legality and ethics of systems which are sometimes described as *fully autonomous*. There appears to be a consensus that current systems are not yet fully autonomous, but there is a belief that they will be developed during the next few years.

Recent papers[1,2] from Human Rights Watch (HRW) discuss autonomous weapon systems, calling for bans on their development and use. Meanwhile, Schmitt (2013),[3] Anderson and Waxman (2013),[4] and Anderson et al. (2014),[5] criticise the concept of a ban from the standpoint of international lawyers. The former argues that HRW do not distinguish between weapons that are unlawful per se and the unlawful use of otherwise lawful weapons. Anderson and co-authors argue that the law will evolve as technology evolves, and so a legal framework will develop which will constrain their use.

This chapter examines highly automated and autonomous weapon systems from an engineering standpoint, covering technology evolution, defence engineering processes, and economic pressures. Weapon system designs are reviewed formally with reviews under Article 36 of Additional Protocol 1 (API) of the Geneva Conventions (called Article 36 Reviews in this chapter for brevity). The legal reviewers expect to see evidence giving measured performance for the system, which will usually come from the engineering teams. Military users will also provide evidence, if the system is at the stage of operational evaluation prior to in-service operation.

New weapon concepts and systems are almost always the result of a steady evolution of technologies in several fields. The term 'drone' is often used by the media to describe unmanned air vehicles (UAVs), with an implication that they are novel and represent a fundamentally new approach to war. In fact they have a history dating back to the 1920s.[6]

It is essential that there is clarity in the terminology used by the law-makers, the engineers designing them, and the military and intelligence agency users. The use of imprecise language or terms that have different meanings among these professions can only lead to confusion and potentially poor legislation and contracts. Several definitions of autonomy have been produced over the last few years and are discussed below. It will also be shown below that autonomy is so closely linked to higher levels of automation that the concepts are indistinguishable.

States have processes for Article 36 reviews which rely on technical data in order to predict performance and reliability in use. Some guidance for the type of evidence required has been published by the International Committee of the Red Cross (ICRC).[7] The engineering profession has developed a large range of techniques which can be, and are, applied to generate appropriate evidence for the legal advisors to make their recommendations. This chapter reviews some of these and shows that they can be applied to autonomous systems. It is shown further that it is possible for procurement authorities to impose contractual specifications on weapon system suppliers that ensure compliance with legal constraints.

Automation, autonomy and control

Throughout the industrial revolution there has been progress in devising systems that do not rely on humans for their operation. Eighteenth-century steam pumping engines required humans to operate the valves until the automatic valve gear was developed. Then speed was controlled by the operator until governors were developed to limit the speed. Now cruise control for most forms of transport is ubiquitous. Steady evolution of technologies has led to modern industry becoming reliant on high levels of automation. General acceptance of the term 'robot' in this context allows advertisers to extol their benefits, for example in car advertisements. However, it should be noted that automated industrial processes execute well-defined operations under strict control regimes.

Increasing public reliance on automatic control of complex systems is a feature of modern society. Aircraft of all types, except the simplest, rely on automated flight control systems; and air traffic control for commercial aircraft in congested airspace would be impossible without automation. The failure of a UK air traffic control computer in December 2014 and subsequent flight delays due to the fall-back solution of human control illustrates this clearly.[8] Despite, or perhaps because of, the rare occurrence of this type of incident, there is public acceptance of these complex systems. It should be noted that the systems produced by the aviation industry at an economic price can only produce high levels of safety because airspace is highly regulated both nationally and internationally and only correctly equipped aircraft can enter it. The use of automatic gearboxes and cruise control in cars is an analogous development to those in the aircraft industry. The step from cruise control to driverless cars is now being discussed widely, with an assumption that the technology will become widespread at some time in the future. Acceptance of their potential can be seen by governments, such as those of the US and UK, passing legislation to allow their use for experimental purposes. It may, however, be many years before driverless cars are accepted for general use.

Weapons have become more automated: the revolver was an early example, with progress through heavy machine guns to automated detect-and-fire systems such as Phalanx and Iron Dome. There do not appear to be moral or legal objections to machine guns, but some are raised about systems such as Phalanx. HRW have argued[9] that these latter systems are automatic rather than autonomous, as they are defensive and only execute pre-programmed instructions more rapidly than a human can perform them. The latter argument would appear sound, but describing a system as automatic or not, based on its use, does not.

Increased sophistication of automated systems has come from both the widespread implementation of algorithms on computers and microprocessors and the combination of subsystems to form a more complex whole. For example, the Phalanx Close-In Weapon System is described by its manufacturer[10] as:

> a rapid-fire, computer-controlled, radar-guided gun system designed to defeat anti-ship missiles and other close-in air and surface threats. A self-contained package, Phalanx automatically carries out functions usually performed by multiple systems – including search, detection, threat evaluation, tracking, engagement, and kill assessment.

It is implicit in the terms *threat evaluation* and *kill assessment* that the system decides whether the object detected is a legitimate target or not, using pre-programmed rules. It is reasonable to assume that the system's use in operations has been subject to a legal review by the United States. It is also reasonable to argue the necessity of a lethal response to a heavy object coming at high speed directly towards the defended platform. Therefore, a legal view of the balance of at least military necessity and proportionality has been made and implemented in the system's use, if not in its design. (The original design was in the 1980s.) Due to the nature of sensor systems and the random noise levels in them, the Phalanx system will have statistically based threshold levels at which it declares an object is a threat. Therefore, the conclusion is that this automated system can make decisions without human intervention, but they are based on a balance of mathematical probabilities derived from human judgement, i.e., it senses its environment, detects a change, assesses it, and changes behaviour as a result. Is this the same as human decision-making? Does this decision-making ability justify calling Phalanx an autonomous system? Are these questions irrelevant in practice as the ethical question is: will this system, as deployed, breach the principles of API?

The defence engineering community addressed the problem of differentiating between automation and autonomy by developing the concept of autonomy levels. These give a guide to the decision-making capability of unmanned air system. These originated in the UAV community when there was a future concept of swarms of UAVs acting in close collaboration. They were published in 2001.[11] As technology limitations and practical issues have emerged, the definitions of autonomy levels have changed. Figure 15.1[12] gives a more modern and generally applicable version used by the ASTRAEA programme in its work on civilian use of UAVs. The implication is that the defence industry recognises that there is not a clear division between automated and autonomous systems.

The arguments in the last paragraph can be summarised as: there is a spectrum of automated capabilities which has the simplest control system at one end and futuristic autonomous weapon control systems at the other. Interpretation of this automation spectrum as a continuum of weapon control systems does not follow; the spectrum is for the individual control subsystems which together make up the whole weapon control system. This point is noted by the ICRC (2014),[13] who state:

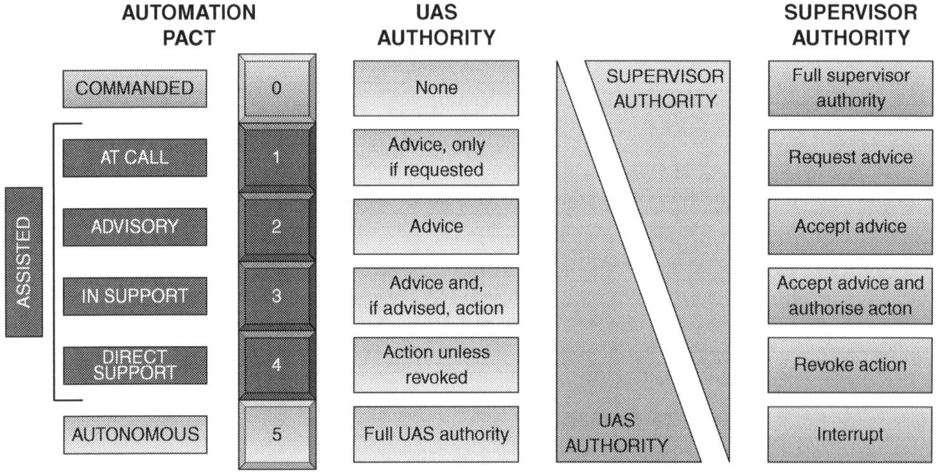

Figure 15.1 Pilot Authorisation and Control of Tasks (PACT) and UAS authority levels.

> Therefore, for a discussion of autonomous weapon systems, it may be useful to focus on autonomy in critical functions rather than autonomy in the overall weapon system. Here the key factor will be the level of autonomy in functions required to select and attack targets (i.e. critical functions), namely the process of target acquisition, tracking, selection, and attack by a given weapon system. Indeed, as discussed in parts B and C of this paper, autonomy in these critical functions raises questions about the capability of using them in accordance with international humanitarian law (IHL) and raises concerns about the moral acceptability of allowing machines to identify and use force against targets without human involvement.

Later in this chapter we show that a system engineering approach with functional partitioning and specifications offers a way to assess the lawfulness of a chain of decisions for an autonomous system in the same way as is carried out currently for human operators.

Automatic and autonomous systems all respond to inputs of various types. The associated engineering problem in weapon design is to design the control system, which is carried out using tools from the discipline of control engineering. A control system is designed to give an output response, which remains within well-defined limits, to its inputs. The simplest one is a single feedback loop such as the one shown in Figure 15.2(a), with feedback as part of the engineered system. The control system design will define the output response to changes in input. Figure 15.2(b) shows the same system with a human providing the feedback and hence the system output.

Traditionally, the output from the control loop is fed into the next part of the whole system which acts on it. The legal question is whether the whole system acts on the output from the control loop or not. The simplest way to separate this action is to put a switch in the output path. This is shown in Figure 15.2(c) as a manual veto, but it could be an input from another system which responds to a different set of inputs such as remote sensors rather than the sensors on the weapon system.

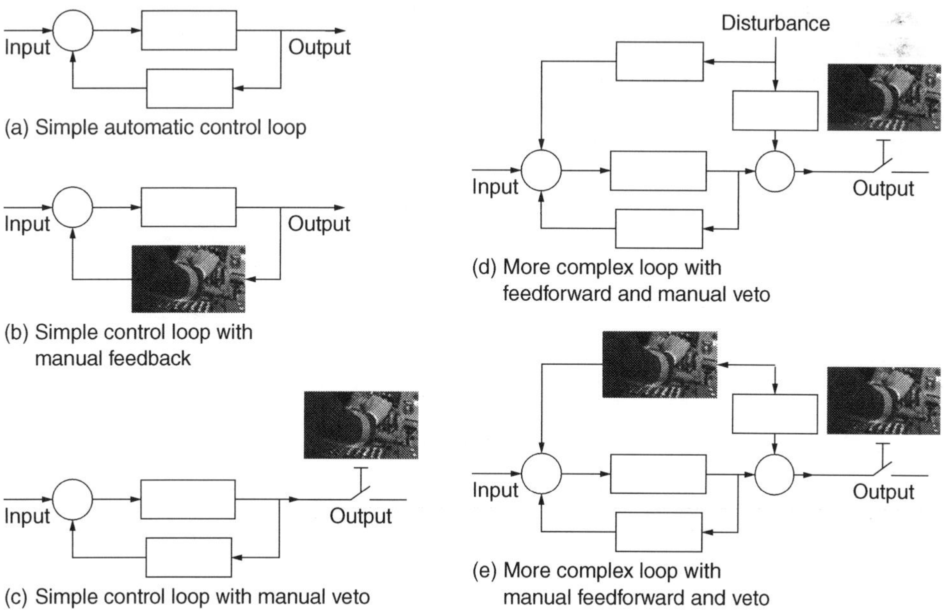

Figure 15.2 Illustrative control loops.

In practice, control systems are more complex. Figure 15.2(d) shows a system with a feed-forward loop to cope with large external disturbances. Figure 15.2(e) shows how a human can intervene in two parts of the control system, providing both a veto and an input to allow for large external disturbances. This is probably the position of the commander in any complex weapon system. (The term 'complex weapon system' here covers both complicated systems and the defence engineering term 'complex weapon system'.)

An important parameter in control system design is the time constant (τ) of each part of the system. This has a strict mathematical definition, but can be thought of as indicating the response time of a control system. Although it would be unwise directly to relate this to times which are not part of the automated system, it is important to consider the times taken for information to flow into and around the military command chain as well as the time constants inside automated control systems.

Economic trends in weapon development

It is well known that most governments have reduced expenditure on military materiel of all types. In the context of autonomous systems, this has had two effects: existing platforms have their lifetimes extended beyond their planned out of service date (OSD); and technologies developed for civil use are very capable and used in military systems wherever possible.

Life extension programmes include processing upgrades and increased levels of automation to enable the military user to operate in an increasingly complex environment. Legacy platform design usually assumed a high level of operator interaction with it, but little interaction with the command chain and other platforms. Modern operations now take place with extensive data networks linking platforms and individual warfighters, increasing the workload on the individual. This leads directly to the development of Information Technology (IT) support tools which are used to inform lethal decisions on the battlefield. One example is the US announcement[14] in 2012 that they hope to extend the life of 300 F-16 aircraft for 15 years with a Structural Service Life Extension Programme (SLEP) and a Combat Avionics Programmed Extension Suite (CAPES) upgrade. (Note that the F-16 started production in 1976 so the basic design is already 50 years old.) CAPES will include sensor fusion and a range of decision aids for the pilot. Increased levels of automation will be included such as its automated electronic warfare suite.

The revolution in IT has been accepted in civil applications; and the technology has then been adapted for military use if this is both possible and economic. The drive to use civil technology is now part of UK procurement policy, with a UK Government White Paper statement[15] that:

> We will also seek to minimise the costs of obtaining operational advantage and freedom of action by, wherever possible:
>
> - integrating advanced technologies into standard equipment purchased through open procurement;
> - sharing and developing appropriate technologies with our key allies;
> - seeking the best and most advanced civilian technology that can be adapted and incorporated into defence and security equipment to give us operational advantage; and
> - making the greatest possible use of synthetic training and simulation to reduce the cost of training personnel, particularly when applying advanced technologies to new capability needs.

One simple example is seen in the role of the Forward Air Controller (FAC) who calls up and commands lethal air support for ground forces. This was once only carried out by a few specialists. Recent campaign pressures and the development of robust data-links make this role more widespread in all land operations. New portable equipment incorporating civil technologies has been developed for the soldier and the new electronic systems added to existing fast-jet aircraft. There is also a complementary procedural and training requirement.

Human and system safety can be compromised by failure modes which will occur in any engineering system. Their study, with the use of statistical methods to reduce their occurrence, is a well-developed field of engineering. Regulating authorities produce engineering standards for the design and safe performance of systems.[16]

Engineers and regulators see a clear difference between automated systems whose failure will be catastrophic and/or lead to loss of life and those whose failure may lead to problems, but which are not catastrophic. The former systems are defined as 'safety-critical'. Their design imposes rigorous testing and extensive procedural processes compared with those for non-safety-critical systems. Designs include increased safety features such as 'multiply-redundant' control systems which have parallel independent control systems with a voting system to decide if one has failed and should be ignored. There was an accompanying reluctance to rely on any software except the most basic designs until the start of the twenty-first century. Progress has been made in the last decade to certify more complex software for aviation applications, but widespread applications are still in the development phase.[17] An unofficial rule of thumb in the industry is that the cost of safety-critical software is at least ten times that of normal, rigorously designed software. This economic factor places constraints on the overall cost of an integrated system and leads to minimising the number of safety-critical subsystems in a larger system; a point which is important for decision aids used to assist human decision-makers and autonomous systems. A decision aid is usually classified as *advisory* not *safety-critical* even if it is *mission-critical.*

Military command chains have well-controlled authorisation procedures for information passed to local commanders. These are straightforward when the information originates from within the coalition carrying out the campaign. The combination of pressure on military budgets and limitations on both data gathering and analysis equipment and reduced staff numbers limits the available information. This contrasts with the expanding capabilities of information gathering, processing and dissemination for commercial purposes such as news agencies and images accompanying maps. The engineering challenge to link the two, military command and commercial data, is a straightforward one of network protocols, standards and firewalls, but this does not address the issue of integrity, age, and provenance of data coming from a range of external sources of variable reliability. The military network may also have increased risk from cyber attack due to the connection to the internet, if the two are linked.

An additional consequence of constrained military budgets is the pressure on a nation's manufacturers to export their products to other nations provided they are acceptable to the former nation's government. There is a risk of proliferation, but limitations are set by international treaties such as the 2013 Arms Trade Treaty (ATT) and the Missile Technology Control Regime (MTCR).

Definitions

Participants in a constructive debate must have a shared understanding of key words, or at least agreement on where their usage and meanings diverge. This assumes a common language, which may not be correct in an international context. Questions about autonomous weapon systems are complicated by terminology used by three separate professions: lawyers, military personnel, and engineers. One example shows the potential for ambiguity: radar designers use the word 'target' to describe any object detected by a radar system whether there is a military context or not; this leads to using terms such as target discrimination, target detection range, false targets, etc., with both radar meanings and military ones.

Although the term 'automatic' is well understood, the same cannot be said about 'autonomy', which has many definitions. These are often defined in a specific context for logical consistency in a discussion about a particular problem.

The US DoD give the following definitions in Directive 3000.09 in an attempt to establish policy and assign responsibilities for development and use, as well as establishing guidelines to minimise failures and their consequences[18]:

Autonomous weapon system

A weapon system that, once activated, can select and engage targets without further intervention by a human operator. This includes human-supervised autonomous weapon systems that are designed to allow human operators to override operation of the weapon system, but can select and engage targets without further human input after activation.

Human-supervised autonomous weapon system

An autonomous weapon system that is designed to provide human operators with the ability to intervene and terminate engagements, including in the event of a weapon system failure, before unacceptable levels of damage occur.

Semi-autonomous weapon system

A weapon system that, once activated, is intended to only engage individual targets or specific target groups that have been selected by a human operator. This includes:

> Semi-autonomous weapon systems that employ autonomy for engagement-related functions including, but not limited to, acquiring, tracking, and identifying potential targets; cueing potential targets to human operators; prioritizing selected targets; timing of when to fire; or providing terminal guidance to home in on selected targets, provided that human control is retained over the decision to select individual targets and specific target groups for engagement.

'Fire and forget' or lock-on-after-launch homing munitions that rely on TTPs to maximise the probability that the only targets within the seeker's acquisition basket when the seeker activates are those individual targets or specific target groups that have been selected by a human operator.

The UK MOD[19] used the following definitions for Unmanned Air Systems (UAS), managing to avoid the difficulties DoD had with targeting seekers:

Automated system

In the unmanned aircraft context, an automated or automatic system is one that, in response to inputs from one or more sensors, is programmed to logically follow a pre-defined set of rules in order to provide an outcome. Knowing the set of rules under which it is operating means that its output is predictable.

Autonomous system

An autonomous system is capable of understanding higher-level intent and direction. From this understanding and its perception of its environment, such a system is able to take appropriate action to bring about a desired state. It is capable of deciding a course of action, from a number of alternatives, without depending on human oversight and control, although these may still be present. Although the overall activity of an autonomous unmanned aircraft will be predictable, individual actions may not be.

These definitions, like many others, put a decision-making capability as a key attribute of an autonomous system. At the simplest level, any weapon with a sensor which triggers a decision to fire or not to fire, such as in Phalanx, or even an anti-tank mine gives the weapon a capability to make a decision, i.e., a level of autonomy.

The conclusion from this discussion and the section above on *Automation, autonomy and control* is that there is no clear distinction between the terms automation and autonomy. They are part of a spectrum of weapon system capability. Any discussion about the use, engineering design or legality of a weapon system can only take place in the context of that particular system and its environment. This does not preclude debate about proposed or conceptual systems, but there must be logical consistency in the concept.

Human Rights Watch in *Losing Humanity* use a different approach to definitions:

> Robotic weapons, which are unmanned, are often divided into three categories based on the amount of human involvement in their actions:
>
> - Human in the Loop Weapons: Robots that can select targets and deliver force only with a human command;
> - Human on the Loop Weapons: Robots that can select targets and deliver force under the oversight of a human operator who can override the robots' actions; and
> - Human out of the Loop Weapons: Robots that are capable of selecting targets and delivering force without any human input or interaction.

In this report, the terms 'robot' and 'robotic weapons' encompass all three types of unmanned weapons, in other words everything from remote-controlled drones to weapons with complete autonomy. The term 'fully autonomous weapon' refers to both out of the loop weapons and those that allow a human on the loop, but that are effectively out of the loop weapons because the supervision is so limited. A range of other terms have been used to describe fully autonomous weapons, including 'lethal autonomous robots' and 'killer robots.'

The three bulleted terms (Human, on, in and out of the loop) are widely used in defence, but it should be clear from Figure 15.3 and the accompanying description that there is not a simple loop for any sophisticated weapon system. The term *loop* is probably best described as the local command chain which has commands propagated down it and obtains feedback from the lethal subsystem and its immediate sensors. A typical military command chain is shown schematically in Figure 15.3 where the double-headed arrows indicate feedback or feedforward loops. There will be additional inputs from human warfighters, sensors or political decisions. This would make Figures 15.2(d) or (e) the most appropriate illustration of the command chain and weapon system based on control system theory. Classifying robots as encompassing all three categories is nonsense, as all weapons with any level of automation will then be considered to be a robot.

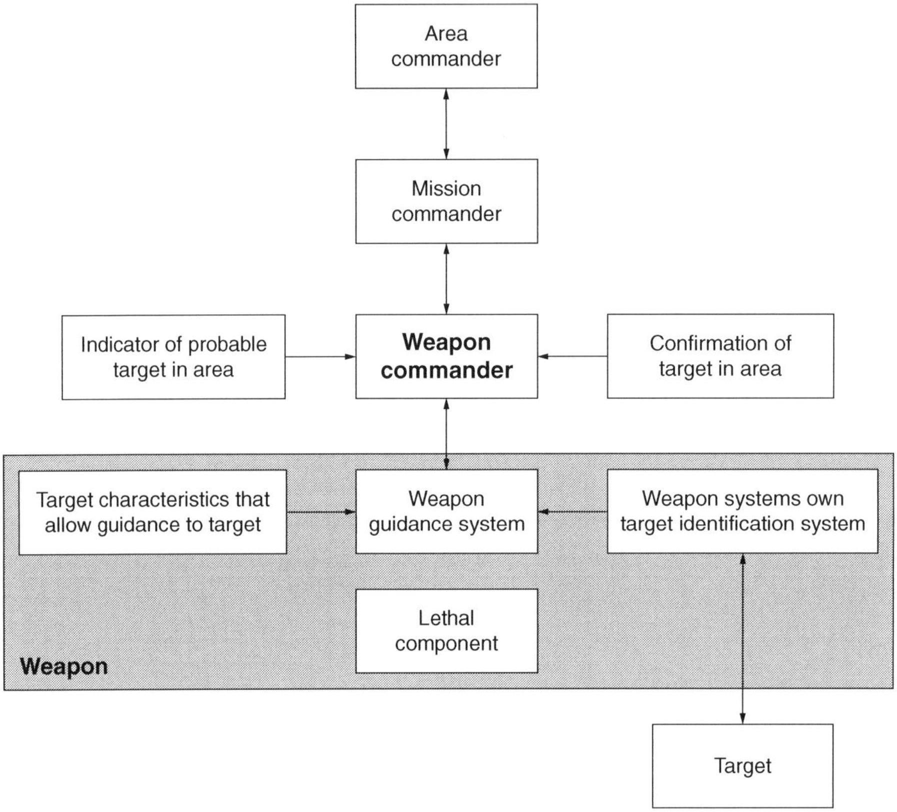

Figure 15.3 Schematic military command chain.

As stated earlier, it is important to consider the time taken for information to flow between and around the human and machine subsystems as well as the time for the whole system to respond to inputs. The phrase above, 'supervision is so limited', can then be interpreted accurately as meaning that the humans in the command chain do not have time or are unable to respond correctly to inputs from a change in the operational area. Even a laser-guided bomb operating perfectly reaches a point in its trajectory when it is too late to guide it away from the designated target point, even if the controller wants to.

Engineering developments

Technologies

Moore's Law predicting a steady, rapid increase in processing power of computers[20] has been accurate for about 50 years. This underpins much of the publicly perceived technology revolution of the last few decades. This chapter assumes that this will continue and there will be a steady influx of civil developments into military equipment, especially in IT. Many of these developments, although important if not essential for military operations, will be irrelevant for discussions about the legality of autonomous weapons. However, the subset of all technology developments affecting lethal decision-making are relevant. These can be the obvious ones such as radar or optical object recognition systems, but could also include facial recognition algorithms and systems that extract information from large amounts of data.

Decision-making technologies for civil applications have evolved over several decades. Automated decisions on mortgage applications were an early example. Developments now extend into the realm of artificial intelligence (AI), with recent public debate about whether machines represent a threat to humanity. As with autonomy, there is not a clear-cut divide between automatic generation, automatic processing, presentation of options to the user, and AI.[21] (Is a car satnav system that offers the driver a quicker route due to changing traffic conditions automatic or intelligent?) Again, there should not be a problem if the military command chain is clear and the boundary and interfaces between decision aids and decision-making is clearly defined in the system design. If the interface is not clear, there is the potential for ambiguity in responsibilities.

Although speculative, it is possible to foresee the scenario of the public expecting 'their' forces to use the full capabilities of highly automated systems and decision aids that will be in common civil use. Military procurement authorities are used to the engineering problems of technology transfer from civil to military domains due to different applications, environmental conditions, and life cycles. Any decisions for the military to use civil products and capabilities which would then come under international humanitarian law (IHL), as well as consumer laws, will add to the complications, and hence cost, of such systems and design processes.

Image and data-processing techniques are developing for many civil and military applications. Their application to automatic target recognition (ATR) has already reached the stage of established conference series[22] and the production of textbooks.[23] The techniques are well in advance of those implied for the Phalanx system discussed above, but some could be in the Iron Dome system. Even if they are not used by Iron Dome, it is highly likely that they will enter service in the next decades. The step from ATR to automatic target identification (ATI) is large, but elementary systems can be developed for ATI of aircraft using radar techniques such as high range resolution profiling (HRRP) and jet engine modulation (JEM). These are both discussed by Tait (2005),[24] which can be used as an introduction.

It is possible that ATR and ATI techniques will be developed for difficult targets in complex scenes. Provided these are used as decision aids giving timely advice to a human commander governed by extant rules of engagement (ROE), there would not appear to be a legal problem. However, linking a positive result from an ATI algorithm directly to a weapon system and firing it may be a small technical step, but will be a large legal and ethical one. This is a similar argument to one discussed by the Center for Strategic and Budgetary Assessments (CSBA).[25] The same argument will apply if the information is presented to the operator, but they do not have time to assess it.

Electronic warfare (EW) is a specialist military area which, of necessity, is not widely discussed. Missile warning systems, for example, rely on probabilistic interpretation of sensor information to make automated responses. If the response is a lethal one, such as releasing an anti-radiation missile, they are subject to Article 36 reviews and their use is strictly controlled. They do show that probabilistic arguments can be used for ROE and review purposes.

Government dependence on open-system technologies such as the internet have opened up the whole problem of cyber warfare. This is a subject which has been seen to be as much of a threat as autonomous weapons by government, military and legal authorities for several years.[26]

Engineering design developments

Chapter 4 of this volume[27] discusses the defence procurement process and how it relates to Article 36 reviews. It describes how system engineering techniques decompose capability requirements into subsystem specifications which can be used in the procurement process.

Design engineering is the process of responding to a specification with a design that is a compromise between delivering a product on time, meeting or exceeding the specification, and cost requirements. The balance between these factors for the final design is a complex function of the internal and external pressures on the design team. These time, quality, and cost constraints are the minimum set relevant to modern designs. There are also a range of implicit ones: including consumer laws, liabilities for product safety, and environmental legislation. These are accepted as business-as-usual by the engineering profession and built into working practice, enabling responsibility for failure to meet broad legal requirements to be allocated to one supplier or part of the supply chain.

A significant development in defence procurement over the last two decades is due to the military need for capabilities which can only be supplied by combinations of equipment and services. This has resulted in the growth of contracting for capabilities in long-term support contracts i.e., initially contracting for a capability at a fixed price with the supplier having the flexibility to offer a solution made up of systems that they can obtain in the most cost-effective manner. This does not absolve the suppliers from their legal obligations as equipment manufacturers. A well-known example of a contract for capability is the UK MOD satellite communication contract with Paradigm, now with Airbus, for a link capacity, whether supplied using MOD satellites or commercial ones. This is intended to reduce wastage due to over-specifying individual pieces of equipment which may not work together in operations. Although weapon systems are still procured as equipment items, such as a missile, the overall capability is specified. The design teams still have to work to detailed specifications, but there may be more flexibility in their interpretation. These specifications now better reflect the overall intended use of the product. Wider requirements such as 'fit for purpose' can be incorporated more readily.

Commercial pressures on the defence industry, in common with the civil sector, have led to a separation of subsystem suppliers and system integrators. The complicating factor in the defence sector is the relatively small number of defence contractors. Most large contractors contain smaller divisions which are effectively fully owned subsidiaries. (Sometimes this is explicitly recognised in the companies' accounting systems and sometimes not.) There are also jointly owned companies such as MBDA, a supplier of missiles and missile systems. It is a multinational group with three major aeronautical and defence shareholders: Airbus Group (37.5 percent), BAE Systems (37.5 percent) and Finmeccanica (25 percent). This can lead to companies interchanging roles for different contracts. An aircraft manufacturer may place a subcontract on a missile supplier to fit an existing missile to it. When a new missile type is ordered by governments from the missile supplier, they could place a sub-contract on the aircraft manufacturer for the interfaces and control systems. Alternatively, the government can place separate contracts on both companies for these items and a third contract, not necessarily on either, for the integration of the two.

National security issues, intellectual property rights (IPR), non-disclosure agreements, and national shareholder interests lead to a complex network of contracts and *internal confidentiality walls* inside design teams. (An internal confidentiality wall is where valuable and useful information is known to some members of a team from one contract, but they cannot pass it to members working on another contract.) The result is that design teams work to hierarchies of internal contracts (work packages) with management partition of the workforce into smaller teams respecting these complexities.

When weapon system design starts, most of the complicating factors discussed above have been solved in a way that is acceptable to the customer or customers. Chapter 4 by Gillespie in this volume discusses the type of evidence that the design and test engineers present for Article 36 reviews. Additionally they provide evidence which is used for guidance when campaign planners are producing rules of engagement (ROE) for the specific campaign. They may also be called on to provide more detailed evidence for ROE guidance for specific weapons at different times in the campaign.

Adherence to IHL may not have been explicitly considered as one of the design constraints for a weapon system, but design information and test results form a large part of the evidence considered at each Article 36 review. Historically, military procurement has been based on buying materiel which can be specified in a relatively complete way and testing for adherence to the specification. Article 36 reviews at the start of the procurement process establish whether the equipment is banned explicitly or if it is capable of being used lawfully. The design and test process then follow the conventional procedures described above or variations on them. With the advent of complex, integrated systems which include more automation in decision-making, IHL will gradually become a standard engineering requirement if that has not already happened.

Controlling the development of highly automated weapons

It is generally accepted by all parties that it is unethical, if not necessarily unlawful, to have weapon systems that can be despatched to an operational area with a high-level instruction which it interprets, with no possibility of human intervention before it acts. It is probably also accepted, as HRW do, that such a system is not yet at the design stage. This does not mean that research and development (R&D) programmes are not underway that could lead there, but these should be part of state-controlled procurement systems using legally acceptable processes. Chapter 4 shows how standard procurement processes relate to Article 36 reviews. Therefore the problem is one of ensuring that Article 36 reviews will be carried out for all R&D programmes that have an aim of increasing automation of a weapon system. Putting this another way, Article 36 reviews are required whenever there is a programme to raise the autonomy level of a state's weapon system, whether through a new system or just incrementally increasing the automation of existing systems or decision aids. The test for a programme would be the interpretation of Article 36's phrase 'new weapon, means, or method of warfare'.

It is argued above that engineering evolves in response to the pressures on it. Anderson, et al., argue that IHL will evolve with technology. If we accept both arguments, the question for this chapter is to ask whether design processes for increased automation will produce designs that are lawful. There are three responses:

1. Current processes will always produce lawful weapon systems;
2. Current processes are inherently incapable of producing lawful weapon systems; or
3. Current processes have the potential to produce lawful weapon systems, but may need to be changed as technology and the law evolve.

HRW argue that response 2 is the valid one. Others could argue that as Phalanx, Iron Dome, and similar systems are lawful and meet the test of having a level of autonomous operation, then response 1 is valid. Both positions would appear to be idealistic, so a more pragmatic approach is taken here. Chapter 4 discussed the type of evidence that the design and test engineers may produce for Article 36 reviews. This is for current systems which are considered to be lawful. Can this type of evidence be produced for autonomous systems and will it satisfy a review?

The first step in any system design process is to agree on a set of specifications and test methods to be used to verify that the delivered system meets the final version of the contractual specifications. (Weapon requirements nearly always change during procurement.) IHL requirements appear to be very generic to an engineer. However, if they are considered to be a set of capability requirements, they can be interpreted and used to derive a set of system requirements and functional specifications for a specific system or system concept. Standard system engineering procedures lead to a logical and consistent set of functions which are executed in one or more defined subsystems, each with an unambiguous specification and test criteria.

Decision-making and decision advice using AI techniques can be considered to be functions within the system and hence their specification, operation, boundaries, and test criteria will be defined through standard processes and procedures. Clearly this is a strong statement and will need R&D by the defence community to ensure that the definitions for AI functions will lead to appropriate Article 36 review evidence.

The first steps in applying the system design process, using API as a capability requirement, has been carried out for autonomous unmanned air vehicles by Gillespie and West.[28] They used a three-component model of the human decision-making process and the legal requirements of necessity, humanity, distinction, and proportionality to derive 25 design requirements. These can be used as a basis for procurement specifications. Twenty-seven functions were identified in the decision-making process with their inputs and outputs. These can be used as a basis for test plans aimed at providing the evidence of actual performance required for Article 36 reviews.

Rules of engagement are treated as one of the inputs to the first stage in the three-part model. These must be expressed in functional terms such as defining safe areas and probability levels for identification of targets or uncluttered blast areas round an aim point. Defining ROE in this way puts pressure on:

- The legal process to produce ROE which are unambiguous and can be expressed in a way which an automated system can interpret. Should the ROE fail this test, then it can be concluded that the system cannot be used lawfully for that mission; and
- The design process to specify the inputs that are needed for automated decisions and the interfaces with external systems. If the interfaces cannot be specified unambiguously, then the system will almost certainly be capable of unlawful use in a range of circumstances. This provides evidence both for Article 36 reviews and guidance for setting ROE in every campaign where the system will be used.

The process developed by Gillespie and West is an example of applying systems engineering to the problem of establishing weapon system lawfulness and ROE guidelines. It should be possible to apply the techniques to other automated systems and generate evidence for Article 36 reviews and produce ROE guidance.

It is concluded from this discussion that response 3 to the above question is the valid one, with potentially the added bonus that modifying current processes may also help identify the unlawful use of new autonomous systems.

Weapon reviews for autonomous systems

The logical phases in the lifetime of a new weapon system are: Concept, Assessment, Development, Manufacture, In-service, and Disposal (the CADMID cycle). Article 36 is interpreted as requiring reviews for new systems in the CADM phases and significant upgrades in the In-service phase. These must have been applied for systems currently in service or well advanced in the CADMID cycle. It should be noted that the long lead times in defence procurement can lead to significant changes in requirements and technical solutions throughout the D and M phases. These should be captured during the review at release to service if not before.

It is necessary for procurement authorities to review new concepts as the first step in the assessment phase for investment decisions. A system that will be unlawful in all circumstances would also be a bad investment as it could never be used. Identifying potential unlawful use of a lawful system will assist effective balance of investment (BOI) decisions in all further steps and when choosing one weapon system from several offered capabilities.

Proposed weapon systems usually offer techniques which may appear to be novel, but they will be based on extant research, whether from military or civil sources. There must be a concept which can be expressed in engineering terms even if at a high level. Similarly there will be a preliminary Concept of Use (CONUSE), which is different from a Concept of Operations (CONOPS) which is used in setting ROE guidance. The CONUSE and weapon system concept provide a basis for an initial review by the relevant legal authorities. This argument must apply to all weapon systems no matter how high their level of automation or autonomy. Weapons that are not under control, or as HRW put it, *out of the loop weapons*, will immediately fail the Article 36 review under both proportionality and necessity criteria.

Accepting that we have an outline concept and a CONUSE, it is possible for engineers and military staff to develop them using representations. The requirement imposed on them by the previous section, and good practice is to identify: which decisions are made by humans; which ones are made by machines; the timescales involved for the necessary information to be presented to the decision-maker; and the time taken for the decision to be made. An essential further step is to examine the sources of information for each decision process and their provenance. The result should be a body of evidence for an initial review.

The results of this initial review should provide further guidance for the next steps. It can also be assumed that the lawyers conducting the review will give opinions on the type of evidence that will be needed to assess measured performance at later reviews. This can then be incorporated in the test philosophy and trials planning during development and manufacture (D and M phases). It is self-evident that it is cheaper and quicker to consolidate these trials into those needed for performance purposes than to add an extra phase of trials due to a legal review.

Complex weapon systems cannot be exhaustively tested under all conditions due to cost and time constraints. This is already recognised; and reviews have to make decisions based on statistical analysis of sometimes low numbers of results. Higher levels of autonomy in decision-making will be no different, but may need new approaches to the tests and the analysis. If research is carried out into the required techniques, it should be possible for Article 36 reviews to be carried out using current frameworks.

Discussion

It is essential to separate marketing hype and journalistic speculation from the hard reality of engineering reliable military systems. Technology developments such as UAVs, robotic weapons, or the Internet are often called game-changing for the conduct of warfare. In practice, these have almost always been the subject of R&D for many years and this is certainly the case with automation and autonomy. Speculation about AI applied to robots mainly draws on extant research and research proposals, so it should be possible to engage in a rational debate about how they would be applied to weapon systems and where the ethical limits lie.

The *Tallinn Manual*[29] gives an example of applying current legal structures to a new technology, in that case cyber warfare. There are some parallels between cyber and autonomous technologies as cyber entities such as computer viruses can be considered to be fully autonomous, but the similarities should not be taken too far. However, we could look at a review of IHL applied to automated decision-making.

This chapter acknowledges the increased level of automation for peaceful purposes. It is assumed that this will lead to public pressure for equally capable systems to be used both to save combatants' lives and as weapons against hostile forces, whether state or non-state organisations. A refusal to pre-emptively consider their legal and defence engineering aspects will reduce the flexibility and increase the timescales of an effective reaction to such threats used against us.

The design and performance-testing problems for fully autonomous systems are, in principle, no different from those encountered in other complex systems. Specific issues that will need to be addressed are: clear separation of decision-making functions into human and machine ones; test philosophies; statistical analysis of performance measures for legal assessment; and definition of functions and behaviours for autonomous decision-makers including AI.

The engineering techniques described in this chapter have also been shown to have the potential for close alignment with requirements from lawyers who are assessing new weapon systems. Although mainly applied in the air domain, there should be no fundamental reason why they cannot be applied in all domains. Therefore it will be possible to provide the technical evidence required under Article 36 for new autonomous weapon systems for the foreseeable future. This mirrors the view that the laws of war will evolve with the technology.

Conclusion

Autonomy has no clear definition in the context of weapon systems, but is part of a capability spectrum with automatic systems. It is more productive to discuss the level of autonomy of the subsystems within a weapon system than to discuss the autonomy level of the latter. It is then possible to identify which decisions are under human control and which are made by an autonomous subsystem. The decision-making process can then be examined for its behaviour in operational conditions. Evidence from trials programmes and design proving will then provide a basis for Article 36 reviews at all stages of procurement.

Technological developments to produce more highly automated systems are underway in many spheres of human endeavour. Therefore it is highly unlikely that any new technology, including AI, could be designated as one that supports the development of autonomous weapon systems without it having widely accepted applications. This makes it impossible to define what could be banned by treaty and renders meaningless any calls for a ban on the 'development, production, and use of fully autonomous weapons', such as the one by HRW.[30]

All weapon systems have humans in both their command and their control chains. What drives the level of automation is the time constant for human intervention in the control chain. The control chain for an automated system is always part of the command chain until the weapon is released from it. The weapon always has clear and unambiguous interfaces with the command chain whilst it is in communication with it; its design defines its behaviour after release. The times between release, stopping communication, and hitting the target are very different for different weapon systems.

In order to review a new technology under Article 36, there must be at least a weapon concept and high level CONUSE. Logically this must give generic descriptions of the weapon, its target set, and the military command and control chain. Without these components there is nothing to review. With these components it is possible to carry out a review of the weapon concept against the well-established principles of military necessity, humanity, distinction, and proportionality. The legality of the proposed autonomous system can then be judged against relevant criteria.

There are existing guidelines for the type of evidence required for an Article 36 review, but these make assumptions about the level of human control over the system. It is suggested that international efforts should be directed at producing guidelines for the evidence required for the review of highly automated systems. These guidelines should evolve as more fully autonomous systems are developed in all spheres.

Notes

1 Human Rights Watch, *Losing Humanity: The Case Against Killer Robots*, at www.hrw.org/reports/2012/11/19/losing-humanity-0.

2 Human Rights Watch, *Shaking the foundations, the human rights implications of killer robots*, at www.hrw.org/reports/2014/05/12/shaking-foundations.

3 Michael N. Schmitt, 'Autonomous Weapon Systems and International Humanitarian Law: A Reply to the Critics', *Harvard National Security Journal* Features Online, 5 February 2013, available at http://harvardnsj.org/2013/02/autonomous-weapon-systems-and-international-humanitarian-law-a-reply-to-the-critics/.

4 K. Anderson and M. C. Waxman, *Law and ethics for autonomous weapon systems, why a ban won't work and how the law of war can*, American University Washington College of Law Research Paper 2013–11, Obtained from www.unog.ch/80256EDD006B8954/(httpAssets)/702327CF5F68E71DC1257CC2004245BE/$file/LawandEthicsforAutonomousWeaponSystems_Whyabanwontworkandhowthelawsofwarcan_Waxman+anderson.pdf.

5 K. Anderson, D. Reisner and M. C. Waxman, 'Adapting the Law of Armed Conflict to Autonomous Weapon Systems', *International Law Studies*, Vol. 90, September 2014, pp. 386–411.

6 Werrel, K. P. *The evolution of the cruise missile* (Alabama: Air University Press, 1985), pp. 17–20.

7 J. M. McClelland, 'The Review of Weapons in Accordance with Article 36 of Additional Protocol I', *International Review of the Red Cross*, No. 850 (2003).

8 One typical report for this incident can be seen at: www.telegraph.co.uk/news/aviation/11290412/Flights-grounded-at-all-London-airports-in-air-traffic-control-computer-meltdown.html.

9 http://icrac.net/2014/06/banning-lethal-autonomous-weapon-systems-laws-the-way-forward/ accessed 20 February 2015.

10 www.raytheon.com/capabilities/products/phalanx/ accessed 20 February 2015.

11 DoD Office of the Secretary of Defense, Unmanned Aerial Vehicles Roadmap 2000–2025, April 2001, www.hsdl.org/?view&did=705358.

12 R. M. Taylor, S. Abdi, R. Dru-Drury and M.C. Bonner, *Engineering Psychology and Cognitive Ergonomics*, Vol. 5, 'Aerospace and transportation systems' (Aldershot UK: Ashgate, 2001), Ch. 10, pp. 81–8.

13 ICRC Background paper in: ICRG, *Autonomous weapon systems: Technical, military, legal and humanitarian aspects*. Expert meeting, Geneva, Switzerland, 26–28 March 2014, 59–73.

14 www.flightglobal.com/news/articles/usaf-details-f-16-life-extension-programme-375914/ accessed 20 February 2015.

15 UK MOD White Paper CM 8278: 'National Security Through Technology: Technology, Equipment, and Support for UK Defence and Security', February 2012, paragraph 13.

16 See, for example, MIL-STD-882E, Department of Defense Standard Practice, System Safety (Washington, DC: DoD, May 2012), which contains many definitions for use in military procurement.

17 The UK Industrial Avionics Group (IAWG) is recognised as one of the leaders in this field; the most prominent of their activities is described at: www.amsderisc.com/related-programmes/.

18 Department of Defense Directive Number 3000.09, *Autonomy in Weapon Systems* (Washington, DC: USDOD, November 2012).

19 UK MOD Joint Doctrine Note JDN 2/11, 'The UK approach to unmanned aircraft systems'.

20 Strictly the axiom by Gordon Moore, the founder of chip manufacturer Intel, is that transistor density in a processor chip doubles every two years.

21 See, for example: Kevin Warwick, *Artificial Intelligence, The Basics*, 1st edn (Routledge, 2012).

22 See, for example, SPIE Automatic Target Recognition XXV, at: http://spie.org/DEF/conferencedetails/automatic-target-recognition.

23 See, for example: D. Blacknell and H. Griffiths, *Radar Automatic Target Recognition (ATR) and Non-Cooperative Target Recognition (NCTR)*, The IET Radar Sonar and Navigation Series, 2013.

24 P. Tait, *Introduction to Radar Target Recognition*, The IET Radar Sonar and Navigation Series 18, 2005.

25 See: http://csbaonline.org/2014/08/21/the-legal-and-moral-problems-of-autonomous-strike-aircraft/.

26 ICRC, 31st International Conference of The Red Cross and Red Crescent, October 2011.

27 Tony Gillespie, 'A defence technologist's view of international humanitarian law', Chapter 4 in this volume.
28 Tony Gillespie and Robin West, 'Requirements for Autonomous Unmanned Air Systems Set by Legal Issues' *The International C2 Journal*, Vol. 4, 2010, pp. 1–32.
29 NATO guidance in the *Tallinn Manual* made clear that cyber attacks must be complaint with international humanitarian law in the same way as any other weapons system, available at: www.ccdcoe.org/tallinn-manual.html, accessed 30 October 2014.
30 Human Rights Watch, *Losing Humanity* (see note 1 above).

PART IV

Synthetic biology

16

BIOTECHNOLOGICAL INNOVATION, NON-OBVIOUS WARFARE AND CHALLENGES TO INTERNATIONAL LAW

Christopher Lowe

Biotechnology provides an ideal platform from which to tackle twenty-first century challenges since it is based on continuously renewable low-energy resources and is able to provide both food and non-food products from managed industrial, agricultural, aquacultural, and forestry ecosystems. In the future, biotechnology will pervade all aspects of human activity – the so-called bioeconomy represents a new route to the sustainable production of biomass, either as a product per se or as a raw material; producing a plethora of food, health, and other material inputs; supplying the industrial, security, and energy sectors – in order to address major environmental, social, and economic challenges and create a safe, healthy and prosperous biosphere for current and future generations. A mature bio-based economy will deliver global food security, improve nutrition and health, create bioproducts and bioprocesses in multiple industrial sectors, contribute to *green* energy provision, and help agriculture, forestry, aquaculture and other ecosystems to adapt to climate change.

The term biotechnology has its modern origins in the early twentieth century and is currently defined as 'the application of science and technology to living organisms, as well as parts, products and models thereof, to alter living or non-living materials for the production of knowledge, goods and services'.[1] However, the repercussions for the twenty-first century of the fundamental breakthroughs in biotechnology that have occurred over the last six decades or so have proven truly revolutionary, particularly in our understanding of the nature, structure, and function of the genetic material, DNA, and our ability to engineer it for various societal and economic purposes. Advocates for biotechnology argue passionately that, if allowed to develop unfettered over the next 50–60 years, the technology is likely to develop thousands of novel genetically modified viruses, bacteria, animals, and plants for sustainable applications in the pharmaceutical, agricultural, food, medical, environmental and energy sectors. These developments should contribute billions of dollars to the global economy. The global biotechnology market was valued at USD270 billion in 2013 and is expected to grow at a compound annual growth rate (CAGR) of 12.3 percent up to 2020.[2] In the field of agriculture alone, without such developments, it is difficult to see how current plant-breeding techniques could increase the world's food supply enough to feed a population which is expected to reach 9.4 billion by 2050.[3] Modern biotechnology can assist farmers in yielding healthy, plentiful harvests with a reduced environmental footprint by providing pest and disease protection, higher yields, new options for weed control, less soil erosion, and higher-quality water and feed stocks, improved grains, and enhanced nutrition. A new area of precision agriculture is being ushered in by using multiplexed sensor technologies to provide geographical and temporal feedback to where crops are being eroded by pests, soil deficiencies, or microclimatic environments. The global area of biotech crops continued to increase for the 18th year in succession at a sustained growth rate of three percent, reaching 433 million acres in 2013.[4] This is very encouraging news, especially for the developing world, where approximately 790 million people are chronically undernourished and millions more suffer from undernutrition.[5] Similarly in human healthcare, advances in genomics are providing greater understanding of diseases and thus better therapies and preventative measures. The World Health Organization (WHO) believes that 'given the burden of infectious diseases in developing countries, biotechnology has the potential to changes the lives of millions of people'.[6] In the industrial sector, the factory farm and biorefinery of the future will encompass an increasing demand for biomanufacturing in sectors as diverse as agri-food, biofuels, energy, and bulk and fine chemicals in order to relieve reliance on petroleum-based technologies, thereby enhancing environmental quality, national security, and sustainability. One specific area poised for exponential growth in the factory farm of the future is bio-sourced industrial products using renewable plant resources to make a variety of industrial products.

Concerns about modern biotechnology

However, while there is a general consensus that biotechnology is set to shape the future of mankind into the foreseeable future, there are growing ethical, political, and societal concerns that pose substantial challenges to international law and transnational institutions. Issues over whether the genetic resources of the world, once manipulated, could be confined within a limited number of intellectual property rights concentrated in a few companies of technologically rich states, thereby allowing them to control access to food, medical, and other resources essential to the health and welfare of billions of people are widely known. Further concerns arise if genetic engineering is abused – either by rogue states to destabilise the biodiversity or biosphere of a target state or its human population with catastrophic consequences[7] – or by more local bioterrorist events promulgated by knowledgeable lone wolves. The use of biology as a weapon is not new.[8] Warring parties have long employed natural bioweapons: poisoned wells, tipped arrowheads with natural toxins, catapulted plague victims into besieged cities, gifted smallpox-infected blankets to Native Americans (by British settlers), and maintained a malaria-infected Terai forest as a natural barrier against invaders of Nepalese-ruled areas from the Ganges plains.[9] Pathogens such as smallpox, plague, and anthrax are deadly enough in their native state without recourse to genetic engineering. However, there are fears that recent progress in the life sciences in genomics, epigenomics, systems biology, and synthetic biology could be used for malicious purposes with much greater effect than possible with the use of wild type organisms alone. There are several reasons for this increased potential risk of novel biowarfare scenarios: First, increased knowledge, capability, and facilities in the medical and pharmaceutical sectors has ubiquitised the technology and raised fears about 'dual-use' biology. Most relatively advanced countries now have the ability to culture pathogenic organisms safely at any scale. The deliberate release of existing pathogens such as the causative organisms of typhoid, anthrax, or smallpox could cause fear, disease, and death in a target population. Second, classical biowarfare agents can be manufactured more efficiently and more virulent with the most rudimentary of genetic manipulations. The former USSR's 'invisible anthrax' displayed altered immunological properties created by the introduction of an alien gene into *Bacillus anthracis*.[10] Creation of novel biowarfare agents for use in conjunction with their complementary vaccines could offer an attractive approach to a potential aggressor. Third, geologically dormant organisms from adverse environments such as the permafrost in Siberia, deep well samples, or, potentially, from extra-terrestrial sources, have been, or could be, resurrected and cultured to create new threats. Fourth, the use of genomics data readily available within the public domain combined with modern methods of synthetic biology allows the resurrection or creation of completely new bioweapons, including ethnic and racially specific bioweapons. Fifth, the production of biological agents could infect the agricultural or industrial infrastructure or harm the natural or built environment. Sixth, new approaches to stabilising, weaponising, and delivering natural or genetically altered agents could evade current detection technologies. Seventh, the ready availability of inexpensive biological building blocks, DNA synthesisers, internet expertise, and the mushrooming bioscience laboratories in the public domain – so-called DIY Biology – may place this know-how in the hands of the lone wolf bioterrorist. Finally, entirely novel bio-agents created with chemical genomics could directly affect human behaviour, consciousness, or fertility or could be incorporated into human genes to adversely affect human evolution itself. This chapter concentrates specifically on the new technologies likely to impact the future of biowarfare and bioterrorism and briefly describes the relevant achievements in these areas, assess to what extent the threats are real, and suggest the associated legal issues which require addressing with the concepts of a 'weapon', 'warfare', the Biological and Toxin Weapons Convention (BWC), and international human rights agreements.

Genomics

Genomics refers to a sub-discipline of genetics that applies recombinant DNA, DNA sequencing methods, and bioinformatics to sequence, assemble, and analyse the function and structure of the complete sets of DNA within a single cell of an organism, i.e., genomes. The first complete genome sequence of a free-living organism to be sequenced was that of *Haemophilus influenzae* (1.8 Mb) in 1995.[11] The following year a consortium of researchers from laboratories across North America, Europe, and Japan announced the completion of the first complete genome sequence of a eukaryote, *Saccharomyces cerevisiae* (12.1 Mb); and since then genomes have continued to be sequenced at an exponentially growing pace,[12] and exponentially decreasing cost. As of 2014, the complete sequences are available for 4,447 viruses, 30,158 archaea and bacteria, and 1,819 eukaryotes. Most of the organisms whose genomes have been completely sequenced are problematic pathogens, such as *Haemophilus influenzae*, well-studied model organisms for medical research such as yeast (*Saccharomyces cerevisiae*), fruit fly (*Drosophila melanogaster*), worm (*Caenorhabditis elegans*), zebrafish (*Brachydanio rerio*) and other fish (e.g., *Takifugu rubripes*), dog (*Canis familiaris*), brown rat (*Rattus norvegicus*), mouse (*Mus musculus*), chimpanzee (*Pan troglodytes*), and plants (*Arabidopsis thaliana*). The first draft of the human genome was completed by the Human Genome Project in early 2001, creating substantial interest.[13] By 2003, the entire genome for one specific person was sequenced, and by 2007 this sequence was declared complete, with an error rate of less than one in 20,000, and with all chromosomes assembled.[14] Since then, the genomes of many other individuals have been sequenced, partly under the auspices of the 1000 Genomes Project, which announced the sequencing of 1,092 genomes in October 2012.[15] More recently, with the consent of participants and the support of the public, Genomics England is creating a lasting legacy for patients, the National Health Service (NHS), and the UK economy through the sequencing of 100,000 genomes[16]; and Google Genomics in 2014 launched a preview of its application programming interface (API) that allows storage, processing, exploring, and sharing of massive genomic datasets of DNA sequences on its cloud infrastructure.[17] A similar initiative by a Harvard Medical School genetics professor, George Church, launched the Personal Genome Project in 2008 to make the genome sequences and medical histories of 1,000,000 people public and searchable.[18] The human genome contains 3,164.7 million chemical nucleotide bases (A, C, T, and G) encoding 30–35,000 genes of average length 3,000 bases. The order of 99.9 percent of the nucleotide bases are exactly the same in all people and the functions are still unknown for over 50 percent of the genes discovered to date. Knowledge of the whole genome sequence may identify the cause of some rare diseases and help point the way to new treatments for these devastating conditions. For example, at least 80 percent of rare diseases are genomic, with half of such new cases found in children, and although rare diseases are individually very uncommon, albeit numbering 5,000–8,000, three million or six to seven percent of the UK population are affected in total.[19] The cost of sequencing per human genome has fallen from just under USD100 million to about USD1000 in the space of 15 years or so; this lowered cost of genome sequencing will allow mutations to be identified.

Ethnic bioweapons

However, despite the benefits, the continued analysis of human genomic data has profound political, social, and security repercussions for human societies. One potential consequence is the development of ethnic bioweapons which aim to harm only, or primarily, persons of targeted ethnicities or specific genotypes. Numerous unsubstantiated reports on the development of ethnic bioweapons have appeared in respected newspapers and journals pointing the finger at various rogue states. These concerns are not new, as in 2005, the International Committee of the Red Cross (ICRC) concluded, 'The potential to target a particular ethnic group with a biological agent is probably not far off',[20] while the British Medical Association (BMA) in its report entitled *Biotechnology, Weapons and Humanity II* suggested that a combination of human genome studies, the development of vectors capable of introducing harmful materials to cells, and new ways to disrupt genes should raise concerns about potential abuse.[21] The report chronicles the mutations in the human genome, the single nucleotide polymorphisms (SNPs) that differ between specific ethnic groups, and concluded, 'Genomic data in public databases revealed that hundreds, possibly thousands, of target sequences for ethnic specific weapons do exist'.[22] In principle, human populations with prominent phenotypes – such as eye, skin, or hair colour, or those with an above-average prevalence of monogeneic disorders, such as lactose intolerance, haemoglobinopathies, or thalassaemia – could represent potential targets. However, sceptics suggest, for instance:

> Trying to find a weapon that affects quite a few of one ethnic group and none of another is just not going to happen. ... Because all groups are quite similar you will never get something that is highly selective: the best you would probably do is something that kills 20% of one group and 28% of another.[23]

Nevertheless, the use of short RNA interference (siRNA) could inhibit vital gene expression[24]; and if the sequence of the gene was different between two ethnic populations, this could create an ethnically specific weapon, although the use of siRNA can be variable and incomplete.[25] Furthermore, modern genome editing techniques can insert, replace, or remove DNA from a genome using recombinant adenovirus associated vectors (RAAV) or artificially engineered nucleases or 'molecular scissors'.[26] The approach uses nucleases to create specific double-stranded breaks at desired locations in the genome and then harness the endogenous mechanisms of the cell to repair the induced break by natural processes of homologous recombination (HR) and non-homologous end-joining (NHEJ). There are currently four families of engineered nucleases being used: Zinc finger nucleases (ZFNs), Transcription Activator-Like Effector

Nucleases (TALENs), the RNA-guided endonuclease Cas9 from the microbial type II Clustered Regularly Interspaced Short Palindromic Repeat (CRISPR/Cas9), system and engineered meganuclease re-engineered homing endonucleases.[27] Genome editing with nucleases such as ZFN is preferred to siRNA since the engineered nuclease is able to modify DNA-binding specificity and therefore can in principle cut any targeted position in the genome and introduce modification of the endogenous sequences for genes that are impossible to specifically target by conventional RNAi. Furthermore, the specificity of ZFNs and TALENs are enhanced as two ZFNs are required in the recognition of their portion of the target and subsequently direct to the neighbouring sequences. TALENs have been used for targeted genome editing in cell cultures[28] and, more recently, genome editing with a switchable insertion in live zebrafish larvae.[29] Zebrafish are used as models of complex diseases; and such an approach could be used to probe the behavioural implications of specific brain neurons or the network of signals that orchestrate development.

Cas9 can be reprogrammed using RNA guides to create targeted DNA double-strand breaks, which can stimulate genome editing via HR or NHEJ. A unique advantage of the Cas9 system is that it can be combined with multiple single-guide RNAs (sgRNAs)[30] to achieve effective multiplexed genome editing in mammalian cells.[31,32] However, although the Cas9 system has been applied in a variety of cell line- and embryo-based demonstrations, *in vivo* applications still remain challenging, particularly due to its large transgene size. Nevertheless, the CRISPR-Cas9 gene editing system has recently become more convenient to use in animal models with the development of the 'Cas9 mouse' using a Cre recombinase and is expected to simplify *in vivo* gene editing, particularly those in which multiple genes and cell types are being manipulated.[33] The technique also has far-reaching potential implications for human genome editing, including targeting individuals with deficiencies such as low intelligence or inborn errors of metabolism.

Epigenomics

The central dogma of biology articulates that cellular DNA is transcribed to RNA which is translated to proteins in order to perform various functions and processes. However, paradoxically, cells display diverse responses to external stimuli: cells with identical complements of DNA can have a plethora of distinct functions and phenotypes. The mechanisms governing phenotypic plasticity are now believed to operate through the regulation of gene expression via mRNA transcription, processing, and transportation as well as in protein translation, post-translational modification, and degradation.[34] The mechanism for epigenetic gene expression was finally established with the finding that DNA methylation and histone modifications were stable, heritable, and reversible and they influenced gene expression without altering the primary structure of DNA. Epigenetic regulation of gene expression can be achieved by gene silencing and is increasingly used to produce therapeutics to combat cancer, infectious diseases, and neurodegenerative disorders. Gene silencing operates by a gene knockdown mechanism since gene silencers, such as RNAi, generally reduce the expression of a gene by at least 70 percent, but do not completely eliminate it.[35]

In principle, modulation of epigenomic factors could constitute a novel bioweapon; however, the principal limitation to malign use of such approaches is delivery. RNAi reagents can be introduced into cells through different routes: siRNAs can be transfected into mammalian cells, while short hairpin RNAs (shRNAs) can be virally transduced into mammalian (and other) cells, and *Escherichia coli* expressing double-stranded RNAs (dsRNAs) can be fed to or microinjected into living animals. Once in the cells, reagents such as dsRNAs are processed into siRNAs of 21–23 nucleotides in length, which are then incorporated into the

RNA-induced silencing complex (RISC) and mediate gene silencing through target mRNA cleavage (if perfect sequence complementarity exists between the target mRNA and the siRNA) or translational interference (if the complementarity is partial). However, there are several challenges associated with gene silencing techniques, particularly with respect to delivery and specificity. Viral vectors used to deliver siRNA into cells, although efficient, can elicit an immune response against the agents.[36] Specificity can also be an issue in gene silencing since both antisense oligonucleotides and siRNA molecules can potentially bind to the wrong mRNA molecule; thus, more efficient methods to deliver and develop specific gene silencing molecules are required.

Synthetic biology

Over four decades after the first demonstration of recombinant DNA techniques, a new armamentarium of genetic tools is again changing the way humans can manipulate life. Synthetic biology mirrors modern engineering by designing novel life functions and forms with a predictable box of materials and parts. Traditional genetic engineering manipulates the genes of an organism by transferring one gene from a donor organism to a recipient organism using a transfer vehicle or vector. The last decade in particular has seen a tremendous expansion in the variety and ease of use of specific techniques associated with synthetic biology, and this, along with rapidly falling costs for DNA synthesis and the dispersion of experimental approaches once thought to be the domain of elite biologists, has resulted in the socialisation of synthetic biology. On the positive side, synthetic biology is expected to be one of the most disruptive innovations of the twenty-first century and impact the chemicals, energy, agriculture, and healthcare sectors with a market size estimated to reach USD16.7 billion by 2018 according to a market report published by Transparency Market Research.[37] One of the principal drivers for synthetic biology is the economics of DNA sequencing and synthesis which is opening new avenues in genome engineering and design.[38] For example, the productivity of DNA sequencing technologies, quoted in terms of the number of base pairs sequenced per day (on each of increasingly inexpensive machines), has increased more than 500-fold over the past decade, and thus the costs of sequencing have decreased by more than three orders of magnitude to less than USD0.001 per base pair. Over the same time period, the productivity of DNA synthesis technologies has increased 700-fold and the costs have fallen from USD30 to less than USD1 per base pair. Substantial further improvements are expected in both enabling technologies which will lead to an inflection point or step change in the capabilities of synthetic biology over the next decade. The combination of access to genome sequences on the internet[39] and the physical availability of synthetic DNA reduces the hurdles to obtaining or creating pathogens and thus alarms both the security community and policymakers. Increasing concern that life science research could be misused by proponents of biowarfare or bioterrorism has been fuelled by numerous public disclosures. For example, the non-intentional enhancement of the virulence of the mousepox virus by inserting the gene for interleukin-4 into the mousepox genome[40] subsequently led to the deliberate increase in lethality in both mousepox and cowpox viruses.[41] A seminal publication chemically synthesised an artificial polio virus *de novo* by starting with the online sequence, ordering small bespoke DNA sequences and stitching them together to reconstruct the native viral genome.[42] Other potential biowarfare agents with relatively short RNA genomes, including the Ebola (~19kb), Marburg (19kb), and Venezuelan equine encephalitis (11.4kb) viruses, could be amenable to synthesis via this route. A further report described the successful reconstruction of the influenza A (H1N1) virus responsible for the 1918 'Spanish flu' outbreak and provided new information about the properties that contributed to its exceptional virulence.[43] The virus was responsible for the influenza pandemic of 1918–19 which killed an estimated 20 to 50 million people worldwide, many more than the subsequent pandemics of the twentieth century and rivalling the two world wars. The work was justified as being critical to evaluating the effectiveness of current and future public health interventions, should an event like that in 1918 re-emerge, and to understand the pathogenesis of contemporary human influenza viruses with pandemic potential.

Other disclosures that have given rise to biosafety concerns include the transfer of the virulence factor of *variola major* into a virus of much lower virulence, the vaccinia virus,[44] the airborne transmission of influenza A H5N1 virus between ferrets,[45] the experimental adaptation of canine distemper virus (CDV) to the human entry receptor CD150,[46] and the demonstration of transmission of H5N1 hybrid viruses bearing the 2009/H1N1 virus genes to guinea pigs via respiratory droplets.[47] These public disclosures focused the attention of the security community on dual-use technologies, synthetic biology, and chemical genomics.[48]

Key developments in synthetic biology

Current synthetic biology encompasses several sub-fields and goals which combine the toolboxes of biology and engineering to create new opportunities, and potential threats, in biotechnology. Key developments are described in the following subsections.

1 Defining the minimal genome/minimal life and constructing the 'protocell'

Synthetic biology is aiming to produce the hypothetical minimal auto-replicative system or chassis cell either by a bottom-up approach starting with microbial cells with small genomes – so-called protocell design and creation – or by genome streamlining, by removing unnecessary functions from larger genomes using a top-down approach.[49] There is an ongoing search for existing cells with minimal genomes, such as the endosymbiotic bacteria of aphids *Buchnera aphidicola* (450kb, 400 genes), the human pathogen *Mycoplasma genitalium* (583kb, 485 genes), the endosymbiotic *Carsonella rudii* (160kb, 521 genes), *Hodgkinia cicadicola* (144kb, 169 genes), and *Tremblaya princeps* (139kb, 110 genes). Several studies have suggested that the minimum number of genes required to define a minimal living cell; that is, one capable of reproduction, maintenance and evolution in a permissive environment, is 200–250 genes,[50] although others have proposed that 100–150 might be sufficient.[51] Current work is aimed at constructing the 'protocell' which incorporates these minimalistic features and to which defined biological parts or 'biobricks' can subsequently be added.

2 Creating the toolbox of biobricks and biovectors

Drawing inspiration from conventional engineering practice, various methods and tools that support the design and construction of new genetic systems from standardised biological parts or biobricks are being developed.[52,53] The key requirement of the biobrick assembly is that any two can be joined to create a composite which itself can be linked to any other biobrick. The Registry of Standard Biological Parts currently maintains a collection of over 13,400 standard parts maintained as plasmids[54] and the principles of reuse and biobrick vectors have now been established to complement the registry.[55] Recently, multiple stakeholders, including academics, industry and interested members of the community have expressed their enthusiasm for the International Open Facility Advancing Biotechnology (BIOFAB) organisation which was founded in 2009 and aims to supply synthetic biologists with a library of genetic parts, sequences of DNA, that have known and predictable functions.[56] BIOFAB has now made about 3,000 well-characterised parts and released ~500 as a high quality curated collection.[57] However, predictable performance is elusive[58] and is undermined by the cellular milieu, including the genetic context of sequences encoding biobricks, the unexpected interacts between the bricks, and their non-specific interactions with other cellular components. A newly described biological device, termed a 'load driver', improves the performance of biobricks by insulating them from each other.[59]

3 Enhancing DNA synthesis

Next-generation DNA sequencing machines can speed up the reading of genomes and thereby create a more effective means to observe cellular behaviour, which, in turn, can help design better genetic circuits. The creation of a bacterium with an entirely synthetic genome was a high-profile success.[60]

4 Genome reprogramming

Many other genomes have been built and engineered, initially with bacteria and viruses, such as chromosome segments, representing about one percent of the 12M bp genome of *Saccharomyces cerevisiae* have been designed using genome editing software to remove repetitive sequences, by adding tags to illuminate the synthetic segments, and then by demonstrating that the engineered yeast strains are as healthy as the natural one.[61] Furthermore, custom changes with synthetic DNA to parts of the zebrafish (*Brachydanio rerio*) genome have now also been achieved.[62] Can these approaches be applied to human cells? Biologists have constructed a programmable genetic 'circuit' that can rewire human cells to respond to particular stimuli,[63] although such approaches are a long way from realising benefits in the clinic.

5 Chemical genomics

The molecule DNA is made up of four bases – adenine (A), thymine (T), cytosine (C), and guanine (G) linked via a sugar-phosphate backbone – and has existed for billions of years as the genetic material of all known life forms. The last few decades have seen attempts to re-engineer DNA by substituting other exotic bases beyond A, T, C, and G and by tinkering with the backbone. Early work on xeno-nucleic acid (XNA) demonstrated that it was possible to rebuild the backbone with non-ionic methylene sulphone linkers[64] and sugars such as threose, hexitol, and glycol,[65] but subsequent efforts were directed at the bases, using xanthosine and isomeric C and G. It was shown that polymerase enzymes could read DNA containing the unnatural bases and that ribosomes could translate the RNA into protein containing an alien amino acid.[66] Subsequent work replacing the oxygen atoms of the natural base T to create difluorotoluene (designated F) showed that this too could be transcribed, not because of its ability to hydrogen bond with A, but more related to its shape and ability to base stack in the DNA core.[67] Others screened 3,600 combinations of 60 abnormal bases for the pair that copied most faithfully[68] while other groups produced unnatural base pairs, P and Z, which copied with a fidelity of >99.8 percent per replication.[69] It appears that the polymerase cannot copy more than four of the paired bases in a row and thus other groups have engineered the polymerase to convert DNA into XNA and vice versa.[70]

6 Xenobiology

Much of the XNA work has been conducted to date *in vitro*, although an early example has shown that the addition of new letters to the 'alphabet of life' can create viable life forms. Replacement of most of the T bases with chlorouracil in an automated system which introduced the base gradually to a strain of *Escherichia coli* that had no ability to produce thymine on its own, showed that after about 5 months, some of the bacteria could not survive without chlorouracil and had eliminated ~90 percent of T from their genomes.[71] A significant challenge was to coax the cells to accept the foreign bases required to maintain the alien base in DNA through repeated rounds of cell division. This was achieved by engineering the *E. coli* to express a gene derived from a diatom which encoded a transporter protein required to allow the molecules to pass through the membrane of the bacterium.[72] The *E. coli* cells were able to survive on the foreign nucleotides until the supply was exhausted before replacing them with native nucleotides. It should be stressed that the alien bacterium contains only a single pair of foreign DNA bases out of millions; nevertheless, the potential of the approach to create alien life has been demonstrated, and it is worth pointing out that this approach may represent a means to avoid potential interference with naturally evolved DNA while working with biotechnology.

7 Resurrection biology

A novel giant virus, *Pithovirus sibericum*, was recently isolated from a more than 30,000-year-old radiocarbon-dated sample from the Siberian permafrost.[73] The revival of such an ancestral amoeba-infecting virus suggests that the thawing of the permafrost either from global warming or industrial misuse of the circumpolar regions might present a future threat to human or animal health. This report follows a series of earlier reports where giant DNA viruses with large genomes (>1.9–2.5Mbp, >1,100 genes) have been found in freshwater ponds and in coastal waters off Chile. These studies suggest that life could have emerged with a greater variety of pre-cellular forms than previously thought, since the new giant viruses bear little resemblance to the three accepted domains of cellular life, namely eukaryota, eubacteria, and archaea.

Synthetic biology and bioterrorism

Since one of the stated objectives of synthetic biology is to make biology easier to use, this presents a conundrum in that it also makes it easier for those with malicious intent to use. A key issue in assessing the risk that would-be bioterrorists could exploit synthetic biology to recreate or resurrect pathogenic viruses is whether they have, or could acquire, the necessary technical skills.[74] At present, whole genome synthesis requires multiple capabilities in software, hardware, and wetware to be brought together and integrated in a well-founded laboratory environment. Examination of published papers over the last five years shows that most are released with multiple authors and institutions. However, it is also clear that synthetic biology is being 'lego-ised' and de-skilled with much of the genomic data becoming available on the internet and many of the biobricks and process kits becoming commercially available. This trend to 'kitification' of biology follows the way that molecular biology has now permeated all aspects of biological science, where cloning is a matter of choosing the right kit available from manufacturers and the process could be completed in days, whereas, even as short a time ago as two decades, this would require several years' work by a dedicated team. It would now only take a person reasonably skilled in the art with the appropriate degree of intent to create serious issues in biosafety for the policy and security agencies. The security implications of synthetic biology thus need to be taken seriously.

DIY biology, amateur biologists and citizen scientists

With the reductions in the costs of DNA sequencing, synthesis, and chemical genomics, coupled with the universality of the internet and 'kitification' of biological recipes, this type of research is no longer the preserve of government-supported academic institutions or large corporations, but is now in the domain of biohackers who are able to conduct research in their homes. Innovation, supported by crowdfunding, is now well within the reach of individual entrepreneurs. Would-be biohackers, professional scientists, enthusiastic hobbyists, and rank amateurs alike are establishing laboratories in their kitchens or garages, buying used equipment online or modifying kitchenalia, and using their imagination to create a citizen's laboratory for a few hundred dollars.[75] In a seminal example, one member of the do-it-yourself or DIY biology community produced a glow-in-the-dark yogurt, which received around USD0.5 million in crowdfunding, while another more sophisticated self-styled group, DIYgenomics, aim to analyse their genomes and even conduct limited clinical trials. Many DIYbio groups are linking with similar groups with complementary expertise in software, electronics, and instrumentation. For example, the New York DIYbio group has collaborated with an electronics collective, NYC Resistor, to create a PCR machine and other essential pieces of basic molecular biology equipment, while an Irish group have created the DremelFuge, a device which attaches to a power tool to produce an inexpensive centrifuge. Similarly, in the basement of an unremarkable building close to the city centre of Copenhagen, there is an independent workspace called Labitat, crammed with computers, welders, incubators, 3D printers, and microscopes and open to anyone interested in art, design, science, and technology.[76] DIYbio, created in the Boston area in 2008, is an 'Institution for the Amateur Biologist' and has around 2000 members and a website (www.diybio.org). Similar DIY workspaces are springing up in many of the major cities worldwide, including those in Denmark, the UK, Spain, France, Germany, Canada, and India. The emergence of DIY biology is also promulgated by the multitude of protocols, ideas, and materials that circulate via internet and collaborative chat rooms, blogs, open-source tool archives, and forums.

DIYbio has its roots in the open-science movement which encourages unfettered exchange of materials, data and publications and follows the vision of open-source software.[77] Many devotees of DIYbio are already tackling sophisticated projects involving synthetic biology by piecing together biobricks. This can lead not only to greater innovation for the good of all, but also to deliberate misuse by malicious individuals or organisations and could lead to nurseries for budding bioterrorists. However, current DIY biology efforts should be seen from a more nuanced perspective; it has a serious image problem. Is it a movement of enthusiastic amateurs or lone wolf miscreants? A recent survey of DIY biologists by the Woodrow Wilson International Centre for Scholars in Washington DC has found that 92 percent of DIY biologists work in communal workspaces some of the time, are relatively young (78 percent under 45), are more educated than the average population, and, importantly, only six percent conceived that their work could result in proliferation of human disease.[78] It is also worth noting that 28 percent of respondents said that some or all of their principal work was conducted in an academic, corporate, or government laboratory, and that 19 percent had a PhD and were within the mainstream scientific community.[79] The findings also suggest that most DIY biologists are against government regulation, although a sizeable minority, 43 percent, anticipate a change sometime in the future, as and when the movement matures. Nevertheless, despite these reassuring comments from the legitimate DIY biology community, it is the prospect of potential abuse that drives the policy and security agencies' concerns.

Bioinspired engineering

Life-machine hybrids and bioinspired robotic systems are two further concerns related to the abuse of citizen bioscience which may impact the area of non-conventional weapons in the future. Biologically inspired engineering involves exploration into the way that living cells, tissues, and organisms build, control, manufacture, recycle, communicate, and adapt to their environment. Bioinspired engineers leverage this knowledge to create new technologies and translate them into products that meet real-world challenges. This inchoate technology has developed rapidly over the last few years and includes micromachines controlled by insects, autonomous flying and swimming microrobots, swarm robots, soft exosuits, and new methods of manufacturing such as pop-up MEMS. Recent developments in materials science, synthetic biology, stem cell biology, and tissue engineering are beginning to enable scientists to use synthetic materials, microdevices, and computational strategies to manipulate cell function, guide tissue formation, and communicate with and control complex organ physiology. Insight into how living systems form and function using self-assembling nanomaterials, complex networks, non-linear dynamic control, and self-organising behaviour are leading to entirely new engineering principles; as a result, the boundary between living and non-living systems is becoming blurred.

Electronically controllable insects are useful models for micro and nano-air vehicles (M/NAVs) which could serve as payload couriers to locations which are not readily accessible to humans or terrestrial robots.[80] There are reports of the remote control of beetles in free flight by means of a miniature implantable radio-equipped and lithium polymer battery operated neuro-stimulation system.[81] The results confirm the ability to control flight initiation and cessation and modulate throttle and direction via a relatively simple interface. Similar studies have employed a mobile robot 'driven' by a male silkworm moth and demonstrated its adaptive behaviour in locating odour sources.[82,83]

Bioinspired engineering has now developed high power density piezoelectric flight muscles and a manufacturing methodology which is capable of rapidly prototyping articulated, flexure-based sub-millimetre mechanisms to build an 80 mg, insect-scale, 120 times per second flapping 3 cm wing robot, the robo-bee, modelled loosely on the morphology of flies.[84] Tethered but unconstrained stable hovering and basic controlled flight manoeuvres were demonstrated. Meanwhile, a Dutch start-up aeronautics company has created a robotic bird that acts as a protective scarecrow, keeping real wild birds away from dangerous areas, such as crop fields laced with harmful pesticides, waste dumps, and airports. The so-called 'robirds' are 3-D printed from nylon and glass fibre, are battery-powered, and are modelled on predators such as bald eagles and falcons.[85] It is clear that the burgeoning armamentarium of bioinspired engineering techniques is likely to create many other robotic systems which, given a suitable chemical or biological payload, could present a serious and almost undetectable threat in small-scale conflict. Indeed, one could argue that this threat has already been realised via the current crop of commercially available remote controlled helicopter and quadcopter drones for aerial photography.

Notes

1 OECD, *A framework for biotechnology statistics*, Paris: OECD, 2005.

2 Grand View Research, Biotechnology Based Chemicals Market Analysis, Market Size, Application Analysis, Regional Outlook, Competitive Strategies, and Forecasts, 2015 To 2022, undated, abstract available at www.grandviewresearch.com/industry-analysis/biotechnology.

3 UNFAO, *The State of Food Insecurity in the World 1999* (Rome: UNFAO, 1999), pp. 10–11.

4 C. James, Global Status of Commercialized Biotech/GM Crops: 2013, ISAAA Brief No. 46 (ISAAA: Ithaca, NY, 2013).

5 UN Population Fund (1998) *The State of the World Population 1998* (New York, UNPF), pp. 2–3, Fig 16.1.
6 Bloom, D. E., Cafiero, E. T., Jané-Llopis, E., Abrahams-Gessel, S., Bloom, L. R., Fathima, S., Feigl, A. B., Gaziano, T., Mowafi, M., Pandya, A., Prettner, K., Rosenberg, L., Seligman, B., Stein, A. Z. and Weinstein, C., *The Global Economic Burden of Noncommunicable Diseases*, Geneva: World Economic Forum, 2011.
7 S. D. Murphy, 'Biotechnology and International Law', *Harvard Intl Law Journal*, Vol. 41, No. 1 (2001), p. 47.
8 J. van Aken and E. Hammond (2003) 'Genetic engineering and biological weapons: New technologies, desires and threats from biological research', EMBO Reports, Vol. 4 (Suppl 1), S57–60.
9 'The Terai Forests' (2006) Forest Monitor, available at www.forestsmonitor.org/fr/reports/549391/549398.
10 A. P. Pomerantsev et al., 'Expression of cereolysine AB genes in Bacillus anthracis vaccine strain ensures protection against experimental hemolytic anthrax infection', *Vaccine*, Vol. 15 (1997), pp. 1846–50.
11 R. D. Fleischmann et al,. 'Whole-genome random sequencing and assembly of Haemophilus influenzae Rd', *Science*, August 1995, pp. 496–512.
12 A. Goffeau et al., 'Life with 6000 genes', *Science*, October 1996, pp. 563–7.
13 V. McElheny, *Drawing the Map of Life: inside the Human Genome Project,* New York: Basic Books, 2010.
14 Goffeau et al., 'Life with 6000 genes' (see note 12 above).
15 G. A. McVean et al., 'An integrated map of genetic variation from 1,092 human genomes', *Nature*, 1 Nov 2012, pp. 56–65.
16 The 100,000 Genomes Project, available at www.genomicsengland.co.uk.
17 Ibid.
18 K. Gruber, 'Google for genomes', *Nature Biotechnology*, Vol. 32, No. 508 (2014).
19 The 100,000 Genomes Project (see note 16 above).
20 Jacques Forster, 'Preventing the use of biological and chemical weapons: 80 years on', 22 April 2015, available at www.icrc.org/eng/resources/documents/misc/gas-protocol-100605.htm.
21 British Medical Association, *Biotechnology, Weapons and Humanity II*, London: British Medical Association, Board of Science and Education, 2004.
22 A. Fire et al., 'Potent and specific genetic interference by double-stranded RNA in Caenorhabditis elegans', *Nature*, Feb 1998, pp. 806–11.
23 D. Adam, 'This week's science', *The Guardian*, 28 Oct. 2004, available at http://theguardian.com/science/2004/oct/28/thisweeksscience.
24 Z. Li and T.M. Rana, 'Therapeutic targeting of microRNAs: current status and future challenges', *Nature Rev Drug Disc*, Vol. 13, No. 8 (2014), pp. 622–38.
25 Adam, 'This week's science'.
26 K. M. Esvelt and H. H. Wang, 'Genome-scale engineering for systems and synthetic biology', *Mol Syst Biol*, Vol 9, No. 5 (2013), p. 641.
27 T. Gaj, C. A. Gersbach and C. F. Barbas, 'ZFN, TALEN, and CRISPR/Cas-based methods for genome engineering', *Trends Biotechnol*, Vol. 31, No. 7 (2013), pp. 397–405.
28 J. C. Miller et al., 'A TALE nuclease architecture for efficient genome editing', *Nature Biotechnol*, Vol. 29, 2011, pp. 143–8.
29 V. M. Bedell et al., 'In vivo genome editing using a high-efficiency TALEN system', *Nature*, Vol. 491, 2012, pp. 114–18.
30 M. Jinek et al., 'A programmable dual-RNA-guided DNA endonuclease in adaptive bacterial immunity', *Science*, Vol. 337, 2012, pp. 816–21.
31 L. Cong et al., 'Multiplex Genome Engineering Using CRISPR/Cas Systems', *Science*, Vol. 339, 2013, pp. 819–23.
32 P. Mali et al., 'NA-Guided Human Genome Engineering via Cas9', *Science*, Vol. 339, 2013, pp. 823–6.
33 R. J. Platt et al., 'CRISPR-Cas9 Knockin Mice for Genome Editing and Cancer Modeling', *Cell*, Vol. 159, 2014, pp. 440–55.
34 A. Bird, 'DNA methylation patterns and epigenetic memory', *Genes Dev*, Vol. 16, 2002, pp. 6–21.
35 E. Hood, 'RNAi: What's all the noise about gene silencing?', *Environ Health Persp*, Vol. 112, 2004, pp. A224–9.
36 S. Q. Harper, 'Progress and Challenges in RNA Interference Therapy for Huntington Disease', *Arch Neurol*, Vol. 66, 2009, pp. 933–8.

37 Transparency Market Research, Synthetic Biology Market (Synthetic DNA, Synthetic Genes, Synthetic Cells, XNA, Chassis Organisms, DNA Synthesis, Oligonucleotide Synthesis) – Global Industry Analysis, Size, Share, Growth, Trends and Forecast 2013–2019, undated, abstract available at www.transparencymarketresearch.com/synthetic-biology-market.html.
38 R. Carlson, 'The changing economics of DNA synthesis', *Nature Biotechnology*, Vol. 27, 2009, pp. 1091–4.
39 Fleischmann, 'Whole-genome random sequencing' (see note 11 above); Goffeau, 'Life with 6000 genes' (see note 12 above); McElheny, 'Drawing the Map of Life' (see note 13 above); and McVean, 'An integrated map of genetic variation' (see note 15 above).
40 Jackson, R. J. et al., 'Expression of mouse interleukin-4 by a recombinant ectromelia virus suppresses cytolytic lymphocyte responses and overcomes genetic resistance to mousepox', *J Virol*, Vol. 75, 2001, pp. 1205–10.
41 D. McKenzie, 'Revealed: Scientific evidence for the 2001 anthrax attacks', *New Scientist*, 25 February 2009..
42 J. Cello, A. V. Paul, and E. Wimmer, 'Chemical synthesis of poliovirus cDNA: generation of infectious virus in the absence of natural template', *Science*, Vol. 297, 2002, pp. 1016–18.
43 T. M. Tumpey et al., 'Characterization of the reconstructed 1918 Spanish influenza pandemic virus', *Science*, Vol. 310, 2005, pp. 77–80.
44 A. M. Rosengard, 'Variola virus immune evasion design: Expression of a highly efficient inhibitor of human complement', *Proc Natl Acad Sci USA*, Vol. 99, No. 13 (2002), pp. 8808–13.
45 S. Herfst et al., 'Airborne Transmission of Influenza A/H5N1 Virus Between Ferrets', *Science*, Vol. 336, 2012, pp. 1534–41.
46 M. Bieringer et al., 'Experimental Adaptation of Wild-Type Canine Distemper Virus (CDV) to the Human Entry Receptor CD150', *PLoS ONE*, 12 March 2013, e57488.
47 Y. Zhang et al., 'H5N1 Hybrid Viruses Bearing 2009/H1N1 Virus Genes Transmit in Guinea Pigs by Respiratory Droplet', *Science*, Vol. 340, 2013, pp. 1459–63.
48 M. Schmidt and G. Giersch, '*DNA synthesis* and security', Chapter 6, in M. J. Campbell (ed.), *DNA Microarrays, Synthesis and Synthetic DNA,* Hauppauge, NY: Nova Science Publishers Inc. 2011.
49 G. Murtas, 'Question 7: construction of a semi-synthetic minimal cell: a model for early living cells', *Orig Life Evol Biosph*, Vol. 37, 2007, pp. 419–22.
50 R. Gil, F. J. Silva, J. Pereto and A. Moya, 'Determination of the Core of a Minimal Bacterial Gene Set', *Mol Biol Rev*, Vol. 68, 2004, pp. 518–37.
51 Murtas, 'Question 7' (see note 49 above).
52 C. A. Voight, 'Genetic parts to program bacteria', *Curr Opin Biotechnol*, Vol. 17, 2006, pp. 548–57.
53 R. P. Shetty, D. Endy and T. F. Knight, 'Engineering BioBrick vectors from BioBrick parts', *J Biol Eng*, Vol. 25, No. 5 (2008).
54 'Registry of Standard Biological Parts', available at http://parts.igem.org/Main_Page.
55 Shetty, Endy and Knight, 'Engineering BioBrick vectors' (see note 53 above).
56 A. Katsnelson, 'DNA factory builds up steam', *Nature News*, (2010) Online. Available at doi:10.1038/news.2010.367.
57 R. Kwok, 'Five hard truths for synthetic biology', *Nature News*, Vol. 463, 2010, pp. 288–90.
58 Ibid.
59 E. Klavins, 'Lightening the load in synthetic biology', *Nature Biotechnology*, Vol. 32, 2014, pp. 1198–1200.
60 There is a large popular literature by and about Craig Venter, but the basic paper is D. G. Gibson et al., 'Creation of a Bacterial Cell Controlled by a Chemically Synthesized Genome', *Science*, Vol. 329, 2010, pp. 52–6.
61 J. S. Dymond et al., 'Synthetic chromosome arms function in yeast and generate phenotypic diversity by design', *Nature*, Vol. 477, 2011, pp. 471–6.
62 P. D. Hsu, E. S. Lander and F. Zhang, 'Development and Applications of CRISPR-Cas9 for Genome Engineering', *Cell*, Vol. 157, 2014, pp. 1262–78.
63 S. J. Culler, K. G. Hoff and C. D. Smolke, 'Reprogramming Cellular Behavior with RNA Controllers Responsive to Endogenous Proteins', *Science*, Vol. 330, 2010, pp. 1251–5.
64 C. Richert, A. L. Roughton and S. A. Benner, 'Nonionic Analogs of RNA with Dimethylene Sulfone Bridges', *J Am Chem Soc*, Vol. 118, 1996, pp. 4518–31.
65 M. Schmidt, 'Xenobiology: A new form of life as the ultimate biosafety tool', *BioEssays*, Vol. 32, 2010, pp. 322–31.

66 J. D. Bain et al., 'Ribosome-mediated incorporation of a non-standard amino acid into a peptide through expansion of the genetic code', *Nature*, Vol. 356, 1992, pp. 537–9.
67 S. Moran, R. X. Ren and E. T. Kool, 'A thymidine triphosphate shape analog lacking Watson–Crick pairing ability is replicated with high sequence selectivity', *Proc Natl Acad Sci USA*, Vol. 94, 1997, pp. 10506–11.
68 A. M. Leconte et al., 'Discovery, characterization, and optimization of an unnatural base pair for expansion of the genetic alphabet', *J Am Chem Soc*, Vol. 130, 2008, pp. 2336–43.
69 Z. Yang, F. Chen, J. B. Alvarado and S. A. Benner, 'Amplification, mutation, and sequencing of a six-letter synthetic genetic system', *J Am Chem Soc*, Vol. 133, 2011, pp. 15105–12.
70 V. B. Pinheiro et al., 'Synthetic genetic polymers capable of heredity and evolution', *Science*, Vol. 336, 2012, pp. 341–4.
71 P. Marlière et al., 'Chemical evolution of a bacterium's genome', *Angew Chem Int Edn*, Vol. 50, 2011, pp. 7109–14.
72 D. A. Malyshev et al., 'A semi-synthetic organism with an expanded genetic alphabet', *Nature*, Vol. 509, 2014, pp. 385–8.
73 M. Legendre et al., 'Thirty-thousand-year-old distant relative of giant icosahedral DNA viruses with a pandoravirus morphology', *Proc Natl Acad Sci USA*, Vol. 111, 2014, pp. 4274–9.
74 Jonathan B. Tucker, 'Could Terrorists Exploit Synthetic Biology?', *The New Atlantis*, 2011, pp. 69–81.
75 H. Ledford, 'Rare victory in fight against melanoma', *Nature*, Vol. 467, 2010, pp. 650–52.
76 M. Meyer, 'Domesticating and democratizing science: a geography of do-it-yourself biology', Papiers de Recherche du CSI, *CSI Working Papers* Series No. 032 (2013).
77 Ledford, 'Rare victory' (see note 75 above).
78 'The DIY dilemma: Misconceptions about do-it-yourself biology mean that opportunities are being missed', (Editorial), *Nature*, Vol. 503, Issue 7477 (2013), pp. 437–8, available at www.nature.com/news.
79 Ibid.
80 A. Michelsen et al., 'Honeybees can be recruited by a mechanical model of a dancing bee', *Naturwiss*, Vol. 76, Issue 6 (1989), pp. 277–80.
81 V. D. T. Thang et al., Proceeding of the International Conference on Innovations in Engineering & Technology (2013), available at http://dx.doi.org/10.15242/IIE.E1213583.
82 Y. Kuwana et al., 'Synthesis of the pheromone-oriented behaviour of silkworm moths by a mobile robot with moth antennae as pheromone sensors', *Biosens Bioelectron*, Vol. 14, 1999, pp. 195–202.
83 N. Ando, S. Emoto and R. Kanzaki, 'Odour-tracking capability of a silkmoth driving a mobile robot with turning bias and time delay', *Bioinspir Biomim*, Vol. 8, No. 1 (2013).
84 K. Y. Ma et al., 'Controlled flight of a biologically inspired, insect-scale robot', *Science*, Vol. 340, 2013, pp. 603–7.
85 See http://clearflightsolutions.com.

17

SYNTHETIC BIOLOGY AND THE CATEGORICAL BAN ON BIOWEAPONS

Filippa Lentzos and Cecilie Hellestveit

The relatively new science of synthetic biology is raising significant security concerns:[1] Will it enable the creation of dangerous viruses from scratch? Will it enable the design of radically new pathogens not found in nature? Is synthetic biology breaking down the boundary between expert and non-expert to such an extent that anyone can do this? This chapter analyses the security threat posed by efforts to engineer biology by placing the threat in its technical, historical, social, political, and legal contexts. This approach enables a more robust, broad, and deep consideration, which more often than not is lacking in the political and security discourse about the synthetic biology threat.

The chapter opens with a short introduction to the scientific developments that are putting synthetic pathogens within reach at a rapid pace. It goes on to elaborate the security concerns that this raises, highlighting which concerns are legitimate and which less so, based on a realistic understanding of the technology and the scientific practices surrounding synthetic biology. Focusing on both the socio-political context and from whom the potential threat is coming, the chapter next discusses the interests and capabilities of non-state and state actors in applying synthetic biology to bioterrorism and biological weapons attacks. We argue that while the potential for state use is very low, this is where the most significant security threat from synthetic biology is originating – and not from non-state actors, which receive the vast amount of attention in policy discussions. The final section of the chapter goes on to consider the extent to which the legal framework prohibiting biological weapons is calibrated for the new security risks posed by synthetic pathogens, and it argues that, while the Biological and Toxin Weapons Convention (hereafter, BWC) is broad in scope and covers new technologies, synthetic biology exposes the treaty to particular challenges that may erode the categorical ban on biological weapons under international law.

The engineering of biology

The term 'synthetic biology' was first coined in the scientific literature in 1912 with the publication of a monograph of the same title by French chemist Stephane Leduc.[2] However, in its contemporary form, synthetic biology is seen as a new and emerging field that seeks to create a rational framework for manipulating the DNA of living organisms through the application of bioengineering principles. Although the precise labelling of synthetic biology and whether it represents a distinctly novel field have been called into question,[3] its key founding principle is 'to design and engineer biologically based parts, novel devices and systems, as well as redesigning existing, natural biological systems'.[4] In short, the field aims to engineer biology.

Although many characterise it as a twenty-first century science, the history of today's synthetic biology can be traced back to 1979, when the first gene was synthesised by chemical means.[5] The Indian-American chemist Har Gobind Khorana and 17 co-workers at the Massachusetts Institute of Technology took several years to produce a small gene made up of 207 DNA nucleotide base pairs. In the early 1980s, two technological developments facilitated the synthesis of DNA constructs: the invention of the automated DNA synthesiser and the polymerase chain reaction (PCR), which can copy any DNA sequence many million-fold. By the end of the 1980s, a DNA sequence of 2,100 base pairs had been synthesised chemically.[6]

In 2002 the first functional virus was synthesised from scratch: polio virus, whose genome is a single-stranded RNA molecule of about 7,500 nucleotide base pairs.[7] Over a period of several months, Eckard Wimmer and his co-workers at the State University of New York at Stony Brook assembled the polio virus genome from customised oligonucleotides, which they had ordered from a commercial supplier. When placed in a cell-free extract, the viral genome then directed the synthesis of infectious virus particles. The following year, Hamilton Smith and his colleagues at the J. Craig Venter Institute in Maryland published a description of the synthesis of a bacteriophage, a virus that infects bacteria, called φX174. Although this virus contains only 5,386 DNA base pairs (fewer than polio virus), the new technique greatly improved the speed of DNA synthesis. Compared with the more than a year that it took the Wimmer group to synthesise polio virus, Smith and his colleagues made a precise, fully functional copy of the φX174 bacteriophage in only two weeks.[8]

Since then, the pace of progress has been remarkable. In 2004, DNA sequences 14,600[9] and 32,000[10] nucleotides long were synthesised. In 2005, researchers at the US Center for Disease Control and Prevention used sequence data derived from the frozen or paraffin-fixed cells of victims to reconstruct the genome of the 'Spanish' strain of the influenza virus, which was responsible for the flu pandemic of 1918–19 that killed tens of millions of people worldwide; the rationale for resurrecting this extinct virus was to gain insights into why it was so virulent. In late 2006, scientists resurrected a 'viral fossil', a human retrovirus that had been incorporated into the human genome around five million years ago.[11] In 2008, a bat virus related to the causative agent of human SARS was recreated in the laboratory.[12] That same year, the J. Craig Venter Institute synthesised an abridged version of the genome of the bacterium *Mycoplasma genitalium*, consisting of 583,000 DNA base pairs.[13] In May 2010, scientists at the Venter Institute announced the synthesis of the entire genome of the bacterium *Mycoplasma mycoides*, consisting of more than 1 million DNA base pairs.[14] The total synthesis of a bacterial genome from chemical building blocks was a major milestone in the use of DNA synthesis techniques to create more complex and functional products. In 2014, a designer yeast chromosome was constructed – a major advance towards building a completely synthetic eukaryotic genome.[15]

These advances have been complemented by progress in genome editing technology, which is enabling deletions and additions in human DNA sequences with greater efficiency, precision, and control than ever before. 'CRISPR-Ca9' (clustered, regularly interspaced short palindromic repeats) has become the major technology employed for these purposes and has been used to manipulate the genes of organisms as diverse as yeast, plants, mice, and, reported in April 2015, human embryos.[16]

The security threat

The aspirations and pace of advance in synthetic biology have raised a number of security concerns. Some of these concerns are legitimate, others less so.[17] Often lacking in the political and security discourse is a realistic understanding of the technology and the scientific practices surrounding synthetic biology. Equally lacking is a nuanced portrayal of who typifies the potential threat, with the non-state actor threat dominating the discourse. In this section, we relate the principal concerns that are portrayed in security discussions around synthetic biology and detail some of the misleading assumptions underpinning them. In the following two sections, we go on to discuss the interest and capabilities of non-state and state actors in applying synthetic biology to bioterrorism and biological weapons attacks.

One of the main concerns raised in the political and security discourse is that synthetic biology is making it easier to create dangerous pathogens from scratch. The claim is that well-characterised biological parts can be easily obtained from open-source online registries and then assembled, by people with no specialist training and outside of professional scientific institutions, into genetic circuits, devices, and systems that will reliably perform desired functions in live organisms. This narrative, however, rests on misleading assumptions about synthetic biology.

The narrative does not reflect the situation facing people with no specialist training who work outside professional scientific institutions; it does not even reflect current realities in academic or commercial science laboratories: academic and commercial researchers are still struggling with every stage of the standardisation and mechanisation process. More than a decade in, the translation of proof-of-concept designs into real-world applications is still a major challenge. As recently noted in an article surveying progress in synthetic biology: 'The synthetic part is easy, it's the biology part that's confounding'.[18]

However, even *if* the engineering approaches offered by synthetic biology make processes more systematic and more reproducible, skills do not become irrelevant, and all aspects of the work do not become easier. And, importantly, 'easier' does not mean 'easy'. Aeronautical engineering provides a useful analogy: planes are built from a large number of well-characterised parts in a systematic way. But this does not mean that any member of the general public can build a plane, make it fly, and use it for commercial transportation. So advances in synthetic biology do not make it easier for just *anybody* to engineer biological systems, including dangerous ones.

This leads to a second concern raised in the political and security discourse: that synthetic biology is breaking down the expert and non-expert boundary. In other words, the growth of a do-it-yourself (DIY) biology community, along with DNA synthesis becoming cheaper and easily outsourced, could make it easier for terrorists to obtain the basic materials to create biological threat agents.

However, the link between synthetic biology and DIYbio, and the level of sophistication of the experiments typically being performed, is grossly overstated. Do-it-yourself biologists typically comprise a wide range of participants of varying levels of expertise, ranging from complete novices with no prior background in biology to trained scientists who conduct experiments on their own time. Some do-it-yourself biologists work in home laboratories assembled from everyday household tools and second-hand laboratory equipment purchased online; the majority conduct their experiments in community labs or 'hackerspaces'. Studies of scientific practice in community labs demonstrate the challenges that amateur biologists face while trying to conduct even rudimentary biological experiments successfully. These amateurs particularly lack access to the shared knowledge available to institutional researchers, highlighting the importance of local, specialised knowledge and enculturation in laboratory practices.

DNA synthesis is one of the key enabling technologies of synthetic biology. There are now many commercial companies that provide DNA synthesis services, so the process can be outsourced: a client can order a DNA sequence online and receive the synthesised DNA material by post within days or weeks. The price charged by these companies has been greatly reduced over the last 20 years and the service is now within reach of a broad range of actors. This has led to routine statements suggesting that it is now cheap and easy to obtain a synthesised version of any desired DNA sequence.

There are, however, several challenges that need to be taken into account when assessing the potential for misuse that inexpensive DNA sequencing might enable. First, simply ordering online the full-length genome sequence of a small virus (or those of larger bacteria) is not currently possible. The alternative, ordering short DNA sequences and assembling them into a genome, requires specialist expertise, experience, and equipment available in academic laboratories, but not easily accessible to an amateur working from home. Assembling DNA fragments was the major technological feat in the work conducted at the Venter Institute that produced the 'synthetic' bacterial genome in 2010, and the Gibson assembly method developed for that project is now widely used. The description of that work, however, demonstrates how the assembly of smaller fragments into larger ones and eventually into a *functioning* genome requires substantial levels of expertise and resources, including those needed to conduct trouble-shooting experiments to identify and correct errors when assembled DNA constructs do not perform as expected. So, as noted by the US National Science Advisory Board for Biosecurity (NSABB), while the

> technology for synthesizing DNA is readily accessible, straightforward and a fundamental tool used in current biological research [...] the science of constructing and expressing viruses in the laboratory is more complex and somewhat of an art. It is the laboratory procedures downstream from the actual synthesis of DNA that are the limiting steps in recovering viruses from genetic material.[19]

Again, it is the biology, not the synthetic part, that is complicated, and DNA synthesis requires extensive training in basic molecular biology techniques, such as ligation and cloning, including hands-on experience that is not 'reducible to recipes, equipment, and infrastructure'.[20]

A third concern often voiced is that synthetic biology may enable radically new pathogens to be designed: Synthetic biology could be used to enhance the virulence or increase the transmissibility of known pathogens, creating novel threat agents.

Mousepox and bird flu (H5N1) experiments are frequently cited to demonstrate how dangerous new pathogens could be designed. But assessments of this threat tend to overlook a salient fact: in both of these experiments, the researchers did not actually design the pathogens. With respect to H5N1, researchers had indeed been trying to design an air-transmissible virus variant for some time, without success. The ferret experiment was set up as an alternative approach, to see whether natural mutations could generate an air-transmissible variant. The researchers had no influence on the specific mutations induced. In the mousepox experiment, researchers inserted the gene for interleukin-4 into the mousepox virus to induce infertility in mice and serve as an infectious contraceptive for pest control. The result – that the altered virus was lethal to mice – was unanticipated by the researchers; that is, it was not designed.

Moreover, some of the lessons that came out of the extensive Soviet programme to weaponise biological agents involve the trade-offs between improving characteristics that are 'desired' in the context of a bioweapons programme – such as virulence – and diminishing other equally 'desired' characteristics, such as transmissibility or stability. Pleiotropic effects – that is, when a single gene affects more than one characteristic – and genetic instability are common in microorganisms. While it is too simple to say that increased transmissibility will always be associated with reduced virulence, this is often the case for strains produced in laboratories. As other commentators have noted:[21]

> To create ... an artificial pathogen, a capable synthetic biologist would need to assemble complexes of genes that, working in union, enable a microbe to infect a human host and cause illness and death. Designing the organism to be contagious, or capable of spreading from person to person, would be even more difficult. A synthetic pathogen would also have to be equipped with mechanisms to block the immunological defences of the host, characteristics that natural pathogens have acquired over eons of evolution. Given these daunting technical obstacles, the threat of a synthetic 'super-pathogen' appears exaggerated, at least for the foreseeable future.

Non-state actors

Clearly, synthetic biology raises security concerns, but, as outlined above, these are often not as straightforward or as immediate as regularly portrayed in the political and security discourse. Alongside the more technical assumptions, there is a set of assumptions in the discourse about who the threat is coming from, what their intentions are, what capabilities they might pursue, and the level of skills and resources available to them. For example, in one of President George W. Bush's earliest statements following 9/11 and the 'anthrax letter' attacks that drew the American people's attention to the biological weapons threat, he said:

> Since September 11, America and others have been confronted by the evils these [biological] weapons can inflict. This threat is real and extremely dangerous. Rogue states and terrorists possess these weapons and are willing to use them.[22]

Later, he set up a WMD Commission and tasked it with examining the threat posed by the nexus of international terrorism and the proliferation of weapons of mass destruction (WMDs). In its report, this Commission asserted:

> Unless the world community acts decisively and with great urgency, it is more likely than not that a weapon of mass destruction will be used in a terrorist attack somewhere in the world by the end of 2013. The Commission further believes that terrorists are more likely to be able to obtain and use a biological weapon than a nuclear weapon. The Commission believes that the U.S. government needs to move more aggressively to limit the proliferation of biological weapons and reduce the prospect of a bioterror attack.[23]

Bioterrorism became one of the Bush Administration's key security concerns over its two terms in office. One estimate of civilian biodefence expenditure across the federal government since 2001 is that more than USD70 billion have been spent.[24] Despite this, on the ten-year anniversary of 9/11 and the anthrax letter attacks, the former US senators who chaired the WMD Commission, Bob Graham and Jim Talent, released a 'report card' on America's bio-response capabilities that concluded that the US was still unprepared to respond to large-scale biological attacks. It also warned:

> Naturally occurring disease remains a serious biological threat; however, a thinking enemy armed with these same pathogens – or with multi-drug-resistant or synthetically engineered pathogens – could produce catastrophic consequences. A small team of individuals with graduate training in several key disciplines, using equipment readily available for purchase on the Internet could produce the type of bioweapons created by nation-states in the 1960s. Even more troubling, the rapid advances in biotechnology, such as synthetic biology, will allow non-state actors to produce increasingly powerful bioweapons in the future.[25]

Some of the technical assumptions previously discussed – about de-skilling and increased access, and about the ease of designing new dangerous pathogens – can be seen here to underlie concerns that connect the advent of synthetic biology with terrorists' potential ability to launch a mass attack.

The senators were not alone in their assessments. For instance, the US Senate Majority Leader Bill Frist made a similar warning in an earlier speech outlining the global threat of infectious disease and bioterrorism, and the need to better prepare the US and the world to respond to epidemics and outbreaks:

> No intelligence agency, no matter how astute, and no military, no matter how powerful and dedicated, can assure that a few technicians of middling skill using a few thousand dollars' worth of readily available equipment in a small and apparently innocuous setting cannot mount a first-order biological attack [...] Never have we had to fight such a battle, to protect so many people against so many threats that are so silent and so lethal.[26]

Similar messages were reinforced at the highest level. Addressing Biological Weapons Convention members at their five-year review meeting in 2011, Secretary of State Hillary Clinton said:

> The advances in science and technology make it [...] easier for states and non-state actors to develop biological weapons. A crude, but effective, terrorist weapon can be made by using a small sample of any number of widely available pathogens, inexpensive equipment, and college-level chemistry and biology.[27]

She also acknowledged, however, that not everyone in the international community shared the US assessment:

> I know there are some in the international community who have their doubts about the odds of a mass biological attack or major outbreak. They point out that we have not seen either so far, and conclude the risk must be low. But that is not the conclusion of the United States, because there are warning signs, and they are too serious to ignore.[28]

The policy discourse on security concerns raised by synthetic biology has overwhelmingly emphasised the threat of terrorists seeking to produce mass casualty weapons and pursuing capabilities on the scale of twentieth-century state-level bioweapons programmes as an imminent concern. Yet, the 'warning signs' seen by Secretary of State Clinton have been few and far between: (i) the Rajneesh cult's deliberate contamination of salad bars with Salmonella to sicken voters and make them stay away from the polls during Oregon elections in 1984; (ii) the Japanese Aum Shinrikyo cult's sarin attack on the Tokyo underground in 1995 and its (unsuccessful) attempt to spray anthrax spores into the air; and (iii) the 'anthrax letters' containing high-quality dry-powder preparation of anthrax spores that were sent to media outlets and members of US Congress in 2001 resulting in at least 22 cases of anthrax, five of which were fatal. More recent indications of terrorist interest in biological weapons that are often referenced in the discourse include: (iv) 'evidence' in Afghan caves; (v) al Qa'ida's call to arms for 'brothers with degrees in microbiology or chemistry to develop a weapon of mass destruction'; and (vi) the Islamic States' 'laptop of doom' reportedly containing information on developing biological weapons.

Most leading biological non-proliferation experts distance themselves from the US assessment that this 'evidence' indicates widespread terrorist interest and capability to carry out mass biological attack; indeed many would argue that the cases actually signal the opposite: that it is difficult to weaponise bugs and deliberately spread disease. For instance, on closer examination after the 'laptop of doom' story broke, it appears that the files only included copy/paste information from biology textbooks with descriptions of how to cultivate and extract bacterial strains at the most basic level of microbiology.[29] The only biowarfare agent mentioned was *Clostridium botulinum*, which could not be cultivated with the methods mentioned in the same document. Most experts believe that while the risk of a small-scale crude bioterrorism attack is possible and likely, the risk of sophisticated large-scale bioterrorism attacks is very small.[30] This is backed up by historical evidence.[31]

The emphasis in the policy discourse on high consequence, mass casualty attacks also falsely assumes that producing a pathogenic organism equates to producing a 'weapon of mass destruction'. Considerable knowledge and resources are necessary for the processes of developing, scaling up, storing and disseminating a biological weapon; and these processes present significant technical and logistical barriers.[32] But even if, against the odds, a biological weapon is disseminated successfully, the outcome of an attack would be affected by factors like the health of the people who are exposed to the agent, the speed and manner with which public health authorities and medical professionals detect and respond to an outbreak, and public health communication strategies. A prompt response with effective medical countermeasures, such as antibodies and vaccination, can significantly blunt the impact of an attack, both physically and psychologically.

In short, sophisticated non-state actor use of biological weapons with mass consequences is unlikely – the sort of sophisticated capability developed in the US and Soviet bioweapons programme are out of reach for terrorists. Does synthetic biology change that equation? Not really. But, having said that, there are a couple of crude bioterrorism scenarios that warrant particular attention. One is the 'lone operator', such as a highly trained synthetic biologist, who is motivated to do harm by ideology or personal grievance. This appears to have been the case with Bruce Ivins, the FBI-identified military-insider behind the anthrax letters. The second scenario involves a 'biohacker' who does not necessarily have malicious intent but who seeks to create bioengineered organisms out of curiosity or to demonstrate technical prowess – a common motivation of many designers of computer viruses. As synthetic biology training becomes increasingly available to students at the college and even high-school levels, a 'hacker culture' may emerge, increasing the risk of reckless or malevolent experimentation, but where casualties would be in the tens, not in the hundreds or thousands.

State interest

A much greater security threat would come from military interest in using synthetic biology to develop highly sophisticated biological weapons. We know that there has been significant military interest in bioweapons in the past. Is that still the case? Are there state-level biological weapons programmes today? The BWC bans the entire category of biological weapons, so any overt offensive bioweapons programme would be in violation of the Convention. Clearly, bioweapons programmes are not admitted to, but what about cheaters?

Official US government statements repeated for many years that there were four nations in possession of offensive biological weapons programmes in 1972, when the BWC was signed. By 1989 this number had increased to ten, and in July 2001, just prior to 9/11 and the anthrax letters, it was thirteen.[33] Not all of the accused states were named, but they included: China, Cuba, Egypt, Iran, Iraq, Libya, North Korea, Russia, South Africa, and Syria.

Assessment of the state-level threat from countries that were definitely developing biological weapons became markedly more qualified following 9/11 and the anthrax letter attacks in late 2001. Already in early 2002, the Assistant Secretary of State for Intelligence and Research described in some detail in congressional testimony which nations were thought to possess weaponised stocks of biological agents.[34] In 2003, official US intelligence assessments outlined countries with bioweapon 'acquisition activity of concern'.[35] That year, though, it became clear through military intervention that intelligence on Iraq's 'WMDs', particularly on its biological weapons, was wrong; the offensive Iraqi bioweapons programme had been entirely disbanded through UN intervention following the Gulf War. The South African programme had been terminated around the same time, in the mid-1990s, following the change of government and end of apartheid. It also became clear that, although having intentions to acquire equipment and develop capabilities, Libya had never had an offensive bioweapons programme. In 2004, the US administration also withdrew the charge that Cuba was maintaining an offensive programme.

Towards the end of the Bush Administration, accusations of offensive bioweapons programmes started becoming more muted, lacking in specifics and status. The Defense Intelligence Agency Director's congressional testimony to a 2007 hearing on the current and future worldwide threats to the national security of the US is a good example, saturated with phrases like 'that could be used to support a biological warfare program' and 'possesses a sufficiently advanced biotechnology infrastructure to allow it to develop and produce biological agents'.[36] This characterisation could equally apply to the United States and most European countries.

Today the official tally of states with offensive bioweapons programmes has been drastically revised. The most recent annual US State Department report assessing compliance with disarmament treaties notes concerns about bioweapons programmes in North Korea, and possibly Syria.[37] It also notes that Russia, a BWC depository nation and a permanent member of the United Nations Security Council (UNSC), and China, also a permanent member of the Security Council, have engaged in biological activities with dual-use applications, as has Iran, but that it is unclear whether any of those activities are in breach of the Convention. The 'available information' does not indicate that Egypt is engaging in activities prohibited by the BWC. Ditto for Pakistan, also named in these reports in more recent years.

There are also a significant number of states that have not signed up to the BWC, notably Israel. There is no discussion of Israel's bioweapons capability or the status of its bioweapons programme in any public US government report. There is also very little scholarly work on the Israeli programme, though Milton Leitenberg has noted that several scientists from the Soviet bioweapons programme are known to have immigrated to Israel and that 'Israel almost certainly maintained an offensive BW program for many years and may still do so'.[38]

Of course there are many countries that have legitimate *defensive* bioweapons programmes, but the line between offensive and defensive programmes can sometimes be very blurred. In early 2000, for instance, a series of secret projects were reportedly underway in the United States to improve biodefences. The Pentagon was buying commercially available equipment to build a small-scale germ factory to produce anthrax simulants – *Bacillus thuringiensis*, the biopesticide made at the main Iraqi bioweapons centre before it was blown up by United Nations inspectors in 1997. Another project involved genetically modifying anthrax to make a vaccine-resistant superbug. Meanwhile the CIA, in one of its projects, was building Soviet-style bio-bomblets and testing them for dissemination characteristics and performance in different atmospheric conditions.[39]

Pentagon and CIA lawyers said the projects were legitimate defensive activities: building and operating a bioweapons facility helped uncover the tell-tale clues of distinctive patterns of equipment buying; genetically modifying anthrax was essential to check whether the current vaccines administered to soldiers were effective; and building and testing bomblets was a defensive response to specific intelligence about a possible adversary. Others disagreed, saying the projects were not permitted under the BWC, signed and ratified by the United States in 1975, although not implemented until the 1980s.[40]

The treaty permits almost any kind of research in the name of defence. Some of this work is unquestionably justifiable. Other research edges closer to the blurred line between defensive and offensive work. The trouble with distinguishing permitted biodefence projects from non-permitted projects is that it is not just about the facilities, equipment, and activities, but also about the purpose or intent of those activities. What are we to make, for instance, of the huge synthetic biology investments by the Defense Advanced Research Projects Agency (DARPA), which spearheads US military technology and aims to create technologies with potential for extraordinary advances in national security capability? DARPA Director Arati Prabhakar has said: 'Biology is nature's ultimate innovator, and any agency that hangs its hat on innovation would be foolish not to look to this master of networked complexity for inspiration and solutions'.[41]

'Living Foundries' is one of the synthetic biology projects that DARPA funds. The project aims to provide

> game-changing manufacturing paradigms for the DoD [through developing and applying] an engineering framework to biology that decouples biological design from fabrication, develops and yields design rules and tools, and manages biological complexity through simplification, abstraction and standardization of both processes and components. [The result will be] rapid and scalable development of previously unattainable technologies and products [...] leveraging biology to solve challenges associated with production of new materials [...] biological reporting systems, and therapeutics.

By 2014, USD90 million had been spent or allocated to the Living Foundries project.[42] DARPA, and other research arms of the Pentagon, have become heavyweight funders of synthetic biology, and questions are starting to be raised about the burgeoning field's increasing dependence on defence dollars.[43]

There are also signs of military interest in synthetic biology by other states. In February and March 2012, President Putin and Russian Minister of Defence Anatoly Serdyukov publicly referred to 28 tasks they had established for the Russian military to prepare for threats 30–50 years ahead. One of these tasks was the development of weapon systems using different physical principles, 'beam, geophysical, wave, genetic, psychophysical and other types of weapons. ... Such weapon systems will be as effective as nuclear weapons but will be more "acceptable" from the political and military point of view' (emphasis added).[44] Genetic weapons are of course banned by the BWC, and the statement remains troubling. Putin's remarks have not been revoked or clarified.

However, despite this, most bioweapons experts agree that the potential for state use of biological weapons, conventionally or synthetically produced, is extremely low.[45] Various reasons are cited for this: biological weapons are not 'good' weapons; it is difficult to produce sophisticated and reliable biological weapons; and it is not politically viable to use them because of the strong norm against them. Others have highlighted the intense secrecy necessary to hide a state programme – to conceal its laboratories, production facilities, training and testing grounds, modes of transportation, and special troops – and the difficulties of sustaining such secrecy in a world of globalised commerce, travel, and communication. The norm against biological weapons and how it operates to inhibit offensive programmes has also been emphasised:

> Say someone is playing around with [potentially offensive biological weapons]. They go to senior people and say, 'Do you know what guys, we can do X, Y, Z, this has potential for major contribution to operations, this is a game changer'. At some point in time somebody has to make a political decision saying 'We're on board'. The move from potential to capability, to incorporation into doctrine, that takes a very hard political decision, and I think many people when you begin to cross that line would say, 'Hold on a second, what are we actually gaining here? Outside a survival scenario we are potentially not making our military situation or our political situation better.' So I think this is where the norm plays in. This is where the taboo plays in. This is where the various elements against bioweapons play in, in ways that we cannot easily quantify or qualify.[46]

Bioweapons are a 'no win situation'. As the expert, Jez Littlewood, goes on to explain:

> Only under the most extreme set of circumstances of an existential threat – which we will deal with through our nuclear weapons anyway – does the notion of senior level decision-makers signing off on an offensive bioweapons program in Western countries come into play. So we can worry about biodefense quite legitimately, but we equally have to be cold-hearted in realising this is only one part of an overall process of how a weapon gets integrated into operational use.[47]

But what about non-Western states, rogue states in search of clout? Could biological weapons form a poor man's atomic bomb, as many political hawks and their advisors have suggested? For instance, the Assistant Secretary of State for Intelligence and Research highlighted in his 2002 testimony, 'Because biological weapons are relatively cheap, easy to disguise within commercial ventures, and potentially as devastating as nuclear weapons, states seeking to deter nations with superior conventional or nuclear forces find them particularly attractive'.[48] Biological disarmament and non-proliferation experts disagree:

> There's really only a few rogue states who are rogue enough to do that in this day and age – and I don't think that even they'd do it; the repercussions would be just too great.[49] [The norm itself is not what necessarily affects rogue states,] it's the retribution factor that would have the effect. The retribution, if a state uses biological weapons, would be so much greater even than with chemical weapons. I don't even think North Korea would consider bioweapons.[50]

Other experts weigh in on this too:

> If a country wishes to acquire effective, dependable biological weapons [bringing about planned-for and reproducible effects every time they are tested] of the types that

> the Soviet Union attempted to do ... the acquisition process is very difficult and costly to carry out. Although it is true that countries can try to do so cheaply, the end products of inexpensive programs are likely to be ineffective and undependable. Iraq is an example: its biological weapons arsenal contained bombs and missiles of dubious reliability or effectiveness.[51]

While we agree that the norm against biological weapons is strong, and that the potential for state use is very low, we find the blanket rejection of the bioweapons state threat unhelpful. Like discussions of the bioterrorism threat, the state-level threat also lacks nuance. Bioweapons might not have military utility in all contemporary conflicts, but they might well have utility in a small subset. Biowarfare can be compared with cyber warfare in that you know you have been attacked, but not by whom. The 'benefits' of biological attacks can be that victims do not know and are unable to prove that an attack has taken place – the question 'who is behind this' is in all likelihood not even going to be asked. For instance, Saudi Arabia experienced the largest cyber attack against a non-military entity to date in 2012, on a Saudi national oil company. That same year the emergence of Middle East Respiratory Syndrome (MERS) was first identified, but no questions were asked publicly about whether the disease could also have been deliberately introduced. We are not suggesting that it was, or that MERS is a biological weapon, but if it had been, the political risks would depend on transparency in military affairs in the sending country. The silent and invisible nature of biological weapons could make them highly potent means for weakening the legitimacy of enemy regimes within their own populations, for instance, or for just keeping them busy. So while 'political viability' is an important element in containing some kinds of bioweapons, it does not cover all possible uses, and particularly not instances when it is not supposed to be known that an attack has taken place. In the 'best case' scenario you may actually get rid of enemy regimes without anyone realising foul play.

The international legal framework

The security threat of synthetic biology must be put in a technical, historical, social, and political context. Once you do that, as we have shown above, you can tease out the details of the threat: (1) Creating dangerous pathogens using synthetic biology is complicated; (2) there is a low-level security threat of the crude misuse of synthetic biology; and (3) the primary security threat of sophisticated synthetic biology misuse comes from technically advanced and well-resourced military programmes. This perspective is the basis for the following sections, which consider how the international legal framework prohibiting biological weapons is calibrated and fit to counter the challenges represented by synthetic biology.

Biological weapons are subject to the strongest ban among weapons of mass destruction (WMD) under international law. The *use* of bacteriological weapons is prohibited under the 1925 Geneva Gas Protocol, a protocol that enjoys the status of custom and is therefore binding on all states irrespective of ratification.[52] Further, the *effects* of use of biological weapons make them unlawful under general international humanitarian law. In 1969, the UN General Assembly defined biological weapons as microorganisms with the ability to inflict damage or cause disease, which are not used for prophylactic, protective or other peaceful purposes.[53] Thus, biological weapons violate both pillars that underpin the laws of armed conflict – the rule of distinction and the prohibition to cause unnecessary suffering and superfluous injury.[54] These restrictions apply to any party to an armed conflict, and include both state and non-state actors.[55] The main legal instrument addressing biological weapons, the BWC, essentially bans the *weaponisation* of biology. The Convention does not explicitly prohibit use, but rather aims at the activity prior to use – reliance on biology as a weapon in the sense of stockpiling, acquiring biological agents or toxins intended for a certain type of use, or development of platforms designed for dispersing biological agents or toxins.

The BWC was negotiated in 1972, the first treaty to ban an entire class of weapons. In essence it sprang from a consensus between the United States and the USSR in the early 1970s that biological weapons presented 'less intractable problems' compared to chemical and nuclear weapons.[56] The use of nuclear weapons was not unlawful on the same footing as biological weapons.[57] And while chemical weapons were unlawful in war, some covered weapons were extensively relied on for purposes of law enforcement.[58] States therefore concluded that an agreement on banning biological weapons should not be delayed until agreement on a reliable prohibition of chemical weapons could be reached.

The resulting treaty banning weaponisation of biology paints its coverage in very broad strokes. The BWC is far simpler and less sophisticated than subsequent arms ban treaties – a simplicity often ascribed to its negotiating history, which is considerably less complex than that of the Chemical Weapons Convention (CWC). The BWC contains eight key provisions stipulating prohibitions, measures of implementation, and finally protection of peaceful use. The form of the BWC is also ascribed to the fact that it belongs to the childhood of disarmament treaties, suffering from weaknesses that have been solved in subsequent conventions.

Importantly, the BWC does not establish an organisation tasked with verification and enforcement of the treaty similar to the Non-Proliferation Treaty or the CWC (the pre-existing International Atomic Energy Agency, IAEA, and the Organisation for the Prohibition of Chemical Weapons, OPCW, respectively). The BWC requires the complete destruction of all bioweapons and no future production, but it does not establish an independent intergovernmental institution tasked with overseeing and enforcing the Convention. It is based on 'self-verification': each state party to the BWC is required to designate a domestic agency to be responsible for guaranteeing compliance with the treaty's provisions. Member states must consult and cooperate, and they may lodge a complaint before the United Nations Security Council under Chapter VI. However, measures to ensure respect for the Convention essentially boil down to a set of nonbinding confidence building measures (CBMs), self-verification, and criminalisation at the national level. The irony is that while biological weapons are subject to the strongest and most categorical ban among the weapons of mass destruction – and arguably of any class of weapon – the weaponisation of biology suffers from the weakest mechanisms of verification and enforcement.

A frequent explanation for this paradox is the perceived nature of biological weapons, often referred to as 'inherently unverifiable'.[59] A combination of the natural prevalence of agents and toxins with the need for research and stockpiling for prophylactic purposes has placed the BW verification challenges on a different track compared to that of nuclear and chemical weapons. Other WMD regimes rely on technology controls to prevent access to the weapons. This may take the form of verifying the presence (or absence) of material that could only be used for weapons purposes, or process dynamics may be monitored to ensure that material is not being diverted for prohibited purposes. Traditional biological resources are less fit for such an approach. Biomaterials, knowledge, and resources are intrinsically dual-use, making differentiation difficult. Biomaterials are also living organisms, with an ability to reproduce, complicating material accounting. The BWC (as will the later CWC) therefore relies on a General Purpose Criterion that contains a qualitative and quantitative dimension. It prohibits agents or toxins 'of types and in quantities' that cannot be justified for 'prophylactic, protective or other peaceful purposes'.

The advent of synthetic biology, however, is changing this equation. The introduction of engineering techniques that may be used for industrial purposes adds some of the 'intractable problems' associated with chemical weapons into the biological realm. Equally, however, synthetic biology is presumably not so 'inherently unverifiable' as traditional biological weapons, given that its prevalence is less natural and therefore more traceable. In short, synthetic biology is introducing features traditionally associated with chemical weapons into the BWC regime.

A frequent criticism of the BWC is that the role of non-state actors has not yet been fully absorbed into the BWC regime, and globalisation has pushed the gravity point of dynamic interaction from states to a variety of transnational actors, with highly diverging and opposed interests.[60] While it is true that the complexities associated with verification of synthetic biology are intimately linked to private actors and industries, the implicit state-actor bias of the BWC does not seem outdated when dealing with the misuse risks of synthetic biology.

Synthetic biology and the scope of the BWC

The provisions of the BWC essentially express four principles: the prohibition to acquire or retain biological or toxin weapons; the prohibition to assist others to acquire such weapons; the obligation to take necessary measures to ensure that such weapons are prohibited at a domestic level; and finally the commitment to ensure that peaceful use of biological science and technology may nevertheless develop.[61]

While the simplicity of the BWC may offer flexibility and adaptability, a less benign effect is the unclear scope and an unhelpful lack of specific details. In order to address this weakness, the Convention contains a general review clause expressing the intention of state parties to adapt and develop the treaty without having to resort to a formal modification procedure. Review conferences (REVCONs) held by the States Parties to the BWC, organised every five years, therefore hold a general review power by way of consensus. The resulting understandings and agreements influence the interpretation of the BWC.[62] Specific issues under all articles of the Convention have been addressed over the years.[63]

The BWC does not explicitly prohibit use of biological weapons. However, the fourth REVCON (1996) affirmed that 'the use by parties, in any way and under any circumstances, of microbial or other biological agents or toxins, that is not consistent with prophylactic, protective or other peaceful purposes, is effectively a violation of Article I of the Convention'.[64]

An understanding was also reached that even individuals are covered by the Convention's prohibitions.[65] In 2006, state parties unanimously confirmed that the BWC prohibits use of bioweapons 'by anyone, anywhere, at any time and for any purpose'.[66] The BWC is hence understood to prohibit the weaponisation of biology and the use of biological weapons by international actors, states, sub-national entities, and individuals. Its scope unambiguously extends to all relevant actors involved with synthetic biology.

The flexible nature of the Convention is an advantage in terms of covering novel technologies. The obligations of the BWC can apply to components, organisms, and products resulting from synthetic biology techniques as far as they are microbial, or other biological agents, or toxins. BWC Article I states that all agents and toxins are covered, regardless of 'origin or method of production'. Implicitly, changes in the means of producing biological weapons are already explicitly covered by the Convention. New scientific and technological developments are explicitly stipulated as elements to take into account in review conferences. The second REVCON (1986) reiterated:

> [T]he Convention unequivocally applies to all natural or artificially created microbial or other biological agents or toxins whatever their origin or method of production. Consequently, toxins (both proteinaceous and non-proteinaceous) of a microbial, animal or vegetable nature and their synthetically produced analogues are covered.[67]

All scientific and technological developments in the life sciences and in other fields of science relevant to the Convention are covered.[68] So are agents or toxins and their components, whether they affect humans, animals or plants.[69] A general consensus has emerged in favour of an absolute ban on any new biological weapons in whatever form they might emerge. The provision on transfer in Article III was, at the 1991 REVCON, explicitly confirmed to extend to 'all relevant scientific and technological developments, *inter alia* in the fields of microbiology, genetic engineering and biotechnology'.[70] The prohibitions and obligations under the BWC therefore in principle extend to all new development in synthetic biology and manipulation of potential living weapons and their products.

Dual-use

The prohibitions in the BWC are negatively circumscribed by the General Purpose Criterion discussed as part of the Framework, above. During the negotiations of the Convention it was clarified that 'prophylactic' encompasses medical activities such as diagnosis, therapy, and immunization, whereas 'protective' covers the development of protective means, including vaccines, and warning devices, but must not be interpreted as permitting possession of biological agents and toxins for defence, retaliation, or deterrence.[71] Protective purposes may imply offensive intent. Research and development for defences against biological weaponry are highly ambiguous activities and their pursuit may mislead others in the absence of transparency. The term 'other peaceful purposes' was not defined during the negotiations, but may be understood to include scientific experimentation.

It may be suggested that bioengineered threats are more sophisticated and challenging in terms of dual-use than their twin weapons of mass destruction. In several respects, the risks associated with bioweapons resemble cyber warfare rather than other WMDs. They are diffuse and far-reaching, largely falling outside the remit of states' capabilities to monitor, detect, and deter.[72]

The BWC process may serve as a multilateral basis for DURC (dual-use-research-of-concern) related dialogue, providing a platform accessible to almost all authorities globally. However, the process is weak and unable to provide verification akin to that ensured for nuclear and chemical weapons. The 2001 REVCON established an intersessional process of annual meetings. Until this point, BWC member states had undertaken reviews of new scientific and technological developments relevant to the Convention as part of the five-yearly reviews. Following repeated calls for a more robust process, developments in science and technology are now discussed annually at the Meeting of Experts and Meeting of States Parties.[73] A standing agenda item on developments in the field of science and technology related to the Convention was included into the programme for 2012–2015.[74]

The dual-use problems associated with chemical weapons, and which contributed to making their ban technically more complicated in the 1970s, are now being adsorbed into the BWC regime through synthetic biology.[75] Synthetic biology therefore raises new and vexing questions about the appropriate balance between diffusion and appropriate control of such technological advancements.[76]

Fault lines

A less benign effect of such broad strokes in a convention is the uncertainties and cracks that may arise at its edges. In this sense, synthetic biology exposes the existing framework of the BWC to a set of challenges, two of which are highlighted below. First, synthetic biology challenges the fault lines between the BWC and the CWC and the grey area between biological and chemical weapons. Second, synthetic biology exposes the relationship between the BWC and the Rome Statute of the International Criminal Court and is likely to affect the shortcomings of the Statute's regulation of biological weapons.

Biological or chemical weapons?

A key component of the BWC is that it bans the development, production, and possession of an entire class of weapons unconditionally – in *all* circumstances. The BWC contains no exception for weapons intended for specific purposes or assignments short of armed conflict. It prohibits any weaponisation of disease agents or germs, irrespective of context or effects, earning the ban the label of categorical. The absence of intended lethality does not influence the lawfulness of a weapon under the BWC. The prohibition therefore also extends to non-lethal weapons, such as biologically engineered microbes, that could attack an enemy's food or fuel, or germs that could corrupt enemy vehicle tyres or belts.[77] On this point, the prohibitions under the BWC are far more extensive than under the CWC.

However, while the General Purpose Criterion defines the scope of the Convention, the BWC does not offer a specific definition of biological weapons.[78] It extends its general prohibitions to all natural or artificially created microbial or other biological agents or toxins whatever their origin or method of production. Both the BWC and the CWC absorbed prohibitions on toxins from the 1925 Geneva Protocol and, taken together, the joint regime of the BWC and CWC was certainly intended to broadly cover the development of new chemical and biological weapons.

The CWC covers three key areas of importance to synthetic biology: (1) it bans development, production, and possession of chemical weapons; (2) it defines chemical weapons to include delivery systems;[79] and (3) it clarifies that peaceful development of pharmaceutical, medical, and agricultural chemicals is not impacted. Unlike the BWC, the CWC does not subject weapons to a categorical ban, but prohibits the use of chemical weapons as means of warfare.[80] Certain chemical weapons for law-enforcement purposes are therefore lawful, within the 'types and quantities' limits.

A major feature of synthetic biology is that it challenges the grey area between biological and chemical weapons, raising the question about whether synthetic biology may in some circumstances be reframed as chemical rather than biological, falling under the scope of the CWC instead of the BWC. The shared intention in the BWC and CWC of banning the weaponisation of biochemistry does not change the fact that the legal effects under the two Conventions differ. If synthetic biology is reframed, it may fall under the prohibitions of the CWC rather than the BWC. Consequently, weapons based on synthetic biology would arguably be *permitted* for use with law enforcement purposes.[81]

A different challenge is linked to what is understood as a weapon. While traditional biological and chemical agents were used against enemy soldiers or non-cooperative civilians and clearly would classify as weapons, modern agents may be used to 'enhance' the capability of a state's own military forces, i.e., performance enhancement of troops. It is unlikely that such agents would amount to 'weapons' under international law. Armour, for example, is not classified as such. While a liberal interpretation of the BWC in this regard is not straightforward, the CWC would open such possibilities.[82]

Synthetic biology is exposing existing prohibitions in the BWC to unprecedented pressure because it may erase and blur the lines between biological and chemical weapons. While other convergent technologies such as nanotechnology, which represents a considerable cross-over between chemistry, biology, and physics, may escape the definitions altogether,[83] synthetic biology is likely to subject the definitional scope of BWC Article I and CWC Article II to substantial pressure.[84] Reframing would subject the relevant technology or weapon to the inspection regime of the CWC, including both mandatory declarations rather than nonbinding CBMs and also challenge inspections (none as yet carried out), which are now only available for enforcement of the BWC as an action of the UN Security Council. However, the outcome of such a development may end up relativising, and therefore weakening, the categorical prohibition against biological weapons.

War crime?

Under BWC Article IV, member states have committed themselves to take *any* necessary measures to prohibit and prevent activities in contravention of the BWC on their territory. In other words, they are not only to respond to prohibited activities but also to stop them from happening. An important mechanism of enforcement is individual criminal liability, offering both punitive and preventive effects. UN Resolution 1540 urges member states to adopt and enforce effective laws prohibiting the use of biological weapons.[85] The Resolution also establishes an ad hoc committee (the 1540 Committee) to promote and verify compliance by member states. Follow-up resolutions by the UNSC have asked the 1540 Committee to promote individual criminal liability.[86] About one third of States Party to the BWC have criminalised the use of biological weapons.[87]

Criminalisation at the international level, as an international crime or war crime, provides the strongest and most effective measure for individual liability for violations of international law. The strength of the categorical ban on biological weapons under international law, combined with the emphasis on criminalisation under the BWC regime, leads to an expectation of a stringent prohibition on biological weapons also under international criminal law. However, neither weaponisation of biology nor use of biological weapons has been comprehensively criminalised in the Rome Statute of the International Criminal Court (ICC).[88] The use of 'poison or poisoned weapons', a prohibition first codified in 1899, *is* stipulated as a war crime.[89] Another paragraph is derived from the 1925 Geneva Protocol, making the use of asphyxiating, poisonous, or other gases, and all 'analogous liquids, materials or devices' a war crime. The provision notably does *not* refer to the use of bacteriological weapons, which is prohibited in the Geneva Protocol, and the Statute makes no further reference to either chemical or biological weapons. Some commentators maintain that biological weapons are nevertheless included – relying on the premise that the term 'poisoned weapon' was the first prohibition of both chemical and biological weapons.[90] However, most commentators conclude that biological weapons are not included in the Rome Statute.[91]

When the Rome Statute was negotiated in 1998, criminalising the use of non-conventional weapons of war was 'one of the most controversial issues' in the discussions.[92] Controversy over nuclear weapons led to the exclusion of explicit references to chemical or biological weapons. This position evolved from a view that 'if nuclear weapons were not to be included, then the poor person's weapons of mass destruction, chemical and biological weapons, should not be either'.[93] A review of the negotiating process shows that delegates did not openly object to listing biological weapons use as a war crime. When the final result ended up not including it, there was rather a general dismay at its removal.[94] The non-comprehensive regulation of biological weapons under international criminal law is not due therefore to any opposition to the categorical ban on biological weapons but rather to perceived links with other weapons of mass destruction.[95]

At the first review conference of the Rome Statute in 2010, a Belgian proposal suggested expanding prohibitions with direct references to use of biological and chemical weapons.[96] The amendment proposed making the use of 'the agents, toxins, weapons, equipment and means of delivery as defined by and in violation of' the BWC and 'chemical weapons as defined by and in violation of' the CWC a war crime prosecutable under the ICC. The proposal was not adopted. There was also an objection to the reference to the BWC, as this would be 'tantamount to compulsory universalization' of the treaty.[97] The proposed amendment was abandoned during the review conference, as it was deemed too complicated for a swift treatment.[98] However, the conference did extend existing prohibitions related to poison and gases to protracted non-international armed conflict.[99]

The absence of a provision explicitly making the use of biological weapons an international crime under the Rome Statute was an increasingly striking lacuna in the international legal regulation of biological weapons. Prosecution for the use of a biological weapon under the Rome Statute would depend on the effects of its use. The decisive element would consequently be whether the effects of use constituted violations of international humanitarian law. For example, a weapon that may not be directed at, or with effects that may not be limited to, a specific military objective, would be unlawful as an indiscriminate weapon. Faced with the advent of synthetic biology, this shortcoming risked impacting the overall approach to biological weapons. In 2017 Belgium renewed its proposal to amend article 8 of the Rome Statute.[100] This time the proposal was adopted and is pending ratification in 2019.[101] After entry into force, 'employing weapons, which use microbial or other biological agents, or toxins, whatever their origin or method of production' will be a war crime for which the ICC has jurisdiction.

The provision has been introduced in two separate, equally worded paragraphs. The first incriminates use of biological weapons in case of an international armed conflict (article 8, para. 2, b). The second paragraph extends the jurisdiction of the ICC to the employment of such weapons in case of a protracted armed conflict not of an international character (article 8, para. 2, e). While the ban on biological weapons in the BWC is categorical and not restricted to armed conflict, the new prohibition on use in the Rome Statute to the contrary is restricted to international or protracted non-international armed conflicts. As indicated above, certain sub-categories of use of biological weapons may pose risks in situations beyond armed conflicts, in hybrid situations not covered by the jurisdiction of the Court under the new provisions. Moreover, the framing of biological weapons use as an international crime restricted to armed conflict tilts the treatment of biological weapons under international law in a direction common to chemical weapons.

Conclusion

In the near future, it is likely that synthetic biology will make it possible to create dangerous viruses from scratch. However, as we have argued in this chapter, while synthetic biology is 'de-skilling' the science, it is not doing it to the extent that people with no specialist training operating outside of professional scientific institutions can assemble biological parts into circuits, devices, and systems that will reliably perform desired functions in live organisms. And even professionals will have a hard time creating radically new pathogens or synthetic 'super-pathogens'.

The technical context suggests that the most significant security threat from synthetic biology comes from professional and well-resourced institutions such as national militaries. This is backed up by the historical record of both biological weapons development and bioterrorism incidents. The potential for state use of biological weapons, conventionally or synthetically produced, is, however, extremely low. This is primarily because the norm against biological warfare – encoded in law through the Biological Weapons Convention – is so exceptionally strong. Traditionally, biological weapons were also judged to have limited military utility. We argue, however, that there might be a small subset of contemporary conflicts in which bioweapons do have military utility, and this possibility, however small, makes it imperative that we tighten up the international legal framework.

Synthetic biology is recasting the scenery of the BWC. Although the Convention offers a strong and comprehensive ban on the use of synthetic biology to weaponise pathogens, synthetic biology is blurring the lines between biological and chemical weapons, and risks relativising the existing categorical ban on biological weapons. Certain weaknesses in the international legal regime prohibiting biological weapons – particularly in terms of definitions and enforcement – are likely to become more prominent as the engineering of biology becomes more and more of a reality.

Notes

1 Some elements of this chapter have been drawn from previously published work, in particular Catherine Jefferson, Filippa Lentzos and Claire Marris, 'Synthetic biology and biosecurity: Challenging the "myths"', *Frontiers in Public Health*, August 2014.
2 Luis Campos, 'That was the synthetic biology that was', in Markus Schmidt, Alexander Kelle, Agomoni Ganguli-Mitra and Huib de Vriend (eds.), *Synthetic Biology: The technoscience and its societal consequences* (New York: Springer, 2009).
3 Maureen A. O'Malley, Alexander Powell, Jonathan F. Davies and Jane Calvert, 'Knowledge-making distinctions in synthetic biology', *BioEssays*, Vol. 30, No. 1 (2008), pp. 57–65.
4 Royal Academy of Engineering, *Synthetic Biology: Scope, Applications and Implications* (London: Royal Academy of Engineering, 2009).
5 Har Ghobind Khorana, 'Total synthesis of a gene', *Science*, 16 Feb. 1979, pp. 614–25.
6 Wlodek Mandecki, Mark A. Hayden, Mary Ann Shallcross et al., 'A totally synthetic plasmid for general cloning, gene expression and mutagenesis in *Escherichia coli*', *Gene*, Vol. 94, No. 1 (28 Sept. 1990), pp. 103–7.
7 Jeronimo Cello, Aniko V. Paul and Eckard Wimmer, 'Chemical synthesis of poliovirus cDNA: Generation of infectious virus in the absence of natural template', *Science*, 9 Aug. 2002, pp. 1016–18.
8 Hamilton O. Smith, Clyde A. Hutchinson III, Cynthia Pfannkoch et al., 'Generating a synthetic genome by whole genome assembly: φX174 bacteriophage from synthetic oligonucleotides', *Proceedings of the National Academy of Sciences*, Vol. 100, No. 26 (3 Nov. 2003), pp. 15440–45.
9 Jingdong Tian, Hui Gong, Nijing Sheng et al., 'Accurate multiplex gene synthesis from programmable DNA microchips', *Nature*, 23 Dec. 2004, pp. 1050–54.
10 Sarah J. Kodumai, K. G. Patel, R. Reid et al., 'Total synthesis of long DNA sequences: Synthesis of a contagious 32-kb polyketide synthase gene cluster' *Proceedings of the National Academy of Sciences*, Vol. 101, No. 44 (17 Sept. 2004), pp. 15573–8.
11 Martin Enserink, 'Viral fossil brought back to life', *Science Now*, 1 Nov. 2006, available at http://news.sciencemag.org/sciencenow/2006/11/01-04.html.
12 Nyssa Skilton, 'Man-made SARS virus spreads fear', *Canberra Times*, 24 Dec. 2008.
13 Daniel G. Gibson, Gwynedd A. Benders, Cynthia Andrews-Pfannkoch et al., 'Complete chemical synthesis, assembly, and cloning of *Mycoplasma genitalium* genome', *Science*, 29 Feb. 2008, pp. 1215–20.
14 Daniel G. Gibson, John I. Glass, Carole Lartique et al., 'Creation of a bacterial cell controlled by a chemically synthesized genome', *Science Express*, 20 May 2010, p. 1. See also Elizabeth Pennisi, 'Synthetic genome brings new life to bacterium', *Science*, 21 May 2010, pp. 958–9.
15 Narayana Annaluru, Héloïse Muller, Leslie A. Mitchell et al., 'Total synthesis of a functional designer eukaryotic chromosome', *Science*, 4 April 2014, pp. 55–8.
16 Puping Liang, Yanwen Xu, Xiya Zhang et al., 'CRISPR/Cas9-mediated gene editing in human tripronuclear zygotes', *Protein & Cell*, Vol. 6, Issue 5 (May 2015), pp. 363–72.
17 Filippa Lentzos, Catherine Jefferson and Claire Marris, 'The myths (and realities) of synthetic bioweapons', *The Bulletin of Atomic Scientists*, published online 18 Sept. 2014: http://thebulletin.org/myths-and-realities-synthetic-bioweapons7626; Jefferson et al., 'Synthetic biology and biosecurity', p. 115 (see note 1 above).

18 T.S. Gardner et al., 'Synthetic biology: from hype to impact', *Trends in Biotechnology*, Vol. 31, Issue 3 (2013), pp. 123–5, quoted in *Nature Reviews Microbiology*, Vol. 12, No. 5 (2014), available at www.nature.com/nrmicro/journal/v12/n5/full/nrmicro3261.html#ref2.
19 National Science Advisory Board for Biosecurity (NSABB), *Addressing Biosecurity Concerns Related to the Synthesis of Select Agents* (Bethesda, MD: National Institutes of Health, 2006), p. 4.
20 Kathleen Vogel, 'Bioweapons proliferation: Where science studies and public policy collide', *Social Studies of Science*, Vol. 36, No. 5 (2006), p. 676.
21 Jonathan B. Tucker and Raymond A. Zilinskas, 'The promise and perils of synthetic biology', *The New Atlantis*, Spring 2006, p. 38.
22 US Department of State, *President's Statement on Biological Weapons, 1 November 2001*, available from http://2001–2009.state.gov/t/ac/rls/rm/2001/7907.htm.
23 WMD Commission, *World at Risk: The Report of the Commission on the Prevention of WMD Proliferation and Terrorism* (New York: Vintage Books 2008), p. xv.
24 T. K. Sell and M. Watson, 'Federal agency biodefense funding, FY2013-FY2014', *Biosecurity & Bioterror*, Vol. 11, No. 3 (2013), pp. 196–216.
25 WMD Center, *Bio-Response Report Card* (Washington, DC: Bipartisan WMD Terrorism Research Center, 2011), p. 11.
26 'Frist Calls for "Manhattan Project for the 21st Century"', Selections from Senator Frist's Remarks Delivered on 1 June 2005 at the Harvard Medical School Health Care Policy Seidman Lecture, available from http://votesmart.org/public-statement/101572/frist-calls-for-manhattan-project-for-the-21st-century#.U8PCI6jj7sI.
27 US Department of State, *Opening Statement to the BWC Seventh Review Conference*, delivered by Hillary Clinton, Secretary of State, 7 Dec. 2011, Geneva, Switzerland, available from: https://geneva.usmission.gov/2011/12/07/statement-by-secretary-clinton-at-the-7th-biological-and-toxin-weapons-convention-review-conference/.
28 Ibid.
29 See the discussion in Chapter 18, which refers to the initial alarming news stories surrounding this event.
30 Filippa Lentzos, 'The risk of bioweapons use: considering the evidence base', *BioSocieties*, Vol. 9, Issue 1 (2014), pp. 84–93.
31 Mark Wheelis and Masaaki Sugishima, 'Terrorist use of biological weapons', in Mark Wheelis, Lajos Rozsa, and Malcolm Dando (eds.), *Deadly cultures: Biological weapons since 1945* (Cambridge, MA: Harvard University Press, 2006); W. Seth Carus, *Bioterrorism and Biocrimes: The illicit use of biological agents since 1900* (Washington, DC: Center for Counterproliferation Research, National Defense University, February 2001 revision); Jonathan B. Tucker, 'Introduction', in Jonathan B. Tucker (ed.), *Toxic Terror: Assessing terrorist use of chemical and biological weapons* (Cambridge, MA: MIT Press, 2000).
32 Sonia Ben Ouagrham-Gormley, *Barriers to Bioweapons: The challenge of expertise and organization for weapons development* (Ithaca, NY: Cornell University Press, 2014); Kathleen Vogel, *Phantom Menace or Looming Danger? A new framework for assessing bioweapons threats* (Baltimore, MD: Johns Hopkins University Press, 2013).
33 Milton Leitenberg, 'Evolution of the current threat', in Andreas Wenger and Reto Wollenmann (eds.), *Bioterrorism: Confronting a complex threat* (London: Lynne Rienner, 2007), p. 41.
34 Statement by Carl W. Ford, Assistant Secretary of State for Intelligence and Research before the Senate Committee on Foreign Relations hearing on 'Reducing the Threat of Chemical and Biological Weapons', 19 March 2002.
35 US Central Intelligence Agency, 'Unclassified Report to Congress on the Acquisition of Technology Relating to Weapons of Mass Destruction and Advanced Conventional Munitions': 1 January through 20 June 2003, available at www.cia.gov/library/reports/archived-reports-1/jan_jun2003.pdf, accessed 17 April 2015; 1 July through 31 December 2003 available at www.cia.gov/library/reports/archived-reports-1/721report_july_dec2003.pdf (accessed 17 April 2015).
36 Statement by Lt. General Michael Maples, Director, Defense Intelligence Agency, before the Committee on Armed Service hearing on 'Current and future worldwide threats to the national security of the United States', 27 February 2007.
37 US Department of State, 'Adherence to and compliance with arms control, nonproliferation and disarmament agreements and commitments' July 2014, available at www.state.gov/documents/organization/230108.pdf (accessed 17 April 2015).

38 Milton Leitenberg, 'Assessing the threat of bioterrorism', in Benjamin H. Friedman, Jim Harper and Christopher A. Preble (eds.), *Terrorizing ourselves: Why US counterterrorism policy is failing and how to fix it* (Washington, DC: Cato Institute, 2010), p. 5.
39 Judith Miller, Stephen Engelberg and William Broad, *Germs: The Ultimate Weapon* (New York: Simon & Schuster, 2001).
40 Ibid.
41 Arati Prabhakar, Director of the Defense Advanced Research Projects Agency, Department of Defense, testimony before the US Armed Services Committee Subcommittee on Intelligence, 'Emerging Threats & Capabilities hearing on 'Department of Defense (DOD) Fiscal Year 2015 Science and Technology Programs: Pursuing Technological Superiority in a Changing Security Environment' on 26 March 2014.
42 Ibid.
43 Erica Check Hayden, 'Bioengineers debate use of military money', *Nature*, Vol. 479, Issue 7374 (2011).
44 'Being strong: National security guarantees for Russia', 20 Feb. 2012, available at http://rt.com/politics/official-word/strong-putin-military-russia-711/, accessed 23 Jan. 2015.
45 Lentzos, 'The risk of bioweapons use' (see note 30 above).
46 Iris Hunger, Jez Littlewood, Caitriona McLeish, Piers Millett, and Ralf Trapp 'Roundtable: Bioweapons non-proliferation at the crossroads', in Filippa Lentzos (ed.), *Biological Threat in the 21st Century* (London: Imperial College Press, forthcoming).
47 Ibid.
48 Statement by Carl W. Ford (see note 34 above).
49 Filippa Lentzos, 'The risk of bioweapons use', p. 86 (see note 30 above).
50 Ibid., p. 87.
51 Milton Leitenberg and Raymond A. Zilinskas, *The Soviet biological weapons program: A history* (Cambridge, MA: Harvard University Press, 2012), p. 282.
52 Jean-Marie Henckaerts and Louise Doswald-Beck (eds.), *Customary International Humanitarian Law, Volume I: Rules* (Cambridge: Cambridge University Press for the International Committee of the Red Cross, 2005) (originally published 2004), p. 257.
53 UN General Assembly, 'A Question of chemical and bacteriological (biological) weapons', 16 December 1969, UN Doc. A/RES/2603(XXIV)A.
54 International Court of Justice, 'Legality of the Threat or Use of Nuclear Weapons, Advisory Opinion of 8 July 1996', *ICJ Reports*, p. 226.
55 Henckaerts and Doswald-Beck, *Customary International Humanitarian Law* (see note 52 above).
56 US Department of State, 'Narrative on the BWC', reprinted in Thomas Graham Jr. and Damien J. Lavera, 'Cornerstones of Security: Arms Control Treaties in the Nuclear Era' (Washington, DC: Nuclear Threat Initiative, 2003), p. 192, see www.nti.org/analysis/reports/cornerstones-security-arms-control/.
57 Gro Nystuen, Stuart Casey-Maslen, and Annie Golden Bersagel (eds.), *Nuclear Weapons under International Law* (Cambridge: Cambridge University Press, 2014), pp. 394–6.
58 Walter Krutzsch and Ralf Trapp, '*A Commentary on the Chemical Weapons Convention* (Dordrecht: Nijhoff, 1994).
59 The expression can be traced back to a UK working paper presented at the Eighteen Nation Disarmament Committee, precursor to the current Conference on Disarmament, in 1968, see Document ENDC/231, para. 3, reproduced in SIPRI, *The Problem of Chemical and Biological Warfare, Volume IV CB Disarmament Negotiations, 1920–1970* (Almqvist & Wiksell, Stockholm, 1971), pp. 255–6.
60 Jean-Pascal Zanders (ed.), *Multi-stakeholdership in the BTWC: Opportunities and Challenges*, (Paris: Institute for Security Studies, Egmont paper series, 2011), p. 39.
61 Piers Millet, 'The Biological Weapons Convention: From international obligations to effective national action', *Applied Biosafety*, Vol. 15, No. 3 (2010), pp. 113–18.
62 They are considered to be understandings under article 31(3) (a) of the Vienna Convention on the Law of Treaties 1969, 115 U.N.T.S. 331, see e.g. George Nolte, *Treaties and Subsequent Practice* (Oxford: Oxford University Press, 2013), p. 371.
63 For an overview, see Piers Millett, 'The Biological Weapons Convention: Securing biology in the twenty-first century', *Journal of Conflict & Security Law*, Vol. 15, Issue 1 (2010), p. 33.
64 Fourth Review Conference, 25–6 Dec. 1996, *Final Declaration*, BWC/CONF.IV/9 Part II.
65 Ibid.
66 Sixth Review Conference of the States Party to the BWC, Geneva, Switzerland, 2 Nov.–8 Dec. 2006, *Final Document*, at 9, U.N. Doc. BWC/CONF.VI/6, article I(2) (2006).

67 Second Review Conference of the States Party to the BWC, Geneva, Switzerland, *Final Document*, at U.N. Doc. BWC/CONF.II/5.
68 BWC/CONF.VI/6, article I(2) (2006).
69 Ibid.
70 Third Review Conference of the States Parties to the BWC, Geneva, Switzerland, 9–27 Sept.1991, *Final Document*, U.N.Doc. BWC/CONF.III/23 Part II.
71 Jozef Goldblat, 'The Biological Weapons Convention: An overview', *International Review of the Red Cross*, No. 318, 1997.
72 United Nations Interregional Crime and Justice Research Institute, *Security Implications of Synthetic Biology and Nanobiotechnology: A Risk and Response Assessment of Advances in Biotechnology* (Turin, Italy: UNICRI, 2011), 62, available at http://igem.org/wiki/images/e/ec/UNICRI-synNanobio-final-2-public.pdf.
73 Seventh Review Conference of the State Parties to the BWC, *Final Document*, 2011, U.N. Doc. BWC/CONF.VII/7, D.22.
74 Ibid.
75 See Markus Schmidt and Gregor Giersch, *DNA Microarrays, Synthesis and Synthetic DNA* (Hauppauge, New York: Nova Science Publishers, 2011).
76 Gill Bates, 'Introduction to SIPRI Yearbook 2008' (Stockholm: SIPRI, 2008), p. 2; Ronald Sutherland, *Chemical and Biochemical Non-Lethal Weapons: Political and Technical Aspects*, SIPRI Policy paper (Stockholm: SIPRI, 2008), p. 23.
77 David Koplow, *Death by Moderation* (Cambridge: Cambridge University Press, 2010), p. 208.
78 Michael Bothe, Natalino Ronzitti, and Allan Rosas (eds.), *The New Chemical Weapons Convention: Implementation and Prospects* (The Hague: Kluwer Law International, 1998), p. 37.
79 *Convention on the Prohibition of the Development, Production, Stockpiling and Use of Chemical Weapons and on Their Destruction* (hereafter CWC), 1974, U.N.T.S. 45, Article II.
80 CWC, Articles I(5) and II(9).
81 International Committee of the Red Cross, *Toxic Chemicals as Weapons for Law Enforcement; A Threat to Life and International Law?* (Geneva: ICRC, 2012), p. 4, available at www.icrc.org/en/document/toxic-chemicals-weapons-law-enforcement-threat-life-and-international-law.
82 Mark Wheelis and Malcolm Dando, 'Neurobiology: A case study of the imminent militarization of biology', *International Review of the Red Cross*, No. 859 (Sept. 2005), pp. 562–3.
83 See *inter alia* Evan Wallach, 'A Tiny Problem with Huge Implications: Nanotech agents as enablers or substitutes for banned chemical weapons: is a new treaty needed', *Fordham International Law Journal*, Vol. 33, No. 858 (2009), warning that nanotechnology may design materials that act like chemical agents, but are not classed as such under existing protocols at p. 859.
84 Ibid.
85 UN. S/RES/1540, 28 April 2004.
86 UN. S/RES/1673, 27 April 2006.
87 Henckaerts and Doswald-Beck, Chapter 23, 73.IV (see note 52 above).
88 Use of biological weapons will in many cases be covered by *other* provisions of the Rome Statute of the International Criminal Court, such as Article 8(2)b) xx), prohibiting methods and materials of warfare that are of a nature to cause superfluous injury or unnecessary suffering, or are inherently indiscriminate, if and when an annex has been agreed to the provision.
89 Rome Statute of the International Criminal Court, Article (2)(b)(xvii).
90 Malcolm Dando and Kathryn Nixdorff, 'Chapter 1. An Introduction to Biological Weapons', in Kathryn McLaughlin and Kathryn Nixdorff (eds.), *BWPP Biological Weapons Reader* (Geneva: BioWeapons Prevention Project, 2009), 2, available at www.icrc.org/eng/assets/files/other/irrc_859_whelis_dando.pdf, accessed 27 Aug. 2015; Michael Cottier, 'War crimes: Article 5', in Otto Triffterer (ed.),*Commentary on the Rome Statute of the International Criminal Court, Observers' Notes, Article by Article*, 2nd edition (Oxford: Hart Publishing, 2008), 413.
91 Marcus Wagner, 'The ICC and its jurisdiction – myths, misperceptions and realities' in A. von Bogdandy and R. Wolfrum (eds.), *Max Planck Yearbook of United Nations Law*, Vol. 7 (The Netherlands, Koninklijke Brill N.V., 8 April 2003), p. 460.
92 Cottier, 'War crimes: Article 5', p. 415 (see note 90 above).
93 Ibid., p. 376.
94 Ibid., 412. One response was created in 2001 by a group of legal experts organised by Matthew Meselson and Julian Robinson, under the auspices of the Harvard-Sussex Program on CBW Armament and Arms Limitation, as 'A Draft Convention to Prohibit Biological and Chemical Weapons Under International Criminal Law', see www.fas.harvard.edu/~hsp/crim01.pdf, accessed 27 August 2015.

95 See Annie Golden Bersagel, 'Use of nuclear weapons as a an international crime and the Rome Statute of the International Criminal Court', in Nystuen et al., *Nuclear Weapons under International Law*, pp. 221–45.
96 Belgium, Draft Amendments to the Rome Statute on War Crimes, Amendment 2, 29 Sept. 2009.
97 Assembly of States Parties, Eighth Session, *Report of the Bureau on the Review Conference*, ICC-ASP/8/43, 15 Nov. 2009, para 33.
98 Kara Allen, Scott Spence and Rocío Escauriaza Leal, 'Chemical and biological weapons use in the Rome Statute – a Case for Change' *Vertic Brief* 14, Feb. 2011.
99 Rome Statute, Articles 2(e)(xiii) and (xiv).
100 Proposal of amendments, Belgium, Depositary notification 17. July 2017 C.N 480.2017.TREATIES-XVIII.10.
101 Resolution ICC-ASP/16/Res.4, 16th Session of the Assembly of State Parties to the Rome Statute, New York, December 2017.

18

A THREAT ASSESSMENT OF BIOLOGICAL WEAPONS

Past, present and future

Matteo Bencic Habian

In Western culture, the catastrophic consequences on the fabric of society caused by an epidemic of major proportions are well documented in historical and literary accounts. Illustrious authors such as Thucydides, Lucretius, Giovanni Boccaccio, and Alessandro Manzoni, have described the dreadful devastation – both in the body and the mind – suffered by those living under the plague.[1] Indeed, over the course of history, the plague has intermittently ravaged Europe, causing a total of 50 million deaths in the fourteenth century alone, which amounts to approximately 60 percent of the entire European population of that time.[2]

In recent decades, the relentless development of biotechnology introduced a novel dilemma: could the next great epidemic be engineered in the laboratories of a rogue state or terrorist group? Or even on the computer screen of a single individual?

The very nature of biological weapons makes them morally repugnant, since taking advantage of 'the very bacteria, viruses and toxins that have threatened life from the beginning is to deal with the enemies of mankind'.[3] Nonetheless, the use of pathogens as weapons is an ancient military practice, and for centuries armies have tried to weaken their enemies by employing these 'silent' killers.

The nineteenth century is considered to be the turning point for virology. In particular, as a result of the work of Louis Pasteur and Robert Koch, germ theory of disease finally began to be taken seriously by the scientific community and virology emerged as a full-fledged scientific discipline.[4] Tragically, with the turn of the twentieth century and the beginning of World War I, national armies employed scientific discoveries in the field of virology for the purpose of developing biological weapons. In 1925, the international community sought to put a halt to the development of such weapons by signing the *Protocol for the Prohibition of the Use in War of Asphyxiating, Poisonous or other Gases, and of Bacteriological Methods of Warfare* (henceforth the '1925 Geneva Protocol'), which prohibited the use of chemical and biological weapons in international armed conflicts. Notwithstanding the prohibition, over the course of World War II, all the major parties involved in the conflict developed their own biological weapons programmes.

In 1972, the signing of the Convention on the Prohibition of the Development, Production and Stockpiling of Bacteriological (Biological) and Toxin Weapons and on their Destruction (henceforward the 'BWC') provided some optimism in respect of the possibility of the permanent elimination of biological weapons. However, the failure to establish an international body capable of monitoring and enforcing compliance with the BWC and the complicated issue of defining the boundaries of offensive/defensive biological research represented significant setbacks. The BWC therefore remained a convention 'with no teeth'.[5]

If during the twentieth century the development of biological weapons was only conceivable at the state level, the twenty-first century brought new security concerns, which today appear more relevant than ever. The evolution and spread of international terrorism, the increasing accessibility to fast modes of transport (e.g. air travel), and the introduction of novel means of communication (e.g. the internet), as well as breakthrough discoveries in the fields of biotechnology, genetic engineering, virology, and 3D printing, have caused the emergence of unprecedented threats to international peace and security.

This chapter provides an assessment of the possibility of a biological attack from a threefold perspective: a state-sponsored biological attack, bioterrorism and bio-hacking. The research question that this chapter seeks to answer is whether states remain the sole actors capable of developing biological weapons or whether new actors have actually acquired – or may be in the process of acquiring – such capacity.

The threat of the past: state-sponsored biological attacks

The use of pathogens in warfare dates back to ancient times. Early references even appear in the biblical book of Exodus, where it is told that in order to force the Pharaoh to release the Jews from slavery, Yahweh inflicted ten plagues upon the Egyptian population, one of which caused the appearance of 'boils breaking out in sores' on men and animals.[6] In antiquity, the weaponisation of diseases was of course rudimentary and occurred through the poisoning of arrows and spears, as well as the contamination of wells and food supplies. In the Middle Ages, plague-infected cadavers were catapulted into enemy camps or besieged cities.[7] It is also documented that during the colonisation of America, the British army distributed smallpox-infected blankets to the Native American tribes.[8] Table 18.1 lists some significant events in the history of biological warfare.

However, it was not until World War I that 'biological warfare became a science'.[9] The most memorable case is that of Germany, which was accused of infecting enemies' livestock and crops, although it avoided targeting humans directly.[10] Following the end of World War I, the international community signed the 1925 Geneva Protocol, prohibiting the use of chemical and biological weapons. Yet no restrictions on research and stockpiling were included in the treaty. By the time World War II was over, all the major parties involved – including the US, the UK, Canada, France, the Soviet Union, Germany, and Japan – had developed their own biological weapons programmes.[11]

Table 18.1 F. Frischknecht, 'The history of biological warfare' (2003) 4 European Molecular Biology Organization Reports, Special Issue, 47

Year	*Significant event*
1155	Emperor Barbarossa poisons water wells with human bodies in Tortona, Italy.
1346	Mongols catapult bodies of plague victims over the city walls of Caffa, in the Crimean Peninsula.
1495	Spanish mix wine with blood of leprosy patients to sell to their French foes in Naples, Italy.
1650	Polish fire saliva from rabid dogs towards their enemies.
1675	First agreement between German and French forces not to use 'poison bullets'.
1763	British distribute blankets from smallpox patients to native Americans.
1797	Napoleon floods the plains around Mantua, Italy, to enhance the spreading of malaria.
1863	Confederates sell clothing from yellow fever and smallpox patients to Union troops in the USA.

Japan's biological weapons programme was named 'Unit 731'. It was led by Dr Shiro Ishii and was particularly active during the 1937–1945 Sino-Japanese War.[12] Throughout that conflict, Japan was responsible for experimenting with biological agents on prisoners and attacking Chinese villages in Manchuria with weapons spreading the *Yersinia pestis* bacterium (which causes the bubonic plague) as well as typhoid, cholera, and anthrax.[13] It is difficult to determine the exact number of casualties caused by Japanese attacks, as most of the information regarding Unit 731 has been destroyed.[14] Nonetheless, the final death toll has been estimated to be in the tens, if not hundreds, of thousands.[15]

In 1942, the US established its own offensive biological weapons programme, which was set in the Chemical Warfare Service at Camp Detrick in Frederick, Maryland. The US biological weapons programme was kept secret throughout World War II,[16] and was only made public in 1946.[17] The programme was expanded during the Cold War, but in 1969, under Nixon's presidency, the US government took the unprecedented decision to unilaterally renounce its offensive biological weapons programme, thus formally abandoning an entire category of armament.[18] This new attitude of the American administration promoted the negotiations of the BWC, which was signed in 1972, banning the development, production, stockpiling, and transfer of biological weapons.

The Soviet biological weapons programme was established in the late 1920s.[19] While it initially lacked sophistication and expertise, it later surpassed the US efforts and became the world's most comprehensive biological weapons programme.[20] Unlike the US, the Soviet Union continued the development of an offensive biological weapons programme even after the signing of the BWC. The illegal top-secret programme, named 'Biopreparat', operated under civilian cover from 1972 until at least 1992, reaching enormous dimensions by the late 1980s and early 1990s.[21] Indeed, it is documented that at its peak the programme employed more than 60,000 people and stockpiled hundreds of tons of anthrax spores and tens of tons of other pathogens (e.g. smallpox and plague).[22]

With the termination of the Cold War, the international community's concerns shifted towards Saddam Hussein's Iraq. It is reported that Iraq first established a secret biological weapons programme in 1974, at the Ibn Sina Centre.[23] After various setbacks, major research activity began in 1985 and by April 1991, three pathogens – botulinum, anthrax, and aflatoxin – had successfully been selected for weaponisation.[24]

Following the 1991 Gulf War, the UN adopted Security Council Resolution 687 (1991), which, *inter alia*, ordered Iraq to ratify the BWC and unconditionally accept the destruction of all its chemical and biological weapons and the establishment of an international monitoring system. The UN Special Commission ('UNSCOM') started inspecting Iraq's weapons of mass destruction ('WMDs') capacity and discovered that Iraq's biological weapons programme was much more advanced than expected.[25] While Saddam Hussein initially agreed to the monitoring regime set under UNSCR 687 (1991), he pursued a policy of constant deception, obstruction and intimidation towards the international inspectors.[26] The UN eventually withdrew their inspectors in late 1998, shortly before the US and UK launched Operation Desert Fox, targeting suspected biological and chemical weapons facilities. Iraq's non-compliance led to the establishment of the UN Monitoring, Verification and Inspection Commission ('UNMOVIC'), set up under UNSCR 1284 (1999). UNMOVIC's task was to replace UNSCOM and verify Iraq's compliance with UNSCR 687 (1991). Nevertheless, Iraq rejected UNSCR 1284 (1999) and inspectors were not allowed back into the country.

In a final attempt to solve the Iraqi issue, the UN Security Council unanimously passed UNSCR 1441 (2002), which, acting under Chapter VII and recalling UNSCR 678 (1990),[27] asserted that Iraq had been 'in material breach of its obligations under relevant resolutions, including resolution 687'. UNSCR 1441 (2002) therefore offered Iraq 'a final opportunity to comply' with its disarmament obligations. Notwithstanding Iraq's improved cooperation, a coalition formed by the USA, the UK, and Australia – with the support of some forty countries – decided to resort to force and invaded Iraq on 20 March 2003. The coalition justified Operation Iraqi Freedom as a revival of the authority to use force provided in Resolution 678 (1990).[28] In 2005, the Iraq Survey Group published the final version of the so-called Duelfer Report, which stated that Saddam's WMD stocks had effectively been destroyed by 1991.[29]

A recent report by the US Department of State concluded that North Korea continues to develop a biological weapons programme and may consider the use of biological weapons as a legitimate warfare option.[30] The document also raised concerns about Russia's inconsistent compliance with the BWC and the unverified destruction of Soviet bio-agents stockpiles, as well as the possibility that the Assad regime in Syria may consider the employment of biological weapons as a military option.[31]

The ambivalence of the BWC, which on the one side prohibits any kind of offensive biological weapons programme, but on the other hand allows biodefence research, further complicates the picture. Indeed, 'there is a legitimate concern that defensive research undertaken in one country's program could be misperceived as offensive, or potentially offensive, in character and drive other nations to pursue offensive research as well'.[32] More broadly, the ambiguous nature of biological research makes it almost impossible to define the criteria that make a bioweapons programme defensive rather than offensive. As long as the international community lacks an efficient and impartial monitoring system, questions about the nature of specific programmes, as well as the capacity and intention of certain states to produce and employ biological weapons, will be left unanswered.

Regardless of one's conception of the international system (e.g. anarchic structure, international society, etc.), states generally behave in a much more predictable way than non-state actors. This does not mean that it is easy to foresee the way in which any state will act in the medium or even short term. Nonetheless, states will seek to avoid (or at least limit) conducts which have been blatantly condemned by the international community and banned under international law (e.g. the use of WMDs).[33] Indeed, states will necessarily have to take into account the political, diplomatic, and military consequences that a breach of such a fundamental prohibition would have both on the national and international stage. In the overwhelming majority of cases, states (and their political and military leaders) will thus act rationally, avoiding international condemnation, diplomatic isolation, and the potential consequences of an international military intervention.[34] As we shall see, in exceptional cases, non-state actors can be immune to the threat of such consequences.

Current threat: bioterrorism

As a matter of fact, the most extreme terrorist groups (which can be regarded as 'death cults') seem to be impermeable to the strategy of deterrence. This derives from their uncompromising political or religious stance, as well as their lack of concern about retaliation and international condemnation.[35] The use of biological weapons may – in principle – appear especially appealing to them in light of the apocalyptic and indiscriminate nature of such weapons, which can be used to target and spread terror amongst unprepared non-combatants. In that sense, biological weapons represent the ideal terrorist device to be employed in an asymmetric warfare against a stronger opponent (i.e. a traditional army).

In August 2014, a laptop belonging to a fighter from the so-called Islamic State of Iraq and al-Sham (henceforth 'ISIS') was seized by a moderate rebel group in Northern Syria. The so-called 'laptop of doom' allegedly contained among its files an Arabic-written document, describing how to manufacture a biological weapon employing the bubonic plague from infected animals, as well as a video illustrating the procedure needed to produce ricin.[36] The document also suggested how to maximise civilian loss of life, by targeting confined and crowded spaces, such as underground train stations, stadiums or shopping malls. It was later revealed that the author of the document (and owner of the laptop) was a young Tunisian national who, before joining ISIS, studied chemistry and physics at university level.[37]

In the wake of the Paris terrorist attacks of 13 November 2015, the French Prime Minister, Manuel Valls, warned that the risk of a biological attack perpetrated by ISIS against the French population was indeed very real.[38] Nonetheless, there still is a lack of consensus on whether ISIS – or any other terrorist organisation for that matter – has ever been in possession or would be capable of producing bio-WMDs in the short term. While some experts argue the threat of a biological terrorist attack may be imminent,[39] others label the same threat as a very unlikely or remote event.[40] In December 2015, a report commissioned by the European Parliament claimed that the probability that ISIS could carry out a CBRN (chemical, biological, radiological or nuclear) attack on European soil had increased and that the European Union should prepare accordingly.[41]

ISIS is not the first terrorist group which has attempted to obtain biological weapons. Indeed, in the 1990s, the Aum Shinrikyo doomsday cult infamously tried to cultivate pathogens – in particular the botulinum toxin and *Bacillus anthracis* spores – in order to launch mass-casualty attacks against Japanese citizens.[42] The Aum was eventually capable of both producing the botulinum toxin and obtaining anthrax spores (probably with the assistance of someone working in the laboratories of Obihiro University).[43] Fortunately, all biological attacks failed, mainly because the Aum lacked the practical expertise and tools necessary to manufacture an aerosol system capable of effectively dispersing the pathogens in the air.[44]

Aum's failure is representative of the obstacles that non-state actors face when they try to develop biological weapons. Indeed, effective dissemination of the agents produced probably represents the toughest challenge to overcome. Nevertheless, Aum's failure must be put into context. First of all, more than 25 years have passed since the inception of Aum's biological weapons programme. Since then, biotechnology has developed dramatically and has arguably become a much more accessible discipline. Moreover, Aum's programme only lasted five years, was fairly small both in terms of funding and personnel, it was carried out in a rather disorganised fashion and was founded on the principle of total autarky: Aum produced its own pathogens and built its own aerosol dispensers.[45] Furthermore, there was a clear misallocation of resources, as the cult lacked a strategy for the development of WMDs and often committed to bizarre sci-fi weaponry projects.[46]

As Tucker and Zilinskas noted, assembling an effective biological weapon is indeed a much more complicated endeavour than 'simply' producing or obtaining an infectious agent. It involves the manufacturing of an elaborate system which requires: 1) a sufficient quantity of the agent (either in the form of wet slurry or dry powder); 2) a complex mixture of chemical additives, which allows the pathogen to preserve its infectivity and virulence during the storage; 3) a container to store and transport the agent; 4) an efficient dissemination device, which allows the pathogen to disperse in the air in fine particles and infect the targeted population through respiration. Weather conditions are critical as well, as the aerosol cloud can only be released under particularly favourable atmospheric and meteorological circumstances.[47] While this is specifically true with regard to the distribution of pathogens such as anthrax, it is less relevant when it comes to highly infectious diseases transmissible from person to person, e.g. smallpox or Ebola. In these cases, 'even a localised release, especially in a mobile population, could trigger a widespread epidemic'.[48]

In any case, manufacturing a functioning biological weapon requires interdisciplinary expertise, which is indeed not easy to obtain. Some authors correctly highlighted the fact that one of the most significant obstacles to the production of biological weapons is represented by *tacit knowledge*.[49] According to Jefferson et al., tacit knowledge refers to 'skills and techniques that cannot be readily codified but, rather, are acquired through a process of "learning by doing" or "learning by example," and often take considerable time and effort to gain'.[50] Following this line of thought only highly specialised individuals, operating within either governmental or academic laboratories and working in professional teams would be capable of producing biological weapons.

Jansen et al. also pointed out that it is particularly hard for terrorists to develop biological weapons because they 'operate within the borders of a nation that may seek to destroy them'. The paucity of resources – both from a financial and know-how perspective – and the necessity to conduct their work concealed and always concerned about the intrusive action of the hosting state hinder the capacity of non-state actors to produce, store, and use biological weapons.[51] Consequently, terrorists generally tend to opt for more conventional weapons.

Although correct, these interpretations risk depicting a rather anachronistic situation, as they minimise the impact of potential future developments in the fields of life sciences and technology. Mukunda et al. observed that '[t]he advance of technology normally converts tacit to explicit knowledge over time' and in that sense, '[s]ynthetic biology is unique [...] in the extent to which it is explicitly devoted to the minimization of the importance of tacit knowledge'.[52] Moreover, even though it is certainly true that tacit knowledge (or rather the lack of it) serves as a deterrent for aspiring bioterrorists, it does not follow that terrorist organisations would remain idle and not strive to acquire such tacit knowledge. It should be kept in mind that the pilot hijackers involved in the 9/11 terrorist attacks patiently committed to learning how to fly planes, solely for the sake of turning such planes into weapons and maximising the number of civilian deaths. While the complexity of learning how to fly a plane is (at the moment) hardly comparable to the complexity of mastering genetic manipulation and manufacturing biological weapons, new technological developments are inevitably rendering such activities more accessible.

In relation to the production of effective aerosol dispensers, the rise of the 3D printing technology (also known as additive manufacturing) represents a further point of concern for the security community.[53] Additive manufacturing has already been employed to assemble 3D printed firearms and it can be used to design and reproduce virtually any part, gear, or mechanism (such as complex aerosol systems), which have always been extremely hard to manufacture.[54]

After reaching its maximum extension in October 2014, ISIS has gradually lost considerable portions of the territory it once controlled. This retreat has culminated in the fall of the Caliphate's de facto capital Raqqa, in October 2017.[55] Even though ISIS may be beaten, it is not yet defeated. The terrorist quasi-state may indeed want to resort to an extraordinarily brutal attack, with the aim of 're-establishing its brand' and re-gaining support. Speculations aside, it is important to determine what made ISIS particularly dangerous when it came to obtaining biological weapons, as new terrorist organisations possessing equivalent traits may arise and pose similar threats in the future.

ISIS's capacity to acquire biological weapons must thus be analysed in light of the terrorist group's distinctive features: 1) apocalyptic ideology, 2) large territory control, 3) access to weapons arsenals, 4) extraordinary level of funding, and 5) recruiting of highly educated individuals.

The abovementioned 'laptop of doom' not only contained the practical instructions for manufacturing a biological weapon, but also contained the moral and religious justification – in the form of a *fatwa* – for using such weapons against civilian unbelievers. The *fatwa* was issued by Nasir al-Fahd, an Islamic cleric currently jailed in Saudi Arabia, and reads as follow: 'If Muslims cannot defeat the *kafir* [i.e. unbelievers] in a different way, it is permissible to use weapons of mass destruction ... Even if it kills all of them and wipes them and their descendants off the face of the Earth'.[56] In general, religious doomsday cults are unconcerned with international condemnation and have proven particularly attracted to WMDs especially because of the apocalyptic effects that may result from their use against civilians.[57]

ISIS's control over the territory around its de facto capital Raqqa and the city of Mosul may have lasted for a period of time not long enough for the terrorist organisation to develop biological weapons. Nevertheless, between January 2014 and October 2017, ISIS maintained a core area around Raqqa and did not experience a pressure tantamount to that of other terrorist groups, which must constantly deal with the intrusive action of their host states. The concern is that ISIS could have used occupied academic or governmental laboratories, such as the University of Mosul, to carry out research on pathogens.[58] It is further well documented that many of Saddam's former generals and high-ranking military officers – some of whom have been involved in Iraq's biological and chemical weapons programmes – have formed the core of ISIS's military elite.[59]

Because ISIS has expanded its political and geographical control in Syria, Iraq and Libya filling in the power vacuum caused by the collapse (or weakening) of rogue regimes and their military apparatus, it has also seized the opportunity to take control of substantial weapons arsenals.[60] These concerns, combined with the confirmation by Iraqi officials that ISIS took control of the Muthanna former chemical weapons facility and has obtained 'low grade' nuclear material from the University of Mosul,[61] led Wolfgang Rudischhauser, Director of the WMD Non-proliferation Centre at NATO, to warn that '[w]e might soon enter a stage of CBRN terrorism, never before imaginable'.[62]

Compared to any other terrorist organisation, ISIS has also benefited from unequalled levels of funding, which derive from the regular collection of taxes, donation from wealthy international supporters, black-market oil trading, ransom extortions, human smuggling and trafficking, robberies, and other criminal activities.[63] Furthermore, ISIS's capacity to recruit foreign fighters has been particularly alarming. Indeed, some of these foreign fighters have received scientific training from European universities and – under the direction of former Iraqi top officers involved in Iraq's biological weapons programme – may have been employed to manufacture biological weapons.[64]

History has shown that defection from scientists working in Western countries, who later decide to join authoritarian or theocratic regimes and establish WMDs programmes is also possible. The most notorious case is that of Dr Abdul Qadeer Khan, who after stealing the plans for a nuclear centrifuge from his work site at Urenco in the Netherlands, ended up establishing the Pakistani atomic weapons programme in 1976.[65] Not only Qadeer Khan made Pakistan a nuclear power, but between 1987 and 2002, he also set up the so-called 'Khan Network', exporting sensitive nuclear material, technologies and expertise to countries such as Iran, Libya, and North Korea.[66]

Even more worrying is the fact that the fate of the approximately 60,000 Soviet scientists employed in the top-secret Russian Biopreparat biological weapons programme is still essentially unknown. There certainly is a possibility that some of those scientists may have been recruited by terrorist or criminal organisations. Furthermore, the complete destruction of stocks of pathogens produced under Biopreparat, such as the tens of tons of smallpox, has never been verified.[67] It is thus very possible that some of that hazardous material may still be 'active' and potentially available on the black market or ready to be smuggled by transnational criminal organisations.[68]

The spreading of digital technology offers an additional tool to the aspiring bioterrorist from a threefold perspective. First of all, the so-called 'dark web' provides a safe haven for criminals, smugglers, drug lords, and arms dealers (including those wanting to sell dual-use biological material). In addition to the actual hazardous material, terrorists also use the dark web to exchange information in the form of documents, training videos, and tutorials.[69] Secondly, the laboratory equipment (such as polymerase chain reaction machines) and the raw material necessary to produce toxins can already be purchased from biotech companies online, absolutely legally and at a fairly low cost.[70] Thirdly, the use of online hacking for industrial espionage and theft of intellectual property has become a common criminal activity.[71] As such, there is an increasing risk that terrorists may take advantage of the inadequate cyber security infrastructure of pharmaceutical firms, agriculture companies or hospital laboratories and steal the know-how necessary to produce pathogens.[72]

Many are the hurdles that an aspiring bioterrorist must overcome in order to manufacture a functioning biological weapon. However, it is undeniable that the development of new technologies is actually allowing an increasing number of individuals to access information and material to an unparalleled extent. It follows that historical precedents should not play a significant role when analysing the threat of bioterrorism. As M. Wheelis and M. Sugishima rightly put it 'it would be a mistake to assume that the sparse record of bioterrorism to date accurately predicts the future'.[73]

Future threat: bio-hacking

Between September and October 2001, letters containing anthrax spores were sent to newspaper offices and to two US senators. As a result of the inhalation of the spores, 5 people died and 17 were injured. In August 2008, the FBI and the Department of Justice announced that charges were going to be brought solely against Dr Bruce Edward Ivins. Dr Ivins committed suicide before the charges could be filed against him. In 2010, the Department of Justice formally announced the conclusion of the investigations, explaining that the collected evidence found that 'Dr Bruce Ivins acted alone in planning and executing these attacks'.[74] The validity of such evidence, as well as the scientific methods used by the FBI throughout the investigations, have been thoroughly contested and in December 2014, the US Government Accountability Office released a report criticising the genetic testing used by the FBI.[75] Regardless of the identity of the perpetrator, the 2001 anthrax attack brought the threat of bio-hacking on the front pages of newspapers and in national security agendas.

In their report, titled *Globalization, Biosecurity, and the Future of the Life Sciences*, the US Institute of Medicine and the National Research Council claimed that '[s]ooner or later, it is reasonable to expect the appearance of "biohackers," mirroring the computer hackers that repeatedly cause mischief today through the creation of a succession of more and more sophisticated computer "viruses".'[76]

When devising a biological weapon, biohackers will inevitably face even greater challenges than bioterrorists, i.e. reduced number of trained individuals, lesser level of funding, nonexistent (or weaker) links with international criminal organisations. Nevertheless, laboratory equipment and raw material for the production of pathogens are already widely available on the internet and are fairly cheap to buy. In 2013, after purchasing on eBay the equipment necessary to set up a DIY biology lab for approximately GBP 3000, three BBC journalists ordered from a biotech company two tubes containing the beginning and end portions of the DNA for making ricin for approximately GBP 20. In theory, that is all a biohacker would need to produce ricin.[77]

Tucker and Zilinskas distinguished between two possible scenarios with regard to the threat of bio-hacking. The first option is that of a so-called 'lone wolf'; that is, a highly qualified, typically above-suspicion biologist, who has developed some kind of mental illness or obsessive resentment against a specific group of people or against society as a whole.[78] This possibility is particularly alarming since such an individual not only would have easy access to raw material and to state-of-the-art laboratory equipment, but he or she would also be supposed to possess the tacit knowledge necessary to produce pathogens. The abovementioned factors, paired with a tendency to act solo, would make any plan devised by a 'lone wolf' extremely difficult to uncover.

The second scenario is that of a proper 'biohacker'; that is, 'an individual who does not necessarily have malicious intent but seeks to create bioengineered organisms out of curiosity or to demonstrate his technical prowess – a common motivation of many designers of computer viruses'.[79]

As previously stated, biohackers (as intended in the second scenario above) would necessarily encounter critical hurdles of both a logistical and financial nature, not to mention a potentially insurmountable lack of tacit knowledge, which would significantly hinder their capacity to handle pathogens. Nonetheless, initiatives such as the DIYbio movement and the iGEM competition aim specifically at de-skilling and democratising biology.[80]

Although it is very hard to determine the actual magnitude of the risk posed by bio-hacking, the very nature of disciplines such as synthetic biology, as well as the increasing accessibility in terms of information, raw material and laboratory equipment requires thorough monitoring and debate, both within the scientific and the security community. Albeit the threat may seem remote, it should not be underestimated.[81]

Conclusion

Since manufacturing biological weapons requires an interdisciplinary expertise, substantial funding, uncommon raw material and sophisticated laboratory equipment, states remain the principal actors capable of producing such weapons.[82] Accordingly, the international community has established a system, revolving around the BWC, to effectively prohibit the use of biological weapons in interstate conflicts. Any state employing biological weapons would thus face international condemnation and diplomatic isolation.

The situation with bio-terrorism is more complicated. Indeed, death cult terrorist groups appear to be impermeable to any threat of retaliation, international condemnation and isolation. Considering that terrorist groups such as ISIS and al Qa'ida, have demonstrated a keen interest in obtaining biological weapons and that they have also been capable of recruiting individuals trained in a wide variety of scientific disciplines, it appears that in the near future the bio-terrorist threat is more likely to increase, rather than decrease.

The bio-hacking threat appears to be less pressing and the potential for a disastrous attack is more contained. Nevertheless, it is essential to actively engage with the individuals and communities involved in the DIYbio movements and educate them in respect of the security risks involved with the scientific development in fields such as synthetic biology.

Considering the devastating consequences of a biological attack, with the stakes as high as they are, it would be a critical error to underestimate the capabilities of terrorist groups or biohackers to produce biological weapons.

Notes

1 Thucydides (*The History of the Peloponnesian War*, Book II) and Lucretius (*De Rerum Natura*, Book VI, verses 1145–1196) wrote about the Plague of Athens, which struck the city in 430 BC. In his masterpiece, the *Decameron*, Giovanni Boccaccio provided an incisive account of the consequences of the Black Death, which devastated Florence between 1348 and 1350. The final pages of Alessandro Manzoni's *The Betrothed* contain a description of the effects of the plague, which ravaged Milan between 1629 and 1631.

2 O. J. Benedictow, *The Black Death, 1346–1353: The Complete History* (Woodbridge, Boydell Press, 2006), p. 382.

3 T. Mangold and J. Goldberg, *Plague Wars: the terrifying reality of biological warfare* (New York, St Martin's Griffin, 1999) xi.
4 Germ theory of disease was firstly proposed by Girolamo Fracastoro in 1546.
5 J. B. Tucker, 'Putting Teeth in the Biological Weapons Ban' (1 January 1998) *MIT Technology Review*.
6 The Book of Exodus 9:8–9.
7 For an exhaustive historical overview of the development of biological weapons until the Cold War see: E. Geissler and J. E. van Courtland Moon (eds.), *Biological and Toxin Weapons: Research, Development and Use from the Middle Ages to 1945*, Oxford, Oxford University Press, 1999. See also: A. Gillespie, *A History of the Laws of War*, Vol. 3 (Hart Publishing, Oxford and Portland, Oregon, 2011), pp. 102–4. M. Wheelis, L. Rozsa and M. Dando (eds.), *Deadly Cultures: Biological Weapons since 1945*, Cambridge, Harvard University Press, 2006. G. W. Christopher, T. J. Cieslak, J. A. Pavlin and E. M. Eitzen Jr., 'Biological Warfare: A Historical Perspective', *The Journal of the American Medical Association*, Vol. 278, No. 5 (August 1997).
8 J. B. Tucker, *Scourge: The Once and Future Threat of Smallpox*, New York, Atlantic Monthly Press, 2001.
9 M. R. Hilleman, 'Overview: cause and prevention in biowarfare and bioterrorism', *Vaccine*, Vol. 20, August 2002, p. 3056.
10 Ibid. E. Geissler, 'Biological Warfare Activities in Germany, 1923–1945', in E. Geissler and J. E. van Courtland Moon, *Biological and Toxin Weapons: Research, Development and Use from the Middle Ages to 1945*, Oxford, Oxford University Press, 1999. National Research Council ('NRC'), *Biotechnology Research in an Age of Terrorism* (Washington, DC: The National Academies Press, 2004), p. 20, available at www.nap.edu/catalog/10827/biotechnology-research-in-an-age-of-terrorism.
11 NRC, *supra* note 10, at 20.
12 D. Barenblatt, *A Plague upon Humanity: The Hidden History of Japan's Biological Warfare Program*, New York: Harper Collins, 2005.
13 P. Williams and D. Wallace, *Unit 731: Japanese Army's Secret of Secrets*, London: Hodder & Stoughton Ltd, 1989. S. H. Harris, *Factories of Death: Japanese Biological Warfare, 1932–45 and the American Cover-Up* (New York, Routledge, 2002). M. R. Hilleman, 'Overview: cause and prevention', (see note 9 above), p. 3056.
14 According to Barenblatt, *A Plague upon Humanity* (see note 12 above) and Harris, *Factories of Death* (see note 13 above), the US army struck a deal with the scientists in charge of Unit 731, shielding them from accountability in exchange of the results of their research on biological weapons.
15 Hilleman, 'Overview: cause and prevention' (see note 9 above), p. 3056. According to Barenblatt, *A Plague upon Humanity* (see note 12 above), Unit 731 caused the infection of more than 250,000 people, the vast majority of whom died. According to Harris, over 200,000 people died as a consequence of Unit 731's biological attacks.
16 See for instance: B. Bernstein, 'America's biological warfare program in the Second World War', *Journal of Strategic Studies*, September 1988, pp. 292–317.
17 G. W. Merck, 'Official Report on Biological Warfare', *Bulletin of the Atomic Scientists*, Vol. 2, No. 7 (1 October 1946), p. 17.
18 J. B. Tucker and E. R. Mahan, *President Nixon's Decision to Renounce the U.S. Offensive Biological Weapons Program* (Center for the Study of Weapons of Mass Destruction, National Defense University, October 2009), p. 17.
19 For a detailed account on the Soviet biological weapons programme see M. Leitenberg and R. A. Zilinskas, *The Soviet Biological Weapons Programme: a history*, Cambridge: Harvard University Press, 2012.
20 Ibid, p. 698. NRC, *Biotechnology Research in an Age of Terrorism* (see note 10 above), p. 21. R. L. Frerichs, R. M. Salerno, K. M. Vogel, N. B. Barnett, J. Gaudioso, L. T. Hickok, D. Estes and D. F. Jung, *Historical precedence and technical requirements of biological weapons use: a threat assessment* (Sandia National Laboratories, 2004), p. 21.
21 K. Alibek and S. Handelman, *Biohazard: The Chilling True Story of the Largest Covert Biological Weapons Program in the World – Told from Inside by the Man Who Ran it*, New York: Delta, 1999.
22 NRC, *Biotechnology Research in an Age of Terrorism* (see note 10 above), p. 21. E. Croddy, *Chemical and Biological Warfare: A Comprehensive Survey for the Concerned Citizen* (New York: Springer, 2002), p. 235.
23 E. M. Spiers, *A History of Chemical and Biological Weapons* (London: Reaktion Books, 2011), p. 110. A. H. Cordesman, *Iraq's Past and Future Biological Weapons Capabilities*, Washington, DC: Center for Strategic and International Studies, February 1998.

24 G. S. Pearson, 'The Iraqi Biological Weapons Program', in M. Wheelis, L. Rozsa and M. Dando (eds.), *Deadly Cultures: Biological Weapons since 1945* (Cambridge: Harvard University Press, 2006), pp. 173–9. Spiers, *A History of Chemical and Biological Weapons* (see note 23 above), p. 112.

25 Council on Foreign Relations, *IRAQ: Weapons Inspections: 1991–1998*, 29 April 2003, available at www.cfr.org/iraq/iraq-weapons-inspections-1991-1998/p7705#p3.

26 Arms Control Association, *Iraq: A Chronology of UN Inspections*, Special Report, available at www.armscontrol.org/act/2002_10/iraqspecialoct02. Pearson, 'The Iraqi Biological Weapons Program' (see note 24 above), pp. 181–3.

27 UNSCR 678 (1990) acting under Chapter VII and recalling UNSCR 660 (1990), afforded Iraq an *ultimatum* to withdraw its troops from Kuwait before 15 January 1991, and authorised Member States to use 'all necessary means' in case of non-compliance.

28 C. Gray, *International Law and the Use of Force* (Oxford: Oxford University Press, 2008), p. 358.

29 Central Intelligence Agency (CIA), *Comprehensive Report of the Special Advisor to the DCI on Iraq's WMD, with Addendums (Duelfer Report)* (25 April 2005).

30 US Department of State, 'Adherence to and Compliance With Arms Control, Nonproliferation, and Disarmament Agreements and Commitments', (Department of State, Bureau of Arms Control, Verification and Compliance, July 2014), p. 14. A recent study by the Belfer Center for Science and International Affairs at Harvard Kennedy School has concluded that North Korea maintains an interest in developing biological weapons, although it is hard to assess North Korea's biological weapons capability without access to classified intelligence. See: H. K. Kim, E. Philipp and H. Chung, *The Known and Unknown: North Korea's Biological Weapons Program*, Cambridge, MA: Harvard Kennedy School, Belfer Center for Science and International Affairs, October 2017.

31 US Department of State, 'Adherence to and Compliance With Arms Control (see note 30 above), pp. 16–17. In March 2013, the Director of the US National Intelligence James R. Clapper stated that '[b]ased on the duration of Syria's longstanding biological warfare (BW) program, we judge that some elements of the program may have advanced beyond the research and development stage and may be capable of limited agent production. Syria is not known to have successfully weaponized biological agents in an effective delivery system, but it possesses conventional and chemical weapon systems that could be modified for biological agent delivery'. See: James R. Clapper, *Statement for the Record, Worldwide Threat Assessment of the US Intelligence Community*, Senate Committee on Armed Services, 18 April 2013.

32 Institute of Medicine and NRC, *Globalization, Biosecurity, and the Future of the Life Sciences* (Washington DC: The National Academies Press, 2006) p. 59, available at www.nap.edu/catalog/11567/globalization-biosecurity-and-the-future-of-the-life-sciences.

33 This appears to be less relevant when it comes to interstate conflicts, as it is notoriously documented that rogue regimes (e.g. Saddam's Iraq and Assad's Syria) have used chemical weapons against their own citizens.

34 This interpretation is in line with Iraq's decision not to use WMDs against US forces during the Gulf War. Indeed, the threat that the US would retaliate with tactical nuclear weapons appears to have persuaded Saddam not to deploy chemical or biological weapons against US troops. See: R. A. Zilinskas, *Biological Warfare: Modern Offense and Defense* (Boulder CO: Lynne Rienner Publishers, 1999), p. 195. The US former Secretary of State James Baker claimed that in a meeting with the Iraqi Deputy Prime Minister Tariq Aziz, he 'purposely left the impression that the use of chemical or biological agents by Iraq could invite tactical nuclear retaliation' (J. A. Baker, *The Politics of Diplomacy* (New York: G.P. Putnam, 1995), p. 359). According to the aforementioned Duelfer Report (p. 100), when Saddam was asked why he didn't launch a WMD attack against the US troops during the Gulf War, he reportedly replied 'Do you think we are mad? What would the world have thought of us? We would have discredited those who had supported us'.

35 W. Laqueur, 'The New Face of Terrorism', *Washington Quarterly*, Vol. 21, No. 4 (1998). Institute of Medicine and NRC, *Globalization, Biosecurity* (see note 32 above), p. 58.

36 H. Doornbos and J. Moussa, 'Recipes From the Islamic State's Laptop of Doom', *Foreign Policy*, 9 September 2014, available at https://foreignpolicy.com/2014/09/09/recipes-from-the-islamic-states-laptop-of-doom/. See also discussion in Chapter 17

37 H. Doornbos and J. Moussa, 'Found: The Islamic State's Terror Laptop of Doom', *Foreign Policy*, 28 August 2014, available at http://foreignpolicy.com/2014/08/28/found-the-islamic-states-terror-laptop-of-doom/.

38 *The Daily Telegraph*, 'French PM warns terrorists could use chemical and biological weapons', 19 November 2015, available at www.telegraph.co.uk/news/worldnews/europe/france/12005131/French-PM-warns-terrorists-could-use-chemical-and-biological-weapons.html.

39 See for instance: W. Rudischhauser, 'Could ISIL go nuclear?', *NATO Review Magazine*, May 2015, available at www.nato.int/docu/Review/2015/ISIL/ISIL-Nuclear-Chemical-Threat-Iraq-Syria/EN/index.htm. N. Bar-Yaacov, 'What if Isis launches a chemical attack in Europe?', *The Guardian*, 27 November 2015, available at www.theguardian.com/global/commentisfree/2015/nov/27/isis-chemical-attack-europe-public. W. Yeo, 'Salafi Jihadists and Chemical, Biological, Radiological, Nuclear Terrorism: Evaluating the threat', *Risk Management Solutions*, 24 August 2015, available at www.rms.com/blog/tag/terrorism-risk-2/.

40 J. Burke, 'Chemical weapons attack on Europe seems highly unlikely', *The Guardian*, 19 November 2015, available at www.theguardian.com/world/2015/nov/19/chemical-weapons-attack-europe-unlikely-france-isis. D. MacKenzie, 'ISIS chemical terror threat sounds alarming but is unlikely', *New Scientist*, 20 November 2015, available at www.newscientist.com/article/dn28527-isis-chemical-terror-threat-sounds-alarming-but-is-highly-unlikely/. C. Jefferson, F. Lentzos and C. Marris, 'Synthetic biology and biosecurity: challenging the "myths"', *Frontiers in Public Health*, August 2014. J. Parachini 'Combating Terrorism: Assessing the Threat of Biological Terrorism', *RAND*, October 2001.

41 B. Immenkamp, 'ISIL/Da'esh and "non-conventional" weapons of terror', *European Parliamentary Research Service*, December 2015.

42 R. Danzig, M. Sageman, T. Leighton, L. Hough, H. Yuki, R. Kotani and Z. M. Hosford, *Aum Shinrikyo: Insights Into How Terrorists Develop Biological and Chemical Weapons* (Washington, DC: Center for a New American Security, December 2012), pp. 18–28. P. C. Bleek, 'Revisiting Aum Shinrikyo: New Insights into the Most Extensive Non-State Biological Weapons Program to Date', *Nuclear Threat Initiative*, 11 December 2011, available at www.nti.org/analysis/articles/revisiting-aum-shinrikyo-new-insights-most-extensive-non-state-biological-weapons-program-date-1/. See also: A.T. Tu, 'Aum Shinrikyo's Chemical and Biological Weapons: More Than Sarin', *Forensic Science Rev*iew, Vol. 2, No. 26 (July 2014).

43 Danzig et al., *Aum Shinrikyo*, p. 25 (see note 42 above).

44 Ibid., pp. 36–7.

45 Ibid., p. 25.

46 Danzig et al. depict the situation as follows (Ibid., p. 20): 'There did not appear to be a strategy for the choice or use of weapons of mass destruction, but simply a fascination with these tools and an attempt to bring them into reality. The interaction between Asahara [i.e. the leader of the cult] and his scientists has been compared to kids playing in a school yard, excited by the prospect of building and using new technology for its own sake. Leaders often got bored with one "toy" if there were any difficulty involved and went on to the next one with passion … The soundest generalization is that Aum took an erratic course, rather than adopting a methodical research and development program. Different members pursued different projects with widely varying enthusiasms and organizational support.'

47 J. B. Tucker and R. A. Zilinskas, 'The Promise and Perils of Synthetic Biology', *The New Atlantis*, Spring 2006, p. 39.

48 M. Rees, *Our Final Century* (London: Arrow, 2004), p. 51.

49 G. Mukunda, K. A. Oye and S. C. Mohr, 'What rough beast? Synthetic biology, uncertainty, and the future of biosecurity', *Politics and the Life Sciences*, Vol. 28, No. 2 (2009), p. 14. J. B. Tucker, 'Could Terrorists Exploit Synthetic Biology?', *The New Atlantis*, Spring 2011, pp. 73–7. J. E. Suk, C. Bartels, E. Broberg, M. J. Struelens and A. J. Ozin, 'Dual-Use Research Debates and Public Health: Better Integration Would Do No Harm', in J. E. Suk, K. M. Vogel and A. J. Ozin, *Dual-use Life Science Research and Biosecurity in the 21st Century: Social, Technical, Policy, and Ethical Challenges* (Frontiers in Public Health, 2015), p. 47. K. M. Vogel, 'Bioweapons proliferation: Where science studies and public policy collide', *Social Studies of Science*, Vol. 36, No. 5 (2006), pp. 659–90. Jefferson et al., 'Synthetic biology and biosecurity' (see note 40 above).

50 Jefferson et al., 'Synthetic biology and biosecurity' (see note 40 above), p. 22.

51 H. J. Jansen, F. J. Breeveld, C. Stijnis and M. P. Grobusch, 'Biological warfare, bioterrorism, and biocrime', 20 *Clinical Microbiology and Infection*, No. 6, June 2014, p. 490.

52 Mukunda et al., 'What rough beast?' (see note 49 above). For a comment in support of the argument that biotechnology is undergoing a de-skilling process, see also: G. L. Epstein, 'The challenges of developing synthetic pathogens', *Bulletin of the Atomic Scientists*, 19 May 2008, available at https://thebulletin.org/2008/05/the-challenges-of-developing-synthetic-pathogens/.

53 See for instance M. Goodman, *Future Crimes* (London: Penguin Random House, 2015), pp. 461–5.

54 A. Majoran, 'Tech Terror: Understanding the Security Risks Posed by 3D Printed Firearms', The Mackenzie Institute, December 2015, available at http://mackenzieinstitute.com/tech-terror-understanding-the-security-risks-posed-by-3d-printed-firearms/. A. Greenberg, 'How 3-D Printed

Guns Evolved into Serious Weapons in Just One Year', *Wired*, May 2014, available at www.wired.com/2014/05/3d-printed-guns/. T. Campbell, C. Williams, O. Ivanova and B. Garrett, 'Could 3D Printing Change the World? Technologies, Potential, and Implications of Additive Manufacturing', *Atlantic Council*, October 2011.

55 M. Kranz and S. Gould, 'These maps show how drastically ISIS territory has shrunk since its peak', *Business Insider*, 24 October 2017, available at http://uk.businessinsider.com/maps-of-isis-territory-2014-2017-10?r=US&IR=T.

56 E. Stakelbeck, *ISIS Exposed: Beheadings, Slavery, and the Hellish Reality of Radical Islam* (Washington, DC: Regnery Publishing, 2015), p. 83.

57 Laqueur, 'The New Face of Terrorism' (see note 35 above). Yeo'Salafi Jihadists' (see note 39 above).

58 Doornbos and Moussa, 'Found: The Islamic State's Terror Laptop' (see note 37 above). Yeo, 'Salafi Jihadists' (see note 39 above).

59 L. Sly, 'How Saddam Hussein's former military officers and spies are controlling Isis', *The Independent*, 5 April 2015, available at www.independent.co.uk/news/world/middle-east/how-saddam-husseins-former-military-officers-and-spies-are-controlling-isis-10156610.html. S. Nakhoul, 'Saddam's former army is secret of success for Baghdadi's Islamic State', *Reuters*, 18 June 2015, available at http://blogs.reuters.com/faithworld/2015/06/18/saddams-former-army-is-secret-of-success-for-baghdadis-islamic-state/. S. Ackerman, 'Isis weapons engineer killed in airstrike in Iraq, claims US military', *The Guardian*, 31 January 2015, available at www.theguardian.com/world/2015/jan/31/senior-isis-militant-weapons-killed-abu-malik-airstrike-us-mosul-iraq. D. Smith, M. Chulov and S. Ackerman, 'Head of Isis chemical weapons program captured by US in Iraq last month', *The Guardian*, 9 March 2016, available at www.theguardian.com/world/2016/mar/09/isis-chemical-weapons-leader-captured-iraq-us-special-forces.

60 Amnesty International, *Taking Stock: The Arming of Islamic State*, December 2015. H. Doornbos and J. Moussa, 'How the Islamic State Seized a Chemical Weapons Stockpile', *Foreign Policy*, 17 August 2016, available at http://foreignpolicy.com/2016/08/17/how-the-islamic-state-seized-a-chemical-weapons-stockpile/.

61 Associated Press at the United Nations, 'Isis seizes former chemical weapons plant in Iraq', *The Guardian*, 9 July 2014, available at www.theguardian.com/world/2014/jul/09/isis-seizes-chemical-weapons-plant-muthanna-iraq. M. Nichols, 'Iraq tells UN that "terrorist groups" seized nuclear materials', *Reuters*, 9 July 2014, available at www.reuters.com/article/us-iraq-security-nuclear-idUSKBN0FE2KT20140709.

62 Rudischhauser, 'Could ISIL go nuclear?' (see note 39 above).

63 Financial Action Task Force, *Financing of the terrorist organisation Islamic State in Iraq and the Levant (ISIL)*, February 2015. B. Satti Charles, 'Funding Terrorists: The Rise of ISIS, Security Intelligence', *Security-Intelligence*, 10 October 2014, available at https://securityintelligence.com/funding-terrorists-the-rise-of-isis/. T. Brooks-Pollock, 'Paris attacks: Where does Isis get its money and weapons from?', *The Independent*, 16 November 2015, available at www.independent.co.uk/news/world/paris-attacks-where-does-isis-get-its-money-and-arms-a6736716.html. O. Williams-Grut, 'Here's where terrorist groups like ISIS and Al Qaeda get their money', *Business Insider*, 7 December 2015, available at http://uk.businessinsider.com/how-isis-and-al-qaeda-make-their-money-2015-12/#6-scamming-banks-1.

64 World Bank, *Economic and Social Inclusion to Prevent Violent Extremism*, October 2016, p. 16. L. Dearden, 'Isis documents leak reveals profile of average militant as young, well-educated but with only "basic" knowledge of Islamic law', *The Independent*, 21 April 2016, available at www.independent.co.uk/news/world/middle-east/isis-documents-leak-reveals-profile-of-average-militant-as-young-well-educated-but-with-only-basic-a6995111.html. Rudischhauser, 'Could ISIL go nuclear?' (see note 39 above). Yeo, Yeo, 'Salafi Jihadists' (see note 39 above).

65 D. Frantz and C. Collins, *The Nuclear Jihadist: The True Story of the Man Who Sold the World's Most Dangerous Secrets … And How We Could Have Stopped Him*, New York: Hachette, 2007. N. G. Evans, 'Contrasting Dual-Use Issues in Biology and Nuclear Science', in B. Rappert and M. J. Selgelid, *On the Dual Uses of Science and Ethics: Principles, Practices, and Prospects* (Canberra: The Australian National University Press, 2013), p. 265. S. Miller and M. J. Selgelid, 'Ethical and Philosophical Consideration of the Dual-use Dilemma in the Biological Sciences', *Science and Engineering Ethics*, Vol. 13, Issue 4 (Dec. 2007), pp. 524–5.

66 M. Kroening, *Exporting the bomb: Technology Transfer and the Spread of Nuclear Weapons* (New York: Cornell University Press, 2010), pp. 134–5. D. E. Sanger, 'The Khan Network', Paper presented at the Conference on South Asia and the Nuclear Future, Stanford Institute for International Studies (4–5 June 2004). W. Langewiesche, 'The Wrath of Khan', *The Atlantic*, November 2005, available at www.theatlantic.com/magazine/archive/2005/11/the-wrath-of-khan/304333/.

67 M. J. Selgelid, 'A Tale of Two Studies: Ethics, Bioterrorism, and the Censorship of Science', *Hastings Center Report*, Vol. 37, No. 3 (May-June 2007), p. 38. W. Orendt, *Plague: The Mysterious Past and Terrifying Future of the World's Most Dangerous Disease* (New York: Free Press, 2004), pp. 227–8. See also: 'Prepared Statement by Richard Preston', in *Biological Weapons: The Threat Posed by Terrorists – Congressional Hearing*, Serial no. J-105–97, 1998, p. 133.

68 M. Kilger, 'Evaluating technologies as criminal tools', in M. McGuire and T. Holt (eds.), *The Handbook of Technology, Crime and Justice* (New York: Routledge 2017), p. 341. R. Mowatt-Larssen, 'Al Qaeda's Pursuit of Weapons of Mass Destruction', *Foreign Policy*, 25 January 2010, available at https://foreignpolicy.com/2010/01/25/al-qaedas-pursuit-of-weapons-of-mass-destruction/.

69 Goodman, *Future Crimes* (see note 53 above), p. 50, points out that the internet has effectively become a 'terrorist university', where aspiring terrorists can find all the relevant information to make their attacks more efficient and deadlier. The United Nations Under-Secretary-General of Disarmament Affairs, Izumi Nakamitsu, during her submissions at the 7985th meeting of the UN Security Council has stressed that '[t]he global reach and anonymity of dark web provides non-State actors with new marketplaces to acquire dual-use equipment and materials'. See United Nations Security Council, 7985th meeting, 28 June 2017. In July 2015, a man was convicted in the UK of attempting to acquire Ricin on the dark web for unspecified purposes. See BBC, 'Breaking Bad fan jailed over Dark Web ricin plot', 18 September 2015, available at www.bbc.co.uk/news/uk-england-34288380.

70 See the next section of this chapter for further details. See Kilger, *supra* note 68, at 340–341.

71 Kilger, 'Evaluating technologies' (see note 68 above), p 341.

72 A. P. Acharya and A. Acharya, 'Cyberterrorism and Biotechnology. When ISIS Meets CRISPR', *Foreign Affairs*, 1 June 2017.

73 M. Wheelis and M. Sugishima, 'Terrorist Use of Biological Weapons' in M. Wheelis, L. Rozsa and M. Dando (eds.), *Deadly Cultures: Biological Weapons since 1945* (Cambridge: Harvard University Press, 2006), p. 284.

74 Department of Justice, *Justice Department and FBI Announce Formal Conclusion of Investigation into 2001 Anthrax Attacks,* 19 February 2010, available at www.justice.gov/opa/pr/justice-department-and-fbi-announce-formal-conclusion-investigation-2001-anthrax-attacks.

75 The US Government Accountability Office, *ANTHRAX: Agency Approaches to Validation and Statistical Analyses Could Be Improved,* 19 December 2014, available at www.gao.gov/products/GAO-15-80.

76 Institute of Medicine and NRC, *Globalization, Biosecurity* (see note 32 above), p. 50.

77 H. Charisius, R. Friebe and S. Karberg, 'Becoming biohackers: The experiments begin', BBC, 23 January 2013, available at www.bbc.com/future/story/20130123-hacking-genes-in-humble-settings.

78 Tucker and Zilinskas, 'The Promise and Perils' (see note 47 above), p. 40.

79 Ibid., p. 42.

80 The DIYbio movement can be defined as a biotech movement which involves individuals, communities, and small organisations in the study of biology and life science using the same methods as traditional research institutions. The iGEM competition is an annual, global, synthetic biology event in which multidisciplinary teams of undergraduate university students (as well as high school and graduate students) build genetically engineered systems using standard biological parts called biobricks.

81 See for instance J. Wikswo, S. Hummel and V. Quaranta, 'The Biohacker: A Threat to National Security', *CTC Sentinel*, 15 January 2014, available at https://ctc.usma.edu/the-biohacker-a-threat-to-national-security/.

82 Tucker, 'Could Terrorists Exploit' (see note 49 above), p. 77.

19

THE SYNTHETIC BIOLOGY DILEMMA

Dual-use and the limits of academic freedom

Guglielmo Verdirame and Matteo Bencic Habian

Should the dissemination of scientific knowledge and technological innovation be subject to restrictions when there is a risk of such knowledge being misused? And what restrictions, if any, should we put in place when the nature of the knowledge is such that a small group of individuals (or even an individual alone) can cause significant loss of life and damage to property if they are only able to acquire the right, and ever more affordable, equipment? Can these restrictions be compatible with the view that '[s]cience is the search for truth, that is the effort to understand the world: it involves the rejection of bias, of dogma, of revelation, but not the rejection of morality'?[1]

The potential for dual-use is inherent in scientific knowledge. Virtually '[e]very major technology – metallurgy, explosives, internal combustion, aviation, electronics, nuclear energy – has been intensively exploited, not only for peaceful purposes but also for hostile ones'.[2] In the words of the physicist Richard Feynman, scientific knowledge has 'an enabling power to do either good or bad – but it does not carry instructions on how to use it'.[3]

The discovery of nuclear fission, and its weaponisation since 1945, brought the dilemma of dual-use into sharper relief in the modern era. Interestingly, in the years preceding this discovery, as scientists' understanding of nuclear chain reaction developed, an intense debate took place in the scientific community about the responsibilities of scientists. Some – among them Leo Szilard who would later become one of the leading scientists in the Manhattan Project – were of the view that self-censorship about the discoveries was necessary to protect the public. Enrico Fermi, by contrast, maintained that any form of censorship was unscientific; the debate was however cut short when the details of a similar discovery were published in France.[4]

Once the genie was out of the bottle, the risk of dual-use could no longer be addressed by self-censorship – in the words of Martin Rees: 'Nuclear weapons can be dismantled, but they cannot be uninvented'.[5] The debate on the management of the nuclear threat moved on to deterrence, disarmament and non-proliferation. But the concept of dual-use continued to inform important policy choices, most notably the *Treaty on the Non-Proliferation of Nuclear Weapons* (NPT). Concluded in 1968, it is based on a strategic bargain between nuclear-weapon States and non-nuclear weapon States and on the distinction between good use (i.e. atoms for peace) and bad use (i.e. nuclear weapons).

Dual-use is now being discussed in other scientific contexts, including life sciences and biology. Developments in these areas could potentially benefit millions of people, enhancing the quality of life and providing new cures for diseases. But the same developments also create 'new opportunities for inappropriate and malicious use',[6] some of which are discussed by Lowe in his contribution to this book.[7] In the worst case scenario, there is a possibility that 'the life sciences could be transformed into the 'death sciences'.[8]

The objective of this chapter is to examine the tension between scientific development and security in the context of scientific and technological innovation in the life sciences. In the first section, we discuss the concept of dual-use with a focus on synthetic biology. In the second section, we present the debate which rages between the scientific and national security spheres regarding the limitation of certain dual-use research of concern (DURC). We will use the H5N1 controversy as an illustration. In the third section, we discuss academic freedom and explore the legal and moral challenges of imposing limits on grounds of national security and public health. As we shall see, human rights law is relevant to the management of risks of dual-use and, in particular, to the balancing of fundamental rights and national security considerations. In spite of its relevance in principle, however, there is still considerable uncertainty about the proper interpretation of human rights law in this context.

The dual-use dilemma and biology

Synthetic biology still lacks an agreed definition, in part due to the cross-boundary nature of the subject. In this chapter, we use synthetic biology to mean 'the engineering of biology: the synthesis of complex, biologically based (or inspired) systems, which display functions that do not exist in nature'.[9] Some claim that transformations across this field may be so momentous that we are witnessing a shift comparable with the transformation of chemistry following the introduction of the periodic table in 1869.[10] With the rise of synthetic biology, they claim, biology might 'ultimately [...] become a mechanistic science',[11] – an 'engineering-based methodology [...] used to build sophisticated, computing-like behaviour into biological systems'.[12]

While biology and nuclear science may be similar in terms of destructive potential,[13] the former may offer greater good use potential than the latter. True, since its discovery, nuclear technology has been peacefully employed for energy and radioisotope production; but the range of benefits which had been foreseen by the media in the 1950s (e.g, cars and airplanes that would 'run on vitamin-sized nuclear pellets')[14] have not materialised.[15] Many expect synthetic biology to revolutionise day-to-day life from the health sector (with the creation of new enhanced vaccines and drugs) to the energy and chemical industry (with the production of biofuels and new environmentally friendly materials).[16] Some even maintain that it may become as pervasive as modern computer science.[17]

Whether these predictions will come true or not is difficult to tell, but there is one thing that is perhaps already evident. Notwithstanding the importance of nuclear energy, it is still possible for countries such as Japan and Germany to decide to phase out this source of energy without anyone questioning their credentials as socio-economically advanced countries. By contrast, a country that chose to opt out of synthetic biology would almost certainly have those credentials called seriously into question.

There is a further important difference between nuclear energy and life sciences. Whilst dual-use nuclear energy discoveries have historically been kept classified, those in the life sciences have not been subject to an analogous regulation, and knowledge-sharing has essentially been the rule.[18]

This situation poses distinct challenges in the area of life sciences. Although sixty years have gone by since nuclear weapons were first invented, states have managed to restrict access to both information and material, but this same approach offers fewer chances of success in the field of biotechnologies for two reasons. First of all, the vast majority of scientific information regarding biology is readily available on the internet. This is part of a trend that may be described as the *democratisation* of life sciences. *Democratisation* in this context does not mean that synthetic biology is experiencing a *de-skilling* process of such a degree to allow *anybody* to 'engineer biology' over the short and mid-period,[19] but rather that scientific knowledge is accessible and, at the same time, equipment is becoming more affordable to small organisations and even individuals.[20] Indeed, initiatives such as the 'open science' and DIYbio movements, as well as competitions such as the iGEM,[21] aim at allowing any biology enthusiast – regardless of his academic and professional background – to perform scientific experiments and participate in the general growth of life sciences. DNA sequences can be purchased from online suppliers, which deliver their products via the postal service; with many companies now providing synthesis services prices have plunged.[22] Used DNA sequencers and other equipment can also easily be purchased on the internet.

Secondly, the benefits to human welfare which synthetic biology offers make it particularly difficult to place limits on its expansion.[23] As discussed below, the tension between security interests and the advancement of scientific knowledge, and the promotion of human welfare through it, is acute.[24] However, '[m]isuse of dual-use research of concern is [...] a low-probability but potentially high-consequence event',[25] which calls for the establishment of an effective risk-benefit assessment practice.

A parallel can be drawn here with computer science. The digital revolution has allowed essentially anyone with basic software skills to enjoy many services offered by modern computers. However, because computer science has become increasingly complex, not *anyone* who intends to do *digital harm* has the necessary skills and expertise to become a hacker or cyberterrorist. But some determined cyber criminals have been successful in committing cyber crimes.

The debate: academic freedom v security

In life sciences the dual-use risk often arises simply from knowledge. How can we manage risk then? Is there any way we can ensure that knowledge is put to *good use*?

In Western culture at least, knowledge is generally considered to be intrinsically positive and closely linked to human flourishing. As Deborah G. Johnson put it, 'The search for knowledge, in Western thought, is noble and enduring, even to some, the very meaning of life'.[26] There is a concept of *forbidden* in the Western tradition,[27] but in these religious or literary accounts centred on the association of knowledge with sin and even evil the morale seems to be that attempts to forbid knowledge ultimately fail: Adam, Prometheus, and Faust were not restrained.

This idea of inherent worth of knowledge dominates the sciences too. As Michael J. Selgelid pointed out, 'Scientists commonly believe that knowledge is good in itself and that both freedom of inquiry and the free sharing of information are essential to the purity and progress of science'.[28] The risk which derives from this attitude is that of separating science from its social and ethical dimension. In a speech given to the Association of Los Alamos Scientists, Robert Oppenheimer – who is considered to be the father of the atomic bomb – maintained that:

> secrecy strikes at the very root of what science is, and what it is for. It is not possible to be a scientist unless you believe that it is good to learn. It is not good to be a scientist, and it is not possible, unless you think that it is of the highest value to share your knowledge, to share it with anyone who is interested. It is not possible to be a scientist unless you believe that the knowledge of the world, and the power which this gives, is a thing which is of intrinsic value to humanity, and that you are using it to help in the spread of knowledge, and are willing to take the consequences.[29]

Oppenheimer's 'intrinsic value' assertion could be taken as representative of the way of thinking of many within the scientific world.

The problem with a purist idea of 'intrinsic value' is that it can result in scientific work being placed outside of the reach of social or political control,[30] with the pursuit of knowledge becoming an end in itself and a trump against any attempt at regulation. In a moral sense, it seems difficult to justify a purist conception of intrinsic value admitting of no exception. Knowledge must relate to human flourishing. True, knowing and learning are key aspects themselves of human flourishing, but if they end up undermining it in some fundamental way, for example by threatening our physical survival, then there must be a tempering of the otherwise just general principle that knowledge should not be forbidden.

There are two further problems with 'forbidden knowledge'. First, in the words of Johnson, 'Forbidding is resorted to only in special situations – where there are good reasons to counterbalance the presumption of freedom'. As discussed later, nowadays this presumption of freedom takes effect through the human right to free expression, enshrined in international human rights law and in the constitutions of liberal democracies.[31]

Second, there is a practical difficulty: in most cases it is almost impossible to 'predict the consequences of knowledge prior to obtaining that knowledge'.[32] Once we become aware of the risks, it is often too late. Evidently, this creates a challenge for the exercise of an *ex ante* risk-benefits assessment for dual-use research. Some have suggested that research in the life sciences should follow the precautionary principle.[33] This principle has received some international recognition in international environmental law and, through it, in the area of environmental biosecurity, most notably in connection with GM foods.[34] But there is little indication at present of the acceptance of the precautionary principle on a wider basis in life sciences research. An approach based on, or at least informed by, precautionary consideration may go some way towards allaying concerns about security.

These debates are not purely theoretical or speculative. Over the last two decades some experiments involving the creation or modification of highly virulent pathogens have called into question the consensus in the scientific community on the intrinsic value of scientific research, and sparked a confrontation between that consensus and more security-oriented policy-makers. In 2001, an Australian research team worked on the modification of the mousepox virus gene. Their aim was to alter the virus in such a way that it would sterilise mice and consequently restrain the spread of the pest. However, the experiment resulted in the accidental increase of lethality of the mousepox virus, so that it was now capable of killing mice which were either naturally resistant to or vaccinated against the 'regular' virus.[35] The unexpected outcome of the study, which was nevertheless published in the *Journal of Virology* along with a description of materials and methods employed,[36] raised security concerns. Even the leading researcher, Ronald Jackson, felt uneasy, stating, 'It would be safe to assume that if some idiot did put human IL-4 into human smallpox they'd increase the lethality quite dramatically [...]. Seeing the consequences of what happened in the mice, I wouldn't want to be the one who'd want to do the experiment.'[37] As the supply of idiots has never been short, it is fair to ask what stands between humanity and that catastrophic scenario.

A year later, research which artificially synthesised a polio virus was conducted at the State University of New York at Stony Brook. The researchers created the virus from scratch by consulting polio gene information available on the internet and purchasing DNA material via mail order. The study proved that 'it is possible to synthesize an infectious agent by in vitro chemical-biochemical means solely by following instructions from a written sequence'. It was eventually published in *Science*.[38] Dr Eckard Wimmer, leader of the project, justified the publication of his research as a warning that terrorists may now be capable of producing bioweapons without obtaining a natural virus. Dr Craig Venter, the first scientist to sequence the human genome, took a tougher stance and claimed that '[t]o purposely make a synthetic human pathogen is irresponsible'.[39]

In 2002, another dual-use research project caused quite a sensation. The study, which was published in the *Proceedings of the National Academics of Sciences*, revealed the manner in which the SPICE protein produced by the smallpox virus neutralises the human immune system.[40] The research naturally raised concerns about the possibility of creating viruses with increased virulence. These anxieties did not, however, hinder the performance of a new dual-use research in 2005. This time, with the help of synthetic genomics techniques, scientists brought the Spanish Flu virus 'back to life'.[41] The Spanish Flu is estimated to have caused between 20 and 50 million deaths in 1918–19.

Before its publication in *Science*, the 'Spanish flu study' was sent to the US National Science Advisory Board for Biosecurity (NSABB) for review. Established in 2004, the NSABB is composed of a mixed membership representing both the scientific and security sectors. The task of the Board is to advise the US government on matters regarding dual-use in life sciences. Its decisions are not binding. The NSABB voted unanimously in favour of publication. *Science*'s editor-in-chief had, however, explained that the study would have been published even in the event of a negative NSABB vote.[42]

The NSABB took a different stance vis-à-vis the question of the publication of studies on the transmissibility of the H5N1 flu virus. H5N1 is the most highly pathogenic avian influenza, an infectious disease which spreads among birds and can eventually contaminate humans as well. The H5N1 strain first infected humans in 1997, during a poultry flu outbreak in Hong Kong. Ever since, the virus has spread widely from Asia to the Middle East, Africa, and Europe, causing millions of infections and deaths in birds. Nonetheless, H5N1 does not transmit easily from birds to people and even less easily among humans.[43] Indeed, the great majority of H5N1 infections were linked to close contact with contaminated birds. The frightful peculiarity of the H5N1 flu is that the human mortality rate ranges between 50 and 60 percent. For the 2003–2015 period, the WHO reported that of the 844 human H5N1 cases, 449 resulted in death.[44] Should genetic mutations increase its transmissibility between humans, the impact would be disastrous. In September 2011, at an influenza conference, Dr Ron Fouchier announced that his team at the Erasmus Medical Center in Rotterdam had created an airborne strain of the H5N1 virus that was transmissible via aerosol between ferrets. In the report that followed, Fouchier described how a combination of genetic engineering and serial infection of ferrets resulted in the generation of a virus which could spread among mammals without direct contact.[45] A very similar study was carried out by Dr Yoshihiro Kawaoka, at the University of Wisconsin-Madison and the University of Tokyo.

These studies raised concerns among national security experts. The White House National Security Staff reported its concerns to the US Department of Health and Human Services, which then sought the advice of the NSABB.[46] The NSABB recommended that the papers should not be published in full, but rather with appropriate redactions to exclude 'the methodological and other details that could enable replication of the experiments by those who would seek to do harm'.[47] The Dutch government agreed and imposed an export-control restriction on Fouchier's article, applying the terms of EU Regulation No 428/2009 which, by setting up a Community export control regime for dual-use items, aims at limiting the spread of nuclear, chemical and biological weapons.[48] The Erasmus University Medical Center announced that its researchers would also observe the NSABB recommendation and the government's export-control restriction, but also added that 'academic and press freedom will be at stake as a result of the recommendation'.[49] For his part, Dr Fouchier stated, 'By following the NSABB advice, the world will not get any safer, it may actually get less safe'.[50] In February 2012, in a much-anticipated meeting in Geneva, the WHO reviewed the papers and concluded that both studies should be published in full. The manuscripts were revised and presented to the NSABB.[51] Kawaoka's study was unanimously cleared for publication, while Fouchier's paper was recommended for publication by a 12-to-6 vote.[52]

The H5N1 controversy represented also a novelty in that it was the first – and so far only – case in which the NSABB recommended the partial censorship of a particular research. The H5N1 controversy also represents a good example of the polarised intricacies of debates over the risks of dual-use knowledge. Scientists argued that the benefits of the publication outweighed risks associated with the misuse of the dual-use knowledge produced, for anticipating the way in which the H5N1 virus could mutate and become transmissible between humans would allow us to respond promptly and safely to a future avian flu outbreak. Fouchier's accusation that the (partial) censorship of his research would make the world 'less safe' is illustrative of the position of many virologists, who maintain that the real threat to human welfare comes from nature, rather than bioterrorism.[53] Moreover, there were those who defend the absolute value of academic freedom and the pursuit of knowledge, and reject any kind of risk assessment, let alone one carried out by the state: 'academic freedom is to scientists what civil liberties are for citizens' they claim;[54] and that is no doubt a fair comparison, but it should also encompass the consideration that civil liberties are seldom unlimited.

In the H5N1 debate national security experts tended to stress the risks associated with a laboratory accident and the possibility of terrorists using the same information to produce bioweapons.[55] Scientists are not security experts, they argued, and do not have access to the latest classified information about bioterrorist capacities.[56] The uncontrolled publication of research open to *dual-use* provides both fresh ideas and instructions to terrorists on how to produce new bioweapons.[57] The view, widely shared in security circles, is that if dual-use studies must be published, they should at least omit the descriptions of material and methods employed to limit the risk of bad use. The US Presidential Commission for the Study of Bioethical Issues noted that 'scientists and engineers should recognize the potential impact of their research on those who will experience both its benefits and burdens and their responsibility to those who provide the means, directly or indirectly, for their research'.[58]

A fundamental question is whether this recognition of risk should be left to the moral conscience for scientists, or whether there should be processes in place for assessing risk and, in extreme cases, for imposing limits on the dissemination of scientific knowledge. Any such processes would raise questions as regards, in particular, the applicable law and the method of enforcement. The NSABB is an example of a 'soft' recommendation-based mechanism. An alternative is a legal determination which, if conducted on the plane of international law, would have to take into account human rights law, specifically the right to freedom of expression.

Should academic freedom be limited?

As mentioned, when the Erasmus University Medical Center of Rotterdam announced that it would comply with the NSABB recommendation requiring the redaction of the H5N1 manuscripts, it also warned that this action represented a threat to academic and press freedom.[59] But what is the scope of academic freedom in international law? And can it be limited to manage risks in the field of synthetic biology?

Academic freedom has been defined as the 'freedom of members of the academic community, assembled in colleges and universities, which underlies the effective performance of their functions of teaching, learning, practice of the arts, and research'.[60] The right to scientific inquiry is understood to entail the right to decide on both the means and the ends of the research. As Robertson put it, 'In claiming a right of scientific inquiry or a right to do research, the scientist is claiming to be free from government direction or intervention in choosing topics of research and in selecting means to carry out the research'.[61]

The notion that academic freedom and the right of scientific research constitute human rights is not really disputed. According to Rajagopal, in international law academic freedom exists as a right derivative from the freedom of expression and opinion and from the human right to education.[62] Freedom of expression is guaranteed in Art. 19 of the Universal Declaration of Human Rights (UDHR)[63] and in all the major international and regional human rights treaties (e.g., Art. 19 of the International Covenant on Civil and Political Rights, Art. 10 of the European Convention on Human Rights, Art. 13 of the American Convention on Human Rights, and Art. 9 of the African Charter on Human and Peoples' Rights). The human right to education also appears in the UDHR, and it is further provided for in Art. 13 of the International Covenant on Economic, Social and Cultural Rights alongside the right, in Art. 15, of academic and scientific freedom.

Some national constitutions expressly recognise the right to academic freedom as a constitutionally protected right.[64] Those that fail to do so may still protect the right to academic freedom and scientific research as derivative rights on the basis of provisions such as the freedom of thought and expression (e.g., freedom of expression in the First Amendment to the US Constitution).[65] In general, it would appear that

> at a first basic level, this freedom [of scientific research] receives the same protection given to all other fundamental rights included in the genus of freedom of thought and expression; at a second level, we could find a specific and expressed constitutional recognition for such a fundamental freedom; and finally, at a possible third level, the State is engaged in promoting scientific research.[66]

Some people go further and argue that

> even when conceived in the limited sense of a protected liberty or negative right, the right to research is not properly conceived as a fundamental right, such as the right to free expression or the free exercise of religion, because it is not entailed by the principle of human equality. [...] Nor is conducting research plausibly conceived as a necessary component of human fulfillment, a notion which underlies recent developments in the concept of 'human rights'.[67]

Academic freedom, whether it exists as a derivative or self-standing right (or both), is subject to the limitation clauses that apply to non-absolute rights. These clauses typically identify grounds, among them national security, on which it is permitted to limit rights. In addition, where one non-absolute right conflicts with another, a balancing of the two is called for. In the jurisprudence of the European Convention on Human Rights, the application of the limitation clauses has been framed under the principle of proportionality.[68] One difficulty, as Liora Lazarus has pointed out, is that the right to security is 'inherently ambiguous': it 'encapsulates on one hand a commitment to rights, which we commonly associate with absence from coercion, but on the other hand a commitment to coercion in the name of individual and collective security'.[69] In general, 'the right to security entails the establishment of the factual conditions which give rise to the achievement of someone's actual security – namely the absence from threats or risks of threats – which results in her being able to enjoy other rights'.[70] Some legal philosophers, such as Henry Shue, prefer to think of the right to security as a meta-right, a necessary condition for the enjoyment of other rights.[71] Lazarus is critical of this approach and argues that by conceiving security as a meta-right there is a risk of all other human rights being *securitized*.[72]

There is no case law from international human rights courts or tribunals to provide us with guidance on how to balance academic freedom with national security in the area of dual risk and synthetic biology. But there is some case law from the United States. As mentioned, academic freedom is protected under the First Amendment to the US Constitution. In *United States v The Progressive*, a situation similar to that of the H5N1 controversy arose. The US government attempted to restrict the publication of a magazine article written by Howard Morland, a journalist, who, using information gathered from the public domain and obtaining information independently and from government officials, sought to illustrate the design and functioning of the hydrogen bomb.[73] The Department of Energy was called upon to review the article and it concluded that the manuscript included restricted data and the publication of this information would represent a breach of the Atomic Energy Act.[74] Nevertheless, *The Progressive* announced that the article would be published *in toto*, and as a consequence the US government sought a restraint order, claiming that the publication could cause irreparable harm. Judge Robert Warren decided to grant a preliminary injunction, following the reasoning that '[t]he Morland piece could accelerate the membership of a candidate nation in the thermonuclear club', which would consequently increase the possibility of a nuclear holocaust.[75] The case was eventually dropped after information leaked and was independently published.

Conclusion

The *engineering* of viruses and other pathogens represents an obvious risk of misuse. Should rogue states or terrorist groups get their hands on those enhanced agents and turn them into weapons, the consequences would be catastrophic. The right to research and share sensitive and potentially dangerous information can and should be limited in extreme cases, namely when it collides with national security and the right to public health.

But we have to proceed with great caution. Restrictions on academic freedom are a defining feature of authoritarianism. True, as we have seen, human rights law does not protect free speech and academic freedom on an absolute basis, and restrictions could be justifiable in exceptional circumstances. There are, however, difficulties. To begin with, there does not seem to be a clear understanding on the degree of risk that society is prepared to tolerate or that the law as it currently stands may require us to tolerate. Who should decide on the appropriate level of risk? Scientists? Judges? Or democratically representative institutions? As we have seen, human rights law would play an important part in the legal analysis, but it is somewhat illusory to think that it offers clear guidance on these matters. By their nature human rights are not normally content-specific; they are formulated in open-textured and broad language. Where a body of jurisprudence has developed, it is possible to flesh out what human rights law requires in specific circumstances; but no such body of jurisprudence has emerged in the area of synthetic biology.

Notes

1 Linus Pauling, 'Peace on Earth: the position of the scientists', *Bulletin of the Atomic Scientists*, Vol. 47, October 1967.
2 Matthew Meselson, 'Averting the hostile exploitation of biotechnology', *The CBW Conventions Bulletin*, Vol. 48, June 2000, p. 16.
3 Richard P. Feynman, 'The value of science', public address given at the 1955 autumn meeting of the National Academy of Sciences.

4 See W. Lanouette, *Genius in the Shadows: A Biography of Leo Szilard, the Man Behind the Bomb* (New York: Skyhorse, 2013), chapter 13; see also N. G. Evans, 'Contrasting dual-use issues in biology and nuclear science', in B. Rappert and M.J. Selgelid, *On the Dual Uses of Science and Ethics: Principles, Practices, and Prospects* (Canberra: The Australian National University Press, 2013), p. 266. Richard Rhodes, *The Making of the Atomic Bomb*, New York: Simon and Schuster, 1986. S.R. Weart, 'Scientists with a secret', *Physics Today*, February 1976, pp. 23–30.
5 M. Rees, *Our Final Century* (London: Arrow, 2004), p. 2.
6 Institute of Medicine and National Research Council (NRC), *Globalization, Biosecurity, and the Future of the Life Sciences*, (Washington, DC: The National Academies Press, 2006), p. 35, available at www.nap.edu/catalog/11567/globalization-biosecurity-and-the-future-of-the-life-sciences.
7 See Chapter 16 in this volume.
8 Ronald M. Atlas and Malcom Dando, 'The dual-use dilemma for the life sciences: perspectives, conundrums, and global solutions', *Biosecurity and Bioterrorism: Biodefense Strategy, Practice, and Science*, Vol. 4, No. 3 (September 2006), pp. 276–7.
9 European Commission, *Synthetic Biology: Applying Engineering to Biology* (Brussels: European Commission, 2005) Report of a NEST High-Level Expert Group, 5.
10 J. Keasling, 'The promise of synthetic biology', *The Bridge*, National Academy of Engineering, Vol. 35, No. 4 (Winter 2005), p. 20.
11 A. Kelle, 'Synthetic biology as a field of dual-use bioethical concern', in B. Rappert and M. J. Selgelid, *On the Dual Uses of Science and Ethics: Principles, Practices, and Prospects* (Canberra: The Australian National University, 2013), p. 45.
12 A. S. Khalil and J. J. Collins, 'Synthetic biology: applications come of age', *Nature Reviews Genetics*, Vol. 11, No. 5 (May 2010), p. 367.
13 Model-based reconstructions have highlighted that the destruction caused by a biological attack, employing enhanced-virulence pathogens or the smallpox virus, would be comparable to a series of nuclear attacks. See M. J. Selgelid, 'A Tale of Two Studies: Ethics, Bioterrorism, and the Censorship of Science', *Hastings Center Report*, Vol. 37, No. 3 (May-June 2007). Barbara Bullock notes that the 'unleashing of biological agents against an unprotected civilian population [...] in some cases, constitutes the ultimate medical disaster with the capability to completely overwhelm the present healthcare system'. See B. Bullock, 'Surveillance and detection: a public health response to bioterrorism', in J. A. Davis and B. R. Schneider (eds.), *The Gathering Biological Warfare Storm* (Westport, CT: Praeger, 2004), p. 31. In July 2001 a simulation – named 'Dark Winter' – of a series of smallpox attacks on three US shopping malls was carried out by the Johns Hopkins Center for Civilian Biodefense Strategies and resulted – in the worst case scenario – in the infection of three million people. According to smallpox mortality rates, a third of the infected would have died.
14 Institute of Medicine and NRC, *Globalization, Biosecurity, and the Future of the Life Sciences* (Washington, DC, The National Academies Press, 2006), p. 45.
15 In 1958, Ford even developed a scale model concept car. According to the project, the Ford Nucleon would have been powered by a small nuclear reactor. B. K. Sovacool, *Contesting the future of nuclear power: a critical global assessment of atomic energy* (World Scientific Publishing Co., 2011), p. 259.
16 OECD and the Royal Society, *Symposium on Opportunities and Challenges in the Emerging Field of Synthetic Biology*, 2010, pp. 14–22; The Royal Academy of Engineering, *Synthetic Biology: scope, applications and implications* (London, May 2009), chapter 3. Presidential Commission for the Study of Bioethical Issues, *The Ethics of Synthetic Biology and Emerging Technologies* (Washington DC, December 2010), chapter 3; D. Chakravarti and W. W. Wong, 'Synthetic biology in cell-based cancer immunotherapy', *Trends in Biotechnology*, Vol. 33, No. 8 (August 2015), pp. 449–61; D. F. Savage, J. Way and P. A. Silver, 'Defossiling fuel: how synthetic biology can transform biofuel production', *American Chemical Society: Chemical Biology*, Vol. 3, No. 1 (January 2008); T. Landrain et al., 'Do-it-yourself biology: challenges and promises for an open science and technology movement', *Systems and Synthetic Biology*, Vol. 7, No. 3 (2013), p. 115; European Commission, *Synthetic Biology*, pp. 13–17 (see note 9 above); Khalil and Collins, *Synthetic Biology*, pp. 367–79 (see note 12 above).
17 Freeman Dyson, 'Our biotech future', *The New York Review of Books*, 19 July 2007.
18 M. J. Selgelid, 'A Tale of Two Studies', p. 38 (see note 13 above). S. Miller and M.J. Selgelid, 'Ethical and philosophical considerations of the dual-use dilemma in the biological sciences', *Science and Engineering Ethics*, Vol. 13, No. 4 (December 2007).
19 The physicist Freeman Dyson claimed that in the next 50 years, society will witness a process of *domestication* of biotechnology. In his words '[d]esigning genomes will be a personal thing, a new art form as creative as painting or sculpture'. See Dyson, 'Our biotech future' (note 17 above).

20 See Christopher R. Lowe, 'Biotechnological innovation, non-obvious warfare and challenges to international law', Chapter 16 in this volume. See also, for instance, M. Goodman, *Future Crimes* (London: Corgi, 2016).
21 iGEM is the main student competition in the field of synthetic biology. It encourages students to build genetically engineered biological systems. See http://igem.org/Main_Page.
22 It is now possible to buy synthesised DNA material for USD 0.3 per base pair. See C. Jefferson, F. Lentzos and C. Marris, 'Synthetic biology and biosecurity: challenging the "myths"', *Frontiers in Public Health*, August 2014, p. 6.
23 NRC, *Challenges and Opportunities for Education about Dual Use Issues in the Life Sciences* (Washington, DC: The National Academies Press, 2010), 24, available at www.nap.edu/catalog/12958/challenges-and-opportunities-for-education-about-dual-use-issues-in-the-life-sciences.
24 NRC, Challenges and Opportunities for Education, 19 and 23; Atlas and Dando, The dual-use dilemma, 282.
25 National Science Advisory Board for Biosecurity (NSABB), *Proposed Framework for the Oversight of Dual Use Life Sciences Research: Strategies for Minimizing the Potential Misuse of Research Information* (June 2007) 2, available at http://osp.od.nih.gov/office-biotechnology-activities/nsabb-reports-and-recommendations/proposed-framework-oversight-dual-use-life-sciences-research.
26 D.G. Johnson, 'Reframing the question of forbidden knowledge for modern science' (December 1999) 5 *Science and Engineering Ethics* 4, 449.
27 R. Shattuck, *Forbidden knowledge: from Prometheus to Pornography* (New York: St Martin's Press, 1996), p. 224.
28 M. J. Selgelid, 'Governance of dual-use research: an ethical dilemma', *Bulletin of the World Health Organization* , Vol. 87, 2009, p. 36.
29 Robert Oppenheimer, *Speech to the Association of Los Alamos Scientists* (Los Alamos, New Mexico, 2 November 1945) available at www.atomicarchive.com/Docs/ManhattanProject/OppyFarewell.shtml.
30 Johnson, 'Reframing the question', p. 446 (see note 26 above).
31 Ibid., p. 448.
32 J. Kempner, J. F. Merz and C. L. Bosk, 'Forbidden knowledge: public controversy and the production of nonknowledge' *Sociological Forum*, Vol. 26, Issue 3 (Sept 2011), p. 479.
33 F. Kuhlau, A.T. Höglund, K. Evers and S. Eriksson, 'A precautionary principle for dual use research in the life sciences', *Bioethics* Vol. 25, No. 1 (January 2011), pp. 1–8.
34 E.g.: Art 191(2), *Treaty of Lisbon amending the Treaty on European Union* and the *Treaty establishing the European Community; Cartagena Protocol on Biosafety to the Convention on Biological Diversity*.
35 Selgelid, 'A Tale of Two Studies', p. 720 (see note 13 above).
36 R. J. Jackson et al., 'Expression of mouse Interleukin-4 by a recombinant Ectromelia virus suppresses cytolytic lymphocyte responses and overcomes genetic resistance to Mousepox', *Journal of Virology*, Vol. 75, No. 3 (2001), pp. 1205–10.
37 BBC World Service, 'Mouse virus or bioweapon?', 17 January 2001, available at www.bbc.co.uk/worldservice/sci_tech/highlights/010117_mousepox.shtml.
38 J. Cello, A.V. Paul and E. Wimmer, 'Chemical Synthesis of Poliovirus cDNA: Generation of Infectious Virus in the Absence of Natural Template', *Science*, Aug. 2002, pp. 1016–18.
39 A. Pollack, 'Scientists create a live polio virus', *The New York Times*, 12 July 2002, available at www.nytimes.com/2002/07/12/us/traces-of-terror-the-science-scientists-create-a-live-polio-virus.html?pagewanted=all.
40 A. M. Rosengard, Y. Liu, Y. Z. Nie and R. Jimenez, 'Variola virus immune evasion design: expression of a highly efficient inhibitor of human complement', *Proceedings of the National Academy of Sciences*, Vol. 99, No. 13 (2002), pp. 8808–13.
41 T. M. Tumpey et al., 'Characterization of the reconstructed 1918 Spanish influenza pandemic virus', *Science,* October 2005, pp. 77–80.
42 D. Kennedy, 'Better never than late', editorial by the Editor-in-Chief, *Science*, 14 October 2005.
43 World Health Organization, *Avian Influenza: fact sheet* (updated March 2014), available at www.who.int/mediacentre/factsheets/avian_influenza/en/.
44 WHO, *Cumulative number of confirmed human cases for avian influenza A(H5N1) reported to WHO, 2003–2015*, available at www.who.int/influenza/human_animal_interface/H5N1_cumulative_table_archives/en/.
45 R. Roos, 'Fouchier study reveals changes enabling airborne spread of H5N1', Center for Infectious Disease Research and Policy, University of Minnesota, 21 June 2012, available at www.cidrap.umn.edu/news-perspective/2012/06/fouchier-study-reveals-changes-enabling-airborne-spread-h5n1.

46 S. A. Ehrlich, 'H5N1: a cautionary tale', *Frontiers in Public Health*, August 2014, Art. 117.
47 NSABB, 'Press Statement on the NSABB Review of H5N1 Research', 20 December 2011, available at www.nih.gov/news-events/news-releases/press-statement-nsabb-review-h5n1-research.
48 European Council, *Council Regulation (EC) No 428/2009 of 5 May 2009 setting up a Community regime for the control of exports, transfer, brokering and transit of dual-use items* (2009) OJ L134/1.
49 Erasmus University Medical Center, 'Virologists to observe American bioterrorism recommendation', 20 December 2011, available at www.erasmusmc.nl/perskamer/archief/2011/3530639/?lang=en.
50 Ron Fouchier quoted in J. Cohen, M. Enserink and D. Malakoff, 'A central researcher in the H5N1 flu debate breaks his silence' *Science*, 25 January 2012, available at www.sciencemag.org/news/2012/01/central-researcher-h5n1-flu-debate-breaks-his-silence.
51 S. Herfst et al., 'Airborne transmission of influenza A/H5N1 virus between ferrets', *Science*, June 2012, pp. 1534–41; Imai, T. Watanabe et al., 'Experimental adaptation of an influenza H5 HA confers respiratory droplet transmission to a reassortant H5 HA/H1N1 virus in ferrets' *Nature*, May 2012, pp. 420–28.
52 NSABB, *Findings and Recommendations* (29–30 March 2012), available at www.nih.gov/about-nih/who-we-are/nih-director/statements/statement-nsabbs-march-30-2012-recommendations-nih-h5n1-research.
53 R. G. Webster, 'Mammalian-Transmissible H5N1 Influenza: the Dilemma of Dual-Use Research', *mBio* Vol. 3, No. 1(January–February 2012), American Society for Microbiology; R. L. Frerichs et al., *Historical precedence and technical requirements of biological weapons use: a threat assessment*, Sandia National Laboratories, 2004; *The Economist*, 'The world's deadliest bioterrorist', 28 April 2012.
54 S. A. W. Evans and W. D. Valdivia, 'Export controls and the tensions between academic freedom and national security', *Minerva*, Vol. 50, No. 2 (June 2012), p. 173.
55 G. D. Koblentz, 'Dual-use research as a wicked problem', in J. E. Suk, K. M. Vogel and A. J. Ozin, *Dual-use Life Science Research and Biosecurity in the 21st Century: Social, Technical, Policy, and Ethical Challenges* (Frontiers in Public Health 2015), p. 36.
56 Selgelid, 'A Tale of Two Studies', p. 36 (see note 13 above).
57 Selgelid, 'Governance of dual-use research', p. 721 (see note 28 above).
58 Presidential Commission for the Study of Bioethical Issues, *The Ethics of Synthetic Biology*, p. 141.
59 Erasmus University Medical Center, 'Virologists to observe' (see note 49 above).
60 R. F. Fuchs, 'Academic freedom: its basic philosophy, function, and history', *Law and Contemporary Problems*, Vol. 28, 1963, p. 431.
61 J.A. Robertson, 'The Scientist's right to research: a constitutional analysis'' *Southern California Law Review*, Vol. 51, 1978, pp. 1205–6.
62 B. Rajagopal, 'Academic freedom as a human right: An internationalist perspective', *Academe: Journal of the American Association of University Professors*, Vol. 89, No. 3 (2003), p. 29.
63 Article 19 of the UDHR: 'Everyone has the right to freedom of opinion and expression; this right includes freedom to hold opinions without interference and to seek, receive and impart information and ideas through any media and regardless of frontiers'.
64 Article 5(3) of the German Constitution: 'Art and science, research and teaching are free. The freedom of teaching does not release from allegiance to the constitution'; Article 33 of the Italian Constitution: 'The Republic guarantees the freedom of the arts and sciences, which may be freely taught. [...] Higher education institutions, universities and academies, have the right to establish their own regulations within the limits laid down by the law'; Article 42(1) of the Portuguese Constitution: 'Intellectual, artistic and scientific creation shall not be restricted'.
65 A. Santosuosso, V. Sellaroli and E. Fabio, 'What constitutional protection for freedom of scientific research?', *Journal of Medical Ethics* Vol. 33, No. 6 (June 2007), p. 342.
66 Ibid.
67 M. B. Brown and D. H. Guston, 'Science, democracy, and the right to research', *Science and Engineering Ethics issue*, Vol. 15, May 2009, p. 358.
68 G. Verdirame, 'Rescuing human rights from proportionality', in Rowan Cruft, S. Matthew Liao and Massimo Renzo (eds.), *Philosophical Foundations of Human Rights* (Oxford: Oxford University Press, 2014), pp. 341–60.
69 L. Lazarus, 'The right to security – securing rights or securitising rights', in R. Dickinson et al., *Examining Critical Perspectives on Human Rights* (Cambridge: Cambridge University Press, 2014), p. 89.
70 Ibid., p. 100.
71 H. Shue, *Basic Rights: Subsistence, Affluence, and US Foreign Policy*, Princeton: Princeton University Press, 1996.

72 Lazarus, 'The right to security', pp. 98–103 (see note 69 above).
73 H. Morland, 'The H-bomb secret: To know how is to ask why', *The Progressive*, November 1979, 3–12.
74 J. M. Nesse, 'United States v. Progressive, Inc.: The national security and free speech conflict', *William & Mary Law Review*, Vol. 22, No. 1, p. 142.
75 *United States v. Progressive*, Inc., 467 F. Supp. 990 (W.D. Wis. 1979), p. 994.

PART V

New frontiers

20
SPACE ODDITIES
Law, war and the proliferation of spacepower

Bleddyn Bowen

The twenty-first century is witnessing the vertical and horizontal proliferation of spacepower. With more states developing satellite and space infrastructure, more lucrative targets are being deployed in orbit. Space technology has become integral for tactical and operational military capabilities, not only in the US military. The development of space weapons based on Earth has potential consequences for nuclear stability as some tactically relevant satellites are also responsible for providing a warning of nuclear attack. Yet this reality of spacepower's proliferation is unaddressed in the two major legal initiatives in global space governance due to their foundation in a misplaced hostility towards the 'militarisation' of space and to a fear of a false jeopardy from 'space weaponisation'. The reality of spacepower's relevance for war and peace on Earth clashes with popular conceptions of states pursuing space technology in the name of peaceful exploration which are simply not supported by historical facts.[1]

Established space powers are continuing their development of outer space for military, economic and political purposes, while smaller and newer space powers are also developing their initial or niche space-faring capabilities. Space is becoming normalised as part of critical infrastructure in the twenty-first century which underpins finance, transport, agriculture, and security infrastructures.[2] Satellites and their associated infrastructure enable the web of communications that span the globe, authorise financial transactions and bank withdrawals, assist in the planning and building of terrestrial infrastructure, monitor Earth's weather, climate, and topography, guide emergency responders to those in distress, and of course enable the bombing of people and buildings with extreme accuracy regardless of adverse weather conditions. Military space activities continue to defy normalisation or rigid codification in international law. The development of spacepower – the use of outer space for political and strategic purposes – has been 'dual use' since its inception. Dual-use refers to the ability of devices – weapons and other technological systems – that can possess both military and civilian purposes despite declared intentions of use. Dual-use technologies can be turned from civilian to military purposes, and vice versa. This makes the verification of arms control agreements extremely problematic and is a persistent critique of the Treaty on the Prevention of the Placement of Weapons in Outer Space, and of the Threat or Use of Force against Outer Space Objects (PPWT). An example is the American Global Position System, which is managed by the US Air Force to provide precision warfare capabilities, but has become an economically lucrative service for digital commercial applications that require precise and ubiquitous position, navigation, and timing information. Across the board, space technology has many inherently civilian and military uses, like any generic labelling of technology that is defined by its geographic location. Another example is that detecting the depth of the water table is useful for agriculture and infrastructure planning, but can also reveal whether heavy tracked armoured vehicles had recently passed through the area, along with probable speed and directions of travel.

From the dawn of the space age, the distinctions between the civilian and military uses of space had been intertwined in the United States, but distinctly militaristic in the Soviet Union.[3] As satellites become more useful for tactical and operational capabilities, notably in the emergence of precision or reconnaissance-strike regimes, the desire to attack satellites through physical or more discrete means are increasing.[4] Space infrastructure is therefore becoming more useful in a non-nuclear conflict and, consequently, a more lucrative and likely target in warfare. These satellites can be struck with weapons based on Earth, including physically destructive weapons as well as devices and techniques that merely temporarily jam, disrupt, or hijack communications streams. It is in this context that two problematic legal initiatives on governing outer space reside.

The International Code of Conduct (CoC or the Code) is sponsored by the United States and the European Union (EU), while the PPWT mentioned earlier is sponsored by Russia and China. Both rest on flawed assumptions about outer space and do not address the looming spectre of space warfare and the ubiquitous nature of dual-use space technology. These make aggressive and peaceful distinctions of technological systems in space technically impossible. This chapter highlights the problems underlying both legal initiatives, which are deep-rooted in popular misconceptions of the historic and present realities of the military–political uses of outer space, or spacepower. In the process, the key provisions of space law and core strategic realities of outer space are encountered. By critiquing these two initiatives, it is argued that the popular terms of the 'militarisation' and 'weaponisation' of space are disguising the primary forces motivating the space powers of the twenty-first century. The primary motivations driving the exploitation of outer space among the major powers is to enhance the state's capabilities across the board – military, political, economic, and technological. In that sense, space is an unremarkable geography as it mirrors precedents in the exploitation of the sea and the air.

If the flaws of using terms such as militarisation and weaponisation are not acknowledged, analysts and practitioners of space arms control and space security debates will perpetually chase red herrings while decisions that affect the strategic realities of outer space and Earth are made under the smokescreen of using space for 'peaceful purposes' and the misguided aim of preventing the 'weaponisation' of outer space. In short, those who believe space is only *now* being militarised defy the reality that space has been militarised since the dawn of the space age; while those who preoccupy themselves with 'space weaponisation' overstate the potential impact of *space-based* weapons on international stability and space security. Fears of space weaponisation also overlook the potential impact that existing *Earth-based* space warfare technologies and weapons may be having on space security.

The failure to recognise the permanence of the dual-use nature of space technology fatally undermines the Code of Conduct as it is currently drafted, whilst the PPWT does nothing to address the most likely forms of space warfare. Outer space is unremarkable in the way that it, like the land, the sea, and the air, is used for a multitude of peaceful and violent purposes. Legal attempts to curb the use of space for military purposes appear doomed to flounder, as would any attempt to keep the sea or the air entirely free of military vessels and action. Attempting to prevent the militarisation of space is akin to closing the stable door after the horse has bolted, and attempting to prevent the weaponisation of space does nothing to forestall the risks of space warfare that Earth-based space weapons have already created. Space warfare is not inevitable – but if it does come, it will not be the kind of space-based warfare that is feared and targeted by the PPWT. Understanding the strategic rationales behind the continued development and deployment of spacepower – in its military, political, and economic forms – should provide the basis for more useful and successful international legal initiatives to govern outer space and reduce unintentional confusion and alarm in orbit. This requires the normalisation and acceptance of the military nature of space, as well as the existing spread of space warfare and anti-satellite capabilities on Earth.

Space weapons

All space weapons can be divided into three categories: Earth-to-space, space-to-space and space-to-Earth. Space-to-Earth weapons remain the most exotic and non-existent type today and will not be discussed here.[5] Space-based and Earth-based anti-satellite weapons systems can include kinetic energy (ramming) or explosive warheads launched from missiles or satellites, co-orbital satellites that can capture or explode near a target satellite, lasers, radiofrequency weapons, microwave emitters, electronic warfare, and cyber intrusion capabilities. Spoofing is another particular risk, where false data are provided to receivers.[6] Electronic warfare and cyber intrusion can be reversible, while the others generate lasting physical damage. Indeed, the physical methods of space warfare do involve the risks of space debris generation.[7] Earth-based anti-satellite weapons systems are being developed in the United States, Russia, and China, and the reversible and 'disruptive' methods of space warfare, such as electronic warfare and low-powered laser interference, are highly proliferated and are within reach of the smallest determined state. These weapons can be based in almost any terrestrial environment and can be mobile as well as static.

Myriad Earth-based weapons systems already exist to destroy, degrade, disrupt, disable, and deny space systems. China continues with its kinetic energy ASAT testing programme, only one method of many at China's disposal to attack space systems. It should be noted that little or no debris has been created in Chinese ASAT tests since the infamous 2007 test, casting further doubt on another belief among some in the arms control community that kinetic weapons tests *always* produce masses of long-lived debris in orbit.[8] China's series of ASAT tests since the mid-2000s has involved kinetic direct-ascent interceptor launches. An interceptor in 2013 reached the altitude of 30,000 km, which is not far below the orbits of crucial geosynchronous communications and early warning satellites.[9]

Space weaponisation is often defined as placing weapons systems in orbit. Indeed, space-to-space weapons have been deployed before: definitionally, 'space weaponisation' may thus already have happened. The Soviet Fractional Orbital Bombardment System (FOBS) was developed to base nuclear weapons in orbit (its deployment tests did not take nuclear weapons into orbit) so that they could be launched from a southerly direction at the United States to avoid their early-warning radars.[10] The Soviet Union also mounted a 'self-defence' cannon on the *Saluyt-3* space station in the 1970s, which was test-fired before deorbiting.[11] The Soviet Union also developed the *Istrebitel Sputnikov* and a subsequent *Naryad* upgrade, which were space-based satellite interceptors in the 1970s and 1991, respectively.[12] The fact of these past deployments should impose a degree of caution on the alarmism often exhibited in discussions of basing weapons in space.

Space weapons can be deployed on Earth, and the Cold War again has precedents. The United States' Nike-Zeus-based Project Mudflap was designed to intercept satellites and the USSR's FOBS with nuclear-tipped missiles based on Earth.[13] In the 1980s, the US Air Force deployed several F-15s with conventionally armed anti-satellite interceptors capable of shooting down satellites in low-Earth orbit. This was arguably in response to the Soviet Union's development of radar ocean reconnaissance satellites (RORSAT) that could pinpoint the location of US

Navy surface ships for long-range Soviet anti-ship missiles.[14] Today, the United States fields residual, but tested, anti-satellite weapons capabilities in the various forms of the SM-3 missile and Aegis missile defence systems that can be placed on destroyers and in Aegis Ashore facilities.[15] Similar to this system, Anatoly Zak notes that anti-satellite capabilities may be added to Russia's S-400 and S-500 air defence missile systems.[16] Basing weapons on Earth is far cheaper than launching them into low orbit, as access to low-Earth orbit is measured in thousands of dollars per kilogram, and even more so into higher orbits. A space weapon loitering in orbit would likely have to withstand the hostile physical environment of orbit, enjoy minimal or no maintenance on its hardware, and be reliably called upon to act after months or years of inactivity. Basing space weapons on Earth allows for greater deception, maintenance, upgrades, redundancy, economy of force, and hardening of the weapons systems. Therefore, it is reasonable to anticipate that it is not the deployment of Earth- or space-based space weapons that generates a response in other countries to develop such space weapons. Rather, it is whether or not the most likely adversaries are becoming dependent on satellites for military capabilities that drives the acquisition of anti-satellite technology. In the twenty-first century, all permanent members of the UNSC as well as India, Israel, and Japan are either becoming or already acutely dependent on spacepower for military, political, and economic life. This generates a more widespread strategic pressure for some space weapons capability. This pressure may not have been largely present in the Cold War, with the exception of the United States' tactical application of space services and the Soviet RORSAT system by the mid to late 1980s.

Space weapons are diverse, and so will be the character of space warfare. The weapons of space warfare and anti-satellite systems can range from 'hard-kill' systems involving ramming and explosive 'kinetic' weapons such as those above, to those involving 'soft-kill' methods such as radiofrequency jamming, microwave emitters, electromagnetic pulse emitters, and lasers.[17] Lasers, if powerful enough, could be included in the hard-kill category. These can all be placed on Earth to engage in space warfare, or anti-satellite operations, without 'weaponising' space. The PPWT does not address any of the weapons systems based on Earth, which, unlike space-based weapons, have already proliferated. In 1996, Indonesia had successfully jammed a Hong Kong communications satellite located in geostationary orbit leased by Tonga following a dispute over the right to place a satellite in a certain slot.[18] In the 2003 Iraq War, the Iraqi Army used Russian-made GPS jammers with limited and localised successes.[19] Basic laser technology has also proliferated to the point where dozens of states take part in the International Laser Ranging Service to track satellites.[20] Increasing these lasers to weapons or disruption-grade capabilities is a matter of refining target acquisition, power generation, focusing the laser beam, and coordinating dispersed laser emitters to take advantage of their additive effects on the target. This demonstrates that even basic disruptive tools in space warfare are within reach of small powers because Earth-based weapons can create effects in orbit. In this increasingly fraught environment with a long history of military uses of space, the Code of Conduct has been pushed by its detractors into standardising military as well as benign everyday activities in outer space.

The Code and militarisation

The Code of Conduct was first introduced by the European Union in 2008, with several subsequent redrafts. It aims to establish 'a set of principles and best practices designed to promote responsible behaviour in space'.[21] It is a device of soft law, meaning it is a voluntary mechanism that space-faring states can join in order to harmonise and standardise the ways in which rockets are launched, satellites are orbited and managed, and traffic is to be deconflicted. The most recent draft was written in March 2014.[22] The code hopes to build on existing treaties regarding outer space, such as the Outer Space Treaty (OST) of 1967 and the Limited Test Ban Treaty (LTBT) of 1963. These banned the deployment of 'weapons of mass destruction' (WMD) in orbit and on celestial bodies other than Earth and the test-firing of nuclear weapons at high altitudes and outer space, respectively. The LTBT followed the disastrous environmental and infrastructural consequences of American and Soviet high-altitude nuclear testing, most infamously the Starfish Prime test. These Cold War-era treaties do not address contemporary concerns regarding everyday 'responsible' behaviour in outer space and the security of satellites from non-WMD weapons such as lasers, kinetic energy, explosives, particle beams, radiofrequency weapons, orbital hijacking, docking, and ramming. Both the Code and PPWT try to address these concerns, but are ultimately flawed for attempting to prevent conflict that involves satellite infrastructure and its targeting. Due to the pervasiveness of the dual-use nature of satellite technology, it is difficult to verify adherence to such norms. The space technology that is useful in everyday and benign activities only requires the will to turn into makeshift weapons or to support terrestrial military and violent methods to achieve policy objectives.

The Code has faced an uphill battle since its inception, culminating in 2015 when Michael Krepon – one of its more prominent proponents – despairingly lamented the possible end of the Code as he knew it. Krepon bemoaned the fact that the EU conceded that the Code would have to go through the more formal and supranational approach of the United Nations General Assembly (UNGA), rather than pursuing an intergovernmental approach. An intergovernmental non-UN approach would allow a core of space powers to develop soft customary international space law between them. The UN approach could stall any progress by democratising the process, which is exactly what Russia, China, and Brazil desire.[23] Indeed, Russia and China are backing the PPWT as 'their' space treaty of choice, which may reflect more of a case of legal warfare or 'lawfare' and political grandstanding rather than a genuine attempt to sanctuarise outer space from conflict.[24] The Code has a severe weakness in that it relies on states to want to cooperate and behave in an ethical manner in order to achieve its goals.[25] The CoC therefore will do little to stop intentionally aggressive behaviour. Attempting to ensure the 'security, safety, and sustainability of all Outer Space activities' is nothing short of ambitious, and its declared goals are noble.[26] Though the Code is meant to be a voluntary and continually evolving piece of soft law, it seems unlikely to meet its sympathetic goal of 'safeguard[ing] the peaceful and sustainable use of outer space for current and future generations, and in a spirit of greater international cooperation, collaboration, openness and transparency', precisely because it also seeks to 'prevent outer space from becoming an arena of conflict' while also enshrining the right of self-defence in space under Article 51 of the UN Charter.[27] The Code does not need to attempt to make outer space free from conflict in order to achieve its goals of regulating everyday space activities that are in increasing need of regulation and management as orbital traffic and the risks of collision, unintentional jamming, and disputes over orbital slots increase.

After being 'mugged' in New York, the Code now couples the curbing or even prevention of strategic activities that are fundamental to the security and military–economic prowess of the major states of Earth with mundane everyday activities in orbit. Analogically, it would be as if the navies of Earth would not be allowed to operate with their broad immunities as allowed in the United Nations Convention on the Laws of the Sea (UNCLOS).[28] Despite the UNCLOS language also adopting the 'peaceful purposes' tone, the practical reality that the sea is a place that can be controlled by military force and used for violent activities by sovereign states under various conditions is not seriously questioned in security and strategic communities. Unlike the sea, however, outer space is not formally recognised as a *global commons*. This does not alter the reality, however, that military and commercial access to and transit within space, like the sea, is seen as a fundamental strategic necessity.[29] The Code will do nothing to prohibit intentional acts of space warfare, and rules of the road for willingly peaceful actors need not address it, just as UNCLOS does not prohibit naval warfare. The kind of operational transparency promoted by the Code would better suit the United States, as it reduces the scope for surprise behaviour from other signatories (such as codifying 'safe' distances and orbital flight patterns between satellites), as opposed to China and Russia, who may need to rely on surprise to make up for the material disadvantage vis-à-vis the United States in crises. Such soft law would provide a useful hedge against in-orbit weapons which may capture or explode near target satellites; however, such technology remains marginal and expensive given Earth-based alternatives to conduct anti-satellite or space warfare.

Linking the Code to curbing the military and aggressive uses of outer space is a potential death sentence for it. It is remarkable that in some academic and policy quarters outer space is still not treated as 'just another realm' for conflict, similarly also used for everyday economic, commercial, and other activities.[30] In practice, it is already recognised by the US military that in a time of war, states will seek to control areas of the global commons, be it international seas, airspace, or outer space, with a return to free innocent passage when hostilities end.[31] Contemporary writings on Chinese space warfare imitate such language regarding 'space control', and Russian statements echo a similar intent to threaten and defend various satellites in a time of war.[32] Conceptual thought is progressing in that vein within military forces across Earth, even if declared intentions from political leaders do not reflect it.

Divergences in globalist and geopolitical viewpoints about outer space do not undermine the need for norms and regulations to codify 'normal' or acceptable behaviour.[33] Earth orbit is populated by several space powers, each bringing its own space applications and technologies to bear on Earth. China is no exception in utilising space for military purposes,[34] and military and strategic spacepower has proliferated across Earth to other countries such as Brazil, India, Israel, Iran, and Nigeria.[35] This is partly why the United States now refers to Earth orbit as a 'congested, competitive, and contested' environment.[36] In political–strategic terms, there is nothing new under the sun, contrary to the alarmist connotations associated with terms like the 'militarisation' and 'weaponisation' of space that are often associated with the deployment of 'hair-trigger' space-based weapons.[37] The heritage of space technology could not be more militaristic, and satellites have been threatened by weapons systems in the past and in the present without triggering a destabilising 'space arms race' or a third world war.

Humanity's entry into outer space was military in nature. German V-2 rockets – designed by Wernher von Braun's team and constructed by slave labour from concentration camps in the Mittelwerk factory – first probed the highest reaches of Earth's atmosphere during World War II en route to London. In the closing stages of the war in Europe, the United States' Office of Strategic Services secured as many German rocket scientists as possible in Operation PAPERCLIP. A similar operation was undertaken by the Soviet Union as the Red Army swept westwards across Europe. The Soviet and American drives to build rockets in the early Cold War was developed through the technology's potential as carriers of nuclear warheads and strategic communications and reconnaissance satellites – not for exploration and scientific research.[38] Indeed, Lyndon B. Johnson was only full of praise for the value of strategic reconnaissance on Soviet nuclear capabilities from satellites. He had realised that he and the entire leadership were harbouring fears they did not need to harbour because pre-satellite intelligence and public discussion vastly overestimated Soviet nuclear delivery capabilities in the late 1950s and early 1960s.[39] This fundamental aspect of nuclear weapons delivery and reconnaissance from outer space continue to be a part of the rationale for the investments in rocket and satellite technology today. The military origin of the opening of the cosmos betrays the ignorance or disingenuousness of those who are seemingly alarmed that some states today are 'militarising' outer space.[40]

The failure at least to adequately acknowledge the military core and strategic necessity of spacepower means that an obfuscative discursive game of cat-and-mouse is perpetuated in academia and international law regarding both the definitions of military space and space weapons and then how to verify such activities as either peaceful or aggressive, rather than accepting such fundamental and universal activities (in their broadest sense) as part of the 'astroscape' which is essential for modern power politics and security. Rather, specific systems or activities should be singled out for regulation, rather than such broad and unhelpful categories of 'space weapons' or 'non-aggressive' purposes. Such ideals of using outer space for 'peaceful purposes' are employed to disdain the military space activities of another, knowing full well that all major space powers embrace both military and non-military space systems that have a direct military and strategic value. The Code may yet be revived on an intergovernmental basis between states that wish to regulate everyday activities in orbit without curbing their strategic freedom of action, as is already done at sea and in the air. However, this can only be done by embracing and accommodating the military necessity and utility of *outer space infrastructure that can both endanger and bolster 'peace'* in the international system. Often accompanying the misleading conceptions surrounding the militarisation of outer space is the question of space weaponisation. As seen below, space weaponisation – often taken in commentary and literature to mean the placement of weapons in orbit – ignores the fact that space weapons are proliferating *on Earth*. The anti-satellite weapons that the United States, China, and Russia have tested are based on Earth and shoot 'up' into space. This may not 'weaponise' space per se, but that is beside the point. The reality is that the ability to shoot down or disrupt satellites exists regardless of whether outer space is directly weaponised itself. This also raises the spectre that space weaponisation is not the doom-laden future that some argue it to be.

This refusal to abandon the frames of militarisation and weaponisation furthers an obsession involving two red herrings: that space is a 'sanctuary' from war and can be used only for peaceful purposes; and that space-based weapons threaten the peaceful uses and sanctuary of outer space. The effects of a condition of 'space sanctuary', which may have once virtually existed in the early space age,[41] is not contingent upon the non-deployment of space-based weapons or kinetic – or explosive – kill anti-satellite systems based on Earth. In other words, space is not a 'sanctuary' from hostile actions today *despite* the fact that space-based weapons are not deployed. This makes the ambitions of the PPWT – to prevent space warfare by banning space-based weapons – ring hollow. Indeed, to define space weaponisation extremely narrowly as the long-term or permanent placement of weapons in orbit is logically valid, but practically useless because weapons systems based on Earth can generate similar effects and de facto enable space warfare, thereby reducing the need to put weapons *in space* itself. This renders the PPWT impotent if the true intention of its sponsors is to prevent space warfare and support the 'peaceful' uses of outer space.

Peaceful purposes and the PPWT

The principle of self-defence, in conjunction with the economic, strategic, and tactical military advantages derived from space systems, can be employed to necessitate and justify all manner of military space systems. As officials or government agencies refer to comprehensive activities in outer space, the invocation of 'peaceful purposes' is never far away. Neither is the phrase or principle of 'the right to self-defence' kept at a distance. The OST, which entered into force in 1967, and has been signed by all major space-faring states, declares that its signatories recognise 'the common interest of all mankind in the progress of the exploration and use of outer space for peaceful purposes'.[42] However, American parlance has a long tradition of making it clear that 'peaceful purposes' does not mean non-military, as military applications are not necessarily 'aggressive'. This dates back to the very beginning of the space age, to the Eisenhower administration's objective of securing the right of overflight for US strategic reconnaissance satellites whilst also claiming to be promoting the use of space for 'peaceful purposes'.[43] This curbed any potential for the phrase 'non-military' to be employed in treaty language instead.[44] Any claims that the United States is violating or ever has violated the principle of 'peaceful purposes' in outer space will fall foul of this historic terminological coup for the United States. Such claims also belie the fact that other major space-faring states have developed military space systems that resemble the military space capabilities of the United States. Since the dawn of the space age the Soviet Union too developed space technologies for military purposes while citing 'peaceful purposes' as a public cover. In the 1980s it disguised its own missile defence and space interception research and development programmes under the rhetoric of peaceful purposes. The Soviet Union also accused the United States of violating the principle with Ronald Reagan's ambitious but flawed Strategic Defense Initiative.[45]

The principles of 'peaceful purposes' and the right to 'self-defence' are still invoked in contemporary US policy.[46] The tensions between the 'peaceful use of space' and developing military capabilities are apparent *only* if we refuse to recognise the reality that military space is ingrained into modern military planning and strategic capabilities. Developing military space capabilities does not *inherently* challenge the peaceful uses of space, if one adopts a deterrent and peacekeeping frame of thought. It is not only the United States that does this. In 2008 France, on paper, desired the 'demilitarisation' and non-weaponisation of outer space while simultaneously touting the merits of both a joint military space command and also the continued investment in reconnaissance and military force enhancement through space systems.[47] By 2013 the same song only added increased emphasis on the value of space-based reconnaissance and intelligence gathering, its dual-use capabilities, and the necessity to integrate military space capabilities to achieve dominance over adversaries.[48] Worthy of note is the absence of the desire to 'demilitarise' space in the 2013 French White Paper. Australia's 2013 Defence White Paper specifically cites Chinese 'counterspace' – i.e., capabilities to disrupt/destroy space systems[49] – as a concern and Australia's continued access to space systems as a high priority.[50] The Australian government also simultaneously employed rhetoric over the 'peaceful' uses of space while claiming the military benefits of space systems.[51] China, too, is guilty of the same. Its 2010 defence white paper cites how space is to be used for 'peaceful purposes'[52] while the People's Liberation Army is modernising through no small use of space-enabled force enhancement.[53] Russian space policy follows in the same track, with force enhancement and counterspace capabilities being developed while simultaneously proselytising its pacifist rhetoric to prevent an 'arms race in space' and keep space use solely for 'peaceful purposes'.[54] The United Kingdom's recent and first foray in national security space policy employs the same rhetorical devices.[55] This brief discursive survey, which includes the five permanent members of the UN Security Council plus Australia, demonstrates that capable space powers have developed military space systems while paying lip service to the notion of space as being an ostensibly peaceful arena. The notion that outer space is a realm used solely for peaceful, non-aggressive and non-military purposes is a rhetorical tool that does not accurately represent what is occurring in orbit. Space systems are essential links in the kill-chains of the most modern military powers to visit precise and rapid death and suffering on Earth. This doublespeak is not restricted to the United States, which should not be uniquely demonised for its military space activities, as is a hallmark of some polemic space security literature.[56]

The rhetoric on space sanctuary and 'peaceful use' may persist in being taken seriously through strategic ignorance, naiveté, and the political and professional capital invested in such terms among the actors embedded in a space arms control complex. This has an analogical precedent in what is termed the nuclear non-proliferation complex.[57] Of course, showing the hypocrisies or contradictions in international relations and associated think-tanks is unimportant and trivial in and of itself, but it can be a useful point of entry to critique pervasive logical fallacies associated with recurrent arguments that influence policies and international laws, norms and treaties, or are used as rhetorical tools to discredit others in the peanut gallery of mainstream or generalist media. These fallacies need to be challenged when those who use them are urging or developing actual policies and are taking part in public discourse and debate, but therein mislead the public and unfamiliar analysts regarding strategic realities. Such is the case with the PPWT and the conceptual fallacies it perpetuates.

The PPWT,[58] proposed by Russia and China in 2008,[59] attempts to ban all 'weapons' from being deployed in orbit (Article II). It becomes apparent that the PPWT is incriminated as a proposal doomed to failure, if one examines the Russian and Chinese ambassadors to the Conference on Disarmament's letter in response to questions submitted by their American counterpart.[60] Responses to questions 1, 2, 3, and 6 demonstrate the practical difficulties, if not impossibility, of reconciling a duplicitous desire for peaceful purposes of space with a purposefully nebulous right to self-defence that only restricts the future potential strategic freedom of a particular state most likely to develop comprehensive orbital docking and inspection technologies for *many* purposes – the United States.[61] However, Chinese orbital rendezvous and piloting capabilities are progressing, which may provide a residual interest and expertise in some space-based weapons capabilities. Russia has a great heritage in space-based weapons capabilities, as mentioned below, but it lacks the intellectual and financial resources to deploy, maintain, and modernise extensive orbital weapons systems today. Perhaps the PPWT was never intended to succeed – but to provide public diplomacy points at the cost of the United States with a potential reward of slowing American development of space-based weapons technologies, should it ever choose to do so.

Sanctuary and the PPWT's strategic irrelevance

The debate on whether the United States should deploy weapons *in space* is merely a distraction from the real and varied uses of spacepower in the United States and beyond – in developing space services, satellite infrastructure and Earth-based anti-satellite capabilities.[62] In general terms, opponents of space weaponisation advocate 'some variation on the same policy theme' of keeping orbital space free of weapons because of fears of making the international security situation unstable, or that it may ultimately make the United States less secure, as others will develop space-based weapons in response.[63] The PPWT has more political merits by drawing on the flawed popular imagination of outer space as a sanctuary free from conflict and a place used only for peaceful purposes. The strategic reality is that space-based weapons only provide a peripheral and exotic interest if a state is to seriously plan for space warfare.

The idea of space sanctuary, a vision of space as somewhere relatively benign that is proselytised in the face of the military nature of humanity's exploitation of outer space, is used to create a false jeopardy in order to advance intransigent pacifist agendas for the way that Earth orbit should be used.[64] To be fair, there are indeed uncompromising militant texts advocating that outer space should be weaponised as soon as possible which fall foul of the same logical traps, virtually devoid of nuanced political analysis and the uncertainties inherent in making strategic net assessments.[65] The existence of 'counterspace' or Earth-based anti-satellite weapons already threatens, to varying degrees, the safe and reliable operation of space systems, therefore questioning the utility of describing outer space as a sanctuary in the first place. Weapons do not need to be based *in space* to bring about space warfare and destroy the notion of space as a sanctuary from conflict. This strategic reality undermines the entire space warfare prevention purpose given to the PPWT. In addition, the fact that space is not a sanctuary does not mean that the world is doomed to suffer more destabilisation or certain death from above.

The debate as to the possibilities and limitations of anti–satellite weapons are not new, dating back as far as the first US ASAT proposals in the 1950s and the Strategic Defense Initiative (SDI) in the 1980s.[66] Lopez reviewed some of the common assumptions and arguments made by 'space doves', who believe in sanctuarising outer space and propose restrictive arms control regimes. First, the doves believe that there is stability in the present international system with the absence of space-based weapons and overtly deployed Earth-based kinetic- or energy-kill ASAT weapons. Second, other states will be able to respond in kind to an American drive to 'weaponise' space. Third, that allies and potential adversaries would act unilaterally and forgo the possibilities of balancing, or bandwagoning dynamics. Finally, the only way to prevent an arms race in space is through arms control treaties; in the absence of such treaties an arms race is all but inevitable.[67]

Lopez challenges the assumption which underlines the PPWT and the first feature of the space dove argument: space-based weapons would make a presently stable international system unstable. Lopez argues that disproportionate US dependence on space systems relative to other powers for military capabilities, and without any reliable means to protect them, makes the current international order *already* unstable. Similarly, space weapons could be portrayed as stabilising instead of destabilising depending on one's analysis.[68] Whether a strategic relationship is stable or not is a subjective judgment that rests upon political, economic, social, and psychological factors, as well as the character of material capabilities (foremost the status of guaranteed nuclear second-strike capabilities). 'Stability' in and from space is no different and space security analysis must avoid the technocentric and linear strategic thought often exhibited in the debate on space weaponisation. The ability to mutually threaten satellites may impose a measure of calm and caution because China has put many of its own targets in orbit for American Earth-based anti-satellite weapons to shoot down or disrupt.[69]

A final analysis is contingent upon the individual or organisation's socio-political conclusions in two ways: on the real and perceived vulnerability and dependence upon satellites, and on the net assessment of (in)stability in the political–economic status quo. These reflect the nature of peace, its preservation, and war's planning and execution as political activities. Political questions abound and their answers should be treated as such – subjective and open to critique, not iron laws. In this sense, space warfare is the continuation of Terran politics by other means.[70] Stability or instability in space does not occur in isolation from events on Earth: the Code and the PPWT reflect the strategic interests of their sponsors. The view that putting weapons *in* space would be destabilising is open to debate because what happens in orbit is but one part of the grand strategic landscape. Indeed, sometimes space weapons, including space–based weapons, can be envisioned as assisting in crisis stability.[71] The PPWT and its proponents do not consider the effects caused by the deployment of Earth-based weapons systems (or ballistic missile defences) and the integration of vulnerable satellites into modernised military forces. Indeed, serious questions prevail over to what extent nuclear stability is undermined by *Earth–based* space weapons, which may generate a first strike option or added effect in a nuclear war. This is problematic because the satellites used for nuclear attack warnings can also provide tactically relevant information. For example, the American space-based infrared satellite (SBIRS) system can detect the afterburners of jet aircraft, not only strategic missile launches. If future satellites that are essential in conventional wars are not distanced or separated from those needed to calm fears of an incoming nuclear attack, strategic stability may be undermined.[72] It is worth stressing that this discussion does not involve the spectre of space weaponisation or the placement of weapons in orbit; rather it involves Earth-based anti-satellite weapons, the changing capabilities of specific satellites, and conventional space-dependent kill-chains.

Some space-based weapon advocates prescribe the future actions of potential US adversaries in a linear fashion: 'If an adversary is first to deploy or use space weapons, then it may already be too late for the U.S. to do anything about it. Loss of a war in space would be utterly catastrophic for the U.S. and for the world.'[73] However, the space dove community can use that same fatalist logic to portray a doomsday consequence along the lines of a security dilemma for the US if it pursues a certain space-based weapons policy. A critique of such a problematic logical thought train is only then replaced with another equally problematic thought train to support initiatives like the PPWT: banning space-based weapons or choosing not to deploy them will avoid a 'space arms race', improve the security situation for the United States, and, therefore, make satellites more secure. There is no reason to rule out an arms race in the event *and* non-event of a deployment of weapons in orbit because space-based weapons are only one aspect of the strategic picture of spacepower and space warfare. The politics and uncertainties of war and peace may produce a mixed and creative response against any deployment of a particular kind of weapons system. There is no obvious 'threshold', point of no return, or an event horizon, to space weaponisation. A major concern for strategic stability today, which the Code and the PPWT fail to acknowledge, is the consequences of intentionally destroying or disrupting tactically relevant space systems that are also responsible for ensuring nuclear stability and assured retaliatory capabilities. This can be achieved with various Earth-based weapons systems today. It is not the development of extensive space weapons (whether Earth- or space-based) that is driving the development of anti-satellite weapons; rather it is the tactical and operational utility of satellites being developed by most major powers. With more satellites in space forming the political, economic, and military instruments of more states, there are more incentives to engage in space warfare without needing to place dedicated weapons platforms in space, or to 'weaponise space', which makes the PPWT a sideshow.

Conclusion

This chapter has outlined the strategic context of contemporary space security through a critique of two legal initiatives. The International Code of Conduct seeks to enshrine rules of the road and norms to govern normal behaviour in orbit. However, it falls foul of the dual-use nature of space activities as the Code will now have to incorporate clauses about preventing space warfare. The Code, if it is to have an impact, must accommodate the military uses of outer space by not attempting to curb them and by not enabling force by granting legal privileges to military or national security activities. Though most space policies revere the peaceful uses of space, all are exploiting the benefits of spacepower to threaten or impose violence, harm, and destruction on Earth in more efficient, reliable, and effective ways. This is no more cause for alarm than the harnessing of the seas and the air to transport and support military power and political will. Space has always been militarised, much as the air and the sea have been for decades and millennia, respectively. However, any notion of current or previous nuclear stability that has been enabled by satellites may be challenged if tactically relevant satellites also provide nuclear warning.

This chapter has also criticised the notion of space sanctuary and the PPWT as a legal initiative to prevent space warfare. Be it the United States' Aegis-equipped ships or China's maturing Earth-based space weapons, the PPWT would do nothing to curb their development and use.

In the face of strategic reality, the PPWT and Code disguise the fact that more states are relying on space for their ability to impose their will and to protect themselves, which also drives development in anti-satellite capabilities on Earth. Prohibiting space-based weapons does not forestall the coming of space warfare, should such a war erupt between space powers. Given the integration of spacepower into our daily lives and military capabilities, and not only in the Western world, it is time to view outer space and astropolitics beyond the inhibitive lenses of the 'militarisation' and 'weaponisation' of space. It is time to consider spacepower's place *along-side* geopolitics and security concerns on Earth, and not in isolation. Space warfare will be the continuation of Terran politics by other means, and the proliferation of spacepower reflects the continuing grand strategic interests of the major powers.

Notes

1 Robert C. Harding, *Space Policy in Developing Countries* (Abingdon: Routledge, 2013) p. 17; Everett C. Dolman, *Astropolitik: Classical Geopolitics in the Space Age* (Abingdon: Frank Cass, 2002), p. 5.
2 European Commission, 'Towards a Space Strategy for the European Union that Benefits its Citizens', Brussels, 04/04/2011, 6; U.S. Department of Defense (DoD), Washington, DC, Directive 3100.10, 'Space policy', 18 October 2012.
3 Michael Sheehan, *The International Politics of Space* (Abingdon: Routledge, 2007) p. 38.
4 Thomas G. Mahnken, 'Weapons: The growth and spread of the precision-strike regime', *Daedalus*, Vol. 140, Issue 3 (2011), pp. 45–57; Andrew F. Krepinevich, *Maritime Competition in a Mature Precision-Strike Regime* (Washington, DC: Center for Strategic and Budgetary Assessments, 2014).
5 On space-to-Earth weapons, see: Bob Preston et al., *Space Weapons, Earth Wars* (Washington, DC: RAND, 2002).
6 On radiofrequency and navigation spoofing, see Logan Scott, 'Spoofs, proofs & jamming towards a sound national policy for civil location and time assurance', *Inside GNSS*, September 2012, available at www.insidegnss.com/node/3183 (accessed 7 May 2017).
7 On the politics of space debris and removal technologies, see Bleddyn E. Bowen, 'Cascading crises: orbital debris and the widening of space security', *Astropolitics*, Vol. 12, Issue 1 (2014).
8 See Brian Weeden, 'Through a glass, darkly: Chinese, American, and Russian anti-satellite testing in space', 17 March 2014, Washington, DC, available athttp://swfound.org/media/167224/Through_a_Glass_Darkly_March2014.pdf.
9 Ibid, p. 1.
10 Miroslav Gyűrösi, 'The Soviet fractional orbital bombardment system program: Technical Report APA-TR-2010-0101', *Air Power Australia*, January 2010, updated April 2012, available at www.ausairpower.net/APA-Sov-FOBS-Program.html (accessed 12 November 2016).
11 Anatoly Zak, 'Spacecraft: manned: Almaz: OPS-2 (Salyut-3)', *Russian Space Web*, available at www.russianspaceweb.com/almaz_ops2.html (accessed 12 November 2016).
12 Jana Honkova, 'The Russian Federation's approach to military space and its military space capabilities', *Policy Outlook*, November 2013, (Washington, DC: George C. Marshall Institute), p. 35.
13 Weeden, 'Through a Glass…', p. 23 (see note 8 above).
14 On RORSAT, see: Pavel Podvig, 'Russia and military uses of space', in: Pavel Podvig and Hui Zhang (eds.), *Russian and Chinese Responses to US Military Plans in Space* (Cambridge, MA: American Academy of Arts and Sciences, 2008), p. 10; Wade Boese, 'Chinese satellite destruction stirs debate', *Arms Control Association*, March 2007; Forrest E. Morgan, *Deterrence and First-Strike Stability in Space: A Preliminary Assessment* (Washington, DC: RAND, 2010), p. 12.
15 Laura Grego, 'SM-3 as ASAT', Union of Concerned Scientists, 26 May 2012, available at http://allthingsnuclear.org/lgrego/aegis-as-asat (accessed 12 November 2016).
16 Anatoly Zak, 'The Naryad program', *Russian Space Web*, updated 6 May 2016, available at www.russianspaceweb.com/naryad.html (accessed 12 November 2016).
17 On a brief technical overview, see: UK Ministry of Defence, *UK Military Space Primer* (Shrivenham: Doctrine and Concepts Development Centre, 2010) 3-5 – 3-7; US Air Force, *AU-18 Military Space Primer* (Montgomery, AL: Air University Press, 2009), pp. 273–81.
18 David Shiga and Agence France Presse, 'Mysterious source jams satellite communications', *New Scientist*, 26 January 2007.

19 John J. Klein, *Space Warfare: Strategy, Principles and Policy* (London: Routledge, 2006), pp. 59, 95.
20 David A. Vallado and Jacob D. Griesbach, 'Simulating space surveillance networks', Paper AAS 11–580 presented at the AAS/AIAA Astrodynamics Specialist Conference. 31/07/11–04/08/2011, Girdwood, AK, p. 13. Online. Available at: www.agi.com/resources/white-papers/simulating-space-surveillance-networks.
21 Arvind Gupta, 'Foreword', in Ajey Lele (ed.), *Decoding the International Code of Conduct for Outer Space Activities* (New Delhi: Institute for Defence Studies and Analyses, 2012), p. ix.
22 European Union External Action Service (EEAS), 'Code of conduct for outer space activities', available at http://eeas.europa.eu/non-proliferation-and-disarmament/outer-space-activities/index_en.htm (accessed 18 August 2016).
23 For Michael Krepon, writing on ploys to expand the Code's remit to include space weapons, see: Michael Krepon, 'Space Code of Conduct mugged in New York', *Arms Control Wonk*, 4 August 2015, available at http://krepon.armscontrolwonk.com/archive/4712/space-code-of-conduct-mugged-in-new-york (accessed 18 August 2016).
24 On legal approaches to achieving political aims, see Larry M. Wortzel, *The Dragon Extends its Reach: Chinese Military Power Goes Global* (Washington, DC: Potomac Books, 2013) p. 121, Larry M. Wortzel, 'The Chinese People's Liberation Army and space warfare', *Astropolitics* , Vol. 6 (2007), pp. 116–18.
25 Ajey Lele, 'Space Code of Conduct: inadequate mechanism', in Ajey Lele (ed.), *Decoding the International Code of Conduct for Outer Space Activities* (New Delhi: Institute for Defence Studies and Analyses, 2012), p. 6.
26 Ibid., p. 20.
27 EEAS, 'Code of Conduct', sections 1, 28, and 51.
28 See particularly Article 95 in the United Nations Convention on the Law of the Sea, available at www.un.org/depts/los/convention_agreements/texts/unclos/unclos_e.pdf.
29 Scott Pace, 'Space cooperation among order-building powers', *Space Policy*, Vol. 36, 2016, pp. 25–6.
30 Colin S. Gray and John B. Sheldon, 'Space power and the revolution in military affairs: a glass half full?', *Airpower Journal*, Autumn 1999, p. 27.
31 Michael Sheehan, 'Counterspace operations and the evolution of US military space doctrine', *Air Power Review*, Vol. 12, No. 3 (2009), p. 107.
32 For example, see: Ashley J. Tellis, 'China's military space strategy', *Survival*, Vol. 49, No. 3 (2007), pp. 41–72; Jana Honkova, 'The Russian Federation's approach to military space and its military space capabilities', *Policy Outlook*, November 2013 (Washington, DC: George C. Marshall Institute); Pavel Podvig, 'Russia and military uses of space', in Pavel Podvig and Hui Zhang (eds.), *Russian and Chinese Responses to US Military Plans in Space* (Cambridge, MA: American Academy of Arts and Sciences, 2008); Anthony H. Cordesman, Ashley Hess and Nicholas S. Yarosh, *Chinese Military Modernization and Force Development: A Western Perspective* (New York: Rowman & Littlefield, 2013), p. 20.
33 Michael Listner, 'Redux: it's time to rethink international spacelaw', 24 November 2014, available at www.thespacereview.com/article/2647/1 (accessed 18 August 2016).
34 On China's increasing integration of spacepower into its tactical and operational military capabilities, see: Ian Easton, *China's Evolving Reconnaissance-Strike Capabilities: Implications for the U.S.-Japan Alliance* (Washington, DC: Project 2049 Institute, 2014).
35 See, for example: Robert C. Harding, *Space Policy in Developing Countries* (Abingdon: Routledge, 2013).
36 See United States Department of Defense, 'National Security Space Strategy: Unclassified Summary', January 2011, Washington, DC, 1.
37 Joan Johnson-Freese, *Space as a Strategic Asset* (New York: Columbia University Press, 2007), p. 22.
38 On the history of Soviet and American space technology and rocket development, see: Walter A. McDougall, … *The Heavens and the Earth: A Political History of the Space Age* (Baltimore, MD: Basic Books, 1985).
39 James E. Oberg, *Space Power Theory* (Colorado Springs, CO: USAF Academy, 1999), p. 14.
40 For example, see Ian Sample, 'China's Jade Rabbit rover makes crucial tracks in space and on Earth', *The Guardian*, 20 December 2013, www.theguardian.com/science/2013/dec/20/china-jade-rabbit-rover-space-politics (accessed 12 November 2016).
41 Paul B. Stares, *Space and National Security* (Washington, DC: Brookings Institution, 1987), p. 1.
42 United Nations, ST/SPACE/11, 'UNITED NATIONS TREATIES AND PRINCIPLES ON OUTER SPACE', New York, 2002, p. 4, available at: www.unoosa.org/pdf/publications/STSPACE11E.pdf.

43 Peter L. Hays, Struggling towards Space Doctrine: U.S. Military Space Plans, Programs, and Perspectives During the Cold War (PhD Thesis, Tufts University, 1994), p. 63.
44 Ibid., p. 142.
45 See: Alun Chalfont, 'Red Star Wars: the hidden facts', *The Times*, 3 March 1985, p. 14.
46 The White House, United States National Space Policy, Washington, DC, 28 June 2010, available at: www.whitehouse.gov/sites/default/files/national_space_policy_6-28-10.pdf; The White House, National Security Strategy, Washington, DC, May 2010, 31, available at: www.whitehouse.gov/sites/default/files/rss_viewer/national_security_strategy.pdf; US Department of Defense, National Security Space Strategy, 5.
47 French Government, 'The French White Paper on defence and national security', Paris, 2008, Sections 12, 13, 14, available at: www.ambafrance-ca.org/IMG/pdf/Livre_blanc_Press_kit_english_version.pdf.
48 French Government, 'French White Paper: Defence and National Security 2013', Paris, 2013, 44, 70, 81, 118, available at: www.rpfrance-otan.org/IMG/pdf/White_paper_on_defense_2013.pdf.
49 Space systems are made up of four segments – the Earthbound ground stations and users, the satellite(s), the uplink, and the downlink.
50 Australian Department of Defence, 'Defence white paper 2013', Canberra, 2013, pp. 15, 24.
51 Australian Government, 'Australia's satellite utilisation policy', Canberra, 16 April 2013, pp. 12–15, 18–19.
52 Chinese Government, 'China's national defense in 2010', Beijing, 2010, chapter X, available at: www.china.org.cn/government/whitepaper/2011-03/31/content_22263885.htm (accessed 23 May 2014).
53 Chinese Government, China's National, … Chapter III, available at: www.china.org.cn/government/whitepaper/2011-03/31/content_22263445.htm (accessed 23 April 2013).
54 Jana Honkova, The Russian Federation's Approach, esp. 5–9 (see note 12 above).
55 HM Government, UKSA/13/1292, 'National space security policy', London, April 2014, esp. 5–10.
56 For example, see Johnson-Freese, *Space*, pp. 2–5, 19–23, 51, 99 (see note 37 above).
57 Campbell Craig and Jan Ruzicka, 'The nonproliferation complex', *Ethics and International Affairs*, Vol. 27, No. 3 (2013).
58 CD/1839, Draft: 'Treaty on the Prevention of the Placement of Weapons in Outer Space and of the Threat or Use of Force Against Outer Space Objects', 29 February 2008, available at: www.cfr.org/space/treaty-prevention-placement-weapons-outer-space-threat-use-force-against-outer-space-objects-ppwt/p26678.
59 The PPWT is a contemporary revival of an older and even more problematic Soviet space-based weapons ban proposal: CD/274, 'Draft Treaty on the Prohibition of the Stationing of Weapons of Any Kind in Outer Space', 15 January 1982.
60 CD/1872, 'Letter Dated 18 August 2009 from the Permanent Representative of China and the Permanent Representative of the Russian Federation to the Conference on Disarmament addressed to the Secretary-General of the Conference Transmitting Answers to the Principal Questions and Comments on the Draft 'Treaty on Prevention of the Placement of Weapons in Outer Space and of the Threat or Use of Force Against Outer Space Objects (PPWT)' Introduced by the Russian Federation and China and Issued as document CD/1839 dated February 2008', 18 August 2009.
61 The US Geosynchronous Space Situational Awareness Program and the Russian *Naryad* system, as well as the Kosmos 2491 and 2499 satellites, are examples of orbital manoeuvring, inspection, and docking technologies that can provide many useful and varied non-aggressive functions in orbit, but also demonstrate a residual space-based capability.
62 On works that do not engage with the space weaponisation debate, see: Bleddyn E. Bowen, 'From the sea to outer space: The command of space as the foundation of spacepower theory', *Journal of Strategic Studies*, published online 23 February 2017, http://dx.doi.org/10.1080/01402390.2017.1293531; Robert C. Harding, *Space Policy in Developing Countries: The Search for Security and Development on the Final Frontier* (Abingdon: Routledge, 2013); Sheng-Chih Wang, *Transatlantic Space Politics: Competition and cooperation above the clouds* (Abingdon: Routledge, 2013); Tellis, China's military space strategy, pp. 41–72 (see note 32 above); Roger Handberg and Zhen Li, *Chinese Space Policy: A Study in Domestic and International Politics* (Abingdon: Routledge, 2007).
63 Karl P. Mueller, 'Totem and taboo: depolarizing the space weaponization debate', in John M. Logsdon and Gordon Adams (eds.), *Space Weapons: Are They Needed?* (Washington, DC: Space Policy Institute, 2003) pp. 13–16.
64 Emblematic of this is the argument in Michael Moore, *Twilight War: The Folly of US Space Dominance* (Oakland: Independent Institute, 2008), p. xvi.

65 For example: Sterling Pavelec, 'The Inevitability of the weaponization of space: technological constructivism versus determinism', *Astropolitics*, Vol. 10, No. 1 (2012); Howard Kleinberg, 'On War in Space', *Astropolitics*, Vol. 5, No. 1 (2007).

66 For example, see Hays, *Struggling Towards*, pp. 156–61 (see note 43 above); Keith B. Payne (ed.), *Laser Weapons in Space: Policy and Doctrine* (Boulder, CO: Westview Press, 1983); John Tirman (ed.), *The Fallacy of Star Wars* (Washington, DC: Vintage Books, 1984).

67 Laura Delgado Lopez, 'Predicting an arms race in space: problematic assumptions for space arms control', *Astropolitics*, Vol. 10, No. 1 (2012), pp. 53–60.

68 Ibid., pp. 53–5; on space-based weapons as stabilising, see Payne, *Laser Weapons in Space* (see note 66 above).

69 Bleddyn E. Bowen, 'Down to Earth: the influence of spacepower upon future history', paper presented at the International Studies Association Annual Convention, 25 February 2017, Baltimore, MD.

70 Such was the leitmotif in: Bleddyn E. Bowen, 'Spacepower and Space Warfare: The Continuation of Terran Politics by Other Means' (PhD Thesis, Aberystwyth University, 2016).

71 Keith B. Payne, 'Introduction and overview of policy issues', in Payne, *Laser Weapons in Space*, 1–8 (see note 66 above).

72 On anti-satellite capabilities and nuclear stability, see: Forrest E. Morgan, *Deterrence and First-Strike Stability in Space: A Preliminary Assessment* (Washington, DC: RAND, 2010); Bruce W. MacDonald, 'Deterrence and crisis stability in space and cyberspace', in Michael Krepon and Julia Thompson (eds.), *Anti-satellite Weapons, Deterrence and Sino-American Space Relations* (Washington, DC: Stimson Center, 2013).

73 Kleinberg, 'On War', p. 22 (see note 65 above).

21
OUTER SPACE AND PRIVATE COMPANIES
Consequences for global security

Paweł Frankowski

This chapter focuses on sectors, methods, and spheres of the space activity of private companies to provide empirical analysis of space applications and implications for global security.* Special emphasis has been given to the private companies offering access to satellite imagery and satellite remote sensing, as well as companies entering outer space with new and prospective capabilities such as space mining. The chapter explains the rising importance of geo-intelligence, space surveillance, and telecommunication for global security and new kinds of security challenges and vulnerabilities such as environmental problems in outer space or technological challenges to security. The author argues that profit-oriented companies play a crucial role in the new security environment in the US, effectively changing both law and practice. Finally, the new and growing market for subcontractors in space applications raises questions about the growing dependence on private resources in a traditional sphere of state activity, namely security, in this case provided from and through outer space.

Why private space security?

Outer space for years has been an exclusive area of activity of space-faring powers and, due to the strategic importance of outer space, any private activity was just impossible.[1] Geostrategic or geospatial factors for space exploration and the quest for astropolitics,[2] pursued by the United States, the Soviet Union, but also France – were accompanied with an intrinsic aversion to any loss of strategic position in outer space. Politicians and military commanders at the beginning of the space race were intensely opposed to private actors, but with progress in space exploration such opposition has diminished, and private actors offered sector alternatives or were just being more flexible than state-controlled industries. Private space endeavours, initially limited to sub-contracting, have gradually taken an important position in the space sector in both the US and Europe. Finally, the space industry has become a regular part of the economic landscape in the US and Europe, being important not only from an economic but also a strategic point of view. Nevertheless, such liberalisation and discussion over the role and place of private actors in security should be analysed from two concurring and somewhat overlapping perspectives. In the US and Europe discussions over the privatisation of security, and the role of private actors, often run in opposite directions. While in European countries most of the debate is over whether or not some functions of the state, and state resources, should be transferred to private actors, in the

United States most of the arguments revolve around whether such functions should be first and foremost public.[3] This important transatlantic argument on different preferences and ideas also applies to space affairs. However, the effects of privatisation of space affairs are not straightforward. Natural monopolies, established by private space companies in some sectors, and enhanced due to retrenchments and cutbacks, important after the economic crisis of 2008, have become useful carriers of political messages to constituencies and are strongly tied to the logic and dynamics of liberalisation within military affairs. Therefore, there are good reasons for believing that the logic of privatisation in the military has been adopted also in space affairs, where a narrative on the sunk costs of existing public space policies, opposition from lobbies, but also enduring enthusiasm over space exploration, have facilitated the turn to privatisation in space affairs.

Arguing that the logic of space privatisation has been intertwined with the neo-liberal dynamics of privatisation of public services raises a fundamental question: why do some governments seem to be much more willing to support private space actors than others? Deteriorating economic situations could be a key, but that is an insufficient factor to understand such strategic choices over crucial resources for state security. The answer to this salient question may be centred on preferences, alternatives, and benefits (both immediate and indirect), where the costs of failure could be diffusely distributed among numerous private actors as well as public agencies.

As mentioned earlier, while some space assets and services, such as telecommunications services, have been in private hands from the very beginning of space exploration, for other sectors such as space imagery or synchronising services it was not an easy path. However, strategies geared towards more private involvement are intrinsically similar to strategies and justifications in other public services. John Donahue, referring to the privatisation of public services, argues that the political choice between public and private services basically has two dimensions. The first concerns finance, focusing on the questions of whether or not individuals should pay for services individually, or whether perhaps the same services should be provided by the state, with funds raised from taxation. Apart from financing, the second dimension focuses on performance, flexibility, and the ability to adapt to changing circumstances. Here we should analyse whether services should be delivered from the governmental level or be provided by a non-state entity, with a decreased attachment to procedures, red tape, and a managerial style of governing.[4] Nevertheless, the privatisation of security and military services follows a slightly different logic, because even though private companies acquired contracts to provide security services, such services will still be financed by public money. Indeed, the main source of income for the private space industry is public actors, and space companies can hardly find other clients. For example 66 percent of the European space industry is accounted for by the public sector, and in 2015 European companies provided goods worth as much as EUR534 million for military customers.[5]

Privatisation of security often refers to private actors who provide utility services, after acquiring a state-owned company, in most cases 'privatised' in search of better performance or to lower the financial burden on the state. But private security could be also provided with new projects and services, worked out by private companies and then adopted by the state as security measures. As an example, mobile telephony, or a variety of techniques and services connected to face/shape/pattern recognition, was developed by private entities. However, with security the states are natural monopolies as ultimate security providers, and, as a consequence, can regulate market possibilities with regard to services provided by private actors. These regulatory choices and standards usually vary from one state to another; for example, regulations on CCTV in the European Union differ to a great extent, despite regulatory pressures coming from the supranational level. In the space security realm, a clear example of state regulatory choices comes from the satellite imagery industry, where, despite technological possibilities, private providers are restricted from selling high-resolution optical or radar imagery on the open market to customers throughout the world. So called shutter control prohibits the high-quality imaging of particular areas, or companies are forbidden from collecting or selling imagery of a particular country. For example US companies cannot collect or sell imagery of Israel, with any better quality than is available from other commercial sources.

The literature on privatisation of military services has expanded rapidly, especially after 2002 and involvement of private companies in the Iraqi operation. While appreciating the varying outlook of different scholars dealing with private military companies, this chapter is following Prado[6] to argue that either transferring provision of services to private hands or, conversely, acquiring them from private entities without developing an independent system on the state's behalf can both be beneficial for the state for at least four reasons. The first reason is price, as the cost of private provisions could be lower because private companies can provide services with fewer people, with outsourced services also provided to third countries. The price of military services, to a great extent, depends on the costs of trained personnel, since private companies can hire former soldiers who have already been trained. In contrast, the cost of public security services is increased by the benefits accruing to soldiers after their years of service. For example, from the overall military budget of the United States (USD1 trillion), more than USD200 billion is spent for pensioners, veterans' benefits, and retiree health services.

Secondly, the push for private security may result in more efficient usage of financial and human resources, and soldiers may perform more valuable duties.[7] Therefore PMCs can provide better service for the same price or the same services at a reduced price. This will allow financial resources to be earmarked for another public service or support the argument that public money has been better spent. Thirdly, with private security providers, states can avoid lengthy red tape procedures, with for example standardisation of military procurement in terms of the time required for mobilisation and deployment. Such considerations are important during armed conflict and become more and more important during the planning of infrastructure, using assets, and regulating activity. The demand for more flexible and less troublesome activity in the security realm is constantly increasing, both in Europe and in the Western Hemisphere.

Finally, governments may turn to private resources as a lack of choice, when the state does not have necessary technical or material capabilities to provide security services in a timely fashion.[8] However, some authors suggest that looking for private solutions in security cannot be analysed in isolation from pressure coming from political processes on a larger scale.[9] Nevertheless, distinguishing between the economic power of private actors and the lack of capacity on behalf of the state, as driving factors for privatisation of security services, does not necessarily answer the question why space assets, crucial for power of any important state in the world politics, are developed by private actors, being to some extent neglected by governments.

Some authors argue that privatisation of core state functions, such as security, is rather rare, and private security actors do not enjoy the kind of freedom as other actors in public–private partnerships.[10] Moreover the main business of private security satellite providers is not in protection of life and assets or in any other clearly military activity, but rather in commercial services, albeit to a large extent used by governments. For such companies, even if they are clearly involved to security matters and, it is better to retain an image of being 'highly specialized, knowledge-intensive, expertise-oriented providers of security oriented solutions'.[11]

Privatisation of space activities is based on a cost-saving approach as a dominating actor may result in the lack of administrative oversight. With outsourcing space activities, the very question of public scrutiny, as a part of a democratic state, would also be less relevant. But also, when government turns to private operators, because it lacks capacity itself, as is the case for European states, any meaningful competition may be impossible. Privatisation of space security refers to the discussion of a state's responsibility for space affairs. Nevertheless, as some authors point out, this responsibility can be problematic – since definition of responsibility remains vague – from the perspective of possible implementation.[12]

What is space security?

With regard to space security and private activities it is necessary to point out how this very widely used term has been coined, what is the contemporary understanding of space security, and how private actors could be merged into thinking about all of the policies and programmes in terms of their implications for space affairs. Michael Sheehan correctly points out that space security and the role of space for security have been discussed rather widely, and the military dimension is still the most important part of such discussion. Nevertheless other important issues are also included in the discourse on space security as the meaning of 'space security' has come to be understood in more general way. But for Sheehan 'space security' is limited to space actors, namely states, who wish to use it for the socioeconomic benefits of their populations [and to] increase human security' and perceive it as the 'ability to use space as a vital national interest'.[13]

For Jean Francis Mayence, space security is perceived in three interrelated dimensions, as (1) outer space for security; (2) security in outer space; and (3) security from outer space.[14] For the first dimension, all defence purposes and using space for all security issues is analysed; security *in* outer space relates to the space system and the space sustainability of space activities. The third dimension of space security, as a post-Cold War dimension, focuses on greater security of Earth as a whole, where environmental protection, weather forecasts of floods and droughts, but also rescue and disaster management, are central themes of space endeavours. For Mayence security in outer space, dominated by public actors, such as space-faring nations or the ESA, is also complemented by commercial stakeholders. However, private actors, apart from assessing the political risk of possible destruction of their satellites in orbit, rather focus on reducing any financial impact on their space business. A broader understanding of 'space security' encompasses issues concerning 'more than just activities occurring beyond Earth's atmosphere' but all the elements of ground stations and communication channels.[15]

However, since the end of the Cold War the boundaries between the strictly military and civilian approaches to space, as well as the military space sector and civilian actors, have largely been blurred, as military commanders use civilian satellite systems to gain strategic information or use commercial telecommunication links for health services directly from a battlefield. For example, during the Gulf War international commercial satellites provided services to field commanders and leased mobile satellite terminals were used in the theatre of war to connect communications systems with headquarters facilities in Florida.[16]

Nevertheless, apart from positive examples of cooperation between private and public actors, Sheehan argues that outer space can produce both security and insecurity. Therefore private companies active in outer space, apart from providing security for space assets, could generate threats in outer space, from outer space, and also through outer space.[17] Possible threats may include, among other things, disruption of satellite signals, creation of space debris, or a potentially dangerous influence on Earth (destruction of an object on Earth), but also withholding private data gathered from satellite imagery.

The last approach to space security, again multidimensional, but clearly built in specific terms of private activity, has been provided by the Space Security Index and Project Ploughshares, who identify 17 factors – namely orbital debris, radio frequency (RF) spectrum and orbital positions, natural hazards originating from space, space situational awareness, access to and use of space by various actors, space-based global utilities, priorities and funding levels in civil space programmes, international cooperation in space activities, growth in commercial space industry, public–private collaboration on space activities, space-based military systems, security of space systems, vulnerability of satellite communications, reconstitution and resilience of space systems, Earth-based capabilities to attack satellites, space-based negation-enabling capabilities, outer space governance, national space policies, multilateral forums for space governance, and other initiatives also provided by private actors.[18] However, considering the notion of preferences, as well as the possibilities of private space companies, security *in* and *from* outer space, public and private goals are tightly interconnected. Thus, to understand current trends in space security, it is useful to analyse the role of space services and how these services have been governed.

Space governance and space security

Space governance can be studied from different perspectives; however, established links between the predominant public actors with particular space assets and the still-small number of private space companies greatly limits comparative work to two regions, the US and Europe. However the discussion so far has identified one aspect of space governance as being especially puzzling, namely the growing role of private governance, where public actors deliberately and somehow consciously disengage from regulating space security. This international governance by default, when it comes to space security, has been divided into two intertwined models of global commons and strategic stability. While the model of global commons relates to voluntary action, and self-restraint, the model of strategic stability is more sophisticated when commercial benefits are also included.[19] However, neither the global commons nor the strategic stability model looks deeply into relations between private and public approaches to space security. This is mainly due to the fact that almost 80 percent of revenues from overall commercial space activity (USD195 billion) come from satellite television services).[20] Satellite imagery, remote sensing, or future space mining are less important for commercial actors, and to a large extent they cooperate or coordinate their behaviour with national governments.

Depending on how the interests and responsibilities of private actors and governments are defined, there are five types of interaction between the two centres of power: coordination, cooperation, coexistence, conflict, and coalescence. Five types of relationships are derived from changes that are taking place in the space law in accordance with the latest developments in such areas. It should be emphasised, however, that the change observed in the system can give rise to different responses from the participants in the system. In addition, four types of relations reflect an ideal situation, while in reality we have to deal with either a simultaneous buildup and conversion or with denial and buildup. Therefore, the type of change will be of secondary importance, while the type of interaction will determine the function of private actors and states and the hierarchy of preferences on all sides of the political processes.

These types of interaction are determined by two types of variables: dependent – that is, the interests and preferences of the parties involved; and independent – that is, the structure of the system. The key question is therefore how much possible and permanent change will result from this interaction in the global system of space security. Examples analysed later in this chapter demonstrate the durability of the independent variables in the relative volatility of preferences and interests. The last element essential for the understanding of the relationship between the interests of commercial actors and foreign policy is the willingness of both sides of the process to use the opportunities that the relationship gives them for a specific institutional structure and a degree of articulation of preferences. This includes not only an attempt to coordinate private actions from the government level, but also the willingness to cooperate in the framework of the model of cooperation, striving to meet their own preferences.

Nevertheless, modes of governance are, to large extent, based on national regulations, because technology in the hands of commercial actors is so efficient that some states, especially the US and France, have to limit the possible activities of commercial space actors. This national level is important in understanding contemporary space security, where private space services are coordinated by governments and restricted by states' national interests. The model proposed by Adrienne Héritier addresses the question of the role of private actors, who increasingly influence the shape and extent of public policies. Such an approach clearly determines that there is presently no strictly public policy, as all those engaged in policy formation are not private actors, even when dealing with key issues for the existence of the state, such as public security or defence. In sum one should assume that (1) private actors are included in the process of policy formulation; (2) management is still based on public actors; and (3) the management to a small extent is based on legislation but also on a series of adjustments on both sides.[21]

However, sometimes private actors are so powerful that they are able to control legislation in a pre-emptive manner, when technology cannot keep up with proposed legislation. This is certainly the case of the Commercial Space Launch Competitiveness Act (SPACE Act of 2015), signed by the President of the United States in 2015. This act, which is the culmination of efforts by the American private sector, opens up new opportunities for private companies in terms of not only launching heavy lift cargo and humans, but also in 'commercial exploration and recovery of space assets by the citizens of the United States'. Such an act allows US citizens to obtain, possess, transport, use, and sell space assets in accordance with US law and international law. The SPACE Act of 2015 contains a clause that by adoption this act precludes taking possession or exclusive rights to any celestial body. Nevertheless, the importance of the adopted legislation goes far beyond the Outer Space Treaty of 1967, to which the United States is also party. Although currently owned technology does not allow the start of exploration of asteroids by private companies, the SPACE Act should be understood as an attempt to impose legal arrangements from which other powers will follow. Despite the ban on the appropriation of space contained in the Space Treaty, there are no clear international legal regulations concerning the possibility of acquiring space assets.[22] Moreover, commercial entrepreneurs who may decide to explore outer space and start mining activities would not be bound by any specific positive or negative obligations regarding the benefits of such activity. The only limitations for commercial actors will be to ensure that space mining does not interfere with states' rights.[23] Such a legal gap and thus a security gap would allow private companies to regulate, at least at the very beginning, all rules for environmental issues.

This pre-emptive regulation and legislation follows the logic of the Moon Treaty, an international regulation which regulates the activities of states on the Moon and other celestial bodies, adopted in 1979. The Moon Treaty, in fact not signed by such space-faring states as the US or Russia, contains a commitment to establish a regime laying down the rules of international exploration of the Moon, when this becomes possible.[24] The SPACE Adoption Act of 2015 is a response to the plans of two US companies: Deep Space Industries and Planetary Resources plan to start mining asteroids before 2023 using nano-satellites, with the first cargo of raw materials or ore obtained from asteroids to be brought back to Earth in 2018. Supporters of the Commercial Space Act emphasise that this act fills the gap in the uncharted and non-regulated area of space practice, but does not violate international law, and that the turnover of materials derived outside of Earth is a fact. The concept of exploitation of space assets and their transport back to Earth is presently unprofitable from an economic point of view. However, some experts suggest that in the foreseeable future space mining companies will be able to exploit raw materials for the production of fuel in orbit and replenishment satellites whose design allows for such a possibility.[25] In addition, raw materials or partially processed ore may be used for microgravity metallurgy and 3D printing, which opens new prospects for the use of space. The second option which the law opens for American companies could be a recovery of inactive satellites orbiting on the so-called 'graveyard orbit' by repairing and refuelling with fuel extracted from the exploitation of asteroids. Private actors operating from the territory of any state, which would be the first to reach the possibility of using such technology, will gain an advantage that will change the balance of power in orbit and also will have an impact on the privatisation of space security. Since decommissioned satellites are often sold to international consortia or third states without any interests in satellite operations, refuelling of inactive or abandoned satellites should be treated as a kind of refurbishing activity employing elements found in the space scrap yard. Contemporary international law does not have any mechanism to prevent such hijacking; but the idea behind such refuelling is based on environmental benefits, rather than on any parallel with piracy or stealing. Therefore pre-emptive national legislation, supported by private entities, can be characterised as reverse coordination, where commercial players, motivated by prospective benefits, are able to craft and coordinate space activities without tested technology in hand.

Private actors also change the practice of spacecraft safety, which gradually will be provided by private contractors offering not only the possibility to launch a cargo, but also, for example, cheaper, more effective capabilities suited to the needs of a Member State space debris reduction system. With no attempt to regulate this uncharted area of international law, manoeuvres within the gap in international regulations made by powerful commercial actors is clearly an example of the privatisation of security.

Governments also gave private companies *carte blanche* in space mining. Such *carte blanche*, or the naive assumption that contemporary technology can still be controlled by governments, has also been visible in remote sensing and satellite imagery. Historically the United States played a crucial role in remote sensing, and images taken from Earth orbit were used for military purposes. But the decision taken by the French government to create the Satellite Pour l'Observation de la Terre (SPOT) company, operating on commercial terms, selling images to other states and to private customers, altered the position of the United States. Nowadays detailed space imagery provided by commercial providers is accessible to almost everyone – actors involved in security issues, but also such actors as universities, the media, insurance companies, or even individuals who can buy images from commercial sources – so the contemporary understanding of international security must be changed. When private actors, starting from NGOs and human rights activists, to individuals, have sensing capabilities, their possible impact on security would complement traditional security measures rather than focus entirely on space security as such. However, thanks to Google Maps, actors such as terrorists or insurgency forces can plan in more detailed and effective ways and analyse their attacks in the context of spatial relationship in a way unavailable 15 years ago.

Is it really private space security?

Privatisation of space security and outsourcing affects states' regulatory possibilities, while not always leading to more efficient allocation of resources. States may look for private resources to avoid regulatory barriers or democratic scrutiny, but also when they do not have enough resources to perform some duties. Also, governments may turn to private actors to start meaningful competition in the market. But to be effective such competition requires multi–stakeholder participation, while the space market is still specific. For example, Digital Globe has very strong purchasing power as the global market leader for Earth observation services with 63 percent of worldwide market shares,[26] while Airbus DS Geo-Intelligence has 14 percent and Planet Labs has 5 percent. More than one-third of customers are from the defence and intelligence areas. Thus if we want to understand the choices taken by private companies, since even commercial satellite companies operate using market strategy and criteria, we should not forget that they get most of their income from public orders. Therefore the biggest market for private companies is, in fact, the government, and any assumption about privatisation of security regarding image sensing seems to be farfetched. Currently, there are three important arguments against creating a clear divide between public and private interests in space.

First, public policies regarding private space operators seem to conflict with liberal assumptions and with what one would expect of relations between public and private bodies. Private companies, such as DigitalGlobe or Spot Image, can exist because they sell images to governmental agencies. As some authors suggest, from a conventional and rationalist perspective, the answer to this puzzle is based on the fact that some of the image sensing companies were created by governments or largely supported by long-term contracts. Thus public contracts serve as a major source of revenue, when other services, such as Google Maps or Bing, are less important. For example DigitalGlobe's principal customer is the US National Geospatial-Intelligence Agency (NGA), based on the Enhanced View Service-Level Agreement (SLA). Therefore up to 60 percent of its revenues comes from public resources, with total revenue of more than USD700 million.[27] Such dependence on one partner forces other customers, mainly other government agencies, to use limited satellite time, regardless of preferential treatment for ten Direct Access Partners. Even for SpaceX, a private company, aiming to develop a reusable rocket with the first reusable stage (Falcon 9), to compete with public launchers on price, it receives funding from the state-funded agency Space Florida up to USD7.3 million.[28] The company benefited from NASA and Department of Defense contracts as well. Nevertheless, space activities conducted by private operators with imaging capacity, and their relations with governments, cannot be compared with other space activities, such as telecommunication. Being the most profitable part of the satellite market, telecommunication companies with developed space assets are also private security satellite providers working closely with military; and their assets have been used for the defense sectors for years. However, while DigitalGlobe or Spot Image can exist thanks to long-term contracts, without any direct competitor, the big commercial operators such as Eutelsat, Intelsat General, SES Government Solutions, and Inmarsat are selling their services to the US Department of Defense (US Air Force and the Defense Information Systems Agency) on year-to-year contracts. This situation results from competition on the comsat market, but also from different priorities envisaged each budgetary year. Planning overseas operations military operators demands different levels of services every year, and they don't want be bound with long term contracts or other arrangements. While it seems to be beneficial for public funds, in fact the spot market demands higher prices, and long term agreements between governments and private operators may significantly decrease military spending. Some authors suggest that with purchasing commercial satellite bandwidth individually on the spot market European governments might pay more than a USD 60 million premium.[29] But higher income for comsat operators does not necessarily turn into a win-win situation, because demand from governments outpaces satellite capacity, and any investment into new technology made by private operators is not as profitable as it seems to be.

One of the recent examples of merging private and public interests could be the European Data Relay System (ERDS), based entirely on private resources. ERDS provides access to data transmitted from orbit to ground stations with high-speed laser satellite links, based on Eutelsat-9B, placed in orbit on 29 January 2016. This system is the first European telecommunications system based on laser links capable of transmitting information at a rate 1.8Gbit/s. Eutelsat, a private consortium, leased part of the satellite for ERDS, and from January 2016 a distinction between public and private interests in space security blurred. Moreover ERDS has been funded by a public–private partnership between ESA and Airbus Defence and Space, with Airbus operating the service and the DLR German Space Administration deciding to cover the costs of development of the laser terminal.[30] ERDS, unlike existing satellite communication systems, which in order to transmit data must be in the range of ground stations, collects data through a laser link from European Sentinel satellites and then transmits to ground stations. The data transmission will take place in real time, with regular data transmission starting in mid-2016. Another one of the satellites equipped with ERDS will be placed in orbit in 2017, and in 2020 ERDS will cover the entire globe. In the future, the system will be used to transmit data from drones, which will change significantly the prospect of European security, as these flights will concern both the EU's external borders and areas where military operations are conducted. Satellite laser communications is a very important dimension of providing security; it is worth noting that, on 27 January 2016, satellite Intelsat 29e was put in Earth orbit, whose main task is to transmit a real-time video signal from the unmanned American aircraft providing ISR (intelligence, surveillance, reconnaissance) data. It is estimated that the current capacity for transmission (even with providing over 300 video streams for CENTCOM alone) is only 20 percent, and the new satellite Intelsat is expected to depress this ratio. This shows both the scale of the phenomenon of the use of drones for intelligence activities, as well as the potential for those involved in the exploration of outer space, and also the heavy involvement and role of private actors in security services provided from outer space.

The second argument concerns a tendency to focus on a limited set of possible private goals and strategies, when the nature of linkages of interests declared and pursued by public and private actors, operating in the strategic constellation,[31] has been excluded from the discourse. Public and private interests are intertwined, and thanks to lobbies public agencies may build an impression that by buying services from private operators public interests are better governed, without unnecessary financial risk on governmental side. This argument could contribute to understanding the unique process of privatisation of space security, when prices of space products and services have decreased, and space services 'have become part of the essential fabric of the global economy'.[32] Commercial involvement in space services inevitably translates into closer links between lobbies and public administration, and services with high demand on the public side have been outsourced to private operators. This is the case with telecommunication, but also with the aforementioned satellite imagery.

Privatisation of space security has continued, and the space industry has taken a leading role in debates over regulation of future activities, acting through strategic constellations. For example, in December 2015 the Washington Space Business Roundtable hosted a meeting for military decision-makers with experts from satellite industry to discuss integration of commercial satellite communications services into the national security space architecture.[33] Such integration means planning of future infrastructure, long-term contracts, sharing the cost of launching and satellite operation, and hosted payloads aboard commercial satellites. Merging private interests, with truly military goals and means, should be perceived as the long-term strategy for commercial operators, especially when commercial space assets have been used for information sharing, intelligence gathering, and transmission of signals to remotely piloted aircraft operations. But, on the other hand, commercial operators cannot guarantee that their satellites will be resilient enough to withstand intercepting, jamming, disabling, or even destruction of satellites and satellite transponders.

Outsourcing security assets to one private operator, with a long-term contract and a steady market for services, could be mutually profitable, but when public administration relies on a single provider, any turmoil or lack of service can undermine strategic stability for weeks.

Short-sighted plans for privatisation of vital assets are visible in the American launching sector, where United Launch Alliance, due to its hefty deal with NPO Energomash, the Russian producer of the RD-180 rocket engine, changed the strategic balance in space. After the crisis in Ukraine, and with growing tensions between the American and Russian governments, ULA has been unable to provide a constant launching capability for military purposes. A similar situation arose when NK-33 Aerojet engines, produced by Russian manufacturer Kuznetsov, were imported and used in the Anteres rocket.

However a traditional way of thinking about space assets encompasses big satellites (over 500 kg), available only to powerful states and rich corporations, with the necessary heavy lift capabilities. Nowadays, with expansion and continuing process of miniaturisation of electronic parts, less powerful states or even private persons could own their own cubesat or other small satellite, with costs for a payload less than a few million dollars. This dramatic shift in the cost of manufacturing and launching satellites may change relations between the public and private spheres, from coordination and cooperation to coexistence and even conflict, and a good number of new artificial objects in low Earth orbit may generate space debris without any control and responsibility. The UN Convention on International Liability for Damage Caused by Space Objects points out that the launching state should be responsible for any object placed in Earth orbit. But nowadays, when satellites are smaller, without manoeuvring ability, and sometimes belong to a multinational consortium, proving such responsibility would be complicated.[34]

Lyall and Larsen argue that – apart from traditional problems with liability of private actors in outer space and inconsistency of space law in terms of the current strategic and market situation – private commercial users of outer space operate on the basis of private law. Therefore, from construction of satellites through launching, marketing securities, and possible dispute solving, commercial operators are regulated by private as well as by public laws. As a result, without international consent on the shape and content for binding legal provisions for private commercial operators, countries have to look into specialised national space legislation. When it comes to traditional space-faring powers, such as the US or Australia, those states that enjoy fairly well-developed domestic legal standards, but lack specialised national legislation, such countries are at a disadvantage when disputes arise from the performance of contracts related to the space activity.[35] Therefore, without proper international legislation and an inability to overthrow private litigation, disputes among private operators, intergovernmental organisations and states are resolved on the merit of legislation in other states.[36]

The third argument relates to the absence of real competition between major commercial operators, mostly in the launching market. Only 23 out of 75 launchers around the globe face commercial competition,[37] but sound business spaceflight commercial operators should base operations, for the time being, on contracts provided by public agencies. Without long-term flow (or viable prospect of flow) of public money, any reasonable investment in space resources by commercial operators will not be attractive for potentially interested entrepreneurs. On the other hand, public entities, being aware of the role of private entrepreneurs on the satellite market, such as Surrey Satellite Technology Ltd., are in the position to acquire majority shares, to get access to technology and information produced by such companies. Therefore Surrey Satellite Technology has been acquired by the European Aeronautic Defence and Space Company (EADS), and despite originating as a real private entity in space affairs, it became, after 25 years, just a branch of a state-controlled consortium. Setting aside cost and the problem that a public agency decided to control a private actor, this is an example of the long-term tendency against competition in space security between private and public parties. The example of the deal between Surrey Satellite Technology and EADS may persuade other commercial operators to follow the same strategy. Other examples of not-so-full privatisation of space come from Germany, where very high-resolution multi-mode

X-Band SAR satellite TerraSAR-X, launched in 2007, is operated by the German Space Agency and the company EADS ASTRIUM as a Public Private Partnership (PPP). It is worth stressing that EADS ASTRIUM is a company where majority shares belong to EADS so the private component in PPP is rather inconsequential.

While the European space security market is largely dominated by EADS, and market criteria are subdued to states' interests, Fitzsimmons argues that the US security market could be characterised as neoliberal, and private companies provide services to governments on regular basis. Moreover, the US bureaucracy, combined with 'outsourcing and downsizing of U.S. Armed Forces, and a relatively unrestricted legal and regulatory environment' results in the development of the market for security providers.[38] Whereas the traditional approach to PMCs makes a good deal of sense, it also could be useful for understanding a large-scale involvement of private satellite services providers in the United States. In Europe, however, dominated by regulation, the lack of coherence in the satellite market is more visible and privatisation of satellite services is based on an ostensible public–private partnership. European actors in the position to sell satellite security services are, in fact, controlled by European states through a variety of consortia, where governments are major stakeholders, as in EADS.

Possible scenarios for privatisation of space security

In the terms of privatisation and space security, space remains relatively untapped, but commercial and military benefits from space exploration/exploitation could even lead to the 'privatisation of space'. Such privatisation will result from growing pressure on space-faring countries to defect from cooperation, since it is less viable with the high number of actors that have entered the space.[39] However, space policy and space research are characterised by very high costs, which are often not feasible for private companies limited by economic calculations. As pointed out, under-investment in technological development by private companies is related to the fact that these actors are not focused on profits of a social nature, such as improving the quality of life of the recipient of the product.[40] Thus some technology that is potentially beneficial to society is not developed or introduced into use, because the profit margin is too small to make this viable for commercial players.

This chapter argues that privatisation of space security can develop in unexpected ways, but in today's space environment private actors would rather play the role of security regulators than security providers. When investment in space technologies is less profitable than other areas of economic activity, private actors would rather focus on soft law and conflict prevention in space, and new private initiatives will appear. For example, apart from important space companies, such as SpaceX or Blue Origin which are active in outer space, other private actors, such as Secure World Foundation (SWF), which focuses on space sustainability, will play a more important role in crafting international guidelines for space activities.[41] This path shows the way for future solutions and projects – such as cleaning space debris, extracting resources from asteroids and planetoids, refuelling satellites, and providing payload capabilities for governmental entities on market-based logic – will be based on the activity of non-state actors providing soft law and regulatory solutions where space-faring states are unable to find any compromise. Therefore private companies will be in fact global (or space) regulators, as part of UNCOPUS being involved in space activities.[42]

The final argument for private involvement in space security comes from the common good approach and the resilience of space assets, emphasised by Project Ploughshares, as an important part of space security. As of 2017 there were more than 700,000 man-made objects in Earth orbit bigger than 1 cm, with 17,000 of them bigger than 10 cm.[43] Some of them are tracked by SSA systems, both American and European, but these systems are public-military owned, and private operators are not granted any access to this data. Any collision of a space object with space debris, even with small particles, might result in a chain reaction, called Kessler's syndrome, and not only private but public, and military, assets will be destroyed or impaired. In such conditions, reluctant cooperation between the public and private sectors and unwillingness by public actors to share vulnerable data seem to confirm that private space activity is more than necessary. This is the apparent case when the logic of mistrust among state powers must be overcome by private actors, perhaps by suggesting a common preference for debris mitigation and for space situational awareness. In the case of space debris, the Space Data Association, an initiative supported by the private sector, with its main aim to enhance data sharing between commercial satellite operators, could be an example of nascent public good provided by private actors, for global security.

Notes

* This chapter draws on an earlier version published in *Politeja*, No. 50 (2017), pp. 131–147.

1 Howard Kleinberg, 'On war in space', *Astropolitics*, Vol. 5, No. 1 (2007), pp. 1–27.

2 Everett C. Dolman, *Astropolitik: Classical Geopolitics in the Space Age* (London, Portland, OR: Frank Cass, 2002), p. 157.

3 Simon Chesterman and Angelina Fisher, 'Conclusion: private security, public order', in Simon Chesterman and Angelina Fisher (eds.), *Private Security, Public Order: The Outsourcing of Public Services and Its Limits* (Oxford: Oxford University Press, 2009), p. 225.

4 John. D. Donahue, *The Privatization Decision: Public Ends, Private Means* (New York: Basic Books, 1991), pp. 7–8.

5 ASD-Eurospace, *The European Space Industry in 2015*, p. 13.

6 Mariana Mota Prado, 'Regulatory choices in the privatization of infrastructure', in Simon Chesterman and Angelina Fischer, *Private Security, Public Order: The Outsourcing of Public Services and Its Limits* (Oxford: Oxford University Press, 2009), pp. 110–11.

7 Ibid., p. 110.

8 Ibid.

9 Rita Abrahamsen and Michael C. Williams, *Security beyond the State: Private Security in International Politics* (Cambridge and New York: Cambridge University Press, 2011), pp. 25–6.

10 Thomas Risse and Tanja A Börzel, 'Public-private partnerships: effective and legitimate tools of transnational governance', in Edgar Grande and Louis W. Pauly (eds.), *Complex Sovereignty: Reconstituting Political Authority in the Twenty-First Century* (Toronto: University of Toronto Press, 2005), p. 202.

11 Abrahamsen and Williams, Security beyond the State, p. 41 (see note 9 above).

12 Mathias Fortheau, 'Space Law', in James Crawford, Alain Pellet and Simon Olleson (eds.), *The Law of International Responsibility, Oxford Commentaries on International Law* (New York: Oxford University Press, 2010), p. 904.

13 Michael Sheehan, 'Defining space security', in Kai-Uwe Schrogl, et al. (eds.), *Handbook of Space Security* (New York: Springer, 2015), p. 8.

14 Jean François Mayence, 'Space security: Transatlantic Approach to Space Governance', in Jana Robinson et al. (eds.), *Prospects for Transparency and Confidence-Building Measures in Space Report* (Vienna: ESPI, 2010, 35.

15 Sheehan, 'Defining space security', p. 12 (see note 13 above).

16 Ronald Elliot, 'C3I warfare moves into new era', *Defense News*, January 1991.

17 Sheehan, 'Defining Space Security', p. 15 (see note 13 above).

18 Space Security Index, 'SPACE SECURITY 2015: Space Security Index', 20 October 2015, available at http://spacesecurityindex.org/2015/10/space-security-2015/.

19 Eligar Sadeh, 'Obstacles to international space governance', in Kai-Uwe Schrogl et al. (eds.), *Handbook of Space Security*, p. 24.

20 Cenan Al-Ekabi, 'European space activities in the global context', in Blandina Baranes, et al. (eds.), *Yearbook on Space Policy 2014: The Governance of Space* (Vienna: Springer, 2016), p. 52.

21 Adrienne Héritier, 'New modes of governance in Europe: policy-making without legislating?', in Adrienne Héritier (ed.), *Common Goods: Reinventing European and International Governance* (Lanham, MD: Rowman & Littlefield, 2002), p. 186.

22 Barry Kellman, 'On commercial mining of minerals in outer space: a rejoinder to Dr Ricky J. Lee', *Air and Space Law*, Vol. 6, No. 39 (2014), pp. 411–20; Ricky Lee, *Law and Regulation of Commercial Mining of Minerals in Outer Space*, Springer Science & Business Media, 2012.
23 Kellman, 'On commercial mining', p. 413 (see note 22 above).
24 Jacques Blamont, 'US space exploration strategy: is there a better way?', *Space Policy*, Vol. 28, No. 4 (2012), pp. 212–17; Peggy Finarelli and Ian Pryke, 'A new paradigm for international cooperation in space exploration' *Space Policy*, Vol. 21, No. 2 (2005), pp. 97–9; Thomas Gangale, *The Development of Outer Space: Sovereignty and Property Rights in International Space Law*, Santa Barbara, CA, 2009.
25 James Clay Moltz, *Crowded Orbits: Conflict and Cooperation in Space* (New York: Columbia University Press, 2014), pp. 110–11.
26 European Commission, *Study to Examine the Socio-Economic Impact of Copernicus in the EU. Report on the Copernicus Downstream Sector and User Benefits*, written by PwC (Brussels: European Union, 2016), p. 19.
27 SpaceNews, 'For digital globe, government business steady but commercial disappoints', *SpaceNews.com*, 30 October 2015, available at http://spacenews.com/for-digitalglobe-government-business-steady-but-commercial-disappoints/.
28 Cenan Al-Ekabi, 'European space activities in the global context', in *Yearbook on Space Policy 2014*, p. 67.
29 Cenan Al-Ekabi, 'European space activities in the global context', in Peter Hulsroj, Arne Lahcen and Cenan Al-Ekabi (eds.), *Yearbook on Space Policy 2011/2012. Space in Times of Financial Crisis* (Vienna: Springer, 2014), p. 89.
30 ESA, 'First space data highway laser relay in orbit', 30 January 2016, available at www.esa.int/Our_Activities/Telecommunications_Integrated_Applications/EDRS/First_SpaceDataHighway_laser_relay_in_orbit.
31 Christoph Knill and Dirk Lehmkuhl, 'Governance and globalization: conceptualizing the role of public and private actors, in Adrienne Héritier (ed.), *Common Goods: Reinventing European and International Governance* (Lanham, MD: Rowman & Littlefield, 2002), pp. 85–104.
32 Moltz, *Crowded Orbits*, p. 102 (see note 25 above).
33 GovSat, 'Government space leaders look to COMSATCOM for more resilient communications', 5 January 2016, available at www.ses-gs.com/govsat/defense-intelligence/government-space-leaders-look-to-commercial-satellites-for-more-resilient-communications/.
34 Peter Haanappel, 'Enforcing the liability convention: ensuring the binding force of the award of the claims commission', in Marietta Benkö, Kai-Uwe Schrogl and Denise Digrell (eds.), *Space Law: Current Problems and Perspectives for Future Regulation* (Utrecht: Eleven International Publishing, 2005), pp. 113–20.
35 Francis Lyall and Paul B. Larsen, *Space Law: A Treatise* (Farnham: Ashgate, 2009), p. 468.
36 Alexis Mourre, 'Arbitration in space contracts', *Arbitration International*, Vol. 21, Issue 1 (2005), pp. 37–58.
37 Moltz, *Crowded Orbits*, p. 96 (see note 25 above).
38 Scott Fitzsimmons, 'The market for force in the United States', in Molly Dunigan and Ulrich Petersohn (eds.), *The Markets for Force: Privatization of Security across World Regions* (Philadelphia: University of Pennsylvania Press, 2015), p. 158.
39 James Clay Moltz, *The Politics of Space Security: Strategic Restraint and the Pursuit of National Interests* (Stanford, CA: Stanford University Press, 2008), p. 34.
40 Albert N. Link and John T. Scott, *Public Goods, Public Gains: Calculating the Social Benefits of Public R&D* (New York, Oxford: Oxford University Press, 2011), p. 5.
41 Theresa Hitchens, *Future Security in Space: Charting a Cooperative Course* (Washington, DC: Center for Defense Information, 2004).
42 Gérard Brachet, 'The origins of the "long-term sustainability of outer space activities" Initiative at UN COPUOS', *Space Policy, Highlight: Assuring the Sustainability of Space Activities*, Vol. 28, Issue 3 (2012), pp. 161–5.
43 Bernhard Schmidt-Tedd, Niklas Hedman and Anne Hurtz, 'The 2007 Resolution on Recommendations on Enhancing the Practice of States and International Intergovernmental Organisations in Registering Space Objects', in Stephan Hobe et al. (eds.), *Cologne Commentary on Space Law*: In Three Volumes (Cologne: Carl Heymanns Verlag GmbH, 2015), p. 464.

22
BIOMETRICS AND HUMAN SECURITY

James Gow and Georg Gassauer

Millions of civilians fled the conflict in Syria after 2011; as so often in the past, doing so without the means to prove their identity or their need. Millions of refugees spread into and beyond neighbouring countries, generating a major international crisis, also fuelled by confluence with other streams of migration. There was little new in this, as such. Mass flows of refugees and humanitarian crises are as old as human societies, conflict, and natural disaster. In this context, the Syria-plus crisis after 2011 was only the latest instance in a long history – even if it was one of the largest such maelstroms ever known. Yet, there was also something fairly new – the use of new technologies in international efforts to manage the tides of humanity.[1] The long history of mass migration inevitably also includes human efforts to manage mass movement and collect information on those displaced and seeking safety, whether to protect them and give them status, or to control them and their impact on established communities.[2]

But these efforts to record, more often than not, must have been overwhelmed by the chaos and uncertainty that accompany such events – the fog of mass refugee movement. In the disorder of displacement, identities might be lost, fraudulently assumed, hidden, and the individuals associated with them also lost, or mistreated and generally, dispossessed. Sometimes, officials might want to impose as much order as they can; in others, they might prefer not to have a record, and with it responsibility. Even during the crisis of the 2010s, the different approaches could be seen. In Belgrade after March 2016 for example, not all refugee arrivals in the city were officially recorded, as police officers on patrol had either been instructed not to register refugees or they no longer saw the point of doing so, as most refugees 'disappeared' quickly – probably into informal settlements before leaving the city within a few hours.[3] As a result smuggling operations adapted very quickly and low-tech methods were used to breach EU borders.[4] This had serious consequences more broadly, as it ensured that faulty data was gathered and transmitted to more senior government departments, and from there to international governmental bodies. This was not only typical for cities and towns along the Balkan route in the 2010s, but probably for any such movement at almost any time in history, *mutatis mutandis*. Such practices – the behaviour of the police and of the refugees – undermined efforts to get a clear picture and to understand how many refugees were in transit, their needs and where that need had to be met, and, ultimately, how to allocate funding.[5]

All of this was despite the unprecedented possibilities provided by biometric technologies in the twenty-first century, which were increasingly deployed by governments and international humanitarian missions, in an attempt to establish a full and reliable record. While far from free of problems and risks – including those just noted, and non-use – the emergence of biometric technologies, in principle, offered the chance to create unique individual records, based on biological information to record the presence of an individual and, suitably supplemented with biographical data, to generate understanding of who was where and what needs they had. Inevitably, of course, as with so many technological innovations, the problems were evident, as attempts to use biometric identification increased. In the remainder of this chapter, we set out some of the technological possibilities. We do this with regard to two main contexts, UNHCR use of fixed technologies in Jordan and EU operation of handheld technologies in the Balkan Peninsula; we then assess the challenges faced and the mixture of practical, ethical, and legal issues raised by putting these technologies into use.

Biometric innovation

Biometrics – measures of life (literally) – are measurable human signatures, which include everything from heartbeats, to the shape of ears, and even gait – the way an individual walks. It also embraces fingerprints and facial recognition. Biometrics, then, involves the measurement of any, or every, human biological signature. In terms of official practice and security measures, or the cognate area of refugee and migrant registration, biometrics like fingerprints or iris recognition are particularly valued and more likely to be used because they mark uniqueness. Iris recognition, in particular, has become favoured, because it can provide a unique identifier, but with a minimal degree of physical engagement. Where fingerprints require an individual's applying one more digits to a surface and pressing in the right way, an iris scan merely involves standing still, eyes open towards a lens, for a moment.

Moreover, while any amount biological information will offer an individual signature, as noted, for all the data that might be gained from a full analysis, or a device such as a FitBit, iris recognition offers one of the clearest and fastest ways to record or confirm individual identity. Retinal recognition – sometimes confused with iris recognition in discussion of biometrics because both involve the eye – requires the use of infrared rays projected into and through the eye to enable a print to be made of the pattern of blood vessels on the retina, at the back of the eye. This is a more invasive procedure, making it less easy to use and, consequently, less widely adopted. Because the iris is topical, it is viewable on the surface of the body and can be photographed quite easily.

The iris is the circular diaphragm structure that controls the amount of light entering the pupil – the aperture that is the central part of the normally visible eye structure. It has two layers. The outer has two sets of muscles, which pull the iris radially (with folds perhaps akin to a concertina) to allow more light into the pupil, or to limit the light entering it. The second layer, only two cells deep, is heavily pigmented, which protects the retina from light other than that allowed through the pupil – and also makes the iris the distinctive coloured part of the eye that humans easily recognise. Colour is one part of the structure that gives each iris its uniqueness, and which fine-grained digital imaging can recognise.

Iris recognition was largely made possible by the Daugman algorithm, created by John Daugman at Cambridge University. This involved analysing iris patterns and translating them into a digital barcode. This remarkable achievement was developed into the leading method for iris recognition in the world.[6] Nearly all iris recognition technology uses the Daugman algorithm for analysis. This iris code offers very high levels of reliable identification – even when conditions are not favourable, such as heavy eyelids, or bright glare. Even where parts of the iris might be obscured, this method remains accurate and effective.

The advanced character of processes based on the Daugman algorithm and the relatively small file size of iris images allow easy and almost instant comparison of an image with all records held in databases. This evidently permits prompt and accurate processing. That also means that most identification decisions can be made by computer. This reduces the degree of human involvement in checking records, allowing processes to be largely automated – as the emergence of self-service immigration machines at airports in the early twenty-first century exemplified.

Speed is one of the great advantages of iris recognition over other methods, such as fingerprinting, which also gives highly reliable unique signatures. However, it is quicker to create a new biometric record using iris information than one with fingerprints, because fewer images are required and they can be captured simultaneously. Not only is the process of making a record faster, but iris recognition is also relatively self-contained. By contrast, fingerprint recognition can require qualified latent print expert examiners to provide close analysis needed to make decisions to clarify discrepancies between fingerprint records. Because iris barcodes can provide straightforward information to check matches in a database, the human element is reduced and those humans involved can act quickly with confidence. Because fewer steps and fewer people are required, iris recognition can reduce the overall operating cost of recording and processing information. This, in turn, allows greater numbers of people to be registered and processed in a shorter span of time.

This kind of efficiency was critical to the UN's adoption of biometrics. Other technological means were used to try to tackle falsification, or confusion, both of which were more than possible in purely name-based systems. Following the 2010 earthquake in Haiti, humanitarian assistance organisations sought to remove duplications in their records and to rationalise those millions of records by associating individuals with their personal telephone numbers.[7] However telephone numbers were not particularly reliable. As with names, they could easily be copied, or changed, and could easily be disposed of when no longer useful, or, indeed, linked to false names. Beyond this, telephone numbers could not really confirm, or correlate, an individual's identity for the purpose of travel documents, or applications for support. They offered no record or trace of an individual in a manner equivalent to that in which unique biometric data could.

The UN first used biometrics for humanitarian registration in 2002, in Afghanistan. Subsequently, it developed substantial expertise, finding the use of biometrics in refugee crises to be highly advantageous. Biometrics could significantly reduce fraud in the distribution of aid, cut the time refugees had to wait to receive benefits, and perhaps also because of this, diminished the chances that some vulnerable members of the refugee host might be radicalised. Beyond this, in some cases, biometric registration could be empowering, offering a proof of identity, even without an official document, such as a passport or identity card, issued by a government. Primarily, however, the increasing adoption by the UNHCR and other humanitarian missions, such as the EU's in the Balkans, was driven by the need to register thousands of individuals each day, at times, but also accommodating fluctuations in demand at other times. Using unique digital identification by biometrics meant that those responsible for delivering humanitarian assistance could verify the receipt of relief and services to refugees who might previously have been unaccounted and indistinguishable one from another. The expanded use of biometric technology marked a major development in the management of humanitarian crises.

UNHCR in Jordan

While UN agencies had been using biometric technology for several years by the time the Syrian crisis emerged in 2011, that context took activity to a new level. Jordan represented the first attempt to gather biometric data across a whole country. The embrace of biometric technology by the United Nations High Commissioner for Refugees (UNHCR) in Jordan as part of its humanitarian mission was innovative. It changed the scale and standard of operating procedures for the registration of refugees. It replaced the paper filing systems of the past and went far beyond the use of digital files previously, where it had not been possible to make such extensive records, or to match them so easily, reducing past fallibilities, such as the use of aliases, fraudulent names, or simply the different spellings of names.[8] The scale of the Syrian crisis was clearly beyond measure and its effects could not be quantified. But, the advent of biometric technology allowed a clearer and larger record than had ever been possible in the past – a major development.

UNHCR in Jordan implemented a biometrics registration process through the use of iris recognition technology. In February 2012, the United Nations began to use biometrics technology to register Syrian refugees arriving at UNHCR camps in Jordan.[9] This included the system chosen by UNHCR in Jordan, made by IrisGuard Inc., as could be seen from the Twitter accounts of the UNHCR and Andrew Harper, the UNHCR Representative in Jordan.[10] The IrisGuard IG-AD100 System was used by UNHCR at registration sites and also at the Cairo Amman Bank, using an ATM system. Each of these were fixed points of physical infrastructure, but despite their distribution, were linked to the UNHCR regional repository, located in the capital, Amman.

The extensive use of biometric technology in this case produced a large volume of reliable data, made available at the time by UN officials on social media discussing the challenges they faced and contributing material that could make other sources more reliable.[11] Processing so many individuals was a challenge. However, the introduction of iris scanning enabled the UNHCR to cut an individual refugee's waiting time from 12 months to nothing. Benefiting from ECHO (European Commission Humanitarian Office) and Japanese funding, Irbid was the first registration point in the world to have iris recognition introduced, allowing it to handle 1,300 refugees a day and to clear its vast backlog, completing 100,000 registrations in seven months (ahead of its own 63,000 target). In terms of volume and process, the use of digital biometric processing was transformative.[12]

The approach to unique identity chosen by the UNHCR regarding Syrian refugees in Jordan in some ways echoed the US military experience of using biometric collection platforms outside the US itself. The legal basis on which it operated was both clear, because of agreements to operate in the country, and also ambiguous, because – as so often in conflict and crisis situations – there were grey areas and a degree of fogginess. This highlighted the challenges for any non-sovereign, or non-state organisation – in this case, an international one – when collecting biometric data in a humanitarian capacity without a clear determination of usage beyond the immediate one. But, this was a better situation than would have been the case otherwise.

It is hard for any outsider to imagine how forced migration from Syria affected the individuals involved. They experienced multiple traumas – the conflict in Syria itself, their flight, the experience of arriving in a new country, needing food and shelter and being at the mercy of government and international agencies, often without documentation or money. Part of this was also the reaction to the refugees' arrival in host countries, which found it challenging to cope both with the burden of people, but also the impact on societies, as countries struggled with changing dynamics. In this context, it is salutary to note that the proportion of Syrian refugees who fled to Lebanon was comparable to the entire Mexican population's being absorbed by the United States.[13] Even though hundreds of thousands were living in UNHCR camps in Jordan, this was by no means a majority of the Syrian refugee population even in that country. The majority of them were to be found in among the Jordanian people, not separated in camps. Instead they were competing for jobs and housing, adding two further pressures to the cost of living in the host state and communities. Despite the immense success of the UNHCR in processing hundreds of thousands of refugees using biometric data, that record probably only accounted for a minority of the total number in Jordan. It was certainly far from a complete record.

Although refugee registration was not all-encompassing, it remained remarkably valuable. UNHCR could know where a refugee was housed in a camp at any given moment. This allowed verifiable camp records. Biometric verification stations could be set up at camp entry control points to check that those allowed access into the camp were *bona fide*. Biometric verification stations could also be set up at various points of humanitarian assistance distribution. Cash payments could be made securely, using the digitally recorded biometric data, which could also serve as a way to verify to UNHCR's global donors that aid was indeed reaching those most in need and that those receiving the benefits were not accidentally benefiting twice.

EURODAC and the Balkans

While the impact of the Syrian conflict on the Middle East was enormous, it was also felt strongly in Europe, with the EU and its member states themselves thrown into crisis as refugees from Syria merged with mass unauthorised migration from other parts of the world, challenging the EU's Mediterranean periphery. However, the EU experience with using handheld biometric recording devices was not an overwhelming success, in contrast to the relative and limited success of the UN experience in Jordan. Whatever the EU approach achieved, in the recording of individual data, was lost in a messy set of overlapping systems that lacked coherence.

The EU approach, marked by an agreement between the EU and Turkey in March 2016,[14] was to use Greek officials based in Turkey and Turkish officials based on Greek islands to assist in the mutual processing of people seeking transit from Turkey into Greece as the entry point for the EU. It also limited the right of passage to refugees fleeing Syria and Iraq – so any migrants from other places, including Afghanistan, were not allowed through. These arrangements were reinforced by the introduction of FRONTEX, an EU border agency, which tackled the migrant influx, sometimes with a soft approach, at others with 'Robocop'-like figures, wearing helmets, visors and heavy uniform, as well as carrying firearms. FRONTEX's activity was also aided by the use of new handheld biometric devices for registering those entering Greece – a new technology developed by Crossmatch Biometrics Tech for the US military in Iraq.[15] While this technology might have presented some questions regarding human rights, data control, and sharing of information,[16] it contributed to a concerted effort in which human traffic was limited. In contrast to the more than 1 million souls who passed into Europe in 2015, as 2016 came to an end, the figure had been limited to only 366,350 (with 4,621 dead and 49 missing, in addition).[17] However, with most of these individuals seeking to pass to other destinations, suggesting that they had good cause not to register in their first country of refuge (as the Refugee Convention requires,[18] absent good reason to travel to another state to claim asylum (for example if close family members are there already)), the issue of 75,523 stranded people remained.[19] These were people without papers, without citizenship anywhere. And without credentials, there was nowhere that they could be moved. So, they remained in refugee camp pockets across the Balkans. Beyond this, despite the use of biometric technology, any records were almost meaningless, as data could only have value if stored and calibrated in effective ways.

Identifying how many refugees entered Europe and how many were in transit was difficult by any means, due to the clandestine nature of irregular migratory movements along the Western Balkan route. Identification, which should have been assisted by the introduction of digital biometric records, was made more difficult, however, as the EURODAC and the Schengen

Information System (SIS II) designed to record and identify migrant data became unhinged through the escalation of the refugee crisis in late 2015.[20] Specifically, this meant that the authorities processing new refugee arrivals in Europe were not able to exchange valuable information necessary when refugees settled in their target country and thus transformed from refugee to asylum applicant. This effectively paralysed the Dublin Convention, the EU agreement on how to manage the mass influx of refugees. It should be noted that as most refugees were aware of the Dublin Convention procedures regarding the first country of entry as the country of asylum (confirming the terms of the Refugee Convention), many refused to be registered in Greece, Hungary, or Slovenia, waiting instead until they had arrived in their country of choice: Austria, Germany, or Sweden.

Although these European countries, in due course, accelerated their efforts to sign bilateral re-admission agreements with countries of origin (Afghanistan, Algeria, and Iraq) such agreements involved lengthy negotiating procedures. As a result, a 'back-flow' of failed asylum applicants along the Balkan route started to develop. This was due to the fact that most negative asylum cases, now refugees again (in what Maley calls the 'ordinary language' sense), were deported via their point of entry.[21] As most of the applications made in Central European countries were not registered through either EURODAC or SIS II, it was not uncommon for many deportees to make a new bid for asylum in the country to which they had just been deported. Although this process was not new, what was new was the increased volume of 'back-flow' bids.

As a result, national asylum offices were unable to differentiate between 'new' arrivals from the Balkan route, making their first bid, and those who had just been deported from neighbouring European countries and were making their second, or even third, bid.[22] Additionally, as the EURODAC, SIS II and the Visa-Information System (VIS) were by design inoperable alongside each other, information exchange between national police, border patrol and asylum offices, let alone on an international level, was virtually impossible.[23] Due to the higher volume of cases applying, and the varying nature of applicants after 2015, this led to longer processing times. Invariably this also led to the increased frustration of case-workers who lost access to essential information on applicants due to limitations, such as the 18-month data storage limit on EURODAC servers, or protocol.[24]

Not only did this make data gathering difficult, it also exposed host communities to security risks, as they allowed migrants with potentially questionable motives to assume new identities as they entered Europe.[25] This became evident when reviewing the abundance of falsified Syrian documents in circulation,[26] which could be obtained on the black market by jihadist returnees, war criminals, or regular criminals in Beirut, Gaziantep, or Istanbul.[27] Thus, there was a large degree of ambiguity surrounding the ability of European states to verify an individual's true identity.

The challenges

While the experience of the UNHCR with Syrian refugees in Jordan was relatively successful, in practical terms, the same was not true for the EU's efforts in the Balkans and across its own region, with the EURODAC system failing because of its own self-destruct limitations, as well as a host of organisational and compatibility issues with other EU systems. Those problems, in themselves, offer sufficient challenge to the use of biometric technology in humanitarian crisis management – the technical means are nothing without the knowledge, understanding, and organisation to use them effectively. However, those problems also hint at the broader set of legal and ethical questions that attend the use of biometrics in international humanitarian missions.

While the benefit of gathering biometric data has been shown, its collection gives rise to many difficult questions, both ethical and legal. Where is, or should, this data be stored? Who should have oversight of it? To whom does the information belong? Does it belong to the UNHCR? Clearly the UNHCR invested in the equipment and performed data acquisition. But, is this data stored alongside, in the same location, as data from refugees in, for example, Malawi or Sudan, or elsewhere? Should information relating to the Syrian refugees be stored in the same database as that holding the registration of Somali fishermen?

Or, should the information relating to Syrian refugees in Jordan belong to Jordan, since it was acquired and processed in the country? As it was, the UNHCR operated data sharing agreements with both the Jordanian government and the Cairo Amman Bank, which played an essential role in collecting the data. However, the more data is shared, the greater the vulnerability to leakage, or misuse. Does the data belong to the individual, as it is, in the end, that particular individual's biographical history that constitutes the record? Or, should the information belong to whichever country hosts the database keeping all the information as a safeguard, which is clearly unlikely often to be the same country as that in which the information was collected?

Could there be a case for deeming the data to belong to the government of Syria, given that it was the refugees' home country? Would it be safe to allow such information to be given to a country which is known for its poor application of human rights and its history of reprisal against those who oppose it? Or, should the information belong to the countries where refugees might eventually seek asylum and to take up residence? The evidence from Iraq, albeit a very different situation involving US, rather than international, control, was that the data was used for diverse purposes and to make decisions beyond the immediate situation, well beyond its original purpose (including as a diplomatic tool in relations with the Iraqi government, with a reduced version of the database being tendered as a 'peace' offering in negotiations between Washington and Baghdad).[28]

Should any other state have access to the UNHCR biometric data records – for example, when an asylum seeker is asking for refuge in a particular country? Both the country and the refugee might see benefit in the UNHCR's providing reliable and verifiable information. In a different respect, what if a country, or an international organisation, such as NATO, suspects an individual of being a potential threat, or member of a hostile organisation, and that individual might be seeking safe passage under cover of the UNHCR? Should that country, or the organisation, be allowed access to UNHCR databases to check and cross-reference biometric data as a security provision?

The UN, and any international organisation, represents a new set of questions on the already busy and blurred panorama of ethical and legal challenges surrounding the gathering, processing, storing, and uses of biometric data. The international character of the organisation and its legal standing in relation to sovereign states and other legal jurisdictions is potentially tricky. Some of these challenges were already faced by the US as it pioneered the use of biometrics in Iraq – albeit that the US was a very different actor to the UN, with very different capabilities and mission, in quite different circumstances.

The challenges of operating biometric data collection systems are great. Data synchronisation is an issue. As a registration programme grows, synchronising with multiple sites, remotely dispersed, becomes harder, in terms of time and capacities. However, the UN experience in Jordan proved that it could at least coordinate across that country. The major challenges are not so much practical, as ones of human rights, good practice, and legality. For instance, while the US military benefited in practice from the use of the HIIDE – the Handheld Interagency Identification Device – as it massively speeded up previously slow, desk-based activity, serious flaws in its use were identified by the congressional watchdog on government spending and performance in the US, the General Accounting Office (GAO). The GAO found that the equipment failed to meet FBI (Federal Bureau of Investigation) standards in terms of fingerprint transaction files.[29] The standard concerned a distinction between rolled and 'slap', or flat, finger prints. Thus, the technology itself was limited, reducing the complete credibility and reliability of the data collected. Although the UNHCR had first adopted iris recognition technologies as early as 2002, and clearly had benefited from progress over those years, there was every chance that the data it, or other humanitarian organisations and missions, gathered, would be imperfect and less than wholly reliable.[30]

The challenges to those using biometrics for refugee relief and management are far different than those faced by the US military. By 2005, the US military was using biometrics in Operation Iraqi Freedom and later employed them in Operation Enduring Freedom in Afghanistan to help ensure those given access to US military installations were not known criminals or bomb-makers, and to help isolate known insurgents from the local populations. Sharing data could be a major headache, even with the UN system. This is both a legal-ethical and a technical question, with different UN agencies having different missions. Technically, with different agencies using different systems to different standards, the problems faced by UNHCR in sharing data across different camps could only be amplified. Those questions could be even larger, in view of the prospect of pressures to share outside the UN system. However, functionality can be a driver here. So, in Jordan, when it was in UNHCR's interest to cooperate with the Cairo Amman Bank as part of its programme, ways were found to overcome the challenges. This allowed the international intergovernmental humanitarian agency to work with a private, commercial organisation. However, it was important that this form of sharing did not alter the UNHCR's mission and that it served its purposes. There should be no doubt that function and focus could come under external pressures of different kinds to change, or to allow data to be used for other purposes than those for which it was collected. For example, if the UN were to adopt fingerprint recognition, then this would potentially be useful in other spheres, such as criminal investigation. In this sense, one of the advantages of iris records was that these were not left at crime scenes and so could not be used in the way that fingerprints might. Thus, pressure was avoided.

Among the potential changes in function and purpose that might occur, vetting would be high on the list. Of course, registration in the first instance is a form of vetting – checking a single, individual identity. However, once information exists, it can be hard to resist pressures for it to be used for other purposes.

One of the biggest challenges in putting a biometric recording system into place is having a match of equipment and personnel that enables the registration and processing activity to be effective. Whatever the equipment, it requires operators that have been appropriately trained. A registration scheme requires personnel fully trained in all aspects, including the ethical and legal, of enrolling individuals in a biometric database.[31] It is important that the initial data recording is done well – including correct biographical information and correlation of details, plus any other contextual information that could assist in the future and ensure that appropriate contextual information is available. However, whatever the responsibility on the data gatherer, those in more senior positions also need to fully appreciate and be trained in all aspects of biometric data activity, including its usefulness and the importance of selecting and monitoring those involved in data collection and processing.

A further aspect of concern is retention of data. This has two dimensions. The first involves the potentially indefinite retention of personal data. It is not clear what refugees being enrolled are committing themselves to when they register and have their biometric data recorded. The busy officials involved in collection are not necessarily working to the standards for data collection required by university ethical approval schemes, or the standards of data protection law in

EEA countries. Not only is the duration of retention an issue, but so is its security. Such systems need to be maintained in a manner that preserves their integrity and prevents inappropriate access or theft, or corruption, which means having a constant programme of maintenance across all sites involved in storing and sharing data. Beyond this, there are then issues, including legal agreements, with the countries in which long-term (as well as shorter-term) storage will take place – and ensuring that any databases are compliant with local legal requirements. Simply getting memoranda of understanding, or other legal agreements, agreed might be a challenge, as well as rights of access and all the requirements associated with maintenance and security. Moreover, the formal challenges to be negotiated could well be compounded by cultural ones – different sensibilities and sensitivities might surround the treatment of such data.

The questions of retention and possible further use are among the factors that could make a refugee reluctant to enrol, despite having little alternative if they want to be taken under the protective wing of a humanitarian agency. It is quite understandable that many who were already in positions of vulnerability and hardship, often fleeing persecution, might be reluctant to have their details registered. Indeed, it is possible that, as well as fear and uncertainty, these refugees might also feel that a stigma attaches to having biometric data recorded – records of this kind are usually only maintained in relation to criminal activity or personal health in liberal countries, and otherwise by authorities with the capabilities in police states. While there may be many good operational reasons to record data, there may often also be a need to overcome fears and a sense of stigma.[32]

Conclusion

The collection of identification information among refugees in humanitarian crises is not new. However, in the past, whatever the heroic efforts of officials to make and maintain records, many individuals were lost and not recorded – for better or worse – and fraud and mistakes were not only possible, but probably rife. The advent of biometric technologies, while still not free from problems and risks, is transformative. It makes it theoretically feasible to make unique individual records, which could be matched with other biographical data. A far more reliable record can be created than at any time in the past. Biometrics have been effective in ensuring registration throughput and in clearing up the backlog of individuals waiting to be registered. In addition assured payments have also been a major benefit of the adoption of biometric processing. This is something also evident in India's unique identification authority project, which distributed social security benefits in Afghanistan and to the Afghan National Police.

The data and analytics made possible by the use of biometrics allowed the UNHCR to gain a clearer picture of dynamics in refugee camps. They could study migration between camps, including where there had been returnees. They could map family associations and how many individuals had sought medical attention and for which reasons. The use of biometrics also enables the UN to identify known troublemakers living in a camp or, indeed, entering camps and passing through them.

The advent of biometrics offers transformative opportunities for managing major humanitarian crises However, as we have shown in this analysis, that prospect remains bound in webs of practical, policy, legal, and ethical problems. The issues include the questions of how databases will be used, when created, of ownership, administration and organisation, and of sharing and potential misuse – including, even, either corrupt and illegal, or even legal (though unethical) commercial selling of information (as there appears always to be a market for information). On every level, the potential benefits offered by the new technologies carry with them major concerns of security and protection of the information recorded.

Notes

1 This chapter is informed in part on research papers by Sarah Soliman commissioned for the 'ESRC-DSTL-funded project: 'SNT Really Makes Reality: Technological Innovation, Non-Obvious Warfare and Challenges to International Law', ES/K011413/1 Prof. G. Verdirame, Prof. AJW Gow. and Dr R.C. Kerr) Law. Sarah's commitments with the RAND Corporation, inter alia, mean that she could not contribute directly to this volume. But her pioneering work is acknowledged and the contribution she made, including alerting James Gow and the project team to this particular area. Sarah's research papers are referenced in this study, as is some of her published work. Her published work also includes an article not directly on the topic, but with some points of relevance: 'Identity, Attribution, and the Challenge of Targeting in the Cyberdomain', Col Glenn Voelz, USA, and Sarah Soliman *MCU Journal*, Vol. 7, No. 1, Spring 2016, pp. 9–29. Without the lead she provided, James Gow might not have been alert to the significance of biometrics in humanitarian emergencies – and so too Georg Gassauer's excellent presentations in relation to this topic in the context of the Liechtenstein Colloquium and seminars in support of the Austrian Presidency to the OSCE in 2017, which prompted Gow's suggestion to be co-authors.

2 See the compelling discussion of refugees and displaced humanity, including the efforts of states to control through language and law, by Bill Maley; William Maley, *What is a Refugee?*, London: Hurst and Co., 2016.

3 Only very few refugees passing through Hungary, Serbia, the former Yugoslav Republic of Macedonia (FYROM), or Greece sought protection there. Police and border officers tasked with the physical duty of servicing the policy mechanisms of the 7 March 2016 border closure referred to filling quotas set for them by their governments. Low in morale, they had little incentive to conduct rigorous checks on border crossings that would reach, or exceed, quotas. Alternatively, in March 2016, push-backs of refugees to Bulgaria were common, up to 50 km within Serbia's borders, as a similar approach to push-backs was taken in other Balkan countries and refugees tried repeatedly to enter the EU through these borders. Therefore, between September 2015 and March 2016, the real number of those entering northern Europe could only be an estimate, at best, and was based on the number of refugees apprehended on the Austrian–Slovenian and the German–Austrian borders, and on those that applied for asylum in northern Europe after March 2016.

4 Depending on ethnic origin of refugee groups, smugglers had different *modus operandi*. A majority of Afghan refugees, for example, paid for their journey through the traditional Hawala system, which was paid in full through down payments of either property, opium, or female family members, before commencing on the journey; upon arrival on EU territory, a photograph of the migrant in front of a prominent landmark was sent to a middleman to release the funding. A United Nations (UN) counter-terrorism expert interviewed during research referred to finding detailed contracts in Kandahar regulating the passage to Europe. Syrian and Iraqi migrants, by contrast, paid in segments. Here, a large variety of price-related services were offered. At the top end were those willing to pay thousands of US dollars (standard currency on the Balkan route) who were given false documents and almost guaranteed entry to the Schengen Zone, while, at the bottom end of the scale, refugees were escorted to within 10 km of the Serbian–Hungarian border and handed a set of wire cutters.

5 Movements of migrants out of Turkey, between September 2015 and March 2016, were on such a scale that the Turkish authorities were unable to detect which Syrians with Temporary Protection Status had left Turkish territory and which had remained. As there was no exchange of biometric data between the EU and Turkey, the statistics of those still seeking protection in Turkey was not readjusted to reflect these larger migratory movements. In the wake of the attempted coup in July 2016, a recalibration of these numbers, which was scheduled by Turkish Director General Migration Management (DGMM) for July 2016, did not take place. This was significant, as the EU and its member states based financial assistance to Turkey on these statistics.

6 John Daugman, 'High confidence visual recognition of persons by a test of statistical independence', *IEEE Transactions of Pattern*, Vol. 15 No. 11, November 1993, pp. 1148–61; the following year, Daugman took out US Patent US5291560A to protect his innovation – for details, see https://patents.google.com/patent/US5291560A/en. For a complete summary and evidence of the widespread impact of Daugman's algorithm, see University of Cambridge, 'Iris Recognition' Impact Case Study, UoA 11, REF2014, HEFCE, available at http://impact.ref.ac.uk/casestudies2/refservice.svc/GetCaseStudyPDF/16707 at 31 May 2018.

7 *The Huffington Post*, 12 January 2011 available at https://webcache.googleusercontent.com/search?q=cache:7j1pi_fLvsgJ:www.huffingtonpost.com/2011/01/12/mobile-technology-creates_n_808333.html+&cd=1&hl=en&ct=clnk&gl=uk&client=firefox-b-ab at 2 June 2018.

8 *UNHCR News*, 17 October 2013, available at www.unhcr.org/525fe1569 at 31 May 2018.

9 Maron, Dina F., 'Eye imaging ID unlocks eight dollars of the Syrian Civil War refugees', *Scientific American*, 18 September 2013.

10 twitter.com/Refugees/status/417800185780396032; twitter.com/And_Harper/status/445613787857829888; twitter.com/And_Harper/status/400839321907101696.

11 UNHCR, *Figures at a Glance*, available at www.unhcr.org/uk/figures-at-a-glance at 2 June 2018.

12 *UNHCR News*, 3 October 2013 available at www.UNHCR.org/524d5e4b6 at 2 June 2018.

13 *The Washington Post*, 12 May 2014.

14 *The EU-Turkey Statement, 18 March 2016*, The European Council, available at www.consilium.europa.eu/en/press/press-releases/2016/03/18-eu-turkey-statement/ accessed 14 December 2016.

15 Findbiometrics – Global Identity Management, 1 March 2016, available at http://findbiometrics.com/crossmatch-biometrics-tech-to-id-and-register-migrants-in-greece-303011/# accessed at 12 April 2016.

16 Sarah Soliman, Unpublished Draft Paper, King's College London, for the RCUK Global Uncertainties Science and Security Programme project: *SNT Really Makes Reality: Technological Innovation, Non-Obvious Warfare and the Challenges to International Law* (ES/K011413/1) (2013–15), Professors Guglielmo Verdirame and James Gow, and Dr Rachel Kerr.

17 IOM, *Migration Flows* Europe, 16 November 2016, available at http://migration.iom.int/europe/ accessed 19 November 2016.

18 *Convention and Protocol Relating to the status of Refugees* 1951 available at www.unhcr.org/uk/protection/basic/3b66c2aa10/convention-protocol-relating-status-refugees.html at 3 July 2018; on the *Convention* and discussion of definitions and 'ordinary' usage versus stipulative legal terminology, see Maley, *What is a Refugee?* pp. 15–41.

19 IOM, *Migration Flows* Europe, 16 November 2016, available at http://migration.iom.int/europe/ accessed 19 November 2016.

20 European Commission, *Proposal*, COM(2016)272 final, Brussels, European Commission 2016.

21 This *modus operandi* was meant to hold until the Dublin Convention procedures were reinstated and refugees could be deported directly back to Greece. The European Commission aimed to reinstate the Dublin Convention by March 2017. European Commission, COM(2016)272 final, *Proposal*, Brussels; European Commission Brussels, 2016.

22 'It will also be necessary to store information on illegally staying third-country nationals and those apprehended entering the EU irregularly at the external border for longer than what is currently permitted. A storage period of 18 months is the maximum permitted under the current Regulation for those apprehended at the external border and no data is retained for those found illegally staying in a Member State. This is because the current EURODAC Regulation is not concerned with storing information on irregular migrants for longer than what it necessary to establish the first country of entry under the Dublin Regulation if an asylum application had been made in a second Member State.' European Commission, COM(2016)272 final, *Proposal*, Brussels: European Commission, 2016.

23 European Commission, COM(2016)272 final, *Proposal*, Brussels; European Commission Brussels, 2016.

24 By contrast, in an interview with Turkish DGMM policy analysts, it was revealed that the Turkish authorities had created their own 14-day fast-track programme that would be based on the decisions of the European countries and would also include similar steps. However, as the intended pace of returns to Turkey from Greece was never accelerated from the European side, this was not activated.

25 European Commission, COM(2016)272 final, *Proposal*, Brussels; European Commission Brussels, 2016.

26 Before Aleppo and Homs fell to rebel and Islamist groups, many civil servants fled with the necessary equipment to produce and sell counterfeit passports, and other documents, in Izmir, Istanbul, or Ankara.

27 *Al Jazeera*, 15 November 2015 available at www.aljazeera.com/news/2015/11/easy-buy-syrian-passport-facebook-151121124233394.

28 Sarah Soliman, 'Tracking Refugees With Biometrics: More Questions than Answers', *War on the* Rocks, 9 March 2016 available at warontherocks.com/2016/03/tracking-refugees-with-biometrics-more-questions-than-answers/ at 2 June 2018.

29 GAO, Defense Biometrics: DoD Can Better Conform to Standards and Share Biometric Information with Federal Agencies. 31 March 2011, available at www.gao.gov/products/GAO-11-276 accessed at 30 May 2018.

30 Sarah Soliman, 'Processing Beyond Measure, with Measure: the United National Use of Biometric Technology to Register Syrian Refugees in Jordan', Unpublished Research Paper, King's College London, May 2014, p. 6.

31 GAO *Defense Biometrics*: Additional Training for Leaders and More Timely Transmission of Data Could Enhance the Use of Biometrics in Afghanistan' April 2012 www.gao.gov/products/GAO-12-442.

32 One development that might avoid stigma – but would create new sets of questions otherwise – is stand-off recognition. Whereas the types of equipment used to date in humanitarian missions all involve contact with the subjects, stand-off recognition – which could be used for surveillance – might allow a record to be made without the subject's knowing. However, while security agencies might carry out an activity of that kind, and ignoring the manifold ethical dimension of acting in that way, in practice, the character of any humanitarian action would be undermined by the use of this kind of technology, jeopardising not only a particular mission, but the prospects for others in the future. On stand-off capabilities, see: Larry Anderson, 'CITeR Working To Make Face, Iris And Fingerprint Recognition Systems Better', available at www.securityinformed.com/insights/co-11541-ga-co-11542-ga-co-11540-ga-co-5188-ga.12915.html accessed 31 May 2018.

23

FUTURE WAR CRIMES AND THE MILITARY (1)

Cyber warfare

James Gow and Ernst Dijxhoorn

Issues of wrong and right were always central in the conduct of warfare. During the twentieth century, especially as it came to its close, those issues grew in perceived importance and attitudes towards them shifted, with extensive discussion of wrong and right surrounding all armed conflicts. Discussion of war crimes and calls for prosecution grew, as the changing character of warfare coincided with greater attention to the law – and the boundaries of wrong and right came under ever-greater pressure, while also being harder to distinguish.[1] The wave of technological innovation in the twenty-first century could only serve to blur those boundaries further.

Attention to legality and the conduct of warfare increased greatly in the later twentieth and early twenty-first century. There was ever-greater attention to war crimes issues.[2] The growth of prosecutions (and, even more extensive, demands for prosecutions), above all at the international level, was accompanied by calls for firmer adherence to, and application of, the law. The growing emphasis on war crimes matters was a sign of change not only in the context of warfare, but also in its character. Increased investigation and prosecution certainly reflected a broad international and municipal preparedness to enforce the law, to push for its enforcement, and to encourage more than ever before adherence to it. Yet, there was more at work than this. The character of warfare had changed, creating spaces of uncertainty.[3] In those spaces, both the law itself generally and distinct from narrow legal interpretations were subject to contestation about what was acceptable and that which was not. Whereas war crimes trials in the twentieth century had only ever focused on extreme events that were undoubtedly beyond the scope of anything legitimate, the more 'normal' conduct of armed conflict also came more and more under the spotlight of commentary and legal scrutiny. At some level, every military operation became subject to accusations of criminality.[4] Allegations of war crimes became commonplace (whatever their merits), in part, because the boundaries of warfare were shifting and the new context was one where expectations had also changed.

In this context, soldiers were charged with offences for the first time for ostensibly doing their jobs. At the Yugoslavia Tribunal, Generals Tihomir Blaškić (Croatian) and Stanislav Galić (Serbian), were convicted of, among other offences, excessive and indiscriminate artillery bombardment. Whereas other cases historically had been for actions clearly beyond the pale, these officers were, in a sense, being judged for doing that which they were supposed to do – fire artillery; it was the way in which they were doing it that was in question. This was a situation in which issues needed to be informed by the views of the professionals themselves, to guide courts and non-professionals in assessments of wrong and right – where professional judgement based on a sense of the realities of doing the job became crucial.

In a context in which the character of warfare was changing and lines that might once have been clearer were blurring, the advent of new technologies only presented further potential for ambiguity and accusation. The advent of unmanned aerial combat vehicles was the *avant garde* in this context – with calls for their use to be banned and accusations of war crimes surrounding their use, with similar questions surrounding the prospect of autonomous weapons (also discussed in Chapters 12–15 of this handbook). The impact of synthetic biology largely remained prospective – but also carried with it potentially the most difficult challenges of all (as discussed in Chapters 16–19 of this handbook). These possible future war crimes, or war crimes accusations, would lie at the fuzzy boundaries of complex warfare, where technological innovations could make already non-obvious warfare – warfare that does not involve clear military units, battlefield and combat with blast and destruction, the commonplace – and, perhaps, 'comfortable'[5] – view of warfare – even less obvious and even invisible. In this chapter and the following one, the views of military personnel on some of the potential wrongs and rights surrounding the use of these new technologies in particular situations are offered; material that could inform future war crimes considerations and cases. This is based on research conducted with members of the armed forces, designed to explore these issues hypothetically – though, inevitably, all issues would be case and context specific.[6] In Chapter 24, autonomous weapons and potential synthetic biological weapons are the focus. In this chapter, cyber warfare and war crimes are the topic, following two other sections: one explores war crimes and the military, and the importance of professional perspective in the context of ambiguous, grey area, hybrid or non-obvious warfare; the other introduces the ways in which new technologies might affect future war crimes.

War crimes and the military

Warfare offers many opportunities for accusation. It is riddled with passions and contestation, giving rise to charges of war crimes, both genuine and disingenuous. The importance of violence in warfare makes nasty events inevitable – the questions are about whether those events are justified. In part, the assessment of justification has a broader context in societies at large. However, the specificities of the profession of arms mean that only those with the expertise in the application of restrained coercive armed force can clearly judge, in the practical context, whether particular actions were justified, or not. The values of parent societies have to be a large part of the equation – the military has to be a reflection of society. Yet, it cannot be a pure reflection because of the nature of its business and the demands placed upon it. The military must, in broad terms, be responsive to society and vice versa – including the recognition by the latter that non-specialists are not always qualified to assess the use of force.[7] The civil-military difference is important and has to be well balanced around the hyphen in the middle. Without a clear difference, soldiers will not be prepared to carry out their mission and apply restrained coercive violence. Yet, they cannot do so beyond anything society would see as justifiable. The armed forces cannot be entirely subject to standards in society and the idea that they must completely 'conform to that of civil society, not vice versa'.[8] Certainly, armed forces personnel need to be held to account. But, where accountability is being exercised, professional expertise should still inform examinations of soldiers' conduct and constitute key evidence in trials – a voice so often absent or unheard when accusations of wrongdoing circle.

Of course, this is not to say that wholly egregious actions, such as murder and mass murder, or rape, should not be utterly beyond the pale. It is to say that where there are questions about military conduct, when military personnel are doing that which they are trained for and supposed to be doing – their jobs, the professional perspectives need to be fully appreciated. This pertains to cases where the questions are more about soldiers' 'mis-applying' their trade. It is about their actively, or negligently, using force. This is a different question from deliberate mass murder, or the inhumane treatment of prisoners, and so forth. The allegations, in cases of unlawful attack, involved judgement about the use of artillery, as part of and related to, the conduct of operations.

It is on questions of this kind, irrespective of the detail of cases (but, where necessary, informing them), that what military personnel think must be taken into account – just as the opinions of medical or engineering experts are used in cases involving other practitioners' judgements in their fields. Starting from this understanding, and seeking to illustrate the need to take account of military voices as allegations of war crimes grow in the context of contemporary warfare, this chapter considers three situations, or scenarios, concerning the use of force: the killing of prisoners of war during a special forces mission; the use of cluster munitions in a humanitarian operation; and the use of artillery siege and bombardment.

Technological innovation, non-obvious warfare and the scope of future war crimes

The simultaneous impact of new technologies and (sometimes linked) changes in the character of warfare have presented challenges in the domain of international law.[9] Radical transformations in the technical domain altered the means and methods of warfare, and altered the character of armed conflict, often making it opaque, or 'non-obvious'. Social change – such as transnational insurgencies, or terrorist movements, privatised military and security operations, and networks challenging hierarchical structures – compounded these features.[10] In this context of novelty and non-obvious warfare, the very existence of war was not even recognised by many observers, and, therefore, unknown, or, at best, dimly suspected through layers of ambiguity. The identity of the main actors in warfare, or the specific character of their actions, were not universally recognised, or accepted.

In this complex and obscured environment, the pressures involving international law, itself caught up in the spiral of change, were vast, especially in their prospective impact on international criminal law and future allegations, investigations and prosecutions of war crimes – all matters that could be complex in the context of more conventional contexts of armed conflict.[11] Innovations that revolutionised the future legal and war-criminal landscape included: space and its use, including the notion of space war; cyber warfare and security; electronic warfare, autonomous weapons systems, and drones – or unmanned aerial combat vehicles; and, increasingly disturbingly, weapons of mass destruction, such as nuclear, radiological, chemical and, perhaps most scarily, biological – or, more pointedly, synthetic biological weapons.

The prospect of criminal trials – or even ensuring that the law is aligned (whether through interpretation, or innovation) with technical developments – remains a challenge for the future. But, it is important to understand the changing scenario. Even though it must be recognised that so much from the past has been inadequately addressed, in terms of war crimes – potentially making horizon scanning questionable and a waste of time, at this stage – consideration of the future agenda is imperative. It is vital to look to the future and explore some of the issues that will emerge, before those issues and the controversies they will inevitably bring are upon us. The relationship between science and technology, and warfare, war crimes of the future has never been so critical.

The blending of international humanitarian law and international and human rights law is a further strand of complexity in this changing context, and affects the war crimes sphere. The way that international human rights law has, in some regards, begun to blur with, overlap or even supplant international humanitarian law in grey areas, augments the challenges ahead.[12] In some respects, this is an area in which international humanitarian law (and the international criminal law that embraces it, therefore) is out of touch with realities of the world around it. While the law must be applied and adjusted to new developments, as best it can, it is a trial for all involved. In the absence of authoritative interpretation and established positions, there will inevitably be argument and debate. The way that right and wrong are judged, especially regarding military practitioners and their professional conduct, is subject to immense potential pressure in the context of contemporary warfare involving some of the possible ways in which advancing technological innovations might be applied. The challenges to international law are almost impossible to envisage entirely.

Some of these new technologies generates outrage almost as soon as they emerged – with Human Rights Watch and others calling for autonomous weapons ('killer robots') to be banned, even before they existed, and the use of remotely piloted vehicles ('drones') causing great consternation.[13] There was an apparent common sense that the use of these technologies was intrinsically criminal. That meant that issues surrounding them would clearly be on the war crimes agenda for the future.

Of course the law, in principle, can apply as it is – with appropriate interpretations and adjustments. This is most evident regarding so-called 'drones', or remotely piloted vehicles, otherwise known as unmanned combat aerial vehicles. The simple position, initially, might say they are merely aircraft and that the law should apply exactly the same way as it has to any other use of air force. Objectively, this is clearly the case. Yet, there seems to be something different about it and people feel it is ethically different. The use of these vehicles, therefore, is contested and, where use begins to be contested, allegations of war crimes will circulate. This use raises challenges, perhaps, because people feel a sense of injustice at some of the things this technology allows. From the point of view of the user, drones allow both greater force protection and greater accuracy in targeting. Yet, this reduced risk for the user is perceived (rightly or wrongly) as a 'risk transfer'[14] from those delivering potentially lethal destructive force to those (usually) on the ground, at the receiving end, whether clear and legitimate targets, or those collaterally affected. There is a perception of a greater sense of helplessness regarding those on the ground, whether they are innocents accidentally struck, or belligerents. The apparent disparity reinforces a sense that it must be more difficult to get sense of justice in the use of these weapons. Thus, the depth of this remains to be addressed and tested – albeit that the simple and fundamental position remains that these are aircraft and subject to the law in the way that any other aerial weapons system would be.

Secondly, the use of cyber capabilities in warfare has transformed the landscape and is one of the major motors in the non-obvious character affecting contemporary conflict. In itself, it lies at the source of ambiguities about whether its use actually constitutes warfare – largely based on understandings of warfare limited to combat involving energy transfer weapons of blast and destruction.[15] The very notion of whether, or, if so, when, the use of cyber technology constitutes an act in warfare is contested. It is possible to maintain that the laws of armed conflict were about physical engagement and should only be regarded in that way. In contrast, it is argued that it must be possible to ask at what point can the use of cyber technology can be considered an armed attack – and this has been extensively considered.[16] For example, under Prime Minister Gordon Brown, the UK government carried out an exercise to identify a 13th challenge for the National Security Strategy, which was identified as the cyber threat. This was judged to be highly unlikely to happen, on a broad scale (recognising that it would happen frequently at minor levels), but to be potentially devastating, in terms of impact, were it to occur. The damage that could be done with a successful cyber attack was immense, theoretically – and temporarily limited incidents, such as that afflicting Estonia in 2007, demonstrated the potential. Of course, this was a greater threat to highly cyber-dependent Western societies than to less-developed ones, on one hand – but, equally, a potentially greater one to those with fewer resources and so more vulnerable to attacks on such infrastructure as they possessed. The potential questions of injustice and war-criminality in this context were, again, enormous.

Perhaps the most obscure, and surely, the most challenging of the new technologies is synthetic biology. In some senses, this involves developments that might just not fit any of the existing categories of law, or crime. These include the potential weaponisation of genetic mutation and manipulation. While these technologies have been developed with human improvement in mind, particularly in the contexts of food production and therapeutics, inevitably, their development rests on the conscience, responsibility and good will of people producing these things.[17] But, the same technologies might be developed for use in armed conflicts and, worse, could be highly damaging in the hands of anyone wishing to do harm and perform terrorist acts, who, if they had the money to fund some research, could have the capability to do same. In terms of the law, while the first response might be to look to the 'Biological Weapons Convention' (BWC) or the International Criminal Court (ICC) statute.[18] The BWC is, more accurately, the 'Bacteriological and Toxins Weapons Convention' – thus, it covers bacteriological material (in terms of existing law, this is problematic, as neither of these names clearly covers the emergence of synthetic biology and systems derived from genome technology).[19] While bacteria – living organisms – and toxins (inorganic poisons) are covered, semantically, there is no place for genetics, genome technologies, and genetic mutation and manipulation. Of course, the spirit of the BWC is taken to be that that it should cover all 'biological' capabilities as weapons – though the engineering and fabrication aspects involved in synthetic biology could also be problematic, in this respect. Without clarification, extension, or development of new law, the theoretical possibility of a genome-informed weapon remains (mitigated, legally, in the early stages of development by the necessity of a bacteriological delivery agent – though, again, ambiguities could apply). In terms of the ICC Statute, while the use of chemical weapons is completely prohibited, in line with the Chemical Weapons Convention, there is no corresponding absolute prohibition on the use of any kind of biological weapon, even though the weapons themselves are banned. International criminal law would certainly be tested by these developments. It might be arguable that the complete outlawing of chemical weapons also covers biological weapons because they work, it could be said, as poisons. In terms of using capabilities that could target genetic markers, it might be that the crime of genocide would be relevant. But, that would only be the case, if there were clearly the intent to destroy a group, in part, or in whole – if the genetic marker targeted did not destroy, but simply, in some sense 'disarmed', then the theoretically 'open and shut' case of genocide (by definition, given genetic targeting), would be moot. In addition, if consideration is given to the difficulties experienced by courts dealing with genocide cases, with proof of intention being the highest-level test, but problems evident at other levels, the challenges of bringing genocide cases would be great. And they would be compounded by ambiguities that might rule out any prospect of conviction, given the tests of physical destruction required and the highly conservative treatments of genocide by international tribunals.

It is in this context of ambiguities that this present chapter and the companion one that follows it (Chapter 24) offer what must be a preliminary and indicative empirical material related to the content of potential future war crimes prosecutions. We show how research subjects understood and responded to questions about the use of these new technologies in warfare. This revealed different levels of understanding and concern, regarding each of three areas explored: cyber warfare, autonomous weapons and the prospect of synthetic biological capabilities. The last two are treated in the following chapter. The remaineder of this chapter considers cyber warfare and the prospect of future war crimes questions.

Cyber warfare

Initially, participants' understanding of cyber warfare was explored openly to set the parameters of discussion. While recognising that this was an open issue and that many things could be regarded as cyber warfare, meaning that it could be 'all things to all men' (and that there was therefore 'a lot of complacency' and a 'lack of understanding'), there was a sense that it was simply a matter of conducting warfare in cyberspace, somewhat akin to distinctions between conducting warfare on land, on the seas, and in the air – it was simply a dimension of activity within warfare, in which an opponent's assets were targeted. These assets not might not necessarily be the cyber targets themselves (rather as aircraft might attack ground forces, or other assets), although there would necessarily be some element of attacking the cyber element to have wider effect. This action, it was judged, needed to be carried out by specialists working with computers who would either defend their own capabilities or conduct offensive action to 'deliver a payload to the opponent'. It was clear that cyber warfare involved crossing borders of some kind.

This issue generated sharp dissention from one individual who asserted that 'cyber warfare is a misnomer'. This person insisted that the term was 'used either to generate funding or to justify military activity'. This did not mean that cyber issues were irrelevant, however: 'whether it is warfare or not, there should be interest in protecting military networks'. Otherwise, cyber was 'more a criminal issue and therefore a law enforcement issue than a military issue'. This same participant held the view that 'cyber warfare is massively hyped up'.

Participants initially discussed cyber warfare primarily as a matter of threat to and protection of an actor's own capabilities, saying it was 'not just a defence issue, but a pan-governmental issue [as] all aspects of government depend on systems and are interconnected'. This extended to 'business and society as a whole'. It was felt that 'everyone, if not from a patriotic perspective, then out of self-interest, should be concerned about cyber warfare, or attacks'. There are 'aggressive actors in the cyber world', many of which are states, while many are private actors, but, it was said, 'the threats both pose are underestimated'.

The biggest concern, however, was internal security. One participant, reporting on personal experience of cyber issues, declared:

> The largest threat comes from the inside [and that] people with legitimate access to a network are the greatest threat to a network [either because there is] an insider that willingly corrupts the network [or an individual] makes a cock-up ... usually through a lack of cyber hygiene, like using a USB stick and not changing the default password'.

This meant that the emphasis had to be on defence against internal threats: 'defence against ignorance by education etc. and defence against insider malicious attacks'.

Participants discussed the notion of cyber attacks, considering what would constitute such an attack – whether it meant a system going down or simply the perception of being attacked. The notion of a cyber attack was challenged in one session by the dissident participant, who asked, 'When did those successful cyber attacks happen, except for Stuxnet? ... Both economic and defence attacks happen, but they don't bring organisations down'. This immediately shifted discussion to the domain of effects and the way in which attacks might be gauged by their impact, including issues of 'collateral damage'. This was felt to be an important factor distinguishing a cyber attack from a conventional armed attack: 'The collateral damage is unknowable, as opposed to conventional attacks where the impact and consequences can be calculated', said one officer, explaining that the laws of physics determine predictable effects when a certain amount of explosive material is used against particular materials, but the impact of a cyber attack was judged to be less specific. 'Many attacks are fixable within five minutes', another participant stated, which was complemented by the view that 'cyber is a one-shot weapon; whatever the effect, it can never be repeated again'. Thus, participants underlined repeatability as a quality distinguishing conventional weapons from their cyber counterparts.

The judgement about the distinctive quality of cyber instruments as weapons did not mean that they were not serious. 'Those working with the systems attack feel the seriousness of the attack', participants agreed, with one of them adding, 'When the network goes down in a small country, the effect is felt.' This was a clear reference to the attacks on Estonia in 2007. This was picked up by another participant who commented:

> Estonia was taken down to a point that it no longer had a functioning government – it was assumed to have come from Russia. The country was brought to a halt without a foreign boot on the ground. That is a warning [that] the systems to prevent cyber attacks in the rest of Europe are not more sophisticated than those of Estonia.

This point was related to other contexts by participants, who pointed out that if health service records and patient files were shared, or intensive care systems brought down, or a national grid were to be attacked, so-called *virtual* cyber attacks could 'quite easily turn into something quite physical'. This was underscored: 'These are very real threats.' One participant noted that 'the threat from cyber comes from the ability combined with intent, and many are willing', invoking a conventional reading of what constitutes a threat.

Once again, however, the single doubting voice in one group strongly questioned the discussion, asking (perhaps rhetorically), 'What [...] is it not happening? ... The only time it happened was state on state – not terrorists.' (There is a clear presumption that, although this has not officially been confirmed, Iran had been targeted by the US, constituting a second case.) One response to this was to state that such attacks do happen 'institution to institution' which drew the rebuttal: 'That's all cyber crime.' Attempts were made to take this on board by linking terrorism to organised crime – 'terrorism becomes organised crime' – by pointing to the links between them, with organised crime being used to fund terrorist groups, as had happened in Northern Ireland. Another example involved the suggestion that if Sony PlayStations were to be 'hacked' and all financial information taken this would be equivalent to 'infiltrating defence weapons systems'. However, this only confirmed the dissident's view: 'Since money was invented, people have been looking for ingenious ways to steal it – which is a crime, not an act of war!' The counter that the money might be used to fund terrorism was dismissed: 'That is a crime and a case for police investigation, as they have been doing for years.' Other members of this particular focus group, however, supported the proposition: 'If it is used to fund terrorism, it is part of the war effort – there are interrelated things between crime and war.'

Respondents were presented with a set of military doctrinal terms associated with the cyber sphere and electronic warfare. The discussion focused on distributed denials of service and syntactic attacks as those most relevant to the notion of cyber warfare.[20] One participant said, 'If you are the victim of a DoS [denial of service] or a kinetic attack, the effect is the same'. However, it was clear that 'for an attacker, they work in different ways – Stuxnet is actively interfering in the system of an opponent, DoS is possible – not inside the integrity of the system of the opponent.' Once again, the issue of 'temporal or permanent damage' was relevant.

In terms of a syntactic attack, the issues were viewed as being similar to those of conventional weapons: 'The legitimacy of targets has to be established – is there a difference whether people drown downstream from the bombing of a dam than from opening it by a cyber attack?', it was asked, prompting discussion of the 'Dambusters' raid by British bombers against German dams in 1943. In terms of cyber, it was judged, 'The legitimacy depends on the effects that cyber has brought on'. Others in this same group tended to agree that cyber was a matter of 'engaging the enemy without fighting'. This could be done in different ways, 'whether you place thousands in a cyber army, like China does, or have specialised people that can depend on good intelligence'. Whichever way is followed, however, there was no doubt that 'cyber will have an effect' and that 'like in conventional warfare', militaries and others 'have to prepare for contingencies'.

The idea of a syntactic attack, actually changing code that will affect data, and the extent to which this actually involved physical change was investigated. Respondents discussed whether this was part of the job of the military, with its ultimate uniqueness in managing the application of potentially lethal violence, and whether the law of armed conflict should apply. One soldier said, 'Some sort of justification is needed – this is very hard without some sort of tangible, kinetic, effect'. This was a point of strong agreement in that participant's group, with another voice adding: 'Tangible, yes; ... if 50 billion were lost on the stock market, this would be tangible', while a different voice augmented this by saying, 'The psychological effect of cyber might be bigger than the physical effect'. However there was complete uncertainty among participants as to how this might precisely conform to the notion of an armed attack, or to that of armed conflict, and whether international humanitarian law would apply – or, if it did, at which stage.

The question was posed: when would taking the cyber route be effective if the desired effect were kinetic? One voice said that 'dropping a bomb on a dam is cheaper'. However, even if this were true of cost, there were other aspects to consider – for example, it was said that 'what is useful with cyber is the deniability – at least in the short-term'. A supplement to this was that 'an attack, ideally, via the internet, could be delivered from a hotel room in Caracas to a target in South Africa'. This was clearly a potential advantage in some circumstances, giving cyber 'attractive characteristics'. One contributor pointed out that while deniability would often be an asset, it would be of little value to terrorists, as 'terrorism only works if people know you did it'.

Participants were asked at what point these kinds of attack might warrant invoking the right to self-defence. This discussion began with one member exploring the nature of cyber and its difference from conventional weaponry: 'Why is cyber not kinetic? A bomb says 'boom' – but a cyber attack might also deliver a kinetic effect. The distinction here is what the goal of the attack is.' This sense was generally shared and it was clear that the effect, rather than the act itself was the point at which judgements could be made. It was felt that self-defence would be legitimate if 'most reasonable people say, "Yes, this sounds about right."' It was recognised that various (unspecified) groups in various places around the world might have different views on what would be acceptable, however. Another contributor suggested that this touched on ethics and the question: 'How do you establish proportionality?' This was extended by reflection on what threshold enabled a dispute over cyber attacks to be 'taken to the UN, or another body for arbitration'.

The discussion of the right to self-defence returned to the questions of deniability and attribution. The problem with judging a cyber attack in terms of self-defence 'lies in the deniability of such an attack'. In this situation: 'The victim wants to strike – punch, the victim wants to invoke self-defence – but against who [sic]?' One member of this particular group discussion responded to this by invoking the example of Estonia once more, saying that, even in this instance, NATO had not agreed that it was an attack. However, there was confusion among other members of this group about collective support and collective self-defence. It was – correctly – reported that the Alliance had discussed the attack and how to cooperate under Article 4 of the Washington Treaty. But, some participants, in particular the one who had introduced this topic initially, mistakenly believed that Article 4 related to the commitment to collective self-defence that is, in fact, found in Article 5 of the Treaty. It was clear that, aside from the general issue of attribution, there could also be doubts about establishing 'the causal link' between 'the transmitting of 1s and 0s and a physical event'. It was felt that 'this might be hard to prove'.

Another issue regarding self-defence, was scale. Judging whether or not there might be a right to self-defence was a matter of scale. One participant asked: 'If one set of traffic lights goes out for an hour, is that an act of war? No.' There were clearly issues surrounding assessment of any effect achievement by a cyber attack: 'How do you assess the impact of an attack and how do you decide what is a proportional response?' The answer, it was felt, was a 'political decision', which would depend on how belligerent a population and its government felt following an attack – 'how willing they are to go to war and where we are in the political cycle'. This was compounded, in one participant's view, by the problem that 'there could be a fast or slow setting effect of an attack'. The pattern of an attack and its effects might well make a difference to the way in which the right to self-defence were perceived, as 'attacks may not come at once, and may not constitute enough to warrant an act of war immediately'. Thus, the potential slow-drip character of a cyber attack would make it harder for the victim to be able to respond in self-defence.

Four propositions, or questions, were presented to the groups and their reactions were sought:

1 'Is it reasonable to require adherence to old concepts of what constitutes *use of force* when it is clear that the destructive power of cyberspace operations can threaten the economic integrity of a state?'
2 'Military objectives which could not have been attacked with kinetic means due to the excessive collateral damage in relation to the concrete and direct military advantage gained potentially become liable targets since computer network attacks provide for a means to reduce incidental loss and collateral damage.'
3 'Certain attacks against military objectives, which would be unlawful if executed with kinetic weapons because they are expected to cause excessive incidental civilian damage, may be lawful if conducted by way of disruptive cyber operations.'
4 Can data be seen as the object of attack? Or should an attack only be deemed to have occurred if there are consequences in the analogue world? Should corruption, deletion, or alteration of data (for example in medical records, or operating systems) be enough to be deemed an armed attack irrespective of consequences in the analogue world?

There was overwhelming and vigorous consensus regarding each proposition, although the reservation that it would be impossible to know was expressed regarding the third proposition. This reservation rested on the impossibility of knowing the full effect in most circumstances.

Conclusion

A range of new technical means do not easily fit with existing law – albeit, that in the first instance, there needs to be a need to apply the existing law to them, as far as possible. This creates new fields of concern and the prospect of new types of allegation and problem in the future concerning war crimes. Whether or not the prevailing legal framework can accommodate the new technologies, there are evident pressures both on the law itself and on those who have to conduct warfare. Some of these challenges have been given initial outline in this chapter and subjected to empirical research with military personnel to gauge their understanding of both technological innovation and application with regard to the boundaries of wrong and right in warfare. This material can inform thinking on the factual aspects of future war crimes cases.

For all its novelty and the challenging qualities it brings, research subjects embraced cyber warfare as simply another part of warfare. While participants underlined repeatability as a quality distinguishing conventional weapons from their cyber counterparts – confirming that there were crucial differences – the cyber capabilities were viewed simply an addition to the realm of weapons used in warfare, and issues were generally interpreted through the lens of conventional armed conflict and the contribution that the cyber arms made. One of cyber's biggest assets – and conversely, difficulties, if on the receiving end of an attack – it was judged, was its 'deniability', at least in the short-term. This was very much the epitome of non-obvious warfare. The key to understanding how conventional warfare constitutes the frame for embracing cyber warfare lay in the effects any use of cyber weapons had and their relatively analogous impact in relation to conventional blast and destruction capabilities. In this sense, as with traditional artillery (in part), or other weapons systems, the legitimacy of cyber attacks should be judged by the effects they create – not the act itself and not the intention in making it (in contrast to traditional just war – and legal – thinking). There was a strong sense that social context and community support were and should be decisive factors. Action would be legitimate if most 'reasonable' people deemed it to be so.

There was strong and general agreement that old concepts of what constitutes *use of force* might be relevant and it was reasonable to maintain existing frameworks, but also that the destructive power of cyberspace operations could threaten the economic integrity of a state while potentially by-passing those frameworks. Similarly, there was strong general consensus that cyber capabilities meant that, in principle, military objectives, which could not have been attacked with conventional means due to the excessive collateral damage in relation to concrete and direct military advantage gained, could potentially become liable targets since computer network attacks could reduce incidental damage. There was also a strong sense that attacking data – corruption, deletion or alteration of data (for example in medical records, or operating systems) – should be enough to be deemed an armed attack, irrespective of consequences in the analogue world.

Notes

1 See James Gow, *War and War Crimes: The Military, Legitimacy and Success*, London: Hurst and Co/New York, Columbia University Press, 2013.

2 'War crimes' technically refers either to Grave Breaches of the Geneva Conventions, or to breaches of the Laws and Customs of Warfare. The two categories are subsumed together in the Statute of the International Criminal Court (though they were treated separately in the Statute of the International Criminal Tribunal for the former Yugoslavia). *Rome Statute of the International Criminal Court*, Rome, 17 July 1998, entry into force 1 July 2002 in accordance with Article 126, Registration 1 July 2002, No. 38544, United Nations Treaty Series, Vol. 2187.

3 See Christopher P. M. Waters, 'Is the Military Legally Encircled?', *Defence Studies*, Vol. 8, No. 1 (2008), and W. G. L. Mackinlay, *Perceptions and Misconceptions: How Are International and UK Law Perceived to Affect Military Commanders and Their Subordinates on Operations?*, Defence Research Paper, Shrivenham: Joint Services Command and Staff College, July 2006.

4 Western operations in Libya and Iraq, for example, were surrounded by accusations of war crimes. See the following for examples. Libya accused NATO's airmen of committing war crimes as they prosecuted an air campaign against the Qadaffi regime in 2011. For example, a Qadaffi-loyalist Libyan diplomat, Mustafa Shaban, accused the Alliance of 'crimes against humanity, crimes of war and crimes of aggression' (*Reuters*, 9 June 2011). Disturbingly for the Alliance and its airmen, it was clear that some among the Allies were uncomfortable with the conduct of operations, or even the overall idea of Operation Odyssey Dawn itself, especially when NATO had to concede errors, such as the 'weapons system failure' that resulted in the highly publicised deaths of two babies among nine civilians in one attack on Tripoli. This enhanced concerns about the operation within NATO and outside it, adding to the pressures on the aircrews conducting operations. (BBC News, 20 June 2011.) *The Guardian* also published an article by Ewen MacAskill under the title 'UK should face court for crimes in Iraq, say jurists', *The Guardian*, 21 January 2004.

5 Jan Willem Honig, 'Uncomfortable Visions: the Rise and Decline of Limited War', in Benjamin Wilkinson and James Gow eds., *The Art of Creating Power: Freedman on Strategy* New York: Oxford University Press, 2017, Chapter 2.

6 The research presented in this chapter and the following one was part of a wider programme of research, involving the War Crimes Research Group at King's College London, the Royal College of Defence Studies (RCDS) and the Joint Services Command and Staff College (JSCSC), in the UK, as well as the Humanitarian Law Center in Belgrade and the Center for Interdisciplinary Postgraduate Studies, University of Sarajevo. The bulk of the research used directly in this volume involved the RCDS, which embraces an international mix of defence practitioners, mostly at one-star (Brigadier/Brigadier General) or equivalent levels. Those who took part in the research were an international mix from all continents and a variety of cultures. Participants were from all continents and roughly all parts of all continents. The research involved people from Middle Eastern countries, from all points of the African compass, from different Asian and Australasian countries, and from the Americas. There was a surprising degree of consistency. It is necessary to be careful when describing or referring to the group, because of the confidentiality involved, especially because some of them come from countries where if they could be identified, they might face censure, or punishment. All respondents volunteered freely, with no compulsion or obligation. The agreement to carry out the research with the RCDS involved sophisticated discussions with the Senior Directing Staff there. The type of research that was conducted has two particular strengths. First, it allows the generation of empirical data, simply through the discussion that takes place, with participants offering knowledge and information drawn from experience, as well as giving expression to beliefs, values and attitudes. This provides the data for analysis. Secondly, however, and in contrast to individual research interviews, the group research process, especially when it involves the presentation of specific propositions (or material – part of the research used in this volume draws on research where participants were shown visual material), encourages interaction between participants, leading to discourse developing and the identification of salient beliefs, values and attitudes, where participants may strongly agree, or diverge, on issues.

7 There is an important body of literature on civil-military relations. Classic and key elements include: Samuel P. Huntington, *The Soldier and the State: the Theory and Politics of Civil-Military Relations*, Cambridge, MA: The Belknap Press, 1957; Samuel E. Finer, *The Man on Horseback: the Role of the Military in Politics*, Harmondsworth: Penguin, 1968; Morris Janowitz, *The Professional Soldier: a Social and Political Portrait*, Glencoe: The Free Press, 1960 and *The Military in the Political Development of New Nations*, Chicago: University of Chicago Press, 1964; Amos Perlmutter, *The Military and Politics in Modern Times*, New Haven: Yale University Press, 1977; Martin Edmonds, *The Armed Services and Society*, Leicester: University of Leicester Press, 1988; Peter D. Feaver, 'The Civil-Military Problematique: Huntington, Janowitz and the Question of Civilian Control', *Armed Forces and Society*, Vol. 23, No. 2 (1996); Elliott A. Cohen, *Supreme Command: Soldiers Statesmen and Leadership in Wartime*, Glencoe: The Free Press, 2002.

8 Michael Boethe, 'The Protection of the Civilian Population and NATO Bombing on of Yugoslavia: Comments on a Report to the Prosecutor of the ICTY', *European Journal of International Law*, Vol. 12, No. 3 (2001), p. 531.

9 Lieutenant General Sir Rupert Smith, *The Utility of Force: The Art of War in the Modern World*, London: Allen Lane, 2005, Ch. 7.

10 Martin C Libicki, 'The Specter of Non-obvious Warfare', *Strategic Studies Quarterly*, 2012, pp. 88–101.

11 James Gow, *War and War Crimes: the Military, Legitimacy and Success*, New York: Columbia University Press, 2013.
12 Ibid., pp. 129–33.
13 Human Rights Watch, *Losing Humanity: The Case Against Killer Robots*, at www.hrw.org/reports/2012/11/19/losing-humanity-0 at 18 June 2018.
14 Martin Shaw, *The New Western Way of Warfare: Risk-Transfer War and Its Crisis in Iraq*, Cambridge: Polity Press, 2005.
15 Thomas Rid, *Cyber War Will Not Take Place,* London: Hurst and Co., 2013.
16 These topics are explored elsewhere in this volume, particularly in the chapters on cyber warfare and law by Elaine Korzak and James Gow.
17 For more on the dual use of synthetic biology, please see: Guglielmo Verdirame and Matteo Bencic Habian, 'Challenges in Regulation: The synthetic biology dilemma – dual-use and the limits of academic freedom' (Chapter 19 of this volume).
18 *Convention on the Prohibition of the Development, Production and Stockpiling of Bacteriological (Biological) and Toxin Weapons and on their Destruction*, London, Moscow and Washington, 10 April 1972; Rome Statute of the International Criminal Court, A/CONF.183/9 of 17 July 1998.
19 For more on synthetic biology and the BWC please see: Filippa Lentzos and Cecilie Hellestveit 'Synthetic biology and the Biological Weapons Convention' (Chapter 17 of this volume).
20 The distinction between these two forms of attack is discussed by Elaine Korzak and James Gow in Chapter 6 of this volume; in brief, 'syntactic' refers to attacks that physically change code, where 'denial of service' refers to attacks by attrition to overload a system and stop its functioning.

24
FUTURE WAR CRIMES AND THE MILITARY (2)
Autonomy and synthetic biology

James Gow and Ernst Dijxhoorn

This chapter largely follow on from the previous one. It continues to present empirical research findings that might inform future war crimes discussions, in a context where technological change challenges current international legal norms, focusing on the judgement of military practitioners. This chapter reports parts of this empirical research that focused on two areas: autonomous weapons systems and robotics; and synthetic biology.[1] As noted regarding findings in relation to cyber warfare in the previous chapter, the research explored different levels of understanding and concern, alongside one another.

Autonomy and robotics

Issues of autonomy and robotic weapons were explored in overlapping and discrete research groups, examining understanding of autonomy and concepts of armed attack and armed conflict, as well as discrimination and proportionality. The issues were introduced against the background assumption that the character of warfare had been changing and that this meant there were might be grey areas revolving around human agency and the point at which human agency ceases. Participants were presented with a series of propositions that might apply to autonomous and semi-autonomous systems, including *drones*.

Initially, respondents were asked to give their understanding of the kind of issues identified, what they thought autonomous systems were, what they would define as autonomous, what would qualify as autonomous, and whether they, in their professional life used autonomous systems, as well as how comfortable they were with – or their thoughts on – autonomy. The initial reaction to the question of whether participants used autonomous systems in their professional life was to make a distinction between sensors and systems. Some participants indicated that they had 'been the recipient [of] product from [...] the UAV'.

Among responses on the definition of autonomy, one participant described full autonomy this way: 'I've always thought of full autonomy as replacing the human judgment and decision-making, but clearly there's a degree of autonomy which replaces some functions of human input.' He noted that he, therefore, had 'experience of some degree of autonomy', but added that regarding the degree of autonomy, 'I have a great deal of discomfort with what I consider to be a fully autonomous system'. Full autonomy was further described as a system that 'took whatever the final judgment was away [from the human]'.

The responses that followed made a clear distinction between systems that present the (human) operator with options and those that take final decisions. One participant noted, 'A system that presented you with options for decision through autonomous means, I'd see as something very different from something that made the final decision itself', indicating that he would be more comfortable with the former.

Another respondent argued: 'We've [the Navy] already got systems like Goalkeeper that are autonomous.' This triggered a discussion amongst all participants whether these systems are fully autonomous and how far rule-based systems are fully autonomous. One respondent reported:

> We've got offensive missiles, like the Brimstone missile that you input the signature of the armoured vehicle and they go and hit the vehicle. There are the anti-radiation missiles that hang around in the sky, spot something, says 'That looks like a [whatever] system, I'd better take that out.' But you have to choose to launch it. It's got an element of persistence. An element of persistence, but not a lot.

One participant noted that the discussion moved straight to lethal force, while autonomous systems could entail a wide range of other non-lethal military actions. For instance: 'Autonomous [...] could be very mundane, moving something from point A to point B.' The respondent made clear that he is 'quite happy' with autonomous systems completing tasks like this. He provided further explanation that there are whole rafts of other functions, which do not consist of lethal force, which could operate with some sort of rule-based environment and could be created for autonomy. Other examples included the use of airships instead of using CLBs (combat logistics battalions), and the use of more autonomous UAVs for surveillance as opposed to strikes. He was supported by the assertion that 'Companies like Google are making a big push towards autonomy in terms of delivery for humanitarian aid. Whether it be some sort of UAV that can deliver a container with tents and water to a certain point.'

This respondent however expressed uncertainty about how autonomous these systems are, saying that this depends on the point one chooses to designate a system as 'fully autonomous'. Thus the discussion reverted to the definition of full autonomy and the uncertainty among respondents about when a system becomes autonomous. 'If there's a line, I don't know where that line is', said one. He ruminated: 'An autopilot on a ride, is that autonomous? Because, at that point in time, it is following the way or points you have put in. The fact that you're sitting there and you can flick the switch off doesn't change that.' Another participant remarked: 'We might not see that as autonomous, but we might view the replacement of a driver for a cab for a CLB that has just been replaced by a waypoint system the Google vehicles use – is that autonomy or not?' Most respondents agreed that 'If you just put in a series of waypoints, that does not mean a system is autonomous.'

The respondent that initially doubted the autonomy of systems using waypoints suggested a definition based on

> whether the actual electronics will make a decision on [provided information]. I know it's simplistic and there will be other sub-routines that say: if you see a big hill, try to avoid it; but there's a difference with autonomy in terms of making decisions along the lines of, if you are presented with this information, then do this, as opposed to follow instructions – you know where you need to go, get there, when you get there, you go back. [He based a distinction on] the absence of a human in the loop, that, if it recognises a target, decides to strike on its own without a human saying 'Yes, I confirm ...'

Another participant added that these systems were still following instructions:

> Surely our artificial intelligence is creating those routines to make their own decisions. … It's rule-based. … It's hard to know where one starts and one stops. … I think it's very difficult to work out where rule-based stops and where artificial intelligence starts.

Participants introduced discussion of systems already in use that they considered to have elements of autonomy. 'Patriot … can work in an autonomous mode, and at least one of the guns in the Gulf War was effectively in autonomous mode, or semi-autonomous, whereas the operator has to decide to *not* fire rather than fire.' An air force officer present replied that this was not dissimilar to systems on the Apache helicopter:

> The computer is processing a list of targets and the operator gets presented with 150 targets per second, and he can choose to prosecute the top ten without doing more than flicking one switch. … That's quite close to what I think of intuitively as the limit of comfort in autonomy – the presence or absence of a human override.

The facilitator also asked whether the participants saw autonomous systems being introduced into their day-to-day working life, or operations. Responses regarding what they thought would be introduced differed. One respondent thought that CLP (Combat Logistics Patrol) would probably be the closest we're ever going to get to autonomy, but contemplated: 'It depends if we're going to get to swarming mini-drones with facial recognition and cyanide in their tail. How far is that away? I don't know.' Another suggested, 'Logistics and support is probably quite a good area because in a way it is quite uncontroversial. That could be a link from CLPs to robot dogs that run up to you with small packs of ammunition attached and attack on setting.'

In the discussion of the use of autonomy by support groups, divisional support groups, and security, which would reduce the requirement of tying up whole companies of infantry, examples were given of where autonomous systems would become problematic. For instance: 'If you're a runner from a brigade and you're making your way back to pass a message, or pick something up, and you're not wearing the right ID badge, or something, then the autonomous security starts.' Or: 'Let's say you have Goalkeeper on a ship and you decide to save money by having smaller autonomous service vessels with Goalkeeper on board, if something comes close to the carrier ship …'

While other respondents were coming up with situations in which autonomous systems could make mistakes that a human would be unlikely to make, an army officer shifted the focus to less controversial tasks of autonomous systems focussing on data:

> Seeing it from a slightly different perspective, we would be able to use artificial intelligence to take large dumps of simple, binary data, and dump them into a system where it would pick through and start to make the links without us having to query the system. So we would be pulling out information, so it would be putting business information in a broader context, or information for us to then turn into intelligence. We cannot keep track of all of the information we have, so once we start to feed that into an engine that allows it to think creatively and start to pull some of those links together without the analyst having to do that, especially with the big data.

Another suggested that autonomous systems would be used 'in a context where communications were difficult. [For instance,] in a hostile environment where communications are jammed, and the only possibility is to send off a drone with a certain amount of intelligence to do just what needs to be done on its own.' A respondent added that such a system would 'need to include a form of self-healing' and have 'a self-forming network that is autonomous, in the sense that it knows when it needs to connect'. Such a system would be instructed as to what target to hit: 'It just needs to go find it, identify it itself and strike it itself because we can't talk to it once it's gone past that horizon where the comms are jammed.'

The discussion about self-healing and self-analysing systems was taken even further by another respondent. First, he mentioned systems, using the example of cars, which have the capability to identify and fix their own faults. However, to the amusement of his colleagues, he then took it a step further and suggested autonomous systems that looked at human performance analysis, that check people's blood pressure and metrics to identify what is happening to them on the battlefield. He conceded that this was

> quite a big leap, [but that] it might come along relatively easily in the future. If someone is injured, if nothing else, [he suggested,] get a whop of morphine, anti-coagulant, things like that to keep you alive, preserve you in time for an autonomous drone to come pick you up.

Pressed further, he suggested that non-military technology would 'eventually be militarised'. Autonomous systems could even be used for complex high-end surgery and anaesthesia:

> If you look at any anaesthetist, he sits there, reads the patient, size, weight, age, history, what they're going to do, and once the patient is under, he's always in the loop. He monitors it, but he has a relatively lightly trained individual, who sits there and monitors it. [He developed this point:] Laser surgery, I suppose, has quite a high degree of autonomy, because it makes a decision of how much it's going to burn off of your eye. The machine, I suppose, does the maths and the thinking, the individual has a look at it – it's like a scan, I suppose. Like you can't be put into a scanner and be told what's wrong, the machine does an automatic scan, but there's a person who decides you need the scan, and then there's the person who deciphers it. So you can have a scan on an operation ... and that's communicated back if the signal works to the expert, wherever he is, might be in the UK, might be a couple of miles back, and then the intervention would have to happen.

When asked by the facilitator about the distinction between autonomous systems making a clinical decision and an autonomous weapon, the respondent indicated,

> There is no difference between an autonomous system making a clinical decision and an autonomous weapon system making a targeting decision. [He continued:] I think it's the same thing. It's life and death. ... You're either teaching it to recognise an anomaly against a background, whether that's a cancer cell or a vehicle. I think both depend on the strength of their programming, the strength of their library and what knowledge they have available.

Another respondent also indicated that the level of comfort in using autonomous systems would be dependent on both using 'robots for what they're good at', but also when a system's 'decision-making logic is going to be better than a biased, stressed person trying to make a decision'. The participant who had suggested the use of autonomy in battle-medicine, developed the point, saying:

> 'It is dependent on the outcome rather that the level of autonomy of the system. [He added:] If tele-medicine were your only option, tele-medicine fails and the person dies, they were going to die anyway, so tele-medicine can only improve that person's output; whereas, a drone that gets it wrong, or the software crashes, could do far more damage if it fails. That's why one would proceed far faster than the other I think. … You develop the system, and the effect of one is ultimately to preserve life, in some way; the other is to terminate life. And that starting point is different. Yet, the technology may be incredibly similar.

Another issue regarding the use of autonomous systems was the public perception of collateral damage. One participant argued that

> We forgive *ourselves* collateral damage at the moment to a much greater extent than we would forgive a drone. [Adding:] Ethically it's an awkward situation. Who is accountable for what has happened?' [Underlaying the participant's discomfort,] if it's a pilot making a conscious decision, the chain of accountability is clear, but if it's a drone that's making a decision, whether it's to take clinical, or offensive action, then there's still a question mark over who to hold accountable. [He concluded:] You can't hold a machine accountable.'

In discussion, it was argued that the programmers, the system behind the machine, hold responsibility. Moreover, it was suggested, 'Google Cars – or the litigation that will start as soon as the first autonomous car knocks over a person – will shed light on the liability issue.'

Stemming from all of this was the acceptance that humans are fallible. However, the question was raised of

> whether we allow technology to be fallible to the same percentage that people are? In certain systems we do allow fallibility, a blocked mobile phone or computer that needs to be reset, but we wouldn't seek to accept that in other systems where we would deem it to be safety critical.

Regarding what drives autonomous systems, both civilian and military participants agreed that 'technology, as a whole, could get to a stage where it became better than whatever humans could do'. However, there was also strong agreement that 'military use of autonomous systems will also be driven by the willingness to take casualties'. One participant encapsulated this:

> If our willingness to take casualties drops, or continues along the curve it has been since the Second World War, that might drive innovation faster than it might do otherwise. If, in total war, we have to have a much higher threshold for the taking of casualties, autonomous systems may not be deemed necessary. That is unless it is more economical; as you may be able to build something faster than you could train an individual to achieve it.

Respondents considered the extent to which machines should be able to select their own targets, and the debate over 'human in the loop' and 'on the loop' standards – that is, the level of control the participants are comfortable with. One participant pointed out:

> We are already there, to a point, with the navy torpedoes that select their own targets, such as Stingray, and Goalkeeper. … Moreover, you can set up boxes in the air, and missiles will target anything that goes within that without differentiation.

This led one participant to make an analogy with landmines: 'You don't send friendly forces into [a minefield].' According to another respondent,

> [W]here you cross the line is deliberate targeting, and the prosecution of emerging targets. … That's where I think autonomous systems would struggle. Where there's something we know, there's no real issue. Trying to identify a Taliban guy, I can't see that happening.

The 'human on the loop' discussion focused on the time it takes for humans to take decisions and on whether they were making the difficult decisions. If humans made the difficult decision 'by saying, "We know there are no friendly forces in that box in that area"', all that the autonomous weapon had to do was 'differentiate between the number of different systems in that box'. It was suggested that the acceptability of the decision had to do with the expediency of the decision. However, it was 'the timeliness that autonomous systems have that may give you an edge, because the "human in the loop" just isn't going to be quick enough'.

It was suggested that non-lethal systems would be more prevalent than lethal systems. And in the area of non-lethal judgemental software, that 'may be tried before being used in a lethal system'. It was observed that

> Many, fairly benign autonomous systems are already in place, and seem to be accepted by the public, [but that] self-driving cars have been tested for years without incident, but are not on the market, which suggests that legislators and the public are not comfortable with a system that has the potential to cause casualties.

The discussion was developed by reference to the notion of an autonomous system as a piece of software that thought for itself and spread like a virus. Comparing such a system to smallpox, the potential to lose control was deemed to be high by participants, leading to uncomfortable laughter: 'What could possibly go wrong?' one participant ironically noted. Other red lines with regard to the employment of autonomous systems were also mentioned. First of all, 'the deterrent' was mentioned by a British member of one group, although another argued,

> The kind of autonomy we would be comfortable with is scenario-based, not technology-based. … Arguably, automating the deterrent, it would be an even more powerful deterrent. There is absolutely no way a first-strike capability could take it out.

Another red line discussed concerned living organisms. For instance, one participant asked: 'If you could put a chip in [a fly] so it does exactly what you want it to do, is that ok? … And if you escalate that up to humans, is it OK to automate humans so humans do exactly what you want?' There was uncomfortable laughter at this point, suggesting a deep feeling that this would be going too far, the speaker making this point argued that 'the deterrent would be more effective if autonomous'. He also argued that there was 'no difference between that and putting sensors on an attack dog, that will stop an attack dog doing various things, which we do at the moment'. Other members of the same group found this challenging and pointed out that humans would potentially 'still have families' and be 'part of the community', and they would also have 'basic human rights'.

The last area investigated was the regulation of autonomous systems. The campaign by NGOs to stop robots through a pre-emptive ban on anything that selects targets independently of human beings was deemed to be an 'entirely unrealistic aspiration'. This is because 'every navy and air system is [already] based on a system with elements of that'. This was, therefore, a futile initiative; respondents thought: 'It would be a rollback of many generations of systems to get back to a stage where you are doing it all yourself.' One respondent judged that the initiative revealed 'a lack of understanding on the NGOs' part'. Another participant compared this to the anti-landmine treaty,

> in that regulation will depend on how people abuse the system and on how it proliferates. ... If technology previously discussed, that is already in use, would get into the hands of people that don't have rules and procedures in place, then there might be a drive for non-proliferation.

Another participant argued that the NGO aspiration was:

> We don't want any nation on Earth to create an army of killer robots, programmed sufficiently well to tell the difference between a civilian and a combatant and unleash into a city.' [He clearly agreed, arguing:] I don't think any of us would want any nation to get to that stage where they could place killer robots that they've programmed well enough with an acceptable level of collateral damage and that's how they could choose to fight their war...

Furthermore, he expressed understanding, 'The NGOs aim to pre-emptively set the norm, that if we get to that level of artificial intelligence, internationally we would accept that we won't go there.'

On what the norm is, and should be, some uncertainty existed. It was argued that, in the UK, and for its major allies, 'the norm is for a human decision point in the kill chain'. However, another participant, in a group with no Americans, argued, 'The Americans will go further at some point.' This was amplified with a rhetorical question: 'Do you trust the Americans, if they have got to go re-do Fallujah and they've got a hundred thousand robots, or a hundred thousand marines, do you trust that they would not use the robots?' The previous speaker noted, 'Norms can be ignored, but, that the norm of a human in – or on – the loop should still exist, as norms are heavily influential. ... The Landmine Treaty, for instance, ... everyone could choose to ignore it.' However, he stressed: 'Actually, the very fact that the treaty exists locks democratic societies into that mentality.' It was suggested, augmenting this point, that:

> Even for non-state actors, and deniable state actors, states not using chemical or biological weapons makes using chemical or biological weapons unacceptable. If we were all happy to use chemical and biological weapons, then a terrorist would go – 'Well, states can do, so why can't we?'

The respondent stressed this point: 'It increases the legitimacy of using certain means. If the Americans are using killer robots, why can't the Chinese or Russians do it? And then we quickly get into non-state actors doing it.'

The achievability of a ban on autonomous weapons proposed by NGOs was seriously doubted. However, one voice thought,

> If they made a distinction between human input in the kill chain, or lack of target selection, and between automated systems, they could make that target absolutely

> achievable. If they don't, it's too late. ... We haven't been able to clearly articulate what all those systems will be, so there can be a real reluctance for governments to step forward and legislate the system, because they don't really know what the system is going to be in the future, so, in a way, we are giving away a great advantage, because we hadn't even considered that system when we put the legislation in place.

The risk in this was underlined by reference to problems experienced with use of existing innovative technology, unmanned aerial combat vehicles, or drones: 'And the big danger is there are so many people who don't understand the drone war. Many members of the public have this perception that they open fire over Pakistan and just hit stuff when they see it.' It was clear that respondents were wary of any move – certainly, any rush – for new legislation until more was known about real capabilities and potential uses.

Synthetic biology

The issue of blanket legislation to prohibit use of a capability has resonance in discussions of synthetic biology and the Biological and Toxin Weapons Convention (BWC). The BWC is widely perceived as outlawing any use of biological weapons – a rational position at the time of its adoption, when the chances of developing militarily utile weapons were nonexistent, or negligible. However, the advent of synthetic biology has changed the perspective on some aspects of biological weapons, in certain respects – for example, the availability of targeted weapons to attack stealth paint on warships could be an attractive option for armed forces able to access it, even though most critics view the BWC as outlawing all biological weapons.

Issues of innovative synthetic biology – the prospect of weaponisation and issues of biosecurity, exploring the issues regarding concepts of armed attack and armed conflict, as well as discrimination and proportionality – were examined among both senior and junior-middle-ranking personnel. These were discussed on two discrete levels, albeit with a small amount of inevitable overlap: the low-level, kitchen-sink, or garage, capabilities that individuals, or small groups could develop, albeit more in line with traditional biological weapons; and the high-end novel developments available only to states, or those with the resources of states. Once again, the issues discussed were introduced against the background assumption that the character of warfare had been changing and that this meant there might be grey areas surrounding what was warfare and what was not, and how synthetic biology related to this.

Initially participants' understanding of bio-related terms was explored, recognising differences between concepts of weaponisation, biosecurity, and biological warfare. Members of the first group to engage with these issues identified a range of examples covered by the label 'biosecurity'. Anthrax was identified as the 'classical example' of a biological weapon, precisely because it could be weaponised. Another example was synthesised plague. The participants added that the Ebola virus was of concern and would be extremely dangerous if it could be weaponised (recognising the existence of an international crisis over Ebola at the time the research was conducted). It was judged that a number of biological agents could be weaponised and could be used. Weaponisation was considered to be the key question in terms of biological agents, not the danger and quality of the biological agents themselves.

One particularly striking concern was that the potential of suicide-mass murder operations had significantly altered the framework for thinking about biological attacks. One factor that had made biological weapons hard to use historically was safety and the prospect that their use would also affect – take out, or kill – those seeking to deliver the agents as weapons – or alternatively that they would not survive to be effective, if delivered over longer distances, or timeframes. 'What has changed', said one interlocutor, 'is that the individual meant to deliver the attacks is also meant to be killed in the process.'

In this context, a major concern was the loss of state control in the realm of biological agents and weapons. It was recognised that some agents could be developed, even by lone individuals, in a garage – indeed it was thought likely that such a loner 'would kill himself before Ebola is airborne' (following a discussion about the potential to develop an airborne version of the Ebola virus to increase it salience and epidemiological effect). Even more than the loss of state control, it was 'the loss of effective control' that was of most concern. Even non-state actors, such as Hezbollah, which was said to have chemical weapons obtained from Syria, had to maintain effective control of this capability. However, there was doubt over whether a chemical or biological capability 'in the hands of two individuals loosely connected to ISIS' would, or could, be kept safely and securely.

Respondents discussed the potential combat effectiveness of such weapons. It was acknowledged than an actor might have to cope with an opponent's use of biological agents against a civilian population, in the midst of which a force was operating. This would raise issues of force protection, it was judged, itself raising issues of ethical concern regarding the prioritisation of the force over civilians at risk. An event of this kind would also present 'the need to decide where the theatre is' – and, so, in what area action would need to be taken. This was a real concern for participants, given that the character and context of warfare had changed and it was not possible 'to define armed conflict as clearly as you could in the past'. People – forces or civilians – 'could be attacked in London while the armed forces are fighting somewhere else'. This raised the challenge of 'catering for every threat' – although it might not 'practically be possible to do so'.

Respondents were asked about the extent to which the loss of control identified was a function of the reality (and assumptions) that certain advanced states remained at the forefront of technological advancement. The response to this was that technology could bring key advances in delivery mechanisms. 'In biowarfare, this means overcoming the problem of delivering the weapon without the biological agent dying, or the person delivering it dying', one participant stated. The same participant continued:

> Ebola seems easy. Go to Africa, hug a patient, come back and start kissing as many people as you can before dying. But weaponisation in that way has limited effect with an effective healthcare system. Thus, even where there was the potential for significant harm, this was mitigated where there was strong and appropriate provision to treat the disease.

Participants were concerned that there might be considerable 'scaremongering' in terms of biological weapons capabilities and threats. Participants stressed the importance of 'starting with the science' and getting a 'sound scientific understanding' of exactly what was possible. The facilitator suggested that there might be scientific uncertainty surrounding these issues, but that decisions might have to be made – and asked what could be done in this situation. Some members of the cohort continued, however, to seek certainty and assurance – it was necessary 'to get agreement on the science' and also 'on the way it can be weaponised'. Others, disagreed, saying, 'You have to imagine the unimaginable because someone else is.' On this side of the sharply divided discussion, in one group, it was felt, 'We have to be as proactive as we can be; … when you don't have the knowledge, you have to produce basic principles.' These were judged to be 'principles that you could produce without certainty – and then hold acts against them'. In response to this, a 'three axis matrix' was proposed for evaluating these issues: 'likelihood, time', and in addition the available resources; however, 'at the end of the matrix there may be no money left for this'.

The issue of scientific knowledge was explored in a different way by presenting the respondents with the Fouchier case in the Netherlands, where Fouchier, a synthetic biologist, had developed an airborne strain of the H5N1 virus (avian flu), an already highly infectious and virulent virus that had been a major concern in 2006, but which would be considerably more virulent, if transmission were airborne. Fouchier and his team regarded their achievement as highly dangerous. It was subject to debate about whether, or not, the knowledge of how to create the airborne variant of H5N1 should be published and shared. The Netherlands authorities favoured suppression of the knowledge. However, the scientists argued for publication and the World Health Organization backed publication, on the grounds that knowledge, once known, should be shared, and that the availability of such knowledge was vital to the search for measures to deal with such a variant, should populations become infected by H51N.

While there was a considerable chance that synthetic biology would mean, sooner or later, that small-scale operators could cause damage using a biological agent, which would likely have its widest impact in disseminating fear, it was also evident that this level of action would be limited, would be one governed primarily by criminal law, rather than law relating to armed conflict, and that it would involve bacteriological or viral agents of a relatively familiar kind – even if they were novel in specifics, which would mean their being of limited use. The same was not true for the theoretical possibility of genetically modified weapons, albeit that these could only be developed, or designed, with state-level capabilities.

The issue of genetic modification was both introduced by a facilitator and raised by participants in discussion. One respondent suggested, even before weaponisation of synthetic biology was raised, that it was not impossible that genetically modified organisms could be used (in warfare, or other political action) to destroy crops, undermining a potential opponent in both food supply and the wider economy. This kind of organism could be used as an economic weapon.

The potential of genome technology was clearly the most frightening and the most divisive issue among respondents, with some welcoming potential utility and benefits, while others sharply inferred that any such capability should be outlawed before it had a chance to emerge. Groups were presented with scenarios and predictions of situations in which (blending with robotics) genome stings might be delivered, using so-called Robo-bees, to manipulate targeted genetic features in a target armed force, or in which genetic material could be used to create capabilities that might, for example, strip away stealth paint from military equipment. In either case, the potential effect would transform an armed conflict – even to the extent that the action itself might be 'non-obvious'.

Participants were divided and sometimes confused about these prospects. Although there were divisions within groups, it was also evident that those with greater seniority were broadly more open to discovering the benefits of synthetic biological capabilities than their more junior colleagues. Some were ready to consider and even welcome capabilities that might both give them an advantage and reduce risks to their own force. 'I'll take that', said one respondent, 'if I can gain that advantage with the weapon, not putting my own guys at risk, I'd want that.' This kind of view was supported by some in each group, with participants also seeing benefits in possibly reducing collateral damage, and concurring that such capabilities, should they be feasible, would be 'more discriminating … especially those like the paints'. In contrast, some in each group backed a simple and immediate call 'to outlaw' any synthetic biological capability, even before it had emerged, such was the psychological effect of the prospect of genetic manipulation.

There was a generally pessimistic sense, however, considering the worse end of the potential spectrum of effects. As one participant expressed it, 'someone will turn ploughshares into swords, they always find nefarious uses for technology'. Thus, the 'futuristic, positive human genome benefits for health, which are proactive, are great', but it was vital not to be self- or other-deceiving: 'Don't kid people that someone won't find a way to turn it into something more sinister.' This was supplemented by the suggestion that it might not even need a capability to be fully developed and available in order to be dangerous (as, perhaps evident in the reaction of those respondents, noted above, who immediately judged that a ban on synthetic biological capabilities should be imposed promptly): 'The issue is that if the threat seems real enough to enough people, for example that the armed forces can be attacked with a biological agent that makes bits fall off, or prevent them from having children, that can affect soldiers. It has a real effect on soldiers.' In more conventional terms, another contributor commented that this was like 'flat expanding bullets', to which the response was: 'exactly, the impact on soldiers of dum-dums was such that they agreed not to use them'.

Participants were convinced, even if the science remained uncertain: 'We would ignore this at our peril.' This was the case 'even if we don't know if it is science fiction, or if it will work'. Another participant said, 'We have to be aware and hope, but work with the detail'. Yet another added: 'We have to imagine the unimaginable and what someone might try to do.' It was important to be 'armed with knowledge, then develop and use principles'. In this, 'going to see the scientists, seeing how likely something is' was essential.

As was pointed out, 'The sense of fear depends on the point of reference.' However, there was little doubt that it was 'inevitable that with scientific advance, someone will find a dual use', and this might be 'either by design or not by design'. Thus, 'understanding the range of possibilities' was essential' – but this had to be 'gauged against probabilities as well'. It was important, in this context, to keep in mind the psychological effect that the use of such weapons, or their threat, would have 'in society, on the battlefield, on the soldiers themselves'. This triggered honest discussion about the soldier's experience: 'Soldiers have reconciled themselves with the existence of bullets and bombs, but not with growing three heads, or slowly dying and so on … this is a red line for soldiers.' This perspective was strongly shared by participants, as was the supplemental admission by a member of one group, 'I personally would not like to go into a theatre with biological weapons.' Biological weapons 'would only increase the fear, which is bad enough anyway'.

Conclusion

The understandings of context and ambiguity shown in relation to cyber warfare were also translated into the realms of autonomy and synthetic biology. There was a strong sense that new legislation to regulate autonomous weapons systems, in particular, would be premature, at best. It was necessary to see what the realities were and what was possible before seeking new controls. Existing notions of responsibility could be applied, noting that, however autonomous a machine, ultimately, it is the human who operates, or programmed, it who remains accountable. In this context, motor vehicles operated using Google satellite systems would be guinea pigs for testing issues of responsibility and accountability, well before issues emerged in the sphere of armed conflict.

In terms of autonomy, there was an evident appreciation of its utility and a recognition that it was not possible to keep track of all of the information available. Therefore, feeding information into an engine that allows it to think creatively and start to pull links together without the analyst having to do that, especially with the big data, was a vital step. However, the sense remained that advances in these domains were more likely in areas other than weaponry. In medicine, for example, the prospects for autonomy and robotics were strong, in the view of interlocutors in the research, because the aim, in that sphere, was to preserve life, in some way. Weapons were more associated with ending life. The same technology could be used for creation and preservation or for destruction, as was recognised – but it would be far more easily used for creative and prophylactic purposes than that of armed conflict.

Each of the novel technologies discussed offered chances of protection and were notable for their quiet – not explosive, or booming – nature. The contrast with weapons relying on blast and destruction was strongest regarding the potential for weaponised synthetic biology. Yet, this was also the most contested point of discussion in the research. That which might make it attractive – discriminating targeting of enemies – was also key to making those who would reject the capability completely fearful and rejecting. Weapons that, idealistically, could remove incidental harm and offer great force protection raised major ethical concerns among respondents, concerned that, if such a weapon was used, the potential harm to civilians and others might outweigh the protection offered.

This, in itself, reflected the profound psychological impact that the prospect of using genetic weapons introduced. It was judged that this effect would not only affect the military in the battlespace – wherever the theatre was to be defined – but also society as a whole. This would be a moral and ethical burden potentially too great to consider. At the same time, however, while the striking mental impact, in imagination, of a prospective genetic capability, preying on the practitioners' fear of ghostly capabilities transforming bodies in some monstrous way is powerful, it was one area where divisions among the research subjects were greatest. In contrasting views, some respondents strongly favoured the adoption of genetic weapons, in view of the potential discriminatory benefit they might bring, while others called for a complete ban, even before any deployable weapon had emerged. This call for an outright ban ran counter to the sanguine sense that existing normative frameworks could well embrace both cyber warfare and autonomy. It also confirmed that the strength of division regarding prospective synthetic biological weapons, including the sense that new rules might be required, sets this area of innovation in relation to warfare and the very obviously 'invisible' non-obvious warfare it represents apart from other domains of novel weaponisation, such as autonomy and cyber warfare.

Note

1 Information on the research is contained in Chapter 23.

25

FUTURE WAR CRIMES AND PROSECUTION

Gathering digital evidence

Maziar Homayounnejad, Richard E. Overill and James Gow

The advent of new technologies brought with it the likelihood of new war crimes challenges. The challenges had already been considerable in the quarter of a century of international criminal tribunals and courts that began with the creation of the International Criminal Tribunal for the former Yugoslavia, which was authorised by the UN Security Council in 1993 and began to operate in 1994.[1] The Yugoslavia tribunal had started from scratch, with all aspects of criminal prosecution at the international level to be established. Investigations into atrocity and gross misconduct in the context of armed conflict faced immense challenges – and benefited at times from the advent of new technologies, including DNA identification. The hard work that went into identifying evidence in the human sphere, with painstaking exhumations and analysis of remains was still a largely analogue experience. The new challenges would make new demands, in terms of evidence and linking to elements of crimes to pass the tests of criminal proceedings. Even though much from the past has not yet been adequately addressed, in terms of war crimes, the speed with which new technologies began to have an impact made it imperative to consider 'the future' in the present.

There is much to indicate that legal and technical safeguards could limit the use of new technologies to commit a wilful violation of the law. Experience, indicates, however, that, at some stage, there will be some instance of misuse – and, without doubt, as so much of the discourse around drones and cyber weapons has shown, there will be allegations to be handled. However, in the eventuality of a criminal investigation and trial, in principle, there should be no shortage of electronic forensic evidence to assist the court and allow for trials and punishment – or, indeed, acquittal. This chapter considers four areas in which forensic investigation of war crimes, or international crimes, involving cyber technologies could be possible: the use of black box recorders; access control data; code stylometry; and keystroke analysis.

Black box recorders

Firstly, by mandating that Autonomous Weapons Systems (AWS) must be both equipped *and* deployed with various 'black box' recorders, investigatory authorities would be able to capitalise on the evidentiary value of digital documentation.[2] This suggestion was first proposed by Gubrud and Altmann, in the context of verifying compliance with an AWS ban,[3] and it currently applies in civil aviation (in various international standards and recommendations,[4] as well as in domestic legal obligations[5]), to assist with the investigation of accidents and other unusual incidents.[6]

Here, black box recorders would assist prosecutors in two ways. Firstly, they would provide direct evidence of the *commission of an alleged war crime*. By mandating that AWS be designed with 'always-on' sensory devices – a practice that is commonplace with industrial control systems – such devices would record onto programmable read-only memory[7] digital audio-visual data of everything that the robot 'sees', 'hears' and does while on deployment.[8] Secondly, black boxes can *record the detailed steps involved in programming* an AWS to 'commit' war crimes[9]; in turn, this would assist with forensic analysis of both the executable code instructions and of the precursor source code statements.[10]

Ordinarily, an agreed mechanism would be put in place for decryption in appropriate bona fide circumstances. For example, in civilian contexts, a decryption key is often lodged with a 'trusted third party' (TTP), who is completely independent of the two adversarial parties.[11] Of course, such an approach is likely to be resisted by states, as the classified nature of military operations would typically defeat the idea of a TTP.[12] Accordingly, war crimes investigators may have to rely on, and further develop, standard forensic techniques to gain access to encrypted information.[13]

However, even before the stage of forensic interrogation, there are several potential stumbling blocks. For example, a determined war crimes suspect might order the physical removal and destruction of the 'black box', although this would be a very time-consuming and labour-intensive task, especially if ordered *en masse* in the case of widespread rape, torture and terror. Alternatively, a desperate war crimes suspect sensing the possibility of capture and arrest may order the remote electronic destruction of all black boxes while AWSs are on deployment, using a non-nuclear electromagnetic pulse (NN-EMP) device.[14] These can destroy within a one-mile radius all unprotected electronic devices that rely on integrated circuits[15] – including the AWS units themselves. Thus, there may be a strong disincentive to use an NN-EMP device,[16] save for the most desperate of situations; which, of course, are usually when the enforcement of International Humanitarian Law (IHL), International Criminal Law (ICL), and International Human Rights Law (IHRL) is of greatest importance to the victims and the international community.

Accordingly, and notwithstanding their practical difficulties, the above circumventions may be addressed in two ways. Firstly, data destruction may be *pre-empted* by means of regular, automatic and encrypted back-ups of the log-files to a secure and remote location, using 'hash' functions to ensure the digital integrity of the data.[17] This process can be automatically initiated on a regular basis but also immediately upon evidence of tampering being detected; however, this option is certainly not without its practical problems.[18] Secondly, the undermining effects of data destruction on the credibility of a prosecution can be *mitigated* by recognising a separate international crime resulting from the large-scale removal or destruction of black box devices. At the very least, courts and tribunals may develop presumptions that enable them to take into account deliberate black box destruction as an independent factor in establishing guilt.[19]

Access control data

Secondly, military systems invariably have access control mechanisms in place, to restrict access to authorised programmers only.[20] Moreover, to provide a higher degree of assurance of the identity of those who author new codes or alter existing ones, systems often incorporate 'multi-factor authentication' (MFA) of identity. This technique is very common in the financial services industry,[21] where it 'requires an individual to present a minimum of two separate forms of authentication … before access is granted'.[22] In addition to the unique user ID, there are three categories of authentication, based on something you *know*, something you *have*, and something you *are*.[23] The last is particularly important as it entails biometric authentication, which will potentially leave a non-compromisable trail of digital evidence, of both the timing of a 'war crime algorithm' and its author.[24] Significantly, while MFA has been a hallmark of the commercial sector, recent trends are pointing towards more widespread adoption of this security technique in the intelligence and defence bureaucracies, both as a layer of defence against cyber attacks[25] and as a more general authentication system for access to shared information networks, for example, by the 'Five Eyes' military and intelligence personnel.[26] With these broader developments, it should arguably become easier to expect MFA to be incorporated in AWS access control mechanisms, and to mandate this in legal technical criteria.

De-anonymising war crime algorithms via code stylometry

Thirdly, in the event that the code which gives rise to a war crime is programmed anonymously or fraudulently, stylometric techniques are being developed to identify unnamed coders, given samples of code by known authors,[27] which may be used as initial or corroborating evidence in a criminal trial. However, the phenomenon of 'common culture' amongst programmers should be noted, whereby a whole generation of programmers who learned similar techniques from the same books, manuals or teachers, tends to exhibit very similar stylistic traits in the code they produce. This militates against the ability to identify individual programmers when no unique code samples known to be authored by them are available.[28]

De-anonymising war crime algorithms via keystroke dynamics

Finally, the reliability of code stylometry may be greatly enhanced if the technique is combined with recorded 'behavioural biometrics'; in particular 'keystroke dynamics'.[29] This is also a recognised form of identification in the US *Commander's Guide to Biometrics in Afghanistan*,[30] thereby enhancing its credibility in a military context. Specifically, 'keystroke dynamics' involves building a unique profile of a user based on his detailed typing rhythms, for both identification and verification purposes. This works in four stages.

Firstly, an AWS and all its programming devices are installed with a software-based recorder,[31] which registers a number of vital *keystroke metrics*[32] every time an identified user/programmer uses the system.

Secondly, those metrics data are continuously fed into a *classifier*,[33] which is powered by a machine learning algorithm,[34] to *induce* a unique profile (or 'signature') for each user.

Thirdly, each user's signature is entered on a *signature database*, where it is continuously updated in line with new keystroke data.[35]

Finally, to determine the typist of an anonymous or fraudulently written 'war crime algorithm', the keystroke data recorded alongside the incriminating code is presented to the classifier, which identifies it with an existing user/programmer profile from the signature database,[36] or else it will determine that the algorithm was written by an 'imposter'.[37]

Importantly, the *retrospective identification* described here can complement data on *continuous authentication* of identity that may have been assessed during the writing of the algorithm,[38] potentially as a biometric layer in the system's MFA process. As a final technical point, while the reliability of keystroke dynamics often demands a critical mass of data – many classifiers requiring at least 250 characters for each cycle of continuous authentication – there are newer techniques being refined based on extracting 'nearest character sequences' and Gaussian models, which are intended to verify user identity with as few as 30 characters.[39] Not only is this expected to reduce the verification cycle, but it will also enable a closer comparison of like-with-like and more successive cycles during continuous authentication, thereby increasing the reliability of verification.[40] Arguably, the same logic will apply to the retrospective identification of the typist of a 'war crime algorithm', for which both aggregate data comparison and smaller data samples for nearest character sequences can verify and cross-check the results.

A (qualified) caveat should be made in respect of keystroke dynamics. Namely, the technique identifies the *typist*, who may not always be the *author* of the code. It is not unusual for more senior programmers to author a code, and to delegate its inputting to a typist, who may not be aware of the exact nature of what is being typed in.[41] While this is not invariably the case, such a 'division of labour' may be expected to occur more often in the case of a 'war crime algorithm', especially as a means to hide the true identity of the code's author, and this may potentially defeat the utility of the approach.

On the other hand, keystroke dynamics may still yield investigatory and evidentiary value in two circumstances. Firstly, where a war crime algorithm is *actually typed in* by programmer A, who defeats the system's MFA to fraudulently access it using programmer B's credentials. At the very least, the process of circumventing the MFA is likely to count as a hacking incident, which is one of the classic scenarios where keystroke dynamics will be useful for identifying a perpetrator. Secondly, even if this does not occur, and the offending programmer hands a pre-written code to a typist for code entry, the keystrokes may still uncover an additional witness who was (perhaps unwittingly) involved, and can be later cross-examined in court.

Conclusion

In addition to the above, there have been several recent proposals for protocols and good practice in handling digital evidence at the International Criminal Court (ICC)[42]; thus, it is arguable that over time this form of evidence will become more reliable, and will gain both credibility and institutional support, thereby securing more convictions in war crimes trials.[43] Ultimately, political and military leaders will more easily be held accountable *ex post facto* and, therefore, are more likely be deterred *ex ante* from using AWS to perpetrate war crimes.

It is evident that, while future war crimes investigations relating to cyber misuse will be challenging, in certain respects, forensic investigation will be possible on the technical level. It will still be necessary to have access to the material for investigation and also for technical evidence to be in forms to be admitted in court, that will need to be interpreted in court. And it will be necessary for prosecutors to link them to elements of crimes indictable under a criminal statute, such as that of the International Criminal Court, or some other internationalised, or municipal jurisdiction. As with the problem of attribution for the use of force and international humanitarian law, whatever might be possible technically, a range of other considerations may complicate and compromise the use of evidence, even when it is available. However, the technical possibilities, at a minimum, mean that even if a highly determined, repressive actor were to overcome all legal and technical safeguards that should protect against the gross misuse of autonomous weapons (which, alas, must be anticipated) and to commit a wilful violation of international humanitarian law, international criminal law, or even international human rights law, they would face the possibility of being technically uncovered. The knowledge that this could be the case might, it must be hoped, therefore serve as a deterrent.

Notes

1 On the evolution of international criminal law, see Robert Cryer, Hakan Friman, Darryl Robinson and Elizabeth Wilmshurst, *An Introduction to International Criminal Law and Procedure*, 2nd edn, Cambridge: Cambridge University Press, 2010.

2 On the importance of digital evidence in prosecuting international crimes, as well as the recent history of, current practice in, and some major issues relating to the use of digital evidence at the International

Criminal Court, see Human Rights Center, UC Berkeley School of Law (2014) *Digital Fingerprints: Using Electronic Evidence to Advance Prosecutions at the International Criminal Court*, at: http://citris-uc.org/wp-content/uploads/2014/07/Digital-Fingerprints-Salzburg-Report-2014.pdf.

3 Gubrud, M and Altmann, J 'Compliance Measures for an Autonomous Weapons Convention', *ICRAC Working Paper #2*, May 2013, pp. 6 and 7, discussing both black boxes and a 'glass box' concept, at: https://icrac.net/wp-content/uploads/2016/03/Gubrud-Altmann_Compliance-Measures-AWC_ICRAC-WP2-2.pdf.

4 See Annex 6: Operation of Aircraft, Part I, *Convention on International Civil Aviation*, ¶¶ 3.3.1–3.3.2, which lays down the basic recommendation and standard to be followed regarding the establishment of a flight data analysis program. More detailed technical requirements are provided in other parts of Annex 6.

5 For example, Federal Aviation Administration regulations mandate the use of both cockpit voice recorders (14 CFR 121.359; 14 CFR 125.227; and 14 CFR 135.151) and flight data recorders (14 CFR 91.609; 14 CFR 121.343; 14 CFR 121.344; and 14 CFR 135.152) in certain commercial passenger aircraft operating in the national airspace of the United States.

6 Accordingly, enforceable *legal obligations* that compel the inclusion and use of black box recorders have mostly been achieved with civilian technologies. Expansion of such obligations into the military sphere may require substantial adaptation and political effort.

7 Griffin, J., Matas, B. and de Suberbasaux, C. (1996) *Memory*, Scottsdale: ICE Corp., fn 163.

8 To see how audio-visual evidence can be key to securing a conviction, see the English Courts Martial Appeal Court case, *R* v. *Sergeant Alexander Wayne Blackman and Secretary of State for Defence* [2014] EWCA Crim 1029.

9 Indeed, the specificity with which a 'rape algorithm' or 'torture algorithm' would have to be encoded would leave little doubt as to the intention of the commander who authorised it or the programmer who authored it, and who would be relatively transparently linked with the crime through standard chain-of-command and project ownership protocols.

10 Nikkel, B. (2016) *Practical Forensic Imaging*, San Francisco: No Starch Press.

11 Delfs, H. and Knebl, H. (2017) *Introduction to Cryptography: Principles and Applications*, 3rd edn, Berlin/Heidelberg: Springer, fn 166. See also Bidwell, A 'Private Keys, Trusted Third Parties, and Kerberos', 26 October 2012, at www.ittoday.info/AIMS/DSM/87-20-15.pdf.

12 See Chapter 14 note 102.

13 See, for example, Spruill, A., 'Digital Forensics and Encryption', *Evidence Technology Magazine*, at: www.evidencemagazine.com/index.php?option=com_content&task=view&id=656.

14 Overill, R. E., 'Denial of Service Attacks: Threats and Methodologies', *Journal of Financial Crime*, Vol. 6, No. 4 (1999), p. 351.

15 Adams, J. *The Next World War: Computers are the Weapons and the Front Line is Everywhere* (New York: Simon & Schuster, 2001), Chapter 10.

16 As the destruction of large numbers of AWS units will reduce the military capability of the regime itself.

17 Sommer, P. 'Intrusion Detection Systems as Evidence', *Proceedings on RAID* 98 (1998), at: www.raid-symposium.org/raid98/Prog_RAID98/Full_Papers/Sommer_text.pdf.

18 Above all, armed forces may object to contractors designing such a feature into AWS units, as transmitting radio messages on such a frequent basis would act as a radio beacon indicating the presence of a weapon system, as well as its location and trajectory.

19 Of course, this would have to combine with other independent evidence, such as witness statements from the alleged victims.

20 In particular, military systems are often subject to a combination of 'role-based access control' and 'mandatory access control'. This restricts access and administrator rights to those with the correct role assignment, role authorisation and transaction authorisation. Importantly, and given the security implications of altering military algorithms, the access policy is determined by the system, not the programmer. See Ferraiolo, D. F. and Kuhn, D. R., 'Role-Based Access Control', *15th National Computer Security Conference*, 13–16 October 1992, available at: https://csrc.nist.gov/publications/detail/conference-paper/1992/10/13/role-based-access-controls.

21 For example, in the payment card industry, Requirement 8.3.1, *PCI DSS 3.2*, will (from 1 February 2018 onwards) require all payment card institutions to implement MFA, in addition to assigning a unique user ID, to ensure proper user-authentication of personnel with administrative access to cardholder data. See Payment Card Industry (PCI) Data Security Standard, Version 3.2, April 2016 (hereafter *PCI DSS 3.2*), available at: www.pcisecuritystandards.org/document_library?category=pcidss&document=pci_dss.

22 Requirement 8.3 Guidance, *PCI DSS 3.2*.
23 Requirement 8.2, *PCI DSS 3.2*, provides that, in addition to the unique user ID, the three authentication categories are: a) something you *know*, such as a password, passphrase or a PIN; b) something you *have*, such as a smartcard, a smart key fob, or a particular smartphone, as demonstrated via a one-time password sent by text; and c) something you *are*, namely, a biometric feature, verified through retina scan, iris scan, fingerprint scan, facial recognition, voice recognition, hand and/or earlobe geometry, or keystroke dynamics (see notes 29–40 below). See also PCI Security Standards Council 'Multi-Factor Authentication', *Information Supplement*, February 2017, p. 2, at: www.pcisecuritystandards.org/pdfs/Multi-Factor-Authentication-Guidance-v1.pdf.
24 See Newman, R., *Security and Access Control Using Biometric Technologies*, Boston: Course Technology, 2009.
25 Osborn, K. 'Industry Offers Multiple Authentication Tech for SIPRNet', *Defense systems*, 1 March 2017, at: https://defensesystems.com/articles/2017/03/01/safenet.aspx.
26 The 'Five Eyes' refers to an intelligence alliance between five English-speaking allies: Australia, Canada, New Zealand, the UK and the USA. See Waterman, S., 'DoD Plans to Eliminate CAC Login Within Two Years', *FedScoop*, 14 June 2016, at: www.fedscoop.com/dod-plans-to-eliminate-login-with-cac-cards/, noting the comments of the Pentagon's Chief Information Officer that 'biometric technologies like iris scans and behavioral analytics "are all doable now", and that [Common Access Cards] should be replaced by "some combination of behavioral, probably biometric and maybe some personal data information that's set from individual to individual".'
27 Islam, A. C., Narayanan, A., Harang, R., Voss, C., Greenstadt, R., Liu, A. and Yamaguchi, F., 'De-Anonymizing Programmers via Code Stylometry', (2015), at: www.cs.drexel.edu/~ac993/papers/caliskan_deanonymizing.pdf.
28 Thomas, M., 'Should We Trust Computers?' *1988 BCS/UNISYS Annual Lecture*. BCS Computer Bulletin (1988).
29 Moskovitch, R., Feher, C., Messerman, A., Kirschnik, N., Mustafić, T., Camtepe, A., Löhlein, B., Heister, U., Möller, S., Rokach, L. and Elovici, Y., 'Identity Theft, Computers and Behavioral Biometrics', *Proceedings of the IEEE International Conference on Intelligence and Security Informatics*, ISI 8–11 June 2009, pp. 155–60, also discussing 'mouse dynamics' at p. 157, with key mouse metrics including: mouse movement; drag and drop; point and click; and silence, at: http://ieeexplore.ieee.org/stamp/stamp.jsp?arnumber=5137288.
30 This defines 'biometrics' as any 'measurable biological (anatomical and physiological) and *behavioral* characteristic that can be used for automated recognition' (emphasis added). On the behavioural side, this explicitly includes such subtle characteristics as 'the keystroke pattern on a keyboard'. See US Army Combined Arms Center, 'Commander's Guide to Biometrics in Afghanistan', *Handbook No. 11–25*, April 2011, Appendix A, p. 45, at: https://info.publicintelligence.net/CALL-AfghanBiometrics.pdf.
31 For an example of a keystroke recorder currently used in the commercial sector, see: www.keytrac.net/.
32 Moskovitch et al. fn 251, p. 156, identify four specific metrics: *dwell time* (key A down/key A up); *latency* (key A down/key B up); *flight* time (key A down/key B down); and *interval* (key A up/key B down). In practice, some keystroke recorders measure only 'dwell time' and 'latencies', though many will also measure the overall typing speed, the frequency of errors (via use of the 'backspace' key) and the use of left *versus* right shift key.
33 Ibid., p. 157. A classifier operates on two levels. Firstly, there is a *training phase*, where it learns the unique characteristics of a particular user. Secondly, there is a *classification phase*, where it compares a new sample of keystroke data (in this case, from a programming task writing a 'rape' or 'torture' algorithm) with existing profiles, to classify that data and either match it to an existing user/programmer, or determine that the system was tampered with by an imposter.
34 Ibid. Two of the most popular algorithms being artificial neural networks and support vector machines.
35 Ibid.
36 Along with a score to indicate its statistical confidence level.
37 Clearly, a rational programmer/typist intending to input a 'war crime algorithm' will want to uninstall one or more of these software features (keystroke recorder, classifier, and/or signature database). Hence, some of the design-led safeguards discussed in Chapter 14 may be appropriate here, to maintain the integrity of the digital evidence.
38 Menshawy, D. E., Mokhtar, H. M. O. and Hegazy, O., 'A Keystroke Dynamics Based Approach for Continuous Authentication', in Kozielski, S., Mrozek, D., Kasprowski, P., Małysiak-Mrozek, B. and

Kostrzewa, D. (eds.), *Beyond Databases, Architectures, and Structures* (London: Springer International, 2014), pp. 415–24.

39 Song, X., Zhao, P., Wang, M. and Yan, C., 'A Continuous Identity Verification Method based on Free-Text Keystroke Dynamics', *Proceedings of the IEEE International Conference on Systems, Man, and Cybernetics*, SMC 9–12 October 2016, pp. 206–10.

40 Ibid.

41 Conversely, keystroke dynamics are often most effectively used where the perpetrator and the typist are one and the same, such as when monitoring a hacking incident or a paedophile chatroom.

42 See Umberg, T and Warden, C., 'Digital Evidence and Investigatory Protocols', *Digital Evidence and Electronic Signature Law Review*, Vol. 11, 2014, pp. 128–36, at: http://sas-space.sas.ac.uk/5719/1/2131-3127-1-SM.pdf.

43 See also International Bar Association (2016) *Evidence Matters in ICC Trials*, London: IBA, pp. 19–33, advocating greater use of digital evidence, and making recommendations to the ICC and other parties in preparing for the proliferation of new forms of evidence, at: www.ibanet.org/Document/Default.aspx?DocumentUid=864b7fc6-0e93-4b2b-a63c-d22fbab6f3d6.

PART VI

International perspectives

26

RUSSIAN INFORMATION WARFARE AND ITS CHALLENGES TO INTERNATIONAL LAW

Oscar Jonsson

In Russian military-theoretical debates, information warfare has arisen as the most significant aspect of the changing character of warfare.[1] This appreciation is well reflected in the war in Ukraine, where information warfare has taken centre stage, and Russia demonstrated that it has updated its toolbox accordingly. As well as psychological operations in Russia and Ukraine, the information war has assumed a global focus. Russia's international news network, RT, received a 2014 budget increase of 41 percent to GBP250 million, and will, among other initiatives, start channels broadcasting in the UK, France, and Germany. The importance of information warfare will only continue to rise as our societies grow more connected. This chapter thus asks how Russian information warfare is changing and what challenges this poses for international law.

This chapter argues that, after the rude awakening in the First Chechen War, Russia has successfully adapted the conduct of its information warfare to the realities of the present day. The Russian understanding of information warfare is notably more holistic than the Western understanding. It includes the content of information, the harmony of society, and the stability of the state in its threat definitions, whereas the Western understanding emphasises the free flow of information. The holism has ensured the relevance of the Russian approach to information warfare in the face of the rapid increase in the use of social media whose role was seen in the revolutions in the Middle East and North Africa.

The Russian understanding of information is separated into two domains: the information-psychological and the information-technical. In the information-psychological aspects, Russia has been consolidating control of both traditional and social media to build resilience and achieve hegemony, if not monopoly, of the state's view. Externally, Russia is building its own international news network providing the regime's angle on international affairs, and employs bloggers and commentators to spread the regime's position. This domain dominated in the Chechen wars and now in Crimea and Ukraine. In the information-technical dimension, post-Chechnya, Russia seems likely to have, by proxies, perpetrated two large cyber attacks against Estonia and Georgia (as well as further attacks on the media). The conduct in both of these domains poses formidable challenges to international law, since its conduct has been designed to exploit the grey areas of the law.

Since there is a lack of official ways of translating the Russian concept, this chapter uses the concept of information warfare, which is the most common way to translate both *informatsionnaia voina* and *informatsionne protivborstvo*. Furthermore, information warfare best captures the perception among the Russian elite, the stakes, and the intensity involved. This chapter relies to a certain extent on those news articles and reports on the war in Ukraine that are available at the time of writing. These sources are, however, used with extra caution and their content is triangulated when possible. This is however not perceived as a major problem, because sufficient literature on information warfare per se was available before Ukraine.

First, this chapter will outline the particularities of the Russian understanding of warfare and what has shaped this understanding. It proceeds by reviewing the development in the information-psychological and the information-technical aspects of information warfare respectively. A discussion follows on the legal challenges posed.

The Russian understanding of information warfare

The Russian notion that the importance of information warfare was drastically increasing entered Russian theory and practice after the First Chechen War in 1994–96. In a study of the development of Russian information warfare, Timothy Thomas argues that there were three lessons learned that impacted on this understanding. It was the importance of the information flow to the public, the impact on the soldier's morale from information, and the use of the internet as a new arena where the war was additionally played out.[2] During the First Chechen War, while the Russian forces avoided the press, Russian journalists could fly to Makhachkala, take a taxi into Chechnya, interview Chechen fighters, and then get remuneration for the article and their travels.[3] This contributed to the immense unpopularity of the war in Chechnya, albeit that it was unpopular from its beginning.

The Russian leadership learned these lessons and adjusted themselves to the second war. They put harsh restrictions on what was allowed to be broadcast. The only accepted version this time, was the Russian regime's version. This was made possible by only allowing access to journalists with permission from the government and restricting access only to certain locations, as well as censoring their output.[4] The issue of the soldiers' morale was being remedied by daily briefings on progress, by weekly combat news broadcasts showing soldiers who had displayed courage, and with psychologists helping on the spot.[5] During the Second Chechen War, the use of the internet proved to be crucial for the Chechens while Russian forces were trying to block their voice being heard in the other media. The Chechens used the internet to: receive money; rally the diaspora; disseminate successful attacks on the Russian soldiers; rally support for Jihad; and spread news of Russian atrocities.[6]

It is against this background that the growing Russian understanding of information warfare should be understood. President Putin concluded only later, in 2000 just after becoming president after the Second Chechen War, that to a large degree it was society's morale that led to Russia's defeat in the first war.[7] In the same year, the government approved the Russian information security doctrine to give strategic guidance in the information domain. This doctrine is still in place today. This could seem like a long time, given the vast development in the information and communication technologies (ICT) since, but, arguably, it is a significantly new and wider and broader understanding of war and warfare that has allowed the doctrine to stay relevant. Rather than to stovepipe aspects of a threat within various authorities and disconnect ways of influencing the adversary, the Russian approach is one of holism where the adversary is envisioned as a whole system, so best impacted as a whole. This is reflected in the Russian organisational framework. In December 2014, the National Defence Control Centre was inaugurated. The Russian armed forces are led from the Centre but, in a time of crisis, 49 different agencies will also be subordinated to, and led from, the Centre.[8]

This way of viewing information warfare renders the Russian approach different from the Western approach in three aspects. First, there is an explicit focus on the need to ensure the stability of state authority and social integrity.[9] To this end the doctrine calls for the bolstering and expansion of mass state-media to be able to promptly and reliably convey information to Russian and foreign citizens.[10] This is also reflected in Russia's proto-doctrine for the armed forces' activities in the information space. In it, information war is defined as a confrontation with the aim of damaging information systems, processes, and resources in order to undermine the political, economic, and social system, as well as brainwashing the adversary population aimed at destabilising their society and the state.[11] This is in marked opposition to the Western approach, which focuses on protecting individuals and not governments. Formally, the Russian view contains a balance between the state and individual, but practice suggests that the former is by a wide margin superordinated to the latter.

Second, the area that recently has received most attention in the West, cyber warfare, does not exist in the vocabulary of the Russian security doctrine. The activities in cyberspace fit within the broader concept of information warfare with reference to attempts at hindering the information infrastructure. This holistic approach is not dependent on narrow definitions, but rather incorporates how cyber activities fit in the broader confrontation.

While the Western approach focuses on the protection of the internet infrastructure and the free flow of content, the Russian understanding is viewing the content of information as a potential and actual security threat.[12] In this lies the third notable point: information in itself or the content of information can be seen as a threat. The ICT revolution has enabled information to impact an adversary on the strategic level. This understanding needs to be seen against the backdrop of the Russian media during the First Chechen War of 1994–96 where journalists reported information that resulted in a very negative image of Russia and hindered military operations against the Chechens.

The media image became seen as a crucial element for success and by many was argued to be a weapon. Writing in the mid-1990s during the First War, Tsymbal argued that '[f]rom a military point of view, the use of information warfare against Russia will categorically not be considered a non-military phase of a conflict ... considering the possible catastrophic consequences of the use of strategic information warfare means by an enemy'.[13] This view raises the question of how the perception of being in an information war ties in with allowing for the use of other means of warfare. The question is increasingly pressing, with President Putin talking after the Second War about an information war being waged against Russia, and today the relations are worse.[14]

In the Russian discussions on the changing character of war, the factor that gets the most attention, and is discussed most concretely, is information warfare. For instance, it has been argued to be sufficient to reach strategic objectives.[15] It is also a key component in what Chief of General Staff Gerasimov argues is blurring the boundaries between and war and peace, and making non-military means four times as effective as military ones.[16]

In this sense, it ties into a long-standing appreciation in Russia for the strategy of the indirect approach and of Sun Tzu. It stipulates the best victory in war to be the one where the enemy's army is intact and you have managed to make him act as you want without applying the military instrument. Since information warfare goes straight into the minds of both the domestic and the foreign audience, it directly ties in to the battle of the wills.

More specifically, the Russian military theorists Chekinov and Bogdanov describe information warfare as giving a wealth of opportunities: to disorganise the military, state control, and specifically the aerospace defence system; to deceive the opponent; to impact public opinion; and to organise anti-government protests.[17] These views underline both the primary importance of information warfare in contemporary war, but also both the information-psychological and the information-technical aspects. Let us review them in more detail.

Information-psychological warfare

The information-psychological aspect includes anything that has to do with impacting the minds on both sides of a conflict. Internally in Russia, the consolidation of the media quickened after Vladimir Putin was elected president in 2000. This is in line with the information security doctrine also from 2000. Even if it would be a stretch to suggest that the media was fully free under Yeltsin, the process that started with the presidential change has steadily worsened the situation. The consolidation has, however, not been as obvious as it was during the Soviet Union. Gehlbach argues that the trick of having Gazprom and pro-Kremlin businessmen buying up media outlets gives the regime a certain degree of ambiguity and plausible deniability of how much is controlled and how firmly.[18] In the early 2000s, notable cases included seizure of the oligarch Berezovsky's network ORT (later Channel 1) and Gusinskiy's NTV, after which both of them fled the country.[19] This process of consolidating control continued to the point that Russia was downgraded in Freedom House's report 'Freedom of the Media' to 'Not Free' in 2009.[20]

While international attention was directed to the campaigns in Eastern Ukraine and Crimea, the regime made an accelerated attack on the last of the remaining critical outlets. In February 2015, the radio director of Ekho Moskvy, probably the most critical media outlet, was replaced by an editor coming from Voice of Russia, the regime's mouthpiece.[21] In mid-March, a week before the formal annexation of Crimea, the Editor-in-Chief of Lenta.ru, Galina Timchenko, was fired and replaced by Alexey Goreslavsky. He formerly ran the 'staunchly pro-Kremlin' site Vzglyad.ru; and. after taking over Lenta, he fired half of the employees including the director-general.[22] The crackdown continued with the TV station Dozhd, which was among the first to air the protests against the Russian parliamentary elections in 2011. The channel was excluded from being carried by cable and satellite operators and quickly lost around 90 percent of its viewers and concomitantly a significant amount of its revenues.[23]

Pavel Durov, the founder of the social media network VKontakte, which could be seen as the Russian equivalent to Facebook, was forced to flee the country in 2014. Before doing so, he sold his part of the company to the Kremlin-friendly oligarch Alisher Usmanov, who had been trying to buy him out for some time.[24] During the crisis, the Russian authorities manipulated the social network. The Russian Federal Security Service (FSB) ordered VKontakte to deliver intelligence on pro-Ukrainian groups, while the FSB branch overviewing mass media and IT in general ordered VKontakte to shut down certain pro-Ukrainian groups.[25]

That ability to shut down websites without a court order was given by the adoption of the 2011 federal law 'On Police', covering any sites suspected of providing 'conditions which assist the commission of a crime or administrative violation'.[26] Furthermore, in May, a 'bloggers' law was adopted in Russia. It stipulated that any site with more than 3,000 daily visitors needed to register with the authorities and to observe the same responsibilities as other media outlets. Lastly, in June, the Russian criminal code was amended to include an up to five-year sentence for calls to extremism. This was criticised by the OSCE's High Representative for Freedom of the Media, because 'they threaten free media and compromise online pluralism in the name of fighting the vague concept of extremism'.[27]

Seeking control of the media should not only be interpreted as a way to silence uncomfortable voices. It needs to be understood as a key means of ensuring regime survival. Enemy images rally the Russian people against a common enemy and give a stronger societal resilience. Control also gives the ruling leadership the opportunity to counteract messages that would have a psychologically negative impact on the society and to strictly control messages in times of national crises. Another piece of legislation that confirms this reading is the one approved on 23 September 2014, where foreign ownership is being limited to 20 percent, something also criticised by the OSCE's High Representative.[28]

Media control also significantly increases the regime's ability not only to frame issues, but also to shape a certain version of reality. Since the government controls a critical mass of media outlets, it is able to produce a story in one of them and then reproduce the story across the other main channels, which makes spin control possible. A study by the Levada centre shows that 90–94 percent of Russians have TV as their main source of information, and the vast majority are watching state television. In this respect, it is particularly notable that 70 perccent of the Russian people saw the coverage on Ukraine and Crimea as completely objective or mostly objective.[29] It is in this environment that these actions are highly effective.

An integral element to the information warfare effort is the channel *Rossiya Segodnya/Sputnik*, the government-owned international news agency that was the result of the combining of Voice of Russia and RIA Novosti. Rossiya Segodnya, which means Russia Today, but is not to be confused with the international network RT, is run by Dmitry Kiselyov. He has made himself (in)famous for attacking opponents of the Kremlin and homosexuals, and for trumpeting that Russia is the country that could turn the US to radioactive dust. A specific and quite typical example of his work is his accusations that Carl Bildt, the Swedish Foreign Minister, was a CIA agent, driven by a lust to take revenge on Russia because Sweden lost the major battle of Poltava in 1709 against Russia. This, Kiselyov continued, is perhaps not strange, because Bildt comes from a country 'where sex from the age of nine is the norm'.[30]

This example is interesting because it throws together two of the common themes for demonising someone in Russia. The first is CIA connections and the second is to paint a threat to traditional sexual views. These accusations could look comical, but they are not targeted at outsiders. They are made for people who have a very high degree of belief in TV stations and seldom have alternative ways of acquiring news, have traditional values, and have a tendency for conspiratorial thinking.

An increasingly important arena where the information struggle takes place, both internally and externally, is on social media. To this end, the Russian government has not only recently acquired control over VKontakte, but has also begun hiring masses of commentators and bloggers to spread the regime's message online. The first one to break this story was the Russian journalist Alexandra Garmazhapova who published it in *Novaya Gazeta* in September 2013. She visited the *Agenstvo Internet-Isledovaniy* (the Internet Research Agency) in St Petersburg and found that the agency had personnel required to write 100 comments a day.[31] During that time, the mayoral elections in Moscow took place, and the main task of the agency was to provide positive comments about the sitting mayor Sergey Sobyanin, incidentally a member of President Putin's party, and to write negative comments about his main contender Alexei Navalny.[32] It later turned out that the smear campaign was not sufficient; the popular Navalny was subsequently imprisoned and taken out of politics on very dubious charges, just as were the oligarchs.

A document of the Agency was also leaked to BuzzFeed during the Crimean crisis, which mentioned a budget of USD10 million for employing 600 cyber privateers (recalling the state-condoned pirates of the eighteenth century).[33] Since social media enable instantaneously crowd-sourced new information content, they have become perhaps the main hub of following a conflict in real time, as shown in the war in Ukraine. The addition of the information warfare toolbox represents a significant development in line with how our societies are changing. For instance, during the Georgian War, Twitter had only 2.8 million users and 300,000 tweets daily, while today it has 645 million users and 600 million tweets a day. This is a change that Russia has adapted to the conflict toolbox.

For the external aspect of the information-psychological aspect, the government-owned international news agency, RT, is the most notable achievement of the Russian regime. It was created in 2005 to carry Russia's global message and it broadcasts, in English, Spanish, and Arabic. Russia Today currently broadcasts worldwide in six different languages (Russian, Spanish, French, English, German and Arabic) and has channels in the UK, Spain and the United States. According to its website, it broadcasts to more than 700 million people in over 100 countries worldwide.[34]

RT International has become the second most watched foreign network in the US and it has around 2.5 million viewers in the UK.

The content of RT is often directed to undermine the West's claim to legitimacy based on human rights and democracy. Considerable time and energy are aimed at highlighting the double standards of Western countries, how they are themselves lacking in the values they are criticising in Russia. To this end, conspiracy theorists are frequent guests and it appeals well to anti-establishment oriented individuals. One illustration is an article RT Spanish published on how Western pharmaceutical companies are responsible for creating and spreading Ebola.[35] The article references 'a group of scientists' but the content comes from www.trueactivist.com. This is a quintessential article of RT.

Nonetheless, much light was cast on RT's reporting in the wake of the war in Ukraine. Following the seizure of Crimea, RT's reporter Elizabeth Wahl, in March, quit her job on air, saying that she could not be a part of a network funded by the Russian government and promoting the Russian view when they shamelessly invaded another country.[36] Similarly, the UK correspondent of RT, Sarah Firth, quit over the RT reporting on the crash of the civilian airliner MH17 which was based on strict guidance despite a paucity of known facts.[37] These defections paint an interesting picture of how RT is being micro-managed to favour Russia and attack the West.

One way to exemplify how journalists working for state-controlled media have been used in the information-psychological conflict is through the war against Georgia. Starting from a week before the conflict escalated to a war, journalists from NTV (owned by Gazprom), TV Tsentr (state-owned), Russia Today (state-owned), and TV Zvezda (run by the Russian Ministry of Defence) started arriving in South Ossetia. This enabled them to broadcast the Russian point of view both domestically and globally before the outbreak of the war, whereas their Western colleagues had trouble entering the country to report. As soon as the war started on 8 August, the Russian journalists were joined by colleagues from all the other major Russian news agencies. This represents a continuation from the Second Chechen War and the lessons of having journalists on the scene whose story can be tailored to fit one's own narrative.

In contrast to the way that states' behaviour is often analysed in terms of their foreign policy, they more often perceive themselves in terms of domestic politics. One reason why the information-psychological aspect receives so much attention in Russia is because of its political system. The current Russian regime is reliant upon restrictions in human rights and democracy, which is the foundation of the Western claim to legitimacy. This creates a systemic contradiction. An effect of this, as Blank concludes in a study on Russian threat perception, is a regime that feels vulnerable domestically and presupposes conflict internationally.[38]

It is precisely in this way that Russia perceives the colour revolutions (in Georgia, Ukraine and Kyrgyzstan), the Arab Spring, the unrest in the Russian elections 2011–2012, and, lastly, the protests that started the overthrow of Yanukovich in Ukraine. They are seen as revolutions fomented by the West, cloaked in moral terms, but for geopolitical goals of regime change.[39] Such a campaign is perceived as an efficient use of non-military means by the West, even becoming the weapon of choice instead of conventional military ones.

To conclude, since the First Chechen War the potential for conducting information-psychological warfare has grown significantly with the revolution in information communication technologies and later the advent of social media. This has been mirrored by increasing Russian investments in the means for conducting information-psychological warfare. This type of warfare is particularly effective in societies that are based on free speech and on the rule of, not only the people, but popular opinion. Simultaneously, the Russian government is putting a lot of effort into building resilience against an information-psychological attack that they perceive the West is perpetrating through supporting democracy and human rights. In this conflict, the perpetration and resilience in social media is of particular importance.

Information-technical warfare

The information-technical domain of information warfare is concerned with the machine-driven data components, the means of transmission, and the information infrastructure. This includes the operation of satellites, sensors, computers and their ability to provide information to the armed forces and the state. It is under this category that cyber warfare, which has received so much attention in the West, fits in. Even though the Russian understanding of information warfare is broad, significant attention is directed to cyber aspects. The potential damage of cyber operations is neatly summarised by a group of the most prolific Russian military theorists on cyber warfare:

> The damage done by cyber weapons may include man-made disasters at vital industrial, economic, power, and transportation facilities, financial collapse, and systemic economic crisis. Besides, cyber weapons can cause government and military operations to spin completely out of control, leave the population demoralized and disorientated, and set off widespread panic.[40]

In this way, cyber attacks are seen to be able, in themselves, to disrupt the functioning of a state. Therefore, one of the leading experts on Russian information warfare concludes that information technology will be a key aspect in future war.[41] Not only is cyber warfare seen as potentially achieving strategic outcomes in itself, but since it is not directly violent in the blast and fragmentation sense of the word, the barrier to use is lower than with conventional military means. Of course, cyber-attacks carry the risk of reprisals, but that hinges on their correct identification. President Putin himself stated in his guiding speech on national security just before his re-election in 2012 that new means – of which he gave informational means special attention – 'will be as effective as nuclear weapons but will be more "acceptable" from the political and military point of view'.[42] This statement implies that it can be a way of conducting warfare more cheaply, if not 'on the cheap'.

The very nature of cyber warfare is that attribution is extremely difficult. This makes it hard to evaluate the Russian conduct with a high degree of certainty. At the same time, attribution is always higher than zero. As Libicki pointedly argues, reasoning in terms of opportunity, means, and motives will quite quickly narrow down the field of potential perpetrators.[43] This chapter discusses the most notable cases where indicators point towards Russian involvement in information-technical warfare, or cyber warfare. These are the cyber attacks on Estonia 2007 and on Georgia 2008. For the purposes of this chapter, the cyber attacks domestically against the

Russian opposition will not be dealt with extensively. Thomas, in his 2014 study, includes the techniques of phishing against opposition figures and distributed denial of service (DDoS) attacks against opposition media.[44]

Estonia

On 26 April 2007, unrest unfolded when the Estonian government decided to move the bronze statue of a Soviet soldier erected in memory of the Great Patriotic War (the Second World War). Thousands of mostly ethnic Russians started protesting. When the Estonian government decided to move the statue ahead of the planned date, the protests turned violent and the Russian foreign ministry responded strongly. Sergei Lavrov, the Russian Foreign Minister, stated that the actions were blasphemous and warned of serious consequences.[45] On 27 April, denial of service (DoS) attacks commenced against Estonian governmental institutions and news portals. These were relatively simple and easy to fend off; they were described as 'ineptly coordinated and easily mitigated'.[46]

A second wave of attacks started three days later, on 30 April, and went on for three weeks. These attacks were much more sophisticated and included posting lists with instructions and targets on Russian forums; ready-made electronic files to enable volunteers to join in the cyber offensive; defacements on public websites; and distributed DoS attacks (which is a degree more advanced than simple DoS attacks).[47] The attacks peaked on 8 and 9 of May, which is Victory Day in Russia, one of the most symbolic holidays. The attacks had a large impact due to Estonia's high degree of connectivity.

In analysing these attacks, the Cooperative Cyber Defence Centre of Excellence in Tallinn argued that these attacks required resources that widely exceed what an average motivated hacker could bring to bear.[48] The judgement was based on the evidently centralised command and control structure and the precise coordination of the attacks. The direction of the operation and the initiation came from Russian-language websites that were tracked back to Russian internet addresses and some, allegedly, to institutions of the Russian state.[49] An interpretation that supports this view is that of John Arquilla, the 'inventor' of the term cyberwar, who claimed that 'there was very high confidence that they arose with Moscow's knowledge and approval'.[50] It is appropriate to point out that cyber attacks have the possibility to be redirected through a network of servers, so this is not concrete evidence, but circumstances need to be included in the assessment.

While the protests were going on in Tallinn, the political organisation 'Nashi' had been protesting around the Estonian Embassy in Moscow from the outset of the crisis. The organisation is notable in relation to the cyber attacks since one of its commissars, Konstantin Goloskokov, publicly stated that he was behind the cyber attacks.[51] Furthermore, Sergei Markov, a Duma deputy from Putin's party United Russia, stated that one of his assistants was in control of the attacks.[52] Yet again, this should also be treated with caution, but it is worth mentioning.

Nashi was created by the Russian regime in 2005 following the Orange Revolution in Ukraine. The fear was that when popular unrest started in Moscow, there would be no one who would counter-protest. The connection to the regime could be ascertained for instance when the head of Nashi confirmed in an interview that they were receiving 200 million roubles from the state budget.[53] An additional factor is that Vladislav Surkov, deputy chief of staff (1999–2011) and deputy prime minister (2011–13), was a key organiser for the movement and often met with the leaders both in public and in private. Nashi has been particularly tasked, according to Carr, to ensure 'domination of pro-Kremlin views on the internet', which also ties in to the information-psychological domain.[54] This illustrates the regime's connection to Nashi and the discussion above that Nashi has been active in the crisis.

There is so far no conclusive evidence that the attack was directed from Russia. This is perhaps not too unusual, given that it is precisely the ambiguity in cyber warfare that makes it attractive. However, by using the suggested means, motivations, and opportunity model suggested by Libicki, the indications point in that direction. The Baltic States are the centre of geopolitical contention between a Russia that is seeking to be the hegemon in the former Soviet space and the West that is bound to defend the Baltic States by NATO's Article 5.

More specifically, the bronze statue was moved after a decree banning both Nazi and Soviet symbols, thus implicitly equalising them. This represented a strong symbolic challenge to a Russia where the victory in the Great Patriotic War is still seen today as the Soviet Union saving Europe from fascism as a distinct evil. This is the reason for the demonic image of fascism currently being painted on Ukraine. The explicit warnings of consequences given by Lavrov, the instructions on Russian forums, and Nashi's connections to the perpetrator indicate at least an implicit approval and at most a high degree of involvement.

In analysing the crisis, the cyber security analyst Carr in his book presents a credible hierarchy of the attacks. He proposes a three-tiered organisation with the regime at the top acting through Nashi. Nashi, whose membership includes hackers, were responsible for organising the instructions and making the call for unaffiliated hackers and anyone grabbed by nationalist fervour to join in and conduct the attack; these were the ones that made up the last tier.[55]

Georgia

The cyber operation in Georgia took place in connection with the broader Russia–Georgia conflict, which intensified during the spring and summer 2008, and culminated in a war in August 2008. The first parts of the cyber operation started on 19 July with defacement on government websites, juxtaposing the Georgian president with Hitler.[56] The more advanced attacks commenced the same day as the shooting war, 8 August. The attacks consisted of SQL-injections (code to sabotage website databases), DDoS attacks, as well as defacements aimed against various public web pages and banks.[57]

Just as with the attacks on Estonia, instructions, coordination and prepared packages to join the attack were available on StopGeorgia.ru, StopGeorgia.info, and xakep.ru (hacker in Russian). This again shows a high degree of central control and command, and the SQL-injections which were not seen in Estonia are a degree more sophisticated than DDoS attacks.[58] An additional aspect is that one of Georgia's main outlets for internet traffic is fibre optic cables in the Black Sea. These were severed early on in the war, which very much ties into the information-technical dimension of information warfare.[59]

While analysing the attacks against Georgia, the first notable point is how the *modus operandi* mirrored the cyber attacks against Estonia, including the coordination on Russian forums. It has been argued by Jackson, director of threat intelligence at Secure Works, that the way the attack was combined with the military operations and the swift emergence of the list of targets made it likely that the Russian government was involved.[60] While analysing the cyber operation, Carr found that the origins of the website where the instructions and malicious code were with a company called SteadyHost.[61] It was registered with a bogus name in the US, but it operated from the same building in St Petersburg as the Russian Ministry of Defence's Centre for Research on Military Strength of Foreign Countries, with the GRU's headquarters – the main intelligence directorate of the Russian armed forces – on the same street.[62] Another similarity to the attacks on Estonia was that Nashi had at the start of the hostilities declared an 'information war' on Georgia.[63]

The political contention between Georgia and Russia is even clearer than the one with Estonia. That the first cyber attacks came weeks before the shooting war reflects the longer time perspective of the conflict. The international fact-finding mission concluded that the origins of the war need to be seen as not only the shelling that was the flash point for the war, but, also, the 'impact of a great power's coercive politics and diplomacy against a small and insubordinate neighbour'.[64]

Similarly, then-President Medvedev publicly admitted that the intervention was to halt the possibility of Georgia entering NATO and worsening Russia's geopolitical situation.[65] In this way the motivation for Russia's use of cyber attacks can be established, and the Russians have ready access to well-educated loyal hackers in abundance. The *modus operandi* was the same as in the attacks against Estonia. The capacity of more or less independent hackers was coordinated on Russian forums with readymade packages and target lists.

Ukraine

During the war in Ukraine, the information-technical element has mostly been conspicuous because of its absence. The Crimean operation started with the Russian special forces (popularly and belittlingly referred to as 'little green men') snatching key infrastructure on the peninsula. Priority targets were TV and radio stations, and mobile phone operators.[66] This was, however, the most notable part of the information-technical aspect. The war over Ukraine has had DDoS attacks on both the Ukrainian and the Russian government sites but nothing that approximates the coordination, impact, or intensity of the cases discussed above.

The Swedish Defence Research Agency argues persuasively in its study of the operation that the difference was that Russia needed the information infrastructure to function for enabling political propaganda and thus creating favourable conditions for their military operation.[67] In this way, the interplay between the information-technical and information-psychological aspects of information warfare is shown. Rather than being two different phenomena as they are treated in the West, they are different arenas which are used separately or in tandem when suitable to achieve political influence.

Legal challenges of Russian information warfare

Information-psychological

The possibilities within international law to counter information-psychological warfare are quite meagre. The bulk of its perpetration is reporting of false facts, framing stories to the framer's own benefit, and smear-campaigns. When the Russian state media started to push stories putting the blame on the Ukrainian armed forces for the downing of MH17, there were few things limiting that framing of the event since the investigation of the crash took months to conclude. If we take the example of the above-mentioned show discussing Carl Bildt, it is, rather, classified as libel and the relevant jurisdiction would be Russian domestic law. This epitomises the problem: responsibility lies with the host-government which in this case, if not commands, then encourages this type of reporting. Only a few weeks after the Carl Bildt-story, Kiselyov was appointed head of *Rossiya Segodnya*,

When it comes to *jus in bello* in humanitarian international law, propaganda and misinformation are normally included in the category of 'ruses of war', which is accepted by customary international law. In societies based on free speech, where the tendency is not to prohibit certain types of propaganda, this is particularly hard to counteract. This is so because it goes against the idea of free speech, and, secondly, upholding the distinction could be arbitrary. The trend is the opposite in Russia where new laws are being adopted limiting free speech in different ways to prohibit 'extremist' material, 'treason', or information that can destabilise the state and incite unrest. The laws that could limit the use of propaganda are rather those concerned with hate crimes, inciting genocide or encouraging subversion, but there are few examples of information warfare reaching that intensity.

In the Vilnius regional court in Lithuania, action has been taken against Russian information warfare. Because of the coverage by Russian state-owned RTR Planeta of the conflict in Ukraine, the Lithuanian Radio/TV commission asked the court to suspend it for inciting discord and violence against peaceful people and for fomenting a military intervention.[68] RTR was found guilty by the court of breaching the law on the provision of information to the public. Similarly, NTV Mir (owned by Gazprom) was convicted for airing broadcasts lying about the events of Lithuanian independence. They broadcast a story that it was undercover Lithuanian agents and not the Soviet army who killed 13 people when the TV headquarters had been stormed.[69] Nonetheless, the ban was easier to decide on than to implement. Viasat, which is registered in Sweden and is the biggest TV provider in the Baltics, ignored the ban.[70] The decision of Viasat was also supported by other TV providers such as TeliaSonera, which argued that the decision could conflict with the protection of certain human rights related to the freedom of expression.[71] TeliaSonera then went ahead with 27 other TV providers to appeal the Lithuanian decision on NTV Mir and they later filed an appeal on the ban of RTR Planeta.

This emphasises the difficulty of combating information-psychological warfare in societies based on free speech. If the propaganda reaches an intensity that can be seen as a violation of UN Charter article 2(4) and attributed to the state, then it is a question for the UN Security Council, in which Russia holds a veto. If the incitement is committed by an individual, the theoretical possibility that the International Criminal Court (ICC) – analogous to the International Criminal Tribunal for the former Yugoslavia's prosecution of Vojislav Šešelj, which rests to a large degree on inciting hate crimes – is excluded since Russia has withdrawn from the Rome Statute of the ICC. The opportunities are thus currently meagre to counter Russian information-psychological warfare using international law.

Information-technical

As touched upon above, the nature of cyber warfare is that it is ambiguous due to the possibility to redirect attacks across the world and the limited means needed to perpetrate attacks. In cases where attribution can be ascertained, it is usually a lengthy progress which poses formidable challenges to international law to address the problem. One way to approach it is through state responsibility. The longstanding duty of states to prevent non-state actors from committing cross-border attacks, that states knew about beforehand, has evolved after 9/11 to include a duty to act (including prevention and prosecution) against groups generally known to carry out illegal acts.[72] If a non-state actor is acting on the area of a state by omission or commission, host-state responsibility could be imputed.[73]

In this way and in these cases, responsibility could be imputed upon Russia. If the attacks are classified as amounting to the not clearly defined 'armed attack', self-defence is allowed under UN Charter Article 51. If it does not amount to an armed attack, the relevant legal framework is domestic criminal law. However, in the case of Russia, this has proven unsuccessful. The attacks 'cloaked in nationalism are not only not prosecuted by the Russian authorities, but they are encouraged by their proxies'.[74] This demonstrates how the Russian state has been unable, or unwilling, to prosecute the attacks, which supports the argument that state responsibility could be imputed upon Russia and allow a victim state to respond to protect their citizens.

Rather than concluding on the definite legality of cyber warfare, this discussion highlights the way in which the conduct of Russian information warfare could have been streamlined around the prohibitions of international law. There is currently a vacuum between cyber attacks falling short of being seen as an armed attack and avoiding the enforcement of domestic criminal law. If the host state is sympathetic towards, or responsible for, the attacks, it is easy to get away with the attacks since the host states are the ones responsible for prosecution. This should also be seen in the light of Russia's approach to international information security. Russia is working internationally to limit the use of military capabilities they have not used, while the current state of international law allows them to protect the civilian hackers that they either used or encouraged.[75]

The key goals in the Russian negotiation position, adhered to by China and other countries from the Shanghai Cooperation Organisation, is a 'sovereignisation' of cyberspace which affirms states' rights to protect their own information space from disturbance and to be sensitive to the diversity of social systems in all countries.[76] For instance, the Russian draft convention on international information security wants to prohibit actions undermining political, economic and social systems as well as making 'aggressive information warfare a crime against international peace and security'.[77] What exactly is meant by avoiding 'undermining political, economic and social' is perhaps best illustrated by the official from the Russian Ministry of Defence who suggested at a UN disarmament conference that promoting ideas on the internet in the name of democracy should qualify as aggression and interference in the internal affairs of states.[78] This is perhaps a blunt statement, but the question is how far it diverges from the overall Russian view.

Conclusion

In the war in Ukraine, Russia has showed an effective and well-coordinated effort on the information warfare front. There were surprisingly small-scale instances of cyber warfare, but the information-psychological confrontation reached previously unparalleled proportions. The maintaining of the information-technical infrastructure could thus be mostly seen as a precondition for fighting the information-psychological war. The Russian conduct shows the use of the whole palette of information warfare tools, ranging from physically storming media broadcasting stations, to hiring paid commentators and bloggers online, as well as spreading disinformation and misinformation via traditional media channels.

The information warfare effort did not persuade the world of the justness of the Russian cause, but perhaps it was never meant to. Rather, it managed to generate domestic support for the operation in Russia. In the early stage, it also gave the image of popular support to the uprising both to persuade the Ukrainian army in Crimea that resistance was futile and to sow doubts, locally and internationally, about the identity of the perpetrators of the operations. The information warfare effort further managed to deflect the focus to the potential problem of Ukrainian nationalist extremists seizing power in Kiev.

The success of Russian information warfare is a development resulting from the bitter experiences on the information front during the First Chechen War in 1994–96. During the war, Russian journalists were free to move around Chechnya, and the Chechens paid the expenses for journalists who made the trip to interview them. The essentially free media provided a favourable image of the Chechens, and there was strong opposition to the war in the rest of Russia while the soldiers that fought the war were demoralised. This improved in the Second Chechen War, in which Russia felt the potential of the internet. Thereafter, the Russian information security doctrine was drafted with its broad understanding of information warfare.

There, a position was codified that sees Russia as in a constant information war with the West. The successful conduct in Ukraine should be seen as in opposition to the Georgian war in 2008 where, despite the early arrival of a number of Russian journalists, Russia lost the international struggle for legitimacy, even though Georgia fired the first shots in the shooting war.

The conduct of information-technical warfare is conducted in the absence of agreed international norms and where domestic criminal law is the relevant framework. If Russia then has been commanding or sympathising with the attacks and does not prosecute, the perpetrators are in effect protected. Furthermore, Russia has been working internationally to limit those military cyber capabilities which it has not used, while this conundrum protects civilian hackers within Russia.

A key reason why the Russian conduct of information warfare has been successful is that it has been streamlined around the grey areas in international law and soft spots in ethical claims to achieve maximum impact. The perpetration of information-psychological warfare against Western societies is particularly effective since it strikes against one of the fundamentals in their claim to legitimacy, the freedom of speech.

Notes

1 For an analysis of Russian contemporary warfare, see Oscar Jonsson and Robert Seely, 'Russian full-spectrum conflict: an initial appraisal after Ukraine', *Journal of Slavic Military Studies*, Vol. 28, No. 1 (2015), pp. 1–22.
2 Timothy L. Thomas, 'Information Warfare in the Second (1999-Present) Chechen War : Motivator for military reform?', in Anne C. Aldis and Roger N. McDermott (eds.), *Russian Military Reform 1992–2002* (London: Routledge, 2014), pp. 209–33.
3 Thomas, 'Information Warfare in the Second (1999-Present) Chechen War', 223 (see note 2 above).
4 Ibid., p. 225.
5 Ibid., p. 227.
6 Ibid.
7 Nataliya Gevorkyan, Natalya Timakova, and Andrei Kolesnikov, *First Person: An Astonishingly Frank Self-Portrait by Russia's President Vladimir Putin* (New York: Public Affairs, 2000), p. 2.
8 RT, 'Russia Launches "Wartime Government" HQ in Major Military Upgrade', 1 December 2014.
9 Ministry of Foreign Affairs of the Russian Federation, 'The Information Security Doctrine of the Russian Federation', 2000, Point I.6.
10 Ministry of Foreign Affairs of the Russian Federation, 'The Information Security Doctrine', Point I.1.
11 Ministry of Defence of the Russian Federation, *Conceptual Views Regarding the Activities of the Armed Forces of the Russian Federation in the Information Space*, 2011, Point 1.
12 Timothy L. Thomas, 'Russia's information warfare strategy: can the nation cope in future conflicts?', *Journal of Slavic Military Studies*, Vol. 27, No. 1 (2014), p. 101.
13 Quoted in Timothy L. Thomas, 'Russian views on information-based warfare' *Airpower Journal*, 1996, p. 26.
14 Gevorkyan et al., *First Person*, p. 120 (see note 7 above).
15 Ivan N. Vorobyov and Valery A. Kiselyov, 'The new strategy of the indirect approach', *Military Thought*, Vol. 15, No. 4 (2006), p. 31.
16 Valery Gerasimov, 'Tsennost' nauki v predvidenii', (27 February-5 March, 2013) 8(476) *Voyenno-Promyslennyy Kuryer*, 3.
17 S. G. Chekinov and S. A. Bogdanov, 'Strategy of indirect approach: its impact on modern warfare', *Military Thought*, Vol. 20, No. 3 (2011), p. 5.
18 Scott Gehlbach, 'Reflections on Putin and the media', *Post-Soviet Affairs*, Vol. 26, No. 1 (2010), p. 79.
19 Ibid.
20 Freedom House, *Freedom of the Press*, Lanham, MD: Rowman and Littlefield, 2009.
21 BBC, 'Russian Ekho Moskvy Radio Director Fedutinov Fired', 18 February 2014.
22 BBC, 'Lenta.ru Editor Timchenko Fired in Ukraine Row', 12 March 2014.

23 Gudrun Persson and Carolina Vendil Pallin, 'Setting the scene: the view from Russia', in Niklas Granholm, Johannes Malminen and Gudrun Persson (eds.), *A Rude Awakening: Ramifications of Russian Aggression Towards Ukraine* (Stockholm: Swedish Defence Research Agency (FOI), 2014), p. 27.
24 *The Guardian*, 'Founder of VKontakte Leaves After Dispute with Kremlin-linked Owners', 2 April 2014.
25 Johan Norberg, Fredrik Westerlund, and Ulrik Franke, 'The Crimea operation: implications for future Russian military interventions', in Niklas Granholm, Johannes Malminen, and Gudrun Persson (eds.), *A Rude Awakening: Ramifications of Russian Aggression Towards Ukraine* (Stockholm: Swedish Defence Research Agency, 2014), p. 43.
26 Keir Giles, *Legality in Cyberspace: The Russian View* (London: Conflict Studies Research Centre, September 2013).
27 OSCE, 'OSCE representative criticizes steps to further increase government control', 25 June 2014.
28 OSCE, 'Proposed media ownership requirements could further damage media pluralism in Russia, OSCE representative says', 24 September 2014.
29 Levada Tsentr, 'Rossiyskiy media-landshaf: televideniye, pressa, internet', 17 June 2014.
30 BBC, 'Russia: children's Toilet TV Show drawn into Ukraine-EU row', 4 December 2013.
31 Alexandra Garmazhapova, 'Gde zhivyt trolli. I kto ix kormit', *Novaya Gazeta*, 7 September 2013.
32 Garmazhapova, 'Gde zhivyt trolli'.
33 Max Seddon, 'Documents show how Russia's troll army hit America', *BuzzFeed*, 2 June 2014.
34 www.rt.com/projects/ (accessed 12 December 2018).
35 *RT Actualidad*, '¿Fue el ébola creado proósito por las farmacéuticos occidentales y la ONU?', 23 October 2014.
36 Elizabeth Wahl, 'I was Putin's pawn', *Politico*, 21 March 2014.
37 *Press Gazette*, 'Russia Today London correspondent resigns in protest at "Disrespect for facts" over Malaysian plane crash', 18 July 2014.
38 Stephen Blank, '"No need to threaten us, we are frightened of ourselves": Russia's blueprint for a police state, the new security strategy', in Stephen Blank and Richard Weitz (eds.), *Russian Army Today and Tomorrow: Essays in Memory of Mary Fitzgerald* (Carlisle: Strategic Studies Institute, 2010), pp. 48, 90.
39 Stephen Blank, 'Threats to and from Russia: an assessment', *Journal of Slavic Military Studies*, Vol. 21, No. 3 (2008), p. 500.
40 S. I. Bazylev, I. N. Dylevsky, S. A. Komov and A. N. Petrunin, 'The Russian Armed Forces in the information environment', *Military Thought*, Vol. 21, No. 2 (2012), p. 11.
41 Timothy L. Thomas, 'Russia's information warfare strategy: can the nation cope in future conflicts?', *Journal of Slavic Military Studies*, Vol. 27, No. 1 (2014), p. 103.
42 Vladimir Putin, 'Being strong: National security guarantees for Russia', *RT*, 19 February 2012.
43 Martin C. Libicki, 'The Specter of Non-Obvious Warfare', *Strategic Studies Quarterly*, (2012, Fall), p. 89.
44 Thomas, 'Russia's information warfare strategy', p. 126 (see note 40 above).
45 Vladimir Socor, 'Moscow stung by Estonian ban on totalitarianism's symbols', *Eurasia Daily Monitor*, Vol. 4, No. 19 (2007).
46 Eneken Tikk, Kadri Kaska and Kristel Rünnimeri, 'Cyber Attacks Against Georgia: Legal Lessons Identified' (Cooperative Cyber Defence Centre of Excellence, 2008), p. 19.
47 Ibid.
48 Ibid., p. 23.
49 *The Guardian*, 'Russia accused of unleashing cyberwar to disable Estonia', 17 May 2007.
50 John Arquilla, 'Twenty years of cyberwar', *Journal of Military Ethics*, Vol. 12, No. 1 (2013), p. 82.
51 Jeffrey Carr, *Inside Cyber Warfare: Mapping the Cyber Underworld* (Sebastopol: O'Reilly, 2011), pp. 117–18.
52 Quoted in Libicki, 'The Specter of Non-Obvious', p. 101 (see note 42 above).
53 *Lenta.ru*, 'Poka ne Zagoryatsya Zdaniya', 17 January 2012.
54 Carr, *Inside Cyber Warfare*, p. 164 (see note 50 above).
55 Ibid., p. 119.
56 Tikk et al., 'Cyber Attacks Against Georgia', p. 36 (see note 45 above).
57 Roland Heickerö, *Emerging Cyber Threats and Russian Views on Information Warfare and Information Operations* (Stockholm: Swedish Defence Research Agency, 2010), p. 46.
58 Ibid.

59 R.J. Deibert, R. Rohozinski and M. Crete-Nishihata, 'Cyclones in cyberspace: information shaping and denial in the 2008 Russia-Georgia War', *Security Dialogue*, Vol. 43, No. 1 (2012), pp. 3–24.
60 Quoted in Eneken Tikk, Kadri Kaska, and Liis Vihul, *International Cyber Incidents: Legal Considerations* (Tallinn: Cooperative Cyber Defence Centre of Excellence, 2010), p. 75.
61 Carr, *Inside Cyber Warfare*, p. 108 (see note 50 above).
62 Ibid., p. 109.
63 *Pravda*, 'Rossiya vs Gruziya: voina v seti. Den' pervii', 9 August 2008.
64 Independent International Fact-Finding Mission on the Conflict in Georgia, 'Report, Vol. 1' (2009), p. 31.
65 *RIA Novosti*, 'Russia's 2008 war with Georgia prevented NATO growth – Medvedev', 21 November 2011.
66 Norberg et al., 'The Crimea operation', p. 43 (see note 25 above).
67 Ibid.
68 *Lithuanian Tribune*, 'Lithuanian court bans broadcasts from Russia's RTR Planeta', 7 April 2014.
69 *Reuters*, 'Lithuania bans Russian TV station over "Lies"', 21 March 2014.
70 *Lithuanian Tribune*, 'Regulator asks prosecutors to check Viasat's compliance with legislation', 9 April 2014.
71 TeliaSonera, 'Respecting freedom of expression – blocking of TV in Lithuania', 11 April 2014.
72 Matthew J. Sklerov, 'Responding to international cyber attacks', in Jeffrey Carr (ed.), *Inside Cyber Warfare: Mapping the Cyber Underworld* (Sebastopol: O'Reilly, 2011) pp. 47–8.
73 Ibid., pp. 56–7.
74 Carr, *Inside Cyber Warfare*, p. 29 (see note 50 above).
75 Ibid., p. 170.
76 Thomas, 'Russia's information warfare strategy', p. 111 (see note 40 above).
77 Ministry of Foreign Affairs of the Russian Federation, 'Convention on International Information Security', 2011.
78 Quoted in Giles, *Legality in Cyberspace*, p. 16 (see note 26 above).

27

UNCONVENTIONAL WARFARE AND TECHNOLOGICAL INNOVATION IN ISLAM

Ethics and legality

Ariane Tabatabai

As new means of warfare are developed, questions regarding their morality and, consequently, legality arise. This is especially the case for unconventional means of warfare. In recent years, specifically since the 11 September 2001 terrorist attacks, the use of cyber capabilities and unmanned aerial vehicles (UAVs) or drones as weapons in counter-terrorism operations has been the subject of numerous debates in the Western world.

This chapter investigates whether similar debates exist in the Muslim world and whether they affect law and policy. It concludes that while the nuances that exist in Western debates surrounding the legality of means and methods of warfare are absent in Islamic jurisprudence, the general prescriptions of the faith are very similar to the rules and regulations of international humanitarian law. This absence of such debates in Islam mainly due to the lack of specialisation in the field by those who interpret the sources and make rulings. Most of the legal debate in Islamic jurisprudence is based on such considerations as the distinction between combatants and non-combatants, protection of the environment, collateral damage, and military objectives. The chapter further concludes that the ethical and legal considerations can impact policy, but can also be fashioned by political and strategic goals.

The ethical and legal debate

The Shiite world

As discussed throughout this book, several technological innovations have fundamentally altered the conduct of warfare in recent years. These include the advent of the internet, with the related development of cyber capabilities threatening security, and the increasing reliance on unmanned aerial vehicles (UAVs). While the law generally lags behind technological advances, discussions on the legal and ethical dimensions of their use often follow. In recent years, these discussions have particularly encompassed drone[1] and cyber warfare,[2] with attempts to apply just war theory to these unconventional means and methods of warfare. Despite the proliferation of these technologies to the Muslim world and Muslim state governments embarking on various modern military programmes, the ethical and legal debate is lacking. When there is a debate about these issues, it is often limited and vague. This is mainly due to two factors.

First, the adoption of these technologies by governments is a sensitive issue with great strategic implications. Therefore, there is often a lack of information on the status and scope of these programmes. In Iran, scientific and technological progress and innovation, both civilian and military, of which the nuclear programme, and the nuclear fuel cycle particularly, is a component and the flagship, is presented as a national aspiration. These activities are presented as part of the Iranian nuclear narrative and the general revolutionary narrative, according to which the country's advancement in science and technology cannot be stopped by political pressure, military threats, or economic sanctions. They are thus an inherent part of Iranian politics, foreign policy, economic independence, and defence and military. Various Iranian officials, including the Supreme Leader and different presidents have highlighted the country's technological progress in their speeches. These have been part of an effort to galvanise popular support in light of the political, social, and economic challenges that the nation has faced as a result of Tehran's foreign policy and nuclear programme. The indigenous nature of these innovations and progress has been the main focus of these talking points. This is illustrated by the Supreme Leader's words:

> In the field of technology – petrochemistry, petroleum, steel, defence production and industry – the progress [made in Iran] is amazing. The defence systems that are produced in the country today, one day our country could not even dream of or imagine having these products; but today, these are produced. In high technology, which is talked about in the world with pride, they [the West] have been forced, despite all the animosity, to say that Iran is one of ten countries that has managed to produce a nuclear fuel cycle. This is not a small thing.[3]

Second, in Shiite Islam, morality can only exist within the realm of the faith, and ethical and legal issues are often intertwined. *Shari'a*, the Islamic legal framework, covers issues relating to warfare comprehensively. However, the rulings on the norms of warfare are made by the same scholars who make rulings on a broad variety of issues, with no expertise on these matters. Hence, the terms of the debate often lack precision and depth. What is more, as discussed in the following sections, Islamic jurisprudence relies on a set of general principles to deduce the status and scope of legal rulings on various issues. Some of the few exceptions have been in instances where certain weapons were used and retaliation by using the same category of weapons became an issue. One such case lies in the use of chemical weapons by Baghdad during the Iran–Iraq War. During that war, the Iranian leadership questioned whether chemical weapons could be used to retaliate or counter the Iraqi chemical attacks. Ayatollah Ruhollah Khomeini, the founder of the Islamic Republic, ruled against the use of chemical weapons, on the grounds that the use of indiscriminate means of methods of warfare is prohibited by the faith. Likewise, the issue of the use of another weapon of mass destruction (WMD), nuclear weapons, became topical when Iran's leadership resorted to using what it described as the 'obvious' prohibition of nuclear weapons under *shari'a* as evidence of the peaceful nature of its nuclear programme.

The Sunni world

The ethical and legal debate on warfare in the Sunni world has been overshadowed by the rise of extremism and the formation of such groups as al Qa'ida and the Islamic State in Iraq and Syria (ISIS) in the past three decades. In 1975, noted counter-terrorism expert Brian Jenkins argued that, 'acts aimed at causing thousands or tens of thousands of casualties, for a variety of reasons, may be the least likely'.[4] Al Qa'ida contradicted this understanding of terrorism, by seeking to inflict mass casualties, rather than aiming to have a big audience and few casualties. In this sense, al Qa'ida and ISIS are different from other Islamic groups, such as Hezbollah. Al Qa'ida also distinguished itself from other terrorist groups by developing a very comprehensive religious narrative, justifying its every action. Its leadership issued *fatāwā* or religious decrees (traditionally formulated in response to a question posed by believers in Islamic jurisprudence). Yet, because the organisation's leadership does not have a legitimate religious and juristic authority to interpret the Holy Text, it also sought to legitimise itself by commissioning clerics, mainly in Saudi Arabia, also to issue decrees, supporting al Qa'ida's reasoning.

Al Qa'ida plugs key events, which it describes as targeting Muslims and seeking to weaken the Prophet's legacy and his followers, into its interpretation of Islamic law. By doing so, it seeks to provide a viable justification for its indiscriminate methods. One such effort to justify al Qa'ida position on the targeting of non–combatants can be found in 'The Islamic Front's Declaration to Wage Jihad against the Jews and Crusaders', published in the newspaper *al Quds al Arabi* on 23 February 1998. The document was signed by key figures in al Qa'ida's leadership, including Osama bin Laden, Ayman al-Zawahiri, Ahmad Taha, Sheikh Hamza, and Fazlur Rahman. The declaration highlights three 'crimes and sins committed by the Americans [which] are a clear declaration of war on Allah'.[5] These 'crimes and sins' are the 'occupation' of 'the lands of Islam in its holiest places', the sanctions regime in place throughout the 1990s against Iraq, and the weakening and division of Arabs to 'guarantee Israel's survival'.[6] For these reasons, the organisation issued 'a decree to all Muslims', calling on them to 'to kill the Americans and their allies – civilian and military', formulating it as an 'individual obligation incumbent upon every Muslim who can do it'.[7]

Ultimately, both al Qa'ida and ISIS have used conventional means to inflict as many casualties as possible. Neither has used modern technologies in its 'struggle' against the West. Yet, the reasoning they have offered to justify indiscriminate attacks, would be applicable to the use of technological innovations, as much as traditional means and methods.

The rise of extremism and existence of groups such as al Qa'ida and ISIS has had an impact on the broader debate regarding the ethical and legal dimensions of warfare in Islam. Indeed, not only have they shaped the debate in the West, but they have also forced the Sunni world to phrase its discussions in radical terms. In other words, Sunni leaders and thinkers often discuss Islamic warfare in the context of fundamentalism and the use of indiscriminate means and methods, whether endorsing or denouncing them, rather than pure philosophical and legal discussions. This means that, as the discussions about unconventional warfare in the Shiite world are driven by the Iranian nuclear programme and politicised in that context, in the Sunni world, Salafi-inspired extremism and terrorism provide the framework for the discussions.

According to Amira Sonbol, however, al Qa'ida 'manipulates' jihad. Indeed, 'waging outright and comprehensive war appears to have no basis in Qur'anic text; on the contrary, conflicts were seen as having clear boundaries, waged to ward off aggression and undertaken in a humane way'.[8] The following sections discuss these boundaries and how they may be applied to technological innovation and its use in warfare.

What is the debate based on?

Sources

The Islamic legal system has an extensive body of rules and regulations governing the conduct of warfare. These rules are spread out throughout several key sources, some of which are common between the Sunni and Shiite branches of Islam, and others, proper to the latter. These rules are encompassed in the *shari'a*, the practical framework of Islam, which includes both ethics and jurisprudence (*fiqh*). According to Mohsen Kadivar, the Prophet's main focus was on ethics. After his death, however, there was a paradigm shift, as his followers focused on the law. The focus on the law and lack of attention to ethics, Kadivar argues, is the source of the problem Muslims face today, as the law is not compatible with modern times. This is why, he argues, the majority of Muslim majority states do not implement *shari'a* as a legal framework.[9] Indeed, unlike modern legal frameworks, which are adaptable, the *shari'a* transcends both time and space. Hence, Kadivar suggests that in order for the law to remain compatible with the divine law, society must be contained to avoid progress and thereby maintain its compatibility with the law. However, several tools allow the faith to remain flexible and to address innovations as they present themselves.

The Qur'an is the cornerstone of the Muslim faith and Islamic jurisprudence. The Holy Text is complemented by other key sources: the *sunnah*, the tradition of the Prophet alone in Sunni Islam versus that of Mohammad and also the Imams[10] in Shiite Islam, based on the Prophet's words and deeds (called *ahādīth*),[11] and in both traditions the *'ijma*[12] or consensus among the believers on a given matter. The main method in Islamic law allowing for new means and methods of warfare to be included into this body of law lies in the equivalent of 'precedents' in Islamic law. This method of reasoning is known as deduction (*qiyās*). It is traditionally used to deduce whether a given object or action is tolerated, allowed, obligatory, or prohibited in Islam. These constitute the various 'grades' of prohibition in Islam. The deduction method is used by considering the teachings of the Qur'an and/or deeds and words of the Prophet and the Imams[13] and applying the same reasoning to other actions. Reason (*'aql*) plays a key role in Shiite jurisprudence but is absent from Sunni Islam. Indeed, according to Shiite Islam, good and evil are determined by reason. Indeed, in order for humans to understand God's prescriptions, they must use reason. What is more, the belief in Shiite Islam is that God would not prescribe anything against reason.

Shi'ism, like the early theological and philosophical schools, affirmed the use of rational and intellectual discourse and was committed to a synthesis and further development of appropriate elements present in other religions and intellectual traditions outside Islam.[14]

In Shiite Islam, jurists who have the authority to interpret the *shari'a*'s prescriptions, called *mujtahid* (pl. *mujtahidin*), interpret the law through analogical reasoning. They do so in the absence of the Twelfth Imam.[15] By doing so, they deduce general principles from particular cases. Among *mujtahidin*, *marajeh* have a particular responsibility. Believers follow (*taqlīd*) a particular *marja'* and fashion their behaviour based on his rulings. By contrast, in Sunni Islam, *ijtihad* is practised by the *mujtahid* using *ra'y* or personal opinion.

Principles informing deduction in Islam

An analysis of the legal and ethical positions on a number of tactics and means and methods of warfare allows one to deduce the stance of Islam on modern unconventional warfare. These general principles are used by scholars to deduce the legal status of modern inventions. This is done through analogical reasoning, through which it can be determined whether, based on certain criteria, a certain invention is admissible by the faith. These principles include terrorism, deceit, and the use of indiscriminate attacks. These deductions are based on a rational model of analogy, according to which, in order to establish whether B is prohibited by the jurisprudence, one must return to the religious commands. Hence, if A is prohibited, the use of X, which would lead to A is also prohibited. This is referred to as 'foundation of the rules of religious command' (manāt al-ahkām).[16]

Shiite views

Davood Feirahi has written that 'Although defensive jihad allows any kind of action against aggressors, in Shiism acts of terrorism are forbidden.' Terrorism (*fatk*) is defined as 'an unexpected attack on a civilian in a non-war situation'. This includes 'terror or an unexpected attack as a defensive measure or for deterrence'.[17] This prohibition is based on a *hadith* by Imam Ja'far al-Sadiq (the sixth successor to the Prophet according to Shiites, a key figure in Islamic jurisprudence). According to this *hadith*, Abu-Sabah had asked Imam Sadiq for permission to surprise and kill a neighbour cursing Imam Ali (for Shiites, cursing the first Imam is equivalent to cursing the Prophet and punishable). Imam Sadiq denied the man this permission and indicated that 'this would be an act of *terror* and is prohibited by Prophet of Allah. Beware Abu–Sabah that Islam prohibits terror'.[18]

Likewise, trickery and deceit, 'including any unexpected attack on the armed forces of the enemy at night', are prohibited in Shiite jurisprudence. There is, however, an exception to this rule: it is allowed to use deceit to reciprocate the deeds perpetrated by a deceitful enemy: 'And the recompense of evil is evil the like of it but who so pardons and puts things right, his wage falls upon God; surely He loves not the evildoers.' Likewise, according to another *hadith* attributed to Imam Ali: 'Return the stone they have thrown. Fight fire with fire.'[19] Deceit itself is seen as amounting to a declaration of war in Shiite Islam. Therefore, 'deception is permitted in war because war is a kind of deception. However any deception, even against the unbelievers, is not allowed in no-war situation.'[20]

One of the key precedents used by Shiite scholars to deduce the prohibition of indiscriminate means of warfare, including weapons of mass destruction, lies in that of the use of poison in wartime. In Islam, indiscriminate attacks are those that would result in the death of civilians, primarily viewed as women, children, the elderly, and those suffering from a mental condition, impairing them from taking part in warfare. The protection of non-combatants is of such importance that it is taken for granted and not mentioned explicitly in some cases, including the following verses, which were revealed to give Muslims the permission to fight pagans and idolaters for the first time.

> Fight in the way of Allah those who fight you but do not transgress. Indeed, Allah does not like transgressors.[21]
>
> And kill them wherever you overtake them and expel them from wherever they have expelled you, and fitnah[22] is worse than killing. And do not fight them at al-Masjid al-Ḥarām until they fight you there. But if they fight you, then kill them. Such is the recompense of the disbelievers.[23]
>
> And if they cease, then indeed, Allah is Forgiving and Merciful.[24]
>
> Fight them until there is no [more] fitnah and [until] worship is [acknowledged to be] for Allah. But if they cease, then there is to be no aggression except against the oppressors.[25]

According to the prominent Shiite scholar, Mohammad Hossein Tabatabai (Allameh), these verses from the Qur'an, do not explicitly mention civilian immunity as this is to be taken for granted. These words, he argues, address combatants alone and, since 'women and children' lack the 'power' to engage in combat, there is no need highlight their immunity.[26]

The use of any poison, including the poisoning of bodies of water and the air, as well as cutting off water to towns, was prohibited by the Prophet.[27] According to Sheikh al-Tousi, 'In war with non-Muslims any weapon is approved except poison because if one uses poison one risks the death of women, children and the insane, whose killing is prohibited'.[28] This prohibition is not one merely on poison, but a general religious command. Therefore, the scope of this prohibition also extends to all poisonous weapons and other WMD of indiscriminate nature.[29] However, as noted in the previous cases, the prohibition is also not definitive, as it can be waived under certain circumstances.

As noted previously, targeting civilians is prohibited by Shiite jurisprudence, except in cases of *tatarros*, where the enemy is hiding behind a human shield. This is similar to the notion of collateral damage in modern international humanitarian law. However, as noted by Mohsen Kadivar, *tatarros* can only be a viable rationale for the use WMD if there is absolute certainly that the adversary will be killed during the attack. However, if there is no absolute certainty regarding such an outcome, then *tatarros* falls under a gray area.[30]

Another key point in unconventional means of warfare, especially as relating to WMD, is that of protection of the environment. The burning and cutting of trees (especially those which bear fruits), the destruction of buildings and habitations, and killing and harming of animals fall under this prohibition. According to Grand Ayatollah Khoei, the exceptions to these rules 'must be dealt with as [the case] arises'. Generally, unless there are 'convincing reasons' that the destruction is 'necessary for military consideration', civilian buildings, especially, should not be destroyed.[31] Similarly, animals should only be slaughtered in sufficient numbers to meet the needs of the combatants.[32]

Finally, the development of WMD for deterrence is a grey area. However, according to some Shiite scholars, they are prohibited by the faith, both by the *shari'a* and by *'aql*, given their costly nature and the lack of guarantee that 'irrational' individuals would not use them, leading to the loss of hundreds of thousands of innocent lives.[33]

Sunni views

Similar principles govern warfare in Sunni Islam. They appear in Sunni *ahādīth*. Yet, the different schools of thoughts within Sunni Islam have divergent stances on the conduct of warfare. These include who is provided with a special protection status by being defined as a non-combatant, what constitutes a defensive war, and therefore legitimate and obligatory, what means and methods of warfare are allowed or prohibited. Nevertheless, according to the Persian medieval scholar Tabari, one of the foremost Islamic thinkers, the first Caliph, Abu Bakr, summarised the Prophet's teachings on warfare during the Expedition of Usama bin Zayd (632 AD):

> Oh army, stop and I will order you [to do] ten [things]; learn them from me by heart. You shall not engage in treachery; you shall not act unfaithfully; you shall not engage in deception; you shall not indulge in mutilation; you shall kill neither a young child nor an old man nor a woman; you shall not fell palm trees or burn them, you shall not cut down [any] fruit-bearing tree; you shall not slaughter a sheep or a cow or a camel except for food. You will pass people who occupy themselves in monks' cells; leave them alone, and leave alone what they busy themselves with. You will come to a people who will bring you vessels in which are varieties of food; if you eat anything from [those dishes], mention the name of God over them. You will meet a people who have shaven the middle of their head and have left around it [a ring of hair] like turbans; tap them lightly with the sword. Go ahead, in God's name; may God make you perish through wounds and plague![34]

Another *hadith* prohibits burning combatants alive.[35] This comprehensive view seems to be dominant throughout the Sunni world. Indeed, 'since in Islam, the objective of war is neither the achievement of victory nor the acquisition of the enemy's property, participants of *jihad* are meant to refrain from unnecessary bloodshed and destruction of property when waging war'.[36]

However, some schools hold different views on the various items discussed by Abu Bakr. For instance, according to some scholars, among whom are followers of the Hanafi school of thought,

> [the] inviolability of property is a corollary of the inviolability of its owner. Hence, where life of the owner is not immune, his property cannot possess this quality. This view therefore permits the destruction of enemy property including all fortresses, houses, water supplies, palms and other fruit trees and all other plants and crops [...] It also allows slaughter of any animals belonging to the enemy, including horses, cows, sheep and cattle, poultry of any kind, bees and beehives. Transferring animals and weapons from the enemy back to Islamic territory is also allowed, but if this course of action is not feasible, animals may be slaughtered and burnt, whereas weapons may be destroyed to prevent the enemy from using them.[37]

The diversity of the schools of thought within the Sunni tradition makes it difficult to clearly conclude what means and methods of warfare can be considered as allowed or prohibited by the faith. Nevertheless, most Sunni schools agree on some broad principles, including the existence of a protected group of people, the non-combatants, albeit defined differently within each school of thought, and prohibiting the affliction of unnecessary damage upon the environment (defined as trees, livestock, and so forth).

Application

As noted previously, the general principles of Islamic jurisprudence are taken into consideration to assess the legality of various military innovations. Hence, when there are no specific prescriptions regarding a particular weapon or tactic, scholars refer to these principles to deduce their legal status. In the case of most modern unconventional means and methods of warfare, there is no clear ruling in Islamic jurisprudence legitimising or prohibiting their use. However, by referring to these general principles, scholars can issue rulings on these innovations. The debate has, however, remained fairly limited due to the sensitivity of the matter. Unconventional warfare is an integral part of world affairs with serious strategic implications. In Iran, the ethics and legality of the development and use of WMD have been debated due to the use of chemical weapons by Baghdad during the Iran–Iraq War and the role of the Islamic legal discourse in shaping Tehran's nuclear narrative in recent years. In the words of the Supreme Leader Ayatollah Ali Khamenei:

> According to our faith, in addition to nuclear weapons, other kinds of WMD, such as chemical and biological weapons, also constitute a serious threat to humanity. The Iranian nation, which is a victim of the use of chemical weapons itself, feels the danger of the production and stockpiling of such weapons and is ready to employ all its means to counter them.
>
> We consider the use of these weapons as *haram* [prohibited under Islamic law], and the attempt to immunise human kind from this great disaster the responsibility of all.[38]

Most recently, Grand Ayatollahs Mousavi-Ardabili and Yousof Sanei issued a *fatwa* prohibiting jamming, or the deliberate transmission of radio signals to disturb communications.[39] Radio jamming has historically been used by governments as a tool during both wartime and peacetime. First, during wartime, several methods are used: radio jamming allows governments to control the flow of information, and radar jamming disrupts radars used by the enemy to guide its missiles and aircrafts. Jamming has been used in many conflicts, especially the Second World War. Second, jamming has been used to limit and control access to certain media outlets in specific countries. In Iran, this has been especially the case during periods of crisis, including the contested 2009 presidential elections and their aftermath. A number of television channels and websites have been filtered and jammed. These include websites of major Western media outlets, social media, and television channels. This widespread practice has been widely criticised in Iran due to its implications for the population's health. This has led a number of *marajeh* to take a stance on the topic and condemn it according to the principles of Shiite jurisprudence.

In the Sunni world, much of the debate around warfare in Islam and the ethical and legal questions regarding the use of unconventional means and methods of warfare have been raised due to the rise of Sunni extremism. Groups like al Qa'ida and ISIS have not only stirred the debate in the West, they have also forced Sunni scholars to discuss the issue. Yet, no single clear view has been formed of the use of technological innovations in the context of armed conflict. For instance, while there seems to be a consensus among Iran's Shiite jurists that the destructive use of WMD is inherently prohibited in Shiite jurisprudence, no such consensus exists in the Sunni world. In fact, Pakistan, a self-proclaimed Islamic Republic, has developed a nuclear arsenal. Saudi Arabia has threatened to acquire a nuclear weapon should its northern neighbour, Iran, do so. Also regarding Saudi Arabia, 'after 11 September a Saudi intelligence survey found that 95 per cent of a sample of educated Saudis aged twenty-five to forty-one supported bin Laden's cause'; and, 'in December 2004, CNN reported that a poll in the kingdom had found bin Laden's popularity exceeded that of King Fahd'.[40] This shows that the highly religious and conservative Saudi population believes the methods employed by al Qa'ida not to be in conflict with its religious beliefs. Further, both predominantly Shiite and Sunni governments have used chemical weapons against their own populations or other Muslim populations, including the aforementioned use of chemical weapons by Saddam Hussein's Iraq during the Iran–Iraq War, against both Iran and Iraq's own Kurds, by Egypt in Yemen, and by Bashar al-Assad's Syria against its own citizens.

The author has discussed the legal status of the use of various technological innovations as weapons, including cyber and drone warfare, with a number of scholars. None of the scholars was able to provide a clear and precise response to the questions raised by the author. They all referred the author back to the general principles enumerated above without taking a clear stance on the matter.

Policy implications

During the Iran–Iraq War, the founder of the Islamic Republic, Ayatollah Khomeini, allegedly refrained from reciprocating the Iraqi use of chemical weapons.[41] Khomeini's reasoning was founded on the necessity to distinguish between combatants and non-combatants in Shiite jurisprudence, as discussed in the previous section. He argued that the tactics adopted by Tehran should not 'harm' the Iraqi people.[42] Khomeini's stance on the prohibition of WMD was not shared by the entire Iranian ruling class, however. Key figures in the Iranian leadership urged Tehran to equip itself with different unconventional weapons both during and after the war.[43] Hashemi Rafsanjani had declared in 1988 that 'chemical bombs and biological weapons are poor man's atomic bombs and can easily be produced. We should at least consider them for our defense.'[44] In 2010, he stated that 'curiosity, the need for defence and deterrence, and above all, greed in some human beings and societies have unfortunately led them to step on a path' where they would hurt the health and life of their own kind.[45]

As a result, Iran has embarked on several unconventional weapons programmes since the end of the war. Its nuclear power programme and the IAEA investigation of the 'possible military dimensions' (PMD) of that programme are well known. The country's paramilitary group, the Army of the Guardians of the Islamic Revolution (IRGC), and the Basij militias established by the Islamic Republic to weaken the traditional military forces by creating a balance of power both have cyber capabilities and ambitions. Iran's cyber capability is identified as one of the top five globally, with an important offensive cyber capability. The cyber warfare budget has been increased and is currently estimated at USD1 billion.[46] Two developments have led to this increasing interest in cyber security and warfare. The first one was domestic and lay in the events following the contested 2009 presidential elections, which came to be known as the Green Movement. The second, an external factor, was the Stuxnet incident. Both events essentially coincided, as the roots of Stuxnet were traced back to June 2009,[47] the same month that the presidential campaign, elections, and subsequent unrest took place. As such, the Islamic Republic can and does use its cyber capabilities on two fronts: internal, to track down and persecute the opposition, and external, to hit the United States and Israel. In April 2012, Ilan Berman, the American Foreign Policy Council's Vice-President, appeared before the US House of Representatives Committee on Homeland Security and the Subcommittee on Cybersecurity, Infrastructure Protection, and Security Technologies, as well as the Subcommittee on Counterterrorism and Intelligence, and stated that 'over the past three years, the Iranian regime has invested heavily in both defensive and offensive capabilities in cyberspace. Equally significant, its leaders now increasingly appear to view cyber warfare as a potential avenue of action against the United States.'[48] However, beyond viewing the cyber realm as 'a potential avenue for action against the United States', Tehran also has an interest in investing in cyber capabilities to deter and counter attacks from the United States, and most importantly Israel. This understanding is confirmed by Supreme Leader Ali Khamenei's call for the formation of a Supreme Council of Cyberspace in March 2012.[49]

Similarly, Iran 'has been beefing up its data gathering and transmission systems at air defense sites'.[50] Likewise, Tehran has invested in a UAV programme, which includes missile-equipped drones.[51] The drone programme, one of the oldest in the world, was established during the Iran–Iraq War, of which the first operational product was the Ababil, launched in 1986.[52] Tehran has also invested in missile systems, developing one of the most advanced and sophisticated ballistic missile systems in the Middle East.[53] It has further supported groups that have historically perpetrated deceitful attacks against civilians and led covert operations, including the Harakat al-Muqawamah al-Islamiyyah (HAMAS) and Hezbollah.

Hence, despite much of the attention in the West being focused on Iran's nuclear programme, the nuclear aspirations should be viewed and understood as part of a greater endeavour. Viewed through such a lens, Iran tries to portray itself as a technologically capable and advanced state, which continues to evolve within the context of the *shari'a*. As such, it presents itself as the champion of technological innovation achieved within an ethical framework, set by the divine. Hence, the *shari'a* is presented by the leadership as the single most decisive line of reasoning in its decision-making process. The idea of technological and scientific progress is key to the Islamic Republic's narrative, allowing it to counter arguments presented by both domestic and foreign constituencies, according to which the country has declined since the 1979 Islamic Revolution, due to the rigid rules and regulations of an outdated religious and legal system. Hence, the capabilities developed by Tehran serve to address not only its defence and military needs, but also political ones.

Similarly, across the Sunni world, states have embarked on various defence programmes, dominated by modern technology, ranging from nuclear weapons in Pakistan, to UAV and cyber capabilities in Saudi Arabia. The Kingdom's planned budget for cyber security is USD33 billion for the 2007–2018 period. As part of these efforts, Riyadh is creating a Cyber Security Operations Centre/Network Operations and Security Centre.[54] Likewise, Saudi Arabia has been conducting war games, including a cyber exercise. The 'Sword of Abdullah', a series of military exercises conducted in April 2014, included 'training on electronic warfare'. Indeed, Arab states in the Persian Gulf believe Iran's cyber capabilities to be a threat.[55]

Non-state actors have also attempted to procure and use modern technology to help them effectively fight the west. Yet, al Qa'ida and ISIS have inflicted most of their casualties by using traditional technology and weaponry, such as aircraft, in the case of the 11 September 2001 attacks, or AK-47s. Nevertheless, there is evidence of al Qa'ida seeking a WMD capability, training its recruits to conduct chemical and biological attacks, and conducting crude 'sensible conventional explosive tests' in Afghanistan.[56]

Conclusion

This chapter argued that despite having an extensive body of laws regulating the conduct of warfare, Islamic law cannot adapt to comprehensively address all means and methods of modern warfare. This is due to the lack of specialisation among Islamic scholars, who are trained to have an opinion on all matters, ranging from personal hygiene to governance, but no specialisation allowing them to concretely comprehend these issues. What is more, some of the concepts of key importance in the legal framework can be considered as obsolete in modern warfare. However, the general principles governing the ethical and legal dimensions of warfare are very similar to the rules and regulations in place internationally. The cardinal norms of international humanitarian law, including discrimination, as well as such considerations as the protection of the environment, collateral damage, and military objectives, are also present and discussed in Islamic jurisprudence. These discussions are further often similar to those of international lawyers. They take into consideration the ethical dimension of the means and methods of warfare in relationship to the military necessities and advantages they represent. Islamic jurisprudence, thus, dictates that any technological innovation falling short of distinguishing between combatants and non-combatants, harming the environment, or of a deceitful nature, including (for the purposes of this chapter) WMD and other unconventional means and methods of warfare, should not be utilised as a weapon of war, except under certain circumstances.

In recent years, the Western arms control community has begun to discuss the effects of nuclear weapons beyond the traditional scope of the debate, focused on the central ideas of deterrence and nuclear posture. This focus provides for an abstract discourse on the use of nuclear weapons. The initiative to bring attention to the humanitarian effects of nuclear weapons tries to break away from this trend. Iranian officials, including the Supreme Leader, have also noted the inhumane dimension of WMD a number of times. While the Western initiative does so from a secular perspective, Iranian officials discuss it from a religious perspective, as they see morality only in religion. However, the core argument, according to which due attention should be paid to the impact of these weapons on human life and habitat, remains the same.

Notes

1 Waltzer, Michael, 'Is the military use of drones ethically defensible?', paper presented at an event at the Berkley Center for Religion, Peace and World Affairs, Georgetown University, 13 March 2013,

available at http://berkleycenter.georgetown.edu/events/is-the-military-use-of-drones-ethically-defensible (accessed 15 April 2014).

2 Charles Dunlap, 'The intersection of law and ethics in cyberwar: some reflections', available at http://scholarship.law.duke.edu/cgi/viewcontent.cgi?article=5357&context=faculty_scholarship (accessed 15 April 2014); Patrick Lin, Fritz Allhoff, and Neil Rowe, 'Is it possible to wage a just cyberwar?', *The Atlantic*, 5 June 2012, www.theatlantic.com/technology/archive/2012/06/is-it-possible-to-wage-a-just-cyberwar/258106/ (accessed 15 April 2014).

3 Majles Research Center, 'Stance of the great Supreme Leader of the Revolution regarding sanctions and the Islamic Republic of Iran's nuclear diplomacy' (Tehran: Majles Research Center, 2012), 8, translation by Ariane Tabatabai.

4 Brian Jenkins, 'Will terrorists go nuclear?', Testimony before the Committee on Energy and Diminishing Materials of the California State Assembly, 19 November 1975 (Santa Monica: The Rand Corporation, November 1975), p. 2.

5 Raymond Ibrahim, *The Al Qaeda Reader* (New York: Doubleday, 2007), p. 12.

6 Ibid., pp. 11–12.

7 Ibid., p. 13.

8 Amira Sonbol, 'Norms of War in Sunni Islam', in Vesselin Popovski, Gregory Reichberg, and Nicholas Turner, *World Religions and Norms of War* (Tokyo: United Nations University, 2009), pp. 290–301.

9 Mohsen Kadivar, 'Shariat: Nezam-e hoghoughi ya arzesh-ha-ye akhlaghi?', 19 November 2013, available at http://kadivar.com/?p=12859 (accessed 17 April 2014).

10 The Prophet and the Imams are considered as infallible and their words and deeds are to be emulated by the believers.

11 Similar to '*acta, dicta, et passa Christi in carne*' in Christian theology, *ahādīth* are 'the embodiment of the divine command and an expression of God's law (*shari'a*), [...] preserved by his Companions, in the form of discrete anecdotes [...] transmitted orally through the generations', see Norman Calder, *Studies in Early Islamic Jurisprudence* (Oxford: Clarendon Press, 1993), vi, and note that the legitimacy and accuracy of *ahādīth* are determined through a very intricate and precise process.

12 The Prophet believed that his community would never 'agree on error'. Hence, if the entire *ummah* agrees on a matter, it is settled.

13 In Sunni Islam, the deductions are based on the conduct of the Prophet, but Shi'as also consider those of the Imams.

14 Azim Nanji, 'Islamic Ethics', in Peter Singer, *A Companion to Ethics* (Oxford: Blackwells, 1991), p. 10.

15 In Shiite Islam, the Imam is 'believed to be divinely guided' and acting as 'the custodian of the Qur'an and the Prophet's teaching, and interpreter and guide for the elaboration and systematisation of the Qu'ranic vision for the individual as well as society'. The Twelfth Imam is believed to have 'withdrawn from the world, to reappear physically only at the end of time to restore true justice'; see Nanji, 'Islamic Ethics', 11.

16 Davood Feirahi, 'Norms of War in Shia Islam', in Popovski, Reichberg and Turner, *World religions and norms of war*, p. 273 (see note 8 above).

17 Ibid., pp. 270–71.

18 Ibid., p. 271.

19 Ibid., p. 272.

20 Ibid.

21 Qur'an, 2:190.

22 'Fitnah' is translated as 'chaos'.

23 Qur'an, 2:191.

24 Qur'an, 2:192.

25 Qur'an, 2:193.

26 Seyed Mohammad Hossein (Allameh) Tabatabai, *Tafsir al-Mizan (*Manshurat al-Fajr; 1st edn., 2011), p. 187.

27 Majlesi, *Bihar al-Anwar*, vol. 19, pp. 177–8.

28 Feirahi, 'Norms of War', pp. 272–3.

29 Ibid., p. 273.

30 Mohsen Kadivar, 'Ta'amoli dar hokm-e sharei-ye selaha-ye koshtar-e jamee', 19 November 2009, available at http://kadivar.com/?p=8501 (accessed 14 April 2014).

31 Feirahi, 'Norms of War', p. 274 (see note 16 above).

32 Ibid., p. 275.
33 Kadivar, 'Ta'amoli dar hokm-e sharei-ye selaha-ye' (see note 30 above).
34 Tabari, *The Conquest of Arabia* (State University of New York Press, 1993), p. 16.
35 Shaheen Sardar Ali and Javaid Rehman, the concept of *jihad* in Islamic international law', *Journal of Conflict & Security Law*, Vol. 10, No. 3 (2005), p. 339.
36 Ibid., p. 240.
37 Ibid.
38 Ali Khamenei, 'Message to the first international conference on disarmament and non-proliferation', on 16 April 2010, available at http://farsi.khamenei.ir/message-content?id=9171.
39 'Ayatollah Sanei: Ersal-e parazitha-ye mahvare-i, taadi be solte-ye malekaneh va hoghough-e shakhsi-e mardom ast', Kalameh, 16 April 2014, available at www.kaleme.com/1393/01/27/klm-180690/ (accessed 17 April 2014).
40 Abdel Bari Atwan, *The Secret History of al Qaeda* (Berkeley: University of California, 2008), pp. 150–51.
41 Interview with Mohsen Kadivar cited in Scott D. Sagan, 'Realist perspectives on ethical norms and weapons of mass destruction' in eds. Sohail Hashmi and Steven Lee, *Ethics and Weapons of Mass Destruction: Religious and Secular Perspectives* (Cambridge: Cambridge University Press, 2004), p. 87.
42 Ruhollah Khomeini, *Sahifeh-ye Emam*, Vol. 13, p. 193.
43 Sagan, *Ethics and Weapons of Mass Destruction*, 87 (see note 41 above).
44 'Iran's Chemical and Biological Programmes', *Iran's Nuclear, Chemical and Biological Capabilities: A Net Assessment* (London: IISS, 2011).
45 'Hashemi Rafsanjani: Combat against chemical and biological weapons with deeds not with speeches', JARAS, broadcast 28 June 2010, available at www.rahesabz.net/story/18356/.
46 Yaakov Katz, 'Iran embarks on $1b. Cyber-warfare program', *Jerusalem Post*, 18 December 2011, available at www.jpost.com/Defense/Article.aspx?id=249864 (accessed 7 November 2012).
47 Gregg Keizer, 'Is Stuxnet the "best" malware ever?', *Computerworld*, 16 September 2010, available at www.computerworld.com/s/article/9185919/Is_Stuxnet_the_best_malware_ever (accessed 8 September 2012).
48 Ilan Berman, 'The Iranian cyber threat to the U.S. homeland', testimony before the House Committee on Homeland Security, Subcommittee on Cybersecurity, Infrastructure Protection, and Security Technologies, and Subcommittee on Counterterrorism and Intelligence, 26 April 2012, available at http://homeland.house.gov/sites/homeland.house.gov/files/Testimony%20-%20Berman.pdf (accessed 8 November 2012).
49 Ramin Mostaghim and Emily Alpert, 'Iran's supreme leader calls for new Internet oversight council', *Los Angeles Times*, 7 March 2012, available at http://latimesblogs.latimes.com/world_now/2012/03/iran-internet-council-khamenei.html (accessed 9 November 2012).
50 'Iran to unveil new destroyer, missile defense system on Friday', *Haaretz*, 17 April 2014, available at www.haaretz.com/news/middle-east/1.585966 (accessed 18 April 2014).
51 'Iran unveils "biggest missile-equipped drone"', *Al-Jazeera*, 19 November 2013, available at www.aljazeera.com/news/middleeast/2013/11/iran-unveils-biggest-missile-equipped-drone-201311182223466932.html (accessed 14 April 2014).
52 Arthur Holland Michel, 'Iran's Many Drones', 25 November 2013, available at http://dronecenter.bard.edu/irans-drones/ (accessed 15 April 2014).
53 The Nuclear Threat Initiative, 'Iran', www.nti.org/country-profiles/iran/ (accessed 18 April 2014).
54 Theodore Karasik, 'Saudi Arabia's defense posture is robust', *Al-Arabiya*, 23 September 2013, available at http://english.alarabiya.net/en/views/news/middle-east/2013/09/23/Saudi-Arabia-s-defense-posture-is-robust.html.
55 Abdulmajeed Al-Buluwi, 'What message is Saudi Arabia sending with war games?', *Al-Monitor*, 30 April 2014, available at www.al-monitor.com/pulse/originals/2014/04/saudi-military-maneuvers-sign.html#.
56 Matthew Bunn, Martin Malin, Nickolas Roth and William Tobey, 'Advancing nuclear security: evaluating progress and setting new goals' (Cambridge, MA: The Belfer Center for Science and International Affairs, March 2014), p. 6.

28

CYBER SECURITY, CYBER-DETERRENCE AND INTERNATIONAL LAW

The case of France

Anne-Marie le Gloannec (dec.) and Fleur Richard-Tixier

Over the past ten years, cybernetics has been evolving very rapidly, offering immense possibilities in terms of efficient communications and synergy, economic gains, and, generally speaking, increased capabilities in many areas – military, economic, and social. At the same time, it has opened up new opportunities to non-state actors outside the boundaries of law, such as mafias, phishers, or hackers, and to states engaged in surveillance and intelligence, using cyber instruments to manipulate the environment to their benefit in the security area. We will consider the latter here, i.e., the realm of cyber security, and particularly the use of cyber means in a conflict-ridden environment.

Conceptualising cyber security has made strides, at first in the United States, and later in European countries, especially after the cyber attacks mounted against Estonia in 2007 – as well as probably in major countries endowed with important or increasing military capabilities, such as Russia, China, or India, that will be left aside in this analysis. Two overarching questions guide such a conceptualisation:

1 The nature of the threat or threats, if any. How must they be characterised? Must they be defined as wars, attacks, or risks? In other words, what is the nature of the actions encountered, civilian or military, and what is the degree of maliciousness involved and its consequences?
2 What means must be employed and, if necessary, what strategies must be followed to counter these actions? In other words, what is to be done in case of attacks that have a military dimension, even if a cyber Pearl Harbor has never occurred? Are thoughts devoted to a worst-case scenario, a so-called black swan?

France, which is at the heart of this chapter, offers an interesting case, for several reasons: as is well-known, it is a member of the United Nations Security Council (UNSC); it is a middle-sized military power willing to take risks when sending troops abroad; though not a member of the 'Five Eyes', it is close to them and endowed with intelligence capabilities that are appreciated by its allies; and last but not least, it has nuclear weapons and early on developed its own doctrine on how to deter potential nuclear attacks. To that extent, nuclear deterrence has influenced or, rather, served as a point of departure to the French approach to cyber security – as it also has in some other Western countries, the United States first and foremost. Despite similarities, French thinking on cyber security differs from the American approach, both in terms of deterrence or dissuasion, and in terms of interpretation of international law.

Deterrence and international law constitute a kind a 'meta-language' to frame cyber security and response to attacks of different kinds. In other words, is it possible to define codes of behaviour, one politico-military and the other one legal, that would constrain possible attackers? Regarding the first question, is deterrence and/or dissuasion possible? Regarding the second one, can specific aspects of international law be applied to cyber attacks, and how? After first describing the emergence and evolution of what is understood as cyber security in France, this chapter analyses French debates on the emergence and definition of cyber attacks, cyber deterrence, and approaches of international law, while underlining similarities with and differences from debates among its allies.

Emergence and definition of cyber threats: the evolution in French thinking

Compared with other countries such as the United States, (or even Norway, a leading thinker within NATO), French concern with cyber security emerged relatively late. Certainly, the government voiced concern over the security of information systems as early as the latter part of the 1990s, and even more so in the 2000s, when a single technology became the norm, allowing the first major attacks to take place. The spread of numerical processes and their interconnection became a security liability for French defence.[1] The French defence ministry commissioned reports and the first defence steps were taken. In 2004, Jean-Pierre Raffarin, then Prime Minister, set up a plan to bolster the security of state information systems. Two years later, the security of companies was also taken into account in a report written by representative Pierre Lasbordes, who stressed the fact that France was lagging behind some of its allies. As a French civil servant put it, the French most probably resisted change: 'Cyber seemed reserved to specialists, to cryptanalysts, in short to the geeks.'[2] The cyber attack on Estonia in 2007 was a universal wake-up call and put in evidence that the cyber domain was a 'a confrontational space' (*espace de confrontation*), as the White Book 2013 put it.[3] To passive security (cryptanalysis), a new approach was added: active security.

The ensuing steps are well known: the publication of a White Book in 2008; the creation, in 2009, of the National Defence Authority of Information Systems (Autorité Nationale de Défense des Systèmes Informatiques, ANSSI); the publication, in 2012, of a report on cyber defence, by representative Jean-Marie Bockel; the White Book 2013, that turned cyber defence into a national priority; the seminal speech that French Defence Minister Jean-Yves le Drian delivered in Rennes in June 2013; the enunciation, in 2014, by the minister, of 50 measures to meet cyber attacks, wrapped up in a pact on cyber-defence (Pacte Défense Cyber); and, in parallel, industrial developments, including the creation, in the decade beginning 2010, of several leading providers of cyber products, systems, and services. Thales Communications and Security resulted from the fusion of two Thales branches in 2011: one specialised in military security and the other one in ensuring civilian security; in 2010 Cassidian Security combined with a branch of the European Aeronautic Defence and Space Company (EADS).[4] In other words, the policy that the French government devised and the steps it followed were bottom-up, as often stressed. Or rather, the French government started to address concerns over civilian cyber security before it fully took its military dimensions into account. To that extent, this did not differ much from the approach of other European countries while the United States prioritised the military one.

During this process, a small community of military thinkers and practitioners, along with academics, often connected with one another, e.g., through the creation of chairs in the military academy, as well as senior civil servants pondered the question of a cyber doctrine. To define a doctrine, however, it is necessary to define cyber actions, which is all the more difficult as these actions vary in nature, from spying to sabotage, from disinformation to outright aggression in the framework of a war – for example cyber attacks connected with the Georgia–Russia war of 2008 – and it is necessary to know who is behind these actions, a state, or non-state actors, or a state sponsor behind the latter? Cyber actions involving an economic purpose such as industrial espionage, phishing, and looting will not be considered here, unless they are linked to specific military undertakings. In this chapter, it is the military area that is being scrutinised, the cyberspace as a 'new area of action in which military operations already take place', according to the White Book 2008.[5]

Apart from lone voices, French thinking on cyber threats rules out the notion of a cyber war, outright as it might be, yet limited to the cyberspace.[6] The French espouse Thomas Rid's pithy remark that a '[c]yber war will not take place'.[7] Some, however, have coined the notion of '*cyber-conflictualités*', referring to various forms of hostilities that may eventually involve a war. The notion of war is nonetheless ill defined – and is used very differently in the United States compared to Europe, where analysts are more parsimonious in their use of the word 'war'.[8] As one interviewee put it, 'war is a word, not a fact, for it is possible to wage war without saying it, and say we wage war without doing so'.[9]

This begs two remarks. First, as Martin Van Creveld wrote in 1991, the existence of nuclear weapons has practically ruled out 'major wars' between major powers.[10] Yet, cyberspace and cyber tools may allow different types of actions. These are discrete, hybrid military actions, well beneath the level of wars where cyber is but one instrument among a multiplicity of actions below a certain threshold. Within this context, one may speak of what I call 'sub-liminal wars' – and not only of hybrid wars (i.e., involving many different types of action, and where a cyber attack is a force multiplier). Ukraine is such a case. Hence, according to Olivier Kempf, who was the Saint-Cyr Sogeti Thales cyber defence and cyber security chair and former French Army Chief Digital Transformation Officer and Director of the collection 'Cyberstrategie', of *Editions Economica*, 'cyberspace allows [one] to skirt the taboo of nuclear weapons'.[11] Second, even if cyber attacks do not constitute kinetic wars, they are not necessarily isolated but may constitute a string of stings. As Rear Admiral Arnaud Coustillière, borrowing the words of General Keith Alexander, underlines:

> [a] 'thousand stings' may combine with one another to blunt, damage and impair state functions, disorientate state and society, and sow chaos: as the state cannot protect its citizens any further, the latter lose trust, while the attacker is more difficult to unmask and the actions more difficult to meet than in case of a major action.[12]

Nonetheless, neither actions that support kinetic wars, nor a thousand stings, nor isolated actions, may constitute attacks per se. Analysts underline that to qualify a cyber action as an attack, a certain degree of violence and physical damage must occur. Currently, efforts are being made to define what physical damage means when inflicted; for instance, to hospitals or nuclear power stations. Where there is lethal damage, the cyber attack will be characterised as aggression. France, and its allies, envisage worst-case scenarios involving lethal attacks on vital infrastructure (see below). The White Book 2013 was very clear on that matter, and so is the Wales communiqué that was adopted at the NATO summit in September 2014,[13] according to which a cyber attack can cause damage equal to those created by a conventional attack, and, hence, trigger Article 5.[14]

Conversely, if no lethal damage occurs, a threshold must be defined beyond which re-action will be called for. Unlike certain countries, France does not specify the quantitative and qualitative nature of the threshold beyond which answers will be required. This is the reason why no agreement could be reached over the *Tallinn Manual*.[15] What constitutes a threshold? According to Eric de Beauregard, 'it is necessary to proceed on a case by case basis. As we did regarding nuclear weapons, we do not define an official threshold beyond which we would resort to force – whatever force we might use.'[16] (See also below, in section on international law.)

Nuclear deterrence and cyber deterrence?

Once attacks have been characterised, how do we prevent them and protect state and society against them? In particular, considering France's background and the fact that the country developed its own doctrine of nuclear deterrence, it is readily understandable that French analysts took nuclear deterrence as a point of departure when pondering the question of how to prepare for cyber attacks. The French debate is full of analogies between nuclear deterrence and cyber-deterrence, and questions as to whether nuclear deterrence, that is French nuclear deterrence, may serve as a blueprint for cyber-deterrence. Yet can deterrence as French nuclear strategists conceived it provide a model for a cyber-deterrence *à la française*? Eventually, most researchers and practitioners dismiss any analogy, for several reasons, most of which are well known.

Firstly, very sophisticated and expensive technologies notwithstanding, cyber technology is cheap in a number of cases, easily accessible to all, and perpetually evolves. Since there is no barrier to entry, either technological or financial, it is impossible to establish a control regime – this refers back to the question of threshold, examined earlier.

Secondly, deterrence implies the capacity and willingness to mete out a punishment superior to the damages inflicted. Besides the question of proportionality, delved into in the last part of this chapter, massive reprisals might call for further massive attacks for which our industrial societies are unprepared. Industrial societies are susceptible to profound disruption by cyber attacks, lacking resilience facing such disruptions.

Thirdly, the doctrine, or doctrines of nuclear deterrence were rooted in the use of nuclear weapons. They bore the seal of Hiroshima and Nagasaki. In the absence of trials and nuclear tests, a taboo, albeit fragile, may exist not to go to an extreme, not to cross the Rubicon beyond which lethal cyber attacks would be launched.[17] Yet, in a discrete environment, it is difficult for a state to convince potential aggressors of its capabilities.[18] It is possible to test nuclear weapons and demonstrate nuclear capacities. This is difficult in the case of cyber weapons.[19]

Finally, attribution is most difficult: while it is relatively easy to make a technical attribution, except in the case of thousand stings involving a multiplicity of simultaneous attacks and multiple attackers, where uncertainty may thwart an identification process. This raises the question of whom we are addressing. Kempf speaks of *polylectique*, i.e., multiple dialogues.[20] Yet, what if the attackers are not only numerous, but also different in nature? What if they do not share the same language as we do? Conversely, though we may assume that we may share the same language between peers, i.e. states, these very states hide their actions, and may remain beneath a certain threshold (see above), in particular to avoid retaliation.

For all of these reasons, a comparison between nuclear deterrence and cyber deterrence does not hold. Once it is agreed that nuclear deterrence cannot serve as the blueprint for cyber-deterrence, and that nuclear deterrence is the most credible form of deterrence, what is cyber-deterrence all about? Some analysts, such as Kempf, argue that there is no deterrence *(dissuasion)* but nuclear.[21] Yet the notion of deterrence emerged well before nuclear weapons were invented and nuclear deterrence conceived. Hence, cyber-deterrence rests upon a classical form of deterrence.[22]

Kempf nonetheless puts forth the notion of 'elements of deterrence'.[23] Deterrence may start at a low or very low level, e.g., messages conveyed, including through diplomatic manoeuvres, to a state that is deemed a potential attacker. The latter may be convinced that the damages France will inflict upon it would be more extensive than those incurred by France.[24] In this respect, France's attendance at all major conferences on cyber security, and the announcement, by ANSSI, that the country reinforces its cyber defence capacities, may be considered as a form of deterrence (dissuasion).[25]

In this respect, prevention and attack are two sides of the typical Western strategy. In his 2013 address, the French Defence Minister, Jean-Yves le Drian, called for a combination of security (protection) and of the capacity to answer, including to attack; resorting, if necessary, to kinetic means. Hence, France's strategy is not a cyber strategy per se: cyber attack and defence are embedded within a broad security strategy. To prevent and attack, the French approach relies upon two institutions: ANSSI, which is in charge of protection, while the General Directorate for external security (Direction Générale de la Sécurité Extérieure, DGSE) and the General Directorate for internal security (Direction Générale de la Sécurité Intérieure, DGSI) are responsible for counter-offensive answers. In the US, the National Security Agency (NSA) is responsible for both defence and offence. According to French analysts, separation between responsibilities has the merit of allowing for nuances. In case of cyber attacks, ANSSII will immediately analyse the nature of the attacks and protect providers and the wider public. Intelligence will proceed more cautiously regarding attribution and, if necessary, counter-actions. Both institutions work very closely together; thus, the head of the technical division of ANSSI is a former director of DGSE.

ANSSI works with companies, in particular those that are deemed of a vital importance (Opérateurs d'Importance Vitale, OIV), a list of which was established first for security purposes in the fight against terrorism and served later for cyber security purposes. The last law on military programming[26] commissioned ANSSI to impose security regulations and rules on providers. The list of 200–300 providers of vital importance is classified. An attack on one or several of them would be considered as aggression, yet the nature of France's response would lie in the hands of the sovereign, combining determination and constructive ambiguity.

Thus, France underlines that it has capacities without unveiling their nature. The close cooperation between ANSSI and DGSE and DGSI allows for coherence of the overall approach while the close cooperation between ANSSI and these providers also serves to improve France's status in terms of cyber security, not only within France's borders but also vis-à-vis its allies: a specific French label enhances France's credibility and sovereignty.

Cyber attacks and international law: the French position

If cyberspace is considered as the fifth domain, should international law apply? A response to cyber attacks, launched within the framework of a kinetic war or not, has to respect international law. What is France's position in this regard? Does France espouse a specific legal point of view on certain issues concerning cyber security? As a member of the United Nations Security Council, and of major organisations that deal with security issues (NATO, EU, and the OSCE), and as a well-known advocate of the respect and further elaboration of international law, France plays an important role when it comes down to defining a legal framework regarding cyber security.

International law does not provide a legal definition of cyber attacks or of cyber war.[27] In the late 1990s, some countries, led by Russia, began to devise a legal definition and to work on a convention regarding cyber weapons control. While the UNGA did not endorse a collective definition of cyber war, or attacks, the Shanghai Cooperation Organization (SCO) agreed upon a definition of cyber war, in 2009.[28] Freedom of expression is at stake in this controversy, which opposes those countries, such as France, the United Kingdom or the United States, who advocate a free, open, and little regulated cyberspace, allowing for freedom of expression, and those, like China or Russia, who aim at controlling the internet.[29] China's and Russia's call for international law to classify every cyber attack as malicious and illegal might undermine those cyber actions that promote freedom of expression – the reason why the Western allies, the United States in particular, oppose any such move.[30] An international convention, which several states have been promoting since the 2010 Davos Forum, to regulate cyberspace, might condone the control of the internet by some governments. In this respect, the failure of China to present guidelines at the General Assembly of the United Nations was met with 'relief' by a number of countries.

The Western countries seek to elaborate a doctrine and reach a consensus, in particular in the framework of the UN Group of Governmental Experts, set up in 2010, that was charged with advancing proposals concerning the use of cyber in international security, particularly in kinetic conflicts. The last report, endorsed by the UNGA in Resolution 68/243 of 27 December 2013, reiterated:

> International law, and in particular the UN Charter, is applicable and is essential to maintaining peace and stability and promoting an open, secure, peaceful and accessible Information and Communication Technologies environment. State sovereignty and international norms and principles that flow from sovereignty apply to State conduct of ICT (Information and Communication Technologies)-related activities and to their jurisdiction over ICT infrastructure within their territory, in spite of the dramatic changes that cyber actions may introduce.[31] NATO followed suit at the Newport summit, in September 2014.[32] So does France.

Even though UN, NATO, and France agree to apply the law of war and all international rules to cyber actions, issues such as attribution, the specification or not of thresholds, and, more generally, the recourse to force or the nature of retaliation in the cyberspace require further elaboration of both *jus ad bellum* and *jus in bello* regarding cyber attacks. *The Tallinn Manual on the International Law Applicable to Cyber Warfare*, devised in 2013 by the NATO Cooperative Cyber Defence Center of Excellence (CCDCOE), offers guidelines and legal adjustments on the use of cyber weapons in kinetic conflicts.[33] French experts were not invited to attend the Tallinn Conference, and, regrettably for the French, their approach was not taken into account.[34] To promote the latter, Paris decided, in 2013, to join the Center. As mentioned above, France's participation in the main fora and conferences on cyber security buttresses the government's will to participate in international legal governance in this regard. It is also a form of deterrence per se, show-casing that France has capabilities and is a major actor. As such, legal regulation may dampen the will to launch cyber attacks though it does not amount to coercion.[35]

Most European countries define attack and defence in a restrictive way, in contrast with the US which resorted to the notion of pre-emptive defence to justify the second war in Iraq. In the absence of a clear definition of imminent threat, 'pre-emption' might allow a state to bypass the UN prohibition of use of force. This entails consequences in the cyber domain where attribution in real time is a difficult or even impossible exercise. As Rear-Admiral Coustillière underlines, technical attribution after a cyber attack might take several weeks, while the use of force is legally circumscribed by UN approval of evidence.[36] The US reserves the right to invoke pre-emptive self-defence against state or non-state actors, were a cyber attack deemed imminent. In contrast, France envisages the use of cyber means only as a countermeasure or retaliation after cyber aggression has taken place – the French doctrine speaks of cyber security and cyber defence. However, the French legal framework, while respecting the main international legal principles, has to face several challenges inherent to the cyber area.

Within the remit of *jus ad bellum*, which examines whether it is legal or not to launch a war, the French doctrine assumes that the criteria for aggression have to apply to cyber attacks. While the UN charter failed to provide them, the UN General Assembly did, in a number of resolutions, particularly Resolution 33/14.[37] Accordingly, to qualify as aggression, a cyber attack has to fulfil the criteria of severity, immediacy, directness, invasiveness, measurability, presumptive legitimacy, and responsibility.[38] Both the *Tallinn Manual* and French doctrine espouse this view. However, no major country has ever defined a legal quantitative or qualitative threshold. Law is widely subordinated to sovereignty.[39]

Nevertheless, according to both Rear-Admiral Coustillière and the head of the department in charge of the law of armed conflicts, France would probably qualify a cyber attack as aggression in a case where vital infrastructures were hit and that entailed lethal consequences. Conversely, it would prove difficult to characterise as aggression any non-lethal cyber attack, such as multiple low-intensity disruptions, although 'operating below both the focus of defensive schemes and the legal threshold of States' authority to respond with force, low-intensity cyber attacks may prove to be a future attack strategy of choice in cyberspace'.[40] This is much debated in both military and civil legal circles. Attribution also poses a major challenge when it comes down to qualifying a case as aggression. So does the question of a non-state actor launching a cyber attack in a kinetic conflict. States need to ponder such a question. Since the judgement of the International Court of Justice on Nicaragua (1986), state responsibility over a non-military group is deemed limited. The *Tallinn Manual* fails to specify the legal responsibility of non-state actors.[41] The legal department of the French Defence Ministry follows very carefully the work of the UN Group of Government Experts in New York, who have been asked to clarify this question.[42]

According to *jus in bello*, France has to respect proportionality. The White Book 2013 stated nonetheless that France would answer a cyber attack with all means at its disposal, including military ones, though favouring a diplomatic and legal solution first.[43] Proportionality opens the door to de-escalation.[44] Unlike nuclear weapons, cyber means provide a large gamut of types of responses, starting from low-intensity attacks. This increases the risk of escalation, in particular in the absence of legal constraints to circumscribe the use of cyber instruments, in times of peace or war. Non-state actors also raise a dilemma. In conventional conflicts, the legal distinction between civilian and military has become increasingly blurred. In cyberspace, civilian structures or agents could become a target or might be used for purposes different from their original ones (the dual-use issue).[45] That is why Rear-Admiral Arnaud Coustillière describes cyberspace as the 'fog of war'. The experts who contributed to the *Tallinn Manual* could not reach a common legal definition to distinguish between civilian and military targets (human or physical), a difficulty that French experts recognise. According to the latter, the principle of necessity that *jus in bello* requires must apply to cyber actions, i.e., the imperative to avoid targeting civilians, unless vital goals must be achieved.[46]

Conclusion

Hence both *jus ad bellum* and *jus in bello* need adjustments to answer the characteristics of cyberspace. If France strives to maintain a leading role in the elaboration of cyber regulations, she is also ready to defend her particularities and preserve her sovereignty over a fifth security domain, which is a source of increased vulnerabilities.

The notion of sovereignty permeates the French discourse on cyber security. On the one hand, cooperation with allies is important and necessary, for a number of reasons. Cooperation helps bolster France's security through information received, and it strengthens its political standing through information given. As some experts underline, the latter fact drives the US to take France seriously. Information is a bargaining chip, a power currency, one of the very last in a world where power becomes more diffuse – even more so through the proliferation of non-state actors and of means of nuisance or destruction. Yet, information strengthens security. In this respect, however, it is a multi-edged sword because information sharing also unmasks vulnerabilities that enemies exploit, while erstwhile allies may use that for other purposes – commercial ones for instance. Hence, the idea of cyber alliances encounters limitations.

Notes

1 Interview with Rear-Admiral Arnaud Coustillière, 23 June 2015.
2 Interview with Jean-Baptiste Jeangène-Vilmer, 7 April 2015.
3 Interview with Jeangène-Vilmer and Jean-Marie Guéhenno (ed.), *Livre Blanc [White Book]: Sur la défense et sécurité nationale (Paris: La documentation française*, 2013), 45, hereafter White Book 2013. As early as mid-2008, Roger Romani (Senate) drew up a report ('Cyberdéfense: un nouvel enjeu de sécurité nationale' [2008] *Rapport d'information de M. Roger ROMANI, fait au nom de la commission des affaires étrangères, de la défense et des forces armées*, 449, 8 July).
4 See, in particular, Olivier Kempf, 'Cyberstratégie à la française' (2012) *Revue Internationale et Stratégique*, 87, 121–9. When EADS was created in 2000 a branch dealt with defence and systems electronics, and later, as of 2005, with Defence and Communication Systems. After EADS was turned into Airbus Groups, Airbus Defence and Space was created, incorporating Cassidian, Astrium, and Airbus Military.
5 Jean-Claude Mallet (ed.), *Livre Blanc [White Book]: Défense et sécurité nationale* (Paris: La documentation française, 2008), 53 (hereafter referred to as White Book 2008), original quotation: 'un nouveau champ d'action dans lequel se déroulent déjà des opérations militaires'.
6 For lone voices, see e.g. Bertrand Boyer, 'Cyberguerre, qui franchira le Rubicon?', in Stéphane Dossé, Olivier Kempf, and Christian Malis, *Le Cyberespace. Nouveau domaine de la pensée stratégique* (Paris: Economica, Collection Cyberstratégie, 2013), in particular 172–4.
7 Thomas Rid, 'Cyber war will not take place' (2011) 35 *Journal of Strategic Studies*, 1. For lone voices, see e.g. Bertrand Boyer, *Cyberguerre*, in particular 172–4.
8 On this latter point, see Gilles Andréani 'The War on terror: Good cause, wrong concept' (2013) 49 *Survival: Global Politics and Strategy*, 4, 31–50.
9 Interview with Jeangène-Vilmer.
10 Martin Van Creveld, *The Transformation of War: The Most Radical Reinterpretation of Armed Conflict since Clausewitz* (New York: Free Press, 1991).
11 Original quotation: 'Le cyberespace est l'occasion du contournement de l'interdit nucléaire', Olivier Kempf, *Introduction à la Cyberstratégie* (Paris: Economica, Collections Cyberstratégie, 2012), 104–6.
12 Interview with Coustillière.
13 North Atlantic Treaty Organisation, Wales Summit Declaration (Brussels: NATO, 4–5 September 2014 (hereafter Wales Declaration), available at www.nato.int/cps/en/natohq/official_texts_112964.htm.
14 White Book 2013, 1.
15 Olivier Kempf, *Alliances et mésalliances dans le Cyberespace* (Paris: Economica, Collection Cyberstratégie, 2014). See further discussion of and the reference for the *Tallinn Manual* in the section on 'Cyber attacks and international law: the French position', above.
16 Interview with Commissaire Eric de Beauregard, Armed-conflict Bureau, Legal Department of French Defence Ministry, 24 June 2015.
17 Boyer, 'Cyberguerre'.
18 Interview with Léonard Rolland, French Foreign Affairs Ministry, 5 June 2015.
19 Ibid.
20 Kempf, *Introduction à la Cyberstratégie*, 107.
21 Olivier Kempf, 'Cyberstratégie à la française' (2012) *Revue Internationale et Stratégique*, 87, 121–9.
22 Interview with Coustillière.
23 Kempf, chapter 8, *Introduction à la Cyberstratégie*.

24 Interview with Jeangène-Vilmer.
25 Interview with de Beauregard.
26 L'Assemblée nationale et le Sénat ont adopté, Le Président de la République promulgue la loi dont la teneur suit, 'LOI n° 2013–1168 du 18 décembre 2013 relative à la programmation militaire pour les années 2014 à 2019 et portant diverses dispositions concernant la défense et la sécurité nationale', available at www.legifrance.gouv.fr/eli/loi/2013/12/18/2013-1168/jo/texte.
27 The French legal frame for defence and security is divided between the military programming law, the defence code, and government decrees. The Defence Ministry also has a special department devoted to reconciling warfare and law. They started to work on cyber in 2000 regarding electromagnetic interference. For the past 15 years, they have been working on *jus ad bellum* and *jus in bello* approaches toward cyber attacks and the use of cyber in hostilities. Legal counsellors specialised in cyber will soon be provided to help French officers studying warfare.
28 On the SCO agreement, see the Agreement among the Governments of the SCO Member States on Cooperation in the Field of Ensuring International Information Security, Yekaterinburg, 16 June 2009. For France's role, see interview with de Beauregard.
29 There are differences between the American and the European approaches to a regulated internet, which go beyond the remit of this chapter. To put it briefly, many European countries seek some types of regulation, on the grounds of privacy, to pursue crimes, and to escape American dominance of the internet.
30 Interview with de Beauregard. Concerning the position of the United States, see John D. Negroponte and Samuel J. Palmisano, Chairs, and Adam Segal, Project Director, *Independent Task Force Report No. 70: Defending an Open, Global, Secure, and Resilient Internet* (New York: Council on Foreign Relations, 2013).
31 See the Final Report of UN Group of Governmental Experts on 'Developments in the Field of Information and Telecommunications In the Context of International Security', 7 June 2013, available at www.un.org/ga/search/view_doc.asp?symbol=A/68/98 and published with addition information in issue 33 of *Disarmament*.
32 Wales Declaration.
33 Schmitt, Michaël (general direction), *Tallinn Manual on the International Law Applicable to Cyber Warfare* (Cambridge: Cambridge University Press on behalf of Tallinn, Estonia: NATO Cooperative Cyber Defence Centre of Excellence, 2013), consulted on April 15; the *Tallinn Manual* [as cited herein] is also published on the CCDCOE website. For France's comments on the Manual, see Oriane Barat-Ginies, 'Existe-t-il un droit international du cyberespace?' (2014) *Hérodote*, 152–3 (Jan.), 201–20.
34 Interview with de Beauregard.
35 Interviews with Rolland and with Coustillière.
36 Interview with Coustillière.
37 United Nations General Assembly (UNGA), 'Définition de l'agression' (1974) *Resolution 33/14*, 14 Dec.
38 Criteria proclaimed by the *Tallinn Manual*, rules 11 to 12, 47–52.
39 Ibid.
40 Watts, Sean, 'Low intensity computer network attack and self-defense', in Naval War College, *International Law and the Changing Character of War*, published as (2011) 87 *International Law Studies*, 14 October, 60.
41 Peter Margulies, 'Sovereignty and cyber-attacks: Technology's challenge to the law of state responsibility' (2013) 14 *Melbourne Journal of International Law*, 506.
42 Interview with Rolland.
43 This is fully explained as 'une capacité de réponse gouvernementale globale et ajustée face à des agressions de nature et d'ampleur variées faisant en premier lieu appel à l'ensemble des moyens diplomatiques, juridiques ou policiers, sans s'interdire l'emploi gradué de moyens relevant du ministère de la Défense, si les intérêts stratégiques nationaux étaient menacés', from White Book 2013, 107.
44 Margulies, 'Sovereignty and cyber-attacks', 506.
45 Interview with de Beauregard.
46 See Legal Office of the French Defence Ministry, *Law of War Manual*, 10, available at www.cicde.defense.gouv.fr/IMG/pdf/20130226_np_cicde_manuel-dca.pdf.

29

THE US, THE UK, RUSSIA AND CHINA (1)

Regulating cyber attacks under international law – developments at the United Nations

Elaine Korzak

In light of the challenges that cyber attacks pose to international law on the use of force and international humanitarian law, the need for a new international legal framework to regulate these types of attacks has been continuously raised. Important, yet mostly unpublicised developments have taken place in the framework of the United Nations since the Russian Federation first introduced a proposal directed towards an international agreement in 1998. Since then, cyber security discussions have taken place across various bodies of the UN system.[1] While the Security Council is yet to discuss matters related to cyber security in any depth, the bulk of discussions has proceeded in the various Committees of the General Assembly.[2] However, legal questions have not been explicitly addressed. The Sixth Committee, dealing with questions of international law, has not been tasked to investigate the possibility of an international treaty regulating the use of cyber attacks by states. Instead, questions related to the applicability and adequacy of international law have been embedded in the broader cyber security debate at the United Nations which has been fragmented across the First, Second and Third Committees of the General Assembly.

Initiative for an international treaty

In this context, the First Committee of the General Assembly – dealing with international security and disarmament – has emerged as the main focal point of discussions. In 1998 the Russian Federation submitted a draft resolution entitled 'Developments in the field of information and telecommunications in the context of international security', which has become an integral part of the Committee's annual deliberations since.[3] The resolution's final text highlighted the duality of technological progress. On the one hand, advances in information and communication technologies (ICTs) have brought tremendous benefits to mankind.[4] On the other hand, increasing technological dependency has also created new risks and threats as 'these technologies and means can potentially be used for purposes that are inconsistent with the objectives of maintaining international stability and security', adversely affecting national as well as international security.[5] As Russia argued, 'a fundamentally new area of confrontation in the international arena is in the making, and there is the danger that scientific and technological developments in the field of information and communications technologies might lead to an escalation of the arms race'.[6] The concern over an emerging cyber arms race also helps explain the Russian rationale in bringing this issue to the attention of the First Committee. More importantly, Russia's initiative was ultimately geared towards the negotiation of an international agreement to regulate the military use of information and communication technologies. Contemporary international law was seen as ill-equipped to regulate the development and use of 'information weapons' resulting in an 'obvious need for international legal regulation'. With this, a debate about the need for an international treaty to govern cyber warfare ensued.

However, the Russian initiative was met with considerable scepticism by the United States and European countries. A cyber arms control treaty of sorts was seen as both unnecessary and difficult to operationalise. Technologically advanced countries did not view an international agreement restricting the development of new means and methods to be in their interest.[7] And in contrast to conventional weaponry, the verification of activities such as development, acquisition or stockpiling of prohibited means presented considerable challenges in the cyber context. More importantly, Western opposition to a new international legal agreement resulted in part from concerns over broad definitions of 'information weapons' that could potentially restrict the free flow of information. Russian views exhibited a certain concern with potential damage arising out of the 'uncontrolled transboundary dissemination of information'.[8] Threats in international information security explicitly include the use of information to destabilise a state, such as the 'manipulation of information flows, disinformation and concealment of information with a view to undermining a society's psychological and spiritual environment and eroding traditional cultural, moral, ethical and aesthetic values'.[9] As a result, for Western observers, the Russian notion of 'information weapon' entailed a 'deeper dread of political subversion associated with the free flow of information'.[10]

In rejecting the Russian initiative for an international agreement to regulate a potential arms race in cyberspace, the US and Western European states have focused their attention on different cyber threats. The diplomatic response of the US and the UK has emphasised the significance of criminal and terrorist misuses of information technology, particularly the potential for attacks against critical national infrastructure. Accordingly, resolutions on these topics were subsequently introduced in the Second and Third Committees. In 2000, the United States sponsored a resolution on 'Combating the criminal misuse of information technologies' in the Third Committee responsible for social, humanitarian, and cultural questions.[11] Following this effort, further resolutions were introduced in the Second Committee dealing with economic and financial issues. These sought the '[c]reation of a global culture of cybersecurity' as well as the introduction of elements and voluntary self-assessment tools to protect critical information infrastructures.[12] However, the initiatives in both Committees were comparatively short-lived and discussions were either deferred to other specialised bodies within the UN system or ceased altogether.[13]

Open opposition

In light of the various initiatives in the First, Second and Third Committees, differences in states' positions with regard to the need for a new international legal framework surfaced and openly clashed in 2005. That year the US was the only country to vote against the First Committee's draft resolution.[14] Subsequently, the annual resolution was no longer adopted unanimously and open opposition on the part of the United States foreshadowed a period of stagnation in the Committee's discussions. The work of a Group of Governmental Experts (GGE) that was to report its progress in 2005 was emblematic of this development. The First Committee of the General Assembly tasked the Secretary-General to convene a group of experts to 'consider existing and potential threats in the sphere of information security and possible cooperative measures to address them'.[15] The Group met throughout 2004 and 2005 and was to submit their outcome report in time for the General Assembly session in the fall of 2005. However, the Group comprising 15 states, including the US, the UK, Russia, and China, failed to identify any common ground and was unable to forward any findings or recommendations.[16] The Secretary-General simply noted that 'given the complexity of the issues involved, no consensus was reached on the preparation of a final report'.[17] Curiously, the stalemate of deliberations in the First Committee raised the profile of the Russian initiative, leading to a multilateralisation of the effort in the First Committee. Following the US veto and the failure of the GGE, the annual resolution gained a number of key sponsors, most notably the People's Republic of China.[18] Over the years, a variety of states chose to co-sponsor the Russian resolution, including Kazakhstan, Tajikistan, Cuba, Japan, Serbia, and Brazil.[19]

Unprecedented movement

The situation in the First Committee changed markedly only with the new US administration. Following the election of Barack Obama in 2008, the United States began to re-engage multilaterally on issues of cyber security.[20] From 2009 onwards, the annual resolution in the First Committee was again adopted unanimously.[21] More importantly, progress in discussions was most visible in the proceedings of a second Group of Governmental Experts meeting throughout 2009 and 2010. Its tasking was identical to that of the Group in 2005, namely to consider existing and potential threats in information security as well as concepts aimed at strengthening the security of global information systems.[22] However, in contrast to the first Group, the new GGE was able to arrive at a consensus and produce a final report.[23] Although its findings may not have been path-breaking, the significance of the second GGE was founded in its ability to present a consensus report for the first time. In this way, the report reflected the improved climate of discussions in the First Committee and the fact that discussions were being resumed on an unprecedented scale. The years following 2010 were characterised by a flurry of activities and in many ways the report of 2010 laid the foundation for future developments.

In its report, the second Group of Governmental Experts asserted that 'existing and potential threats in the sphere of information security are among the most serious challenges of the twenty-first century', warning that the 'global network of ICTs [information and communication technologies] has become an arena for disruptive activity'.[24] In particular, the Group highlighted the potential use of information technologies by states, noting that 'there is increased reporting that States are developing ICTs as instruments of warfare and intelligence, and for political purposes'.[25] It provided a general overview of the threat landscape before laying out five recommendations as to the way forward. These recommendations addressed a variety of aspects, including capacity building, confidence-building measures, terminology, and information exchanges.[26] Most importantly, they also addressed the question of norms and rules in cyberspace. The group argued that 'uncertainty regarding attribution and the absence of common understanding regarding acceptable State behaviour may create the risk of instability and misperception'.[27] It thus called for 'further dialogue among States to discuss norms' pertaining to the use of information technologies by states as one of its five recommendations.[28]

With this, the question of norms of responsible state behaviour in cyberspace figured quite prominently in the report. However, the group was not able to agree on more specific recommendations as to the way forward. It did not address the question of whether international law is applicable or adequate in any detail. It simply recognised the issue of norms as an area in need of discussion. Reportedly, the United States had introduced a discussion paper during the deliberations of the second GGE that spoke to the question of applicable international law.[29] According to a US diplomat, 'the U.S. put forward a simple notion that we hadn't said before. The same laws that apply to the use of kinetic weapons should apply to state behavior in cyberspace.'[30] The discussion paper represented the first explicit articulation by the United States that it regards existing international law to be applicable to potential conflicts in cyberspace. In the end, this position did not become a part of the final report of the second GGE due to Chinese opposition. According to accounts, the Russian Federation seemed prepared to go along with a recognition that international laws apply to state actions in cyberspace while the Chinese delegation rejected the idea.[31] The report ultimately reflected this difference in opinion – it highlighted the significance of norms in cyberspace but fell short of acknowledging that existing international law applies to state actions.

This situation changed with the third Group of Governmental Experts installed for 2012 and 2013. In order to capitalise on the momentum created by the second GGE, the third group was quickly tasked to continue examining threats in the sphere of information security and possible measures to address them. A particular emphasis was placed on possible 'norms, rules or principles of responsible behaviour of States'.[32] In what many have regarded as a landmark document, the third group managed to agree on an impressive catalogue of recommendations in three key areas: norms and principles, confidence-building and information exchange, as well as capacity-building.[33] The Group was able to make substantial progress with regard to questions of international law although it recognised that the international dialogue on these questions was still in its early stages. The report of 2013 for the first time acknowledged that international law is in principle applicable to the use of information and communication technologies by states. It laid down the basic rationale as follows:

> The application of norms derived from existing international law relevant to the use of ICTs by States is an essential measure to reduce risks to international peace, security and stability. Common understanding on how such norms shall apply to State behaviour and the use of ICTs by States requires further study. Given the unique attributed of ICTs, additional norms could be developed over time.[34]

Accordingly, international law is found to be applicable to states' actions in cyberspace. However, it remains unclear how specific norms could be implemented with regard to the use of information technologies by states and thus existing laws could also be complemented with additional norms over time. More specifically, the group agreed that '[i]nternational law, and in particular the Charter of the United Nations, is applicable and is essential to maintaining peace and stability and promoting an open, secure, peaceful and accessible ICT environment'.[35] The report subsequently identified several other bodies of law applicable to actions in cyberspace. These include state sovereignty and norms that flow from it, human rights and fundamental freedoms, as well as international law of state responsibility.[36]

With this, the report placed considerable emphasis on issue of norms in cyberspace and was able to explicitly address the question of the applicability of international law. In asserting that international law, in particular the United Nations Charter, is applicable to states' actions, the group put forward a consensus view for the first time. This acknowledgment represents a significant step in resolving the applicability and adequacy of international law in the use of cyber means. As the UN Secretary-General commented,

> I appreciate the report's focus on the centrality of the Charter of the United Nations and international law as well as the importance of States exercising responsibility. The

> recommendations point the way forward for anchoring ICT security in the existing framework of international law and understandings that govern State relations and provide the foundation for international peace and security.[37]

As significant as the outcome of the third Group of Governmental Experts may have been, its report left the most challenging aspect for future deliberations. The question of how international legal norms could be applied in concrete circumstances remains unclear even after it formed the subject for a fourth group of Governmental Experts that met throughout 2014 and 2015. The group was to study 'the use of information and communications technologies in conflicts and how international law applies to the use of information and communications technologies by States'.[38] Although the group adopted a number of voluntary, nonbinding norms, it did not make significant progress on the question of how international law applied to the conduct of states in cyberspace.[39] For the most part, it reaffirmed the balance struck in the 2013 consensus report by recalling the applicability of international law while highlighting the need for further discussion on its implementation.[40]

Conclusion

In light of these developments, a number of observations can be identified relating to the views of states and the need for new or dedicated norms to regulate the use of cyber attacks by states. First, the analysis reveals that the debate at the United Nations has moved away from its initial focus on the negotiation of a new legal framework. Efforts at an international cyber treaty of sorts have not materialised since the Russian Federation first began advocating the concept in 1998. Instead, the debate has broadened into a discussion on norms, rules and principles for the responsible behaviour of states in cyberspace. In this context, the consensus reports of the 2013 and 2015 Group of Governmental Experts have been key developments establishing the applicability of international law, in particular the UN Charter.

Second, with this, the focus of efforts to establish new or dedicated norms to regulate the use of cyber attacks has shifted from a wholesale agreement towards the identification of areas with potential for additional norms. While the Groups of Governmental Experts have recognised the principal applicability of international law, they have also acknowledged the potential need for additional norms in light of the unique attributes of cyber attacks. This point is particularly salient given the continued controversy over the question of adequacy. More specifically, the question of how international law can be applied in specific circumstances has remained largely unresolved.

Third, states' positions with regard to the need for new international laws can be characterised in terms of their position vis-à-vis the legal status quo.[41] States such as the US have sought to preserve the legal status quo by advocating the applicability of existing international law. Following the GGE report of 2013 they have been trying to build upon this consensus to facilitate common understandings and interpretations of current international legal provisions. In contrast, other states, such as Russia have been in favour of altering the legal status quo by advocating the promulgation of wholesale legal agreements and emphasising the need for additional norms even in if existing legal frameworks are applied.

The analysis of cyber security discussions at the UN has shown that Russian efforts at a new international cyber treaty of sorts have not materialised. Instead, a broader debate on norms of responsible state behaviour has emerged. In the course of discussions, states have acknowledged the applicability of international law to state conduct in cyberspace and attention has shifted towards questions of implementation. With this, states in favour of new or dedicated norms to regulate cyber attacks have emphasised the creation of additional norms in lieu of wholesale agreements.

Notes

1 For an overview see Tim Maurer, 'Cyber Norm Emergence at the United Nations: An Analysis of the UN's Activities Regarding Cyber-security', Discussion Paper 2011–11 (Cambridge, MA: Belfer Center for Science and International Affairs, Harvard Kennedy School, September 2011).
2 Maurer, 'Cyber Norm Emergence'.
3 UN General Assembly, Letter dated 23 September 1998 from the Minister for Foreign Affairs of the Russian Federation addressed to the Secretary-General A/C.1/53/3, Appendix.
4 UN General Assembly Resolution A/RES/53/70.
5 Ibid.
6 UN General Assembly, Report of the Secretary-General A/54/213, p. 8.
7 Tom Gjelten, 'Shadow Wars: Debating Cyber "Disarmament"', World Affairs (2010), available at www.worldaffairsjournal.org/article/shadow-wars-debating-cyber-disarmament (accessed 13 May 2014).
8 UN General Assembly, Report of the Secretary-General A/54/213, p. 9.
9 Ibid.
10 Christopher A. Ford, 'The Trouble with Cyber Arms Control', *The New Atlantis – A Journal of Technology & Law* (2010), p. 62.
11 UN General Assembly Resolution A/RES/55/63.
12 UN General Assembly Resolution A/RES/57/239; UN General Assembly Resolution A/RES/58/199; UN General Assembly Resolution A/RES/64/211.
13 UN General Assembly Resolution A/RES/56/121.
14 UN General Assembly Resolution A/RES/60/45. Information regarding voting is available at https://gafc-vote.un.org/, (accessed 13 May 2014).
15 UN General Assembly Resolution A/RES/56/19, para. 4.
16 UN General Assembly, Report of the Secretary-General A/60/202, Annex.
17 Ibid., para. 5.
18 For information regarding sponsorship see https://gafc-vote.un.org/ (accessed 13 May 2014).
19 For information regarding sponsorship see https://gafc-vote.un.org/ (accessed 13 May 2014).
20 See for example John Markoff and Andrew E. Kramer, 'In Shift, U.S. Talks to Russia on Internet Security', *New York Times*, 13 December 2009.
21 For information regarding sponsorship see https://gafc-vote.un.org/ (accessed 13 May 2014).
22 UN General Assembly Resolution A/RES/63/37, para. 4.
23 UN General Assembly, Report by the Secretary-General A/65/201.
24 Ibid., para. 1 and 4.
25 Ibid., para. 7.
26 UN General Assembly, Report by the Secretary-General A/65/201.
27 Ibid., para. 7.
28 Ibid., para. 18.
29 Gjelten, 'Shadow Wars'.
30 Quoted in John Markoff, 'Step Taken to End Impasse Over Cybersecurity Talks', *New York Times*, 16 July 2010.
31 Gjelten, 'Shadow Wars'. See also Eneken Tikk, Developments in the Field of Information and Telecommunication in the Context of International Security: Work of the UN First Committee 1998–2012 (Geneva: ICT4Peace Publishing, 2012), p. 8.
32 UN General Assembly Resolution A/RES/66/24, para. 4.
33 UN General Assembly, Report by the Secretary-General A/68/98.
34 Ibid., para. 16.
35 Ibid., para. 19.
36 Ibid., para. 21–23.
37 UN General Assembly, Report by the Secretary-General A/68/98, p. 4.
38 UN General Assembly Resolution A/RES/68/243, para. 4.
39 UN General Assembly, Report by the Secretary-General A/70/174.
40 Ibid.
41 See Chapter Six.

30

THE US, THE UK, RUSSIA AND CHINA (2)

Regulating cyber attacks under international law – the potential for dedicated norms

Elaine Korzak

This chapter follows on from Chapter 29 and takes a closer look at the emerging interpretative approaches of the US and the UK, and Russia and China, in order to ascertain the potential for the creation of such additional or dedicated norms regulating the use of cyber attacks. The analysis focuses on the emerging views in light of the legal challenges identified above, focusing on the positions of the US and Russia, in particular, as these are aligned with the UK and China, respectively.

With the shift of discussions away from an international treaty towards potential areas of additional norms, states' emerging views and interpretations with regard to the use of cyber attacks become critical points for analysis. The following subsections therefore present and examine the emerging interpretative approaches of the United States and Russia in order to identify if there are any areas where additional or dedicated norms may emerge. The analysis compares and contrasts states' views with regard to the legal challenges identified in the context of the *jus ad bellum* and the *jus in bello*. The first subsection deals with the application of international law on the use of force, specifically with the prohibition on the use of force and the right to self-defence. The second subsection examines states' views with regard to the implementation of the key targeting provisions of international humanitarian law, the principles of distinction and proportionality.

International law on the use of force

As described in the previous section, the United States was an early advocate of the applicability and adequacy of international law to state actions in cyberspace. Although this view was not adopted by Groups of Governmental Experts until 2013, the United States had elaborated on this position with a speech given by Harold Koh, then Legal Advisor at the US Department of State. His address at USCYBERCOM in September 2012 forms the most detailed account of emerging US interpretations with regard to the application of international law on the use of force to date.[1] In particular, it contains references of emerging US views with regard to the *jus ad bellum* thresholds of 'use of force' and 'armed attack'.

In his speech, Koh acknowledged the applicability of the prohibition on the use of force and the right to self-defence to cyber operations.[2] Cyber attacks could, under certain circumstances, constitute a 'use of force' and even an 'armed attack'. In the case of the prohibition on the use of force, the speech appears to adopt a consequence-based interpretation of 'armed force', stating that '[c]yber activities that proximately result in death, injury, or significant destruction would likely be viewed as a use of force'.[3] Thus, if a cyber attack results in physical consequences analogous to the 'physical damage that dropping a bomb or firing a missile' would have, then that cyber attack should be considered a use of force.[4] The support of a consequence-based interpretation of force is also evident in the examples of cyber attacks that would be considered a prohibited use of force: attacks causing a nuclear plant meltdown or disabling air traffic control systems resulting in plane crashes.[5] All these examples unequivocally involve physical consequences; indeed, the level of property damage and human injury would be quite significant in these instances.

With regard to the right to self-defence, Koh's speech acknowledges that cyber attacks could rise to the level of an 'armed attack'.[6] However, no further guidance is given as to precise thresholds. The *International Strategy for Cyberspace* of 2011 simply stated that the US 'will respond to hostile acts in cyberspace as we would to any other threat to our country' and that it 'reserve[s] the right to use all necessary means – diplomatic, informational, military, and economic – as appropriate and consistent with applicable international law'.[7]

With this, the emerging interpretative approach of the United States aligns with its long-held interpretations of 'use of force' as well as 'armed attack'. Since the adoption of the UN Charter, the US has advocated a restrictive reading of the prohibition on the use of force that would exclude political and economic coercion from its purview.[8] Its emerging view with regard to cyber attacks, which appears to be based on a consequence-based approach, continues such a restrictive reading.

In contrast, the emerging position of the Russian Federation points to a broader understanding of the relevant thresholds under the *jus ad bellum*. In its 2011 *Conceptual Views Regarding the Activities of the Armed Forces of the Russian Federation in the Information Space* Russia recognised 'information space' as a domain alongside land, air, sea and space.[9] With this, Russia´s activities in this space are governed by its domestic laws as well as principles of international law.[10] This includes the recognition that the prohibition on the use of force and the right to self-defence apply to actions in cyberspace. According to the *Conceptual Views*, Russia's armed forces will 'use the right for individual and collective self-defence with the implementation of any chosen options and means, which do not contradict the generally recognised regulations and principles of the international law'.[11]

However, the exact threshold for actions that could trigger Russia's right to self-defence remains unclear. Similarly, Russian documents do not provide definite guidance on activities that would constitute a prohibited use of force. Thus, the question which cyber attacks would rise to the level of a 'use of force' or even an 'armed attack' remains unanswered.

On the one hand, Russian official have equated the effects of cyber operations with those of weapons of mass destruction. Already in the 1990s, a Russian official was quoted as saying that Russia retains the right to use nuclear weapons against the means and forces of information warfare, whether there were actual casualties or not.[12] More recently, in a meeting of the General Assembly in 2013, a Russian representative argued that the damage resulting from the use of information and communication technologies 'is comparable with the one inflicted by the most destructive weapons'.[13] Such references are complemented by a concern over attacks on critical infrastructures, listing attacks such as Stuxnet as pertinent examples. A commentary on the *Conceptual Views* highlighted that the 'damage done by cyber weapons may include man-made disasters at vital industrial, economic, power, and transportation facilities'.[14] All these references can be seen as indicators that, at a minimum, cyber attacks which would result in consequences comparable to those of known weaponry would qualify as a prohibited use of force and potentially as an armed attack.

On the other hand, the emerging Russian position potentially extends beyond an interpretative approach that embraces the consequence-based interpretation. Russia's understanding of actions in cyberspace that could constitute a 'use of force' or an 'armed attack' appears to go beyond physical harm. The definition of 'information war' underlying the *Conceptual Views* serves as an illustration. It describes not only a 'confrontation between two or more states in the information space for damaging the information systems, processes and resources, which are of critical importance' but also 'undermining the political, economic and social system, and massive brainwashing of the population for destabilizing the society and the state'.[15] The latter effects would clearly constitute non-physical harm that would fall below the thresholds of 'use of force' and 'armed attack' under a consequence-based interpretation. Similarly, in its submissions of views to the UN Secretary-General, the Russian Federation has shown a preoccupation with a range of non-physical effects that can arise out of the use of information and communication technologies. In particular, uncontrolled information flows within and across national borders are concerning as they can harm a state's psychological and spiritual environment and erode traditional cultural, moral, ethical and aesthetic values.[16] Accordingly, '[s]tates that were previously in a position to ensure a legal regime of information exchange at their own internal level find themselves defenceless, in the new situation, against the transmission from abroad to their territory of information which may be unlawful or destructive'.[17] In discussing the events of the Arab Spring, commentators argue that the effects of uncontrolled information flows could rise to 'a scale comparing well with that of cyberspace cyberwarfare'.[18]

This position is in line with Russia's historically broad interpretative approach towards the scope of the prohibition on the use of force. Since the inception of the United Nations, Russia has been arguing for the inclusion of certain types of economic and political coercion as a prohibited use of force.[19] In the context of cyber operations, observers have similarly highlighted that the consequences of 'information operations' can take physical or non-physical forms, all of them potentially highly damaging:

> Today, a state can be attacked without its territory ever being physically invaded. The damage from such an attack may take different forms, for instance, technical failure of critical industrial, economic, energy and transport facilities, as well as financial collapse and large-scale crisis. Additionally, significant non-material damage could be inflicted as a result of disruption of civil order and military authority, including demoralization or disorientation of the population or mass panic.[20]

Thus, Russia's emerging interpretative approach in this area is highly ambiguous. Its views with regard to activities qualifying as a prohibited use of force or even armed attack are potentially expansive going beyond the physical consequences covered by a consequence-based interpretation. Particularly, the range of non-physical, yet potentially destabilising effects arising out of uncontrolled information flows are of concern. The result is a potentially broad interpretation of the key provisions of international law on the use of force in the context of cyber attacks.

Overall, the analysis of the emerging interpretative approaches of the US and Russia reveals a potential divergence in states' views with regard to the characterisation of cyber attacks under the *jus ad bellum*. States' views indicate a divergence in the interpretation of the scope of the prohibition on the use of force. Whereas the US appears to adopt a consequence-based interpretation in the classification of cyber attacks as a 'use of force' and as an 'armed attack', Russia's approach potentially extends to consequences beyond physical damage. The approach of the

United States, as laid out by the Legal Advisor of the State Department, continues its traditionally restrictive interpretation of the prohibition on the use of force which excludes economic and political coercion. The focus is on the physical consequences of a cyber attack and the question of equivalency. If a cyber attack results in physical destruction or human injury akin to that of known forms of weaponry, then it can be qualified as a prohibited 'use of force' or even an 'armed attack' prompting the right to self-defence. Russia's emerging interpretative approach, on the other hand, appears to be broader – covering certain non-physical consequences as well. A particular concern with the political, economic, social as well as moral effects arising out of uncontrolled information flows is discernible.

This potential divergence in the emerging interpretative approaches of the US and Russia illustrates the legal challenges of uncertainty and ambiguity presented above. It highlights the possibility that cyber attacks might be classified differently under international law by different states. In this case, a range of non-physical, yet potentially destabilising effects are a matter of difference in opinions. The result is increased uncertainty and ambiguity as to the qualification of these new types of attacks.

Interestingly, the US seems to recognise the level of uncertainty and ambiguity in the context of cyber operations. With reference to 'armed attacks' it has acknowledged that 'it may be possible to reach differing conclusion about whether an armed attack has occurred'.[21] However, any controversies are not taken as invalidating the current legal framework as 'such ambiguities and room for disagreement do not suggest the need for a new legal framework specific to cyberspace. Instead, they simply reflect the challenges in applying the Charter framework that already exists in many contexts'.[22] Thus, although there is uncertainty and ambiguity associated with the classification of cyber attacks, the US sees no need for a reconsideration of the current legal framework as these uncertainties also exist in other areas. This is in line with the US position opposing efforts directed towards a new international agreement described above.

More importantly, the apparent divergence in interpretative approaches ultimately limits the potential for additional norms in the area of the *jus ad bellum* and its threshold norms. A closer look at the differences in the emerging views of the US and Russia shows that they are not only diverging but that they are fundamentally incompatible. The apparent broad approach advanced by the Russian Federation places particular emphasis on the political, economic, social and moral effects that can arise out of uncontrolled information flows within and across national borders. Such non-physical effects clearly fall below the threshold of physical consequences required by the consequence-based interpretation advanced by the majority of commentators and seemingly adopted by the United States. The political, economic, or social effects of information flows are not seen as harmful or deleterious consequences. Rather than falling under the use of force paradigm, the question of information flows and their effects constitutes a human rights issue for the US and other Western states. Accordingly, states can only regulate information flows, and with them content, within the boundaries of international human rights law, including the Universal Declaration of Human Rights and the International Covenant on Civil and Political Rights. International human rights law extends to activities online and its application is seen as essential and universal. As the US stated in its *International Strategy for Cyberspace*, '[s]tates must respect fundamental freedoms of expression and association, online as well as off'.[23] The US 'will be a tireless advocate of fundamental freedoms of speech and association through cyberspace; … and will work to encourage governments to address real cyberspace threats, rather than … inappropriately limiting either freedom of expression or the free flow of information'.[24]

Thus, with regard to the issue of information flows and their effects, the emerging views of the US and Russia are not only divergent but fundamentally opposed. As a result, the potential for any agreement over additional norms in this area is severely limited. This is somewhat ironic, since additional norms could help ameliorate the uncertainty and ambiguity associated with the potentially diverging interpretative approaches advanced by the US and Russia. Yet, it is this very divergence that limits the prospects for additional or dedicated norms in this area.

Implementation of international humanitarian law

Following an analysis of emerging views with regard to key provisions of international law on the use of force, this subsection examines the position of the United States and the Russian Federation in relation to the implementation of international humanitarian law and the legal challenges of insufficiency and ineffectiveness.

The US position with regard to the application and implementation of key *jus in bello* norms reveals some initial parameters. The Koh speech at US Cyber Command touched upon not only issues of international law on the use of force but also questions of humanitarian law. In it, the applicability of the *jus in bello* and its key provisions of targeting were acknowledged. Accordingly, Koh explicitly stated that both the principles of distinction and proportionality apply to computer network attacks 'undertaken in the context of an armed conflict'.[25] This leaves open the question whether computer network attacks are also covered by international humanitarian law if they are conducted in isolation from traditional means of warfare. Irrespective of this question, the Koh speech provides only general guidance. The key principles of distinction and proportionality have not been applied in specific cases involving cyber attacks which limits a detailed analysis of whether their use might ultimately result in an increased targeting of civilians and civilian objects.[26] With this, it remains to be seen whether the legal challenges of insufficiency and ineffectiveness might arise and if they do, whether the US would view their development negatively or not.

However, different aspects of the speech provide some clues. With regard to the principle of distinction, Koh stated that it applied to 'cyber activities that amount to an "attack" – as that term is understood in the law of war – in the context of an armed conflict'.[27] This may suggest that the US shares the majority interpretation that restricts the application of the principle of distinction to 'attacks' rather than 'operations' as stipulated in Article 48 of Additional Protocol I. It remains open which cyber activities would qualify as 'attacks' and, more specifically, whether cyber attacks below the threshold of physical consequences would fail to qualify as 'attacks' in the sense of Article 49(1) of Additional Protocol I. The US has traditionally restricted its interpretation of 'attacks' to acts of violence, i.e. those with physical consequences. Although it remains to be seen whether it would support an interpretation that would allow for an increased targeting (but not attacking) of civilian objects in the context of cyber activities, the above statement may be taken as a hint that the US acknowledges a potential legal lacuna created in the use of cyber means of warfare.

Another indicator in this direction can be found in the broad interpretation the US has historically advocated with regard to the definition of 'military objective'. The US assessment is based on the effective contribution an object makes to 'war-fighting or war-sustaining capability' as opposed to 'military action'.[28] This could expose a vast set of objects to attack given societies′ increased reliance of key infrastructures on information and communication technologies. Further, civilian and military networks are pervasively interconnected, increasing the chances that civilian objects could effectively contribute to the war effort in some way. These dependencies, coupled with an expansive interpretation of 'military objective', increase the likelihood that civilian objects may become increasingly affected by the conduct of hostilities.

Finally, potential challenges emerging in the implementation of the principle of proportionality are even less discernible. Although the Koh speech recognises and highlights the problematic issue of dual-use objects and knock-on effects,[29] the lack of specific examples or guidance again limits an assessment whether civilians and civilian objects may become increasingly affected as a consequence of applying international humanitarian norms in the case of cyber attacks.

All in all, the US appears to have endorsed the applicability of the principles of distinction and proportionality to the use of cyber attacks. Yet, it remains to be seen whether the implementation of these principles would lead to an increased involvement of civilians and civilian objects and ultimately to the challenges of insufficiency and ineffectiveness.

In a somewhat similar vein, the position of the Russian Federation with regard to the applicability and implementation of international humanitarian law is also dependent upon future interpretation and development. Some initial guidance is discernible in the *Conceptual Views Regarding the Activities of the Armed Forces of the Russian Federation in Information Space*:

> The Armed Forces of the Russian Federation are guided by the regulations of the international humanitarian law (limitation of indiscriminate use of information weapons; establishment of special protection for information objects, which are potentially dangerous sources of technogenic catastrophes; prohibition of treacherous methods of waging information war).[30]

This brief statement appears to confirm the applicability of international humanitarian law by the Russian Federation. However, it remains unclear which cyber activities could lead to an 'armed conflict' triggering the application of this body of law. As regards the implementation of the principles of distinction and proportionality, the statement does not explicitly address them but highlights a number of relevant aspects. It mentions the prohibition of indiscriminate attacks as well as objects requiring special protection hinting at a particular concern with the protection of civilians and civilian objects. Yet, it remains unclear how this concern would translate in the implementation of the principles of distinction and proportionality. More specifically, it stands to be ascertained whether their implementation might lead to an increased targeting of civilians and civilian objects and whether the challenges of insufficiency and ineffectiveness would be viewed negatively.

In summary, the analysis of states' views with regard to the implementation of international humanitarian law shows that their interpretative approaches are still in their early stages. Very few public parameters have evolved guiding the views of states with regard to the implementation of the principles of distinction and proportionality. It remains to be ascertained whether the implementation of these principles will lead to an increased targeting of civilians and civilian objects. Thus, the emerging interpretative approaches of the US and Russia have not evolved to an extent that would enable a meaningful assessment of states' views with regard to the challenges of insufficiency and ineffectiveness. As a consequence, the potential for the development of additional norms in this area remains open. However, a number of considerations might affect future developments in the use of cyber attacks during 'armed conflict'.

On the one hand, both the US and Russia, alongside an increasing number of states, are developing defensive as well as offensive cyber warfare capabilities.[31] Most prominently, the United States stood up US Cyber Command in 2010 to 'prepare to, and when directed, conduct full spectrum military cyberspace operations'.[32] Although the development of Russian cyber capabilities has not yet resulted in the creation of a dedicated command,[33] these developments indicate the recognition of states that cyber means might be used in future 'armed conflicts'. Western states particularly might be interested in using these capabilities in light of changing goals of warfare. Arguably, objectives have moved away from outright military victory resulting from attrition warfare towards effects-based military campaigns that seek to coerce an opponent into adopting or abandoning certain policies.[34] NATO's Kosovo campaign is an example in this regard as it did not seek control over Serbia but was designed to end Slobodan Milošević's policy of ethnic cleansing. The aim is to effect change at the political level. As a result, '[s]uch a strategy may tempt military commanders to attack any target which will achieve the aims of the operation, and hence end the war, in the most effective manner possible; many times these will be civilian objects'.[35] In this context, cyber attacks may provide an effective tool to achieve these shifting goals of warfare by providing a means to target objectives without the attendant physical damage. The possibility that cyber attacks enable an increased targeting of civilian objects may not be regarded as a negative development but may actually provide a new set of tools highly beneficial from a political and strategic standpoint. Such a development might be further reinforced by the broad interpretative approach taken by the United States with regard to the definition of a 'military objective'. As described above, the US assessment is based on an object's contribution to 'war-fighting or war-sustaining capability' as opposed to 'military action'. In light of modern societies' increased reliance on information infrastructure, a growing number of civilian objects may become targetable through physically less destructive cyber attacks. This would further serve the changing goals of warfare. Thus, these considerations may suggest that even if civilians and civilian objects became increasingly affected by the conduct of hostilities, the challenges of insufficiency and ineffectiveness would not necessarily be viewed entirely negatively.

On the other hand, the very reliance of modern infrastructures on information systems may provide a countervailing incentive against the use of cyber attacks. Whereas past advances in military capabilities have generally offered advantages for the countries pursuing their development, this logic no longer holds true in the case of cyber attacks. States with advanced capabilities are simultaneously rendered the most vulnerable since they offer a plethora of possible targets. This is particularly true for highly networked states such as the US that have so far been able to insulate their dual-use infrastructure from conventional attacks. Western states have thus become particularly concerned with the vulnerability of critical infrastructures as '[h]ackers and foreign governments are increasingly able to launch sophisticated intrusions into the networks that control critical civilian infrastructure'.[36] As the UK *Cyber Security Strategy* pointed out, 'power supply, food distribution, water supply and sewerage, financial services, broadcasting, transportation, health, emergency, defence and government services would all suffer if the national information infrastructure were to be disrupted'.[37] Interestingly, the concern over critical infrastructure extends beyond Western countries, mainly due to (perceived) technological dependencies.[38] In countries such as Russia, an extensive reliance on foreign-built technology has led to a heightened sense of vulnerability with regard to the security of critical infrastructure systems.[39]

All in all, it remains to be seen how these factors will affect future developments. States' interpretative approaches with regard to the implementation of the principles of distinction and proportionality are still in their early stages, limiting an analysis of states' view with regard to the legal challenges of insufficiency and ineffectiveness. Thus, the potential for the development of additional norms in this area is still open. This stands in contrast to the analysis of emerging interpretative approaches in the context of the *jus ad bellum* which has revealed diverging and incompatible viewpoints, ultimately limiting the prospects for dedicated norms in that area.

Conclusion

The emergence of cyber attacks as a new way of warfare has opened up a host of challenging questions for states. Unsurprisingly, questions relating to the applicability and adequacy of international law have been at the forefront of international discussions. More specifically, the question has arisen whether the use of cyber attacks by states necessitates a new regulatory framework under international law given the unique characteristics of cyber attacks. This chapter sought to analyse the challenges posed to current legal frameworks in order to evaluate the potential need as well as prospects for new or dedicated legal norms regulating the use of cyber attacks.

The analysis of the emerging interpretative approaches of the US-UK and Russia-China sought to identify potential areas for the creation of additional norms. While states' emerging views with regard to the implementation of international humanitarian law leave open the possibility that additional norms may be created, interpretative approaches in the context of international law on the use of force show a potential divergence limiting prospects for the development of dedicated norms.

The international debate and development of states' interpretative approaches illustrate the complexity of challenges prompted by the emergence of cyber attacks. An assessment of the adequacy of international law in regulating this new type of warfare will necessarily need to go beyond a debate over the need for a new international legal treaty along the lines of known weapons conventions. However, the nascent stages of states' views with regard to the legal challenges created by cyber attacks indicate that any development of dedicated norms in this area will be subject to numerous factors and their complex interplay. Thus, the need for dedicated norms can only be conclusively ascertained over time. Yet, the preliminary analysis suggests areas with greater and lesser prospects for such developments.

Notes

1 Harold Hongju Koh, *International Law in Cyberspace, Remarks at USCYBERCOM Inter-Agency Legal Conference*, Ft. Meade, MD, 18 September 2012, available at www.state.gov/s/l/releases/remarks/197924.htm (accessed 21 October 2014).
2 Ibid.
3 Ibid.
4 Ibid.
5 Ibid.
6 Ibid.
7 *International Strategy for Cyberspace* (United States, May 2011), p. 14.
8 Matthew Waxman, 'Cyber-Attacks and the Use of Force: Back to the Future of Article 2(4)', *Yale Journal of International Law*, 36 (2011), p. 437. See also Christine Gray, *International Law and the Use of Force*, 3rd edition (Oxford: Oxford University Press, 2008) p. 30.
9 *Conceptual Views Regarding the Activities of the Armed Forces of the Russian Federation in Information Space* (Defence Ministry of the Russian Federation, 2011). Unofficial translation by the NATO Cooperative Cyber Defence Centre of Excellence available at www.ccdcoe.org/strategies/Russian_Federation_unofficial_translation.pdf (accessed 21 October 2014). Document in Russian: Концептуальные взгляды на деятельность Вооруженных Сил Российской Федерации в информационном пространстве (Министерство обороны Российской Федерации, 2011) available at http://ens.mil.ru/science/publications/more.htm?id=10845074@cmsArticle (accessed 21 October 2014).
10 Ibid.
11 Ibid.
12 Vladimir Tsymbal quoted in: Timothy Thomas, 'Russia's Information Warfare Structure: Understanding the Roles of the Security Council, Fapsi, the State Technical Commission and the Military', *European Security*, Vol. 7, No. 1 (1998), p. 161.
13 Vladimir Yermakov, *Statement to the First Committee of the 68th Session of the UN General Assembly*, New York, 30 October 2013, available at www.un.org/disarmament/special/meetings/firstcommittee/68/pdfs/TD_30-Oct_ODMIS_Russian-Fed-(E).pdf (accessed 21 October 2014).
14 S. I. Bazylev et al, 'The Russian Armed Forces in the Information Environment: Principles, Rules, and Confidence-Building Measure', *Military Thought*, Vol. 2, 2012, pp. 11–12.

15 *Conceptual Views Regarding the Activities of the Armed Forces of the Russian Federation in Information Space* (see note 9 above).
16 UN General Assembly, Report of the Secretary-General A/56/164/Add.1, p. 6.
17 Ibid.
18 Bazylev et al., 'The Russian Armed Forces', p. 12 (see note 14 above).
19 Oliver Dörr and Albrecht Randelzhofer, 'Ch.I Purposes and Principle, Article 2(4)', in Bruno Simma, Daniel-Erasmus Khan, Georg Nolte and Andreas Paulus (eds), *The Charter of the United Nations: A Commentary*, 3rd edition, Vol. I (Oxford: Oxford University Press, 2012).
20 Sergei Komov, Sergei Korotkov and Igor Dylevski, 'Military aspects of ensuring international information security in the context of elaborating universally acknowledged principles of international law', *Disarmament Forum*, Vol. 3, 2007, p. 37.
21 Koh, *International Law in Cyberspace*. See also UN General Assembly, Report of the Secretary-General A/66/152, p. 18.
22 Ibid.
23 Ibid., p. 10.
24 Ibid., p. 24.
25 Koh, *International Law in Cyberspace* (see note 21 above).
26 Reportedly, the use of cyber means had been debated in several instances but ultimately abandoned for legal reasons. See for example the potential use during US operations in Libya in 2011. Eric Schmitt and Thom Shanker, 'U.S. Debated Cyberwarfare in Attack Plan on Libya', *New York Times*, 17 October 2011.
27 Koh, *International Law in Cyberspace* (see note 21 above).
28 See Section 8.2 of the US *Commander's Handbook on the Law of Naval Operations*, available at www.usnwc.edu/getattachment/a9b8e92d-2c8d-4779-9925-0defea93325c/ (accessed 13 August 2014).
29 Koh, *International Law in Cyberspace* (see note 21 above).
30 *Conceptual Views Regarding the Activities of the Armed Forces of the Russian Federation in Information Space* (see note 9 above).
31 See for example James Lewis and Katrina Timlin, *Cybersecurity and Cyberwarfare. Preliminary Assessment of National Doctrine and Organization* (Washington, DC: Center for Strategic and International Studies, 2011).
32 For more information see the US Cyber Command fact sheet at www.stratcom.mil/factsheets/2/Cyber_Command/ (accessed 21 October 2014).
33 See Keir Giles, '"Information Troops" – a Russian Cyber Command?', in C. Czosseck, E. Tyugu and T. Wingfield (eds.), *3rd International Conference on Cyber Conflict Proceedings* (Tallinn, Estonia: NATO CCD COE, 2011), pp. 45–60.
34 Heather Harrison Dinniss, *Cyber Warfare and the Laws of War* (Cambridge: Cambridge University Press, 2012), p. 183.
35 Ibid.
36 William J. Lynn III, 'Defending a New Domain. The Pentagon´s Cyberstrategy', *Foreign Affairs*, Vol. 89, No. 5 (2010), p. 100.
37 *Cyber Security Strategy of the United Kingdom: safety, security and resilience in cyber space* (United Kingdom, June 2009), p. 12, available at www.gov.uk/government/uploads/system/uploads/attachment_data/file/228841/7642.pdf (accessed 21 October 2014).
38 Giles, 'Information Troops', p. 47 (see note 33 above).
39 See for example the *Information Security Doctrine* of Russia, *Information Security Doctrine of the Russian Federation* (Ministry of the Interior of the Russian Federation, 9 September 2000), part 4. English language version available at www.mid.ru/bdomp/ns-osndoc.nsf/1e5f0de28fe77fdcc32575d900298676/2deaa9ee15ddd24bc32575d9002c442b!OpenDocument (accessed 21 October 2014).

INDEX

Kolibri 360 GmbH
Ferdinand-Jühlke-Straße 7
99095 Erfurt, Deutschland
produktsicherheit@kolibri360.de